Singular Optics

SERIES IN OPTICS AND OPTOELECTRONICS

Series Editors: **E Roy Pike**, Kings College, London, UK
Robert G W Brown, University of California, Irvine, USA

Recent titles in the series

Singular Optics
Gregory J. Gbur

The Limits of Resolution
Geoffrey de Villiers and E. Roy Pike

Polarized Light and the Mueller Matrix Approach
José J Gil and Razvigor Ossikovski

Light—The Physics of the Photon
Ole Keller

Advanced Biophotonics: Tissue Optical Sectioning
Ruikang K Wang and Valery V Tuchin (Eds.)

Handbook of Silicon Photonics
Laurent Vivien and Lorenzo Pavesi (Eds.)

Microlenses: Properties, Fabrication and Liquid Lenses
Hongrui Jiang and Xuefeng Zeng

Laser-Based Measurements for Time and Frequency Domain Applications: A Handbook
Pasquale Maddaloni, Marco Bellini, and Paolo De Natale

Handbook of 3D Machine Vision: Optical Metrology and Imaging
Song Zhang (Ed.)

Handbook of Optical Dimensional Metrology
Kevin Harding (Ed.)

Biomimetics in Photonics
Olaf Karthaus (Ed.)

Optical Properties of Photonic Structures: Interplay of Order and Disorder
Mikhail F Limonov and Richard De La Rue (Eds.)

Nitride Phosphors and Solid-State Lighting
Rong-Jun Xie, Yuan Qiang Li, Naoto Hirosaki, and Hajime Yamamoto

Molded Optics: Design and Manufacture
Michael Schaub, Jim Schwiegerling, Eric Fest, R Hamilton Shepard, and Alan Symmons

Singular Optics

Gregory J. Gbur

CRC Press
Taylor & Francis Group
Boca Raton London New York

CRC Press is an imprint of the
Taylor & Francis Group, an **informa** business

A TAYLOR & FRANCIS BOOK

CRC Press
Taylor & Francis Group
6000 Broken Sound Parkway NW, Suite 300
Boca Raton, FL 33487-2742

First issued in paperback 2020

© 2017 by Taylor & Francis Group, LLC
CRC Press is an imprint of Taylor & Francis Group, an Informa business

No claim to original U.S. Government works

ISBN-13: 978-1-4665-8077-0 (hbk)
ISBN-13: 978-0-367-78265-8 (pbk)

Visit the Taylor & Francis Web site at
http://www.taylorandfrancis.com

and the CRC Press Web site at
http://www.crcpress.com

Dedicated to
Professor Emil Wolf
Advisor, mentor, and friend

Contents

Series Preface, xiii

Preface, xv

CHAPTER 1 ▪ Introduction: Vortices in Nature   1

CHAPTER 2 ▪ Anatomy of a Vortex Beam   9

2.1	MAXWELL'S EQUATIONS AND PARAXIAL BEAMS	9
2.2	LAGUERRE–GAUSS PARAXIAL BEAMS	14
2.3	PROPERTIES OF PHASE SINGULARITIES	22
2.4	DERIVATION OF HIGHER-ORDER GAUSSIAN BEAMS	23
2.5	EXERCISES	32

CHAPTER 3 ▪ Generic Properties of Phase Singularities   33

3.1	YOUNG'S INTERFEROMETER AND PHASE SINGULARITIES	34
3.2	TYPICAL FORMS OF WAVE DISLOCATIONS	40
	3.2.1 Screw Dislocation	41
	3.2.2 Edge Dislocation	42
	3.2.3 Mixed Edge-Screw Dislocations	44
3.3	CRYSTAL DISLOCATIONS AND WAVEFRONTS	46
3.4	TOPOLOGICAL CHARGE AND INDEX	51

3.5   CREATION AND ANNIHILATION EVENTS   54

3.6   EXERCISES   60

CHAPTER 4 ■ Generation and Detection of
              Optical Vortices   63

4.1   GENERATION   63

  4.1.1   Spiral Phase Plate   64

  4.1.2   Mode Conversion   68

  4.1.3   Computer-Generated Holograms   72

  4.1.4   Direct Laser Generation   77

  4.1.5   Nonuniform Polarization   79

  4.1.6   Other Methods   82

4.2   DETECTION   84

  4.2.1   Interference-Based Methods   84

  4.2.2   Diffraction-Based Method   87

  4.2.3   Shack–Hartmann Method   89

  4.2.4   Computer-Generated Holograms and
           Modans   94

  4.2.5   Geometrical Mode Separation   101

4.3   METHOD OF STATIONARY PHASE   107

4.4   EXERCISES   109

CHAPTER 5 ■ Angular Momentum of Light   111

5.1   MOMENTUM AND ANGULAR MOMENTUM
       IN WAVEFIELDS   112

5.2   ORBITAL AND SPIN ANGULAR MOMENTUM   118

5.3   ANGULAR MOMENTUM OF
       LAGUERRE–GAUSSIAN BEAMS   122

5.4   INTRINSIC AND EXTRINSIC ANGULAR
       MOMENTUM   127

5.5   TRAPPING FORCES   131

5.6   MOMENTUM IN MATTER AND THE
       ABRAHAM–MINKOWSKI CONTROVERSY   136

5.7   EXERCISES   146

CHAPTER 6 ■ Applications of Optical Vortices   149

6.1	MICROMANIPULATION, SPANNING, AND TRAPPING	149
6.2	OPTICAL COMMUNICATIONS	156
6.3	PHASE RETRIEVAL	161
6.4	CORONOGRAPH	166
6.5	SIGNAL PROCESSING AND EDGE DETECTION	171
6.6	ROTATIONAL DOPPLER SHIFTS	176
6.7	EXERCISES	182

CHAPTER 7 ■ Polarization Singularities   185

7.1	BASICS OF POLARIZATION IN OPTICAL WAVEFIELDS	186
7.1.1	Linear Polarization	192
7.1.2	Circular Polarization	193
7.2	STOKES PARAMETERS AND THE POINCARÉ SPHERE	194
7.3	POLARIZATION SINGULARITIES: C-POINTS AND L-LINES	202
7.4	GENERIC FEATURES OF POLARIZATION SINGULARITIES	207
7.5	TOPOLOGICAL REACTIONS OF POLARIZATION SINGULARITIES	212
7.6	HIGHER-ORDER POLARIZATION SINGULARITIES	216
7.7	NONUNIFORMLY POLARIZED BEAMS	218
7.8	POINCARÉ BEAMS	222
7.9	POINCARÉ VORTICES	225
7.10	PARTIALLY POLARIZED LIGHT AND SINGULARITIES OF THE CLEAR SKY	227
7.11	THE PANCHARATNAM PHASE	235
7.12	SECOND-ORDER TENSOR FIELDS AND THEIR SINGULARITIES	244

| | 7.13 | THE HAIRY BALL THEOREM | 252 |
| | 7.14 | EXERCISES | 256 |

CHAPTER 8 ■	Singularities of the Poynting Vector	259
8.1	POWER FLOW IN ELECTROMAGNETIC WAVES	260
8.2	CONCEPTUAL DIFFICULTIES WITH THE POYNTING VECTOR	266
8.3	THE FIRST OPTICAL VORTEX	268
8.4	OTHER EARLY OBSERVATIONS OF POYNTING SINGULARITIES	275
8.5	GENERIC PROPERTIES OF POYNTING VECTOR SINGULARITIES	279
8.6	SINGULARITIES, TRANSMISSION, AND RADIATION	286
8.7	EXERCISES	292

CHAPTER 9 ■	Coherence Singularities	293
9.1	OPTICAL COHERENCE	294
9.2	SINGULARITIES OF CORRELATION FUNCTIONS	305
9.3	GENERIC STRUCTURE OF A CORRELATION SINGULARITY	312
9.4	PHASE SINGULARITIES IN PARTIALLY COHERENT FIELDS	317
9.5	ELECTROMAGNETIC CORRELATION SINGULARITIES	321
9.6	EXPERIMENTS AND APPLICATIONS	324
9.7	TWISTED GAUSSIAN SCHELL-MODEL BEAMS	326
9.8	OAM AND RANKINE VORTICES	331
9.9	EXERCISES	334

CHAPTER 10 ■	Singularities and Vortices in Quantum Optics	335
10.1	QUANTIZATION OF THE ELECTROMAGNETIC FIELD	336

10.2	QUANTUM CORES OF OPTICAL VORTICES	343
10.3	INTRODUCTION TO ENTANGLEMENT	347
10.4	NONLINEAR OPTICS AND ANGULAR MOMENTUM	355
10.5	ENTANGLEMENT OF ANGULAR MOMENTUM STATES	363
10.6	A NONLOCAL OPTICAL VORTEX	375
10.7	BELL'S INEQUALITIES FOR ANGULAR MOMENTUM STATES	380
10.8	VORTICES IN SCHRÖDINGER'S EQUATION	390
10.9	UNCERTAINTY PRINCIPLE FOR ANGULAR MOMENTUM	399
10.10	DIRAC'S MAGNETIC MONOPOLE	416
10.11	EXERCISES	425

Chapter 11 ■ Vortices in Random Wavefields		427
11.1	SPECKLE STATISTICS AT A SINGLE POINT	428
11.2	SPECKLE STATISTICS AT MULTIPLE POINTS	432
11.3	STATISTICS OF VORTICES IN RANDOM WAVEFIELDS	437
11.4	CORRELATIONS OF VORTICES IN RANDOM WAVEFIELDS	442
11.5	QUASICRYSTALLINE ORDER	445
11.6	POLARIZATION SINGULARITIES IN RANDOM WAVEFIELDS	447
11.7	MULTIVARIATE COMPLEX CIRCULAR GAUSSIAN RANDOM PROCESSES	448
11.8	EXERCISES	452

Chapter 12 ■ Unusual Singularities and Topological Tricks		453
12.1	BESSEL–GAUSS VORTEX BEAMS	453
12.2	RIEMANN–SILBERSTEIN VORTICES	465
12.3	FRACTIONAL CHARGE VORTEX BEAMS	473

12.4	KNOTS, BRAIDS, AND LINKED VORTICES	479
12.5	CASCADES OF SINGULARITIES	485
12.6	LISSAJOUS SINGULARITIES	489
12.7	OPTICAL MÖBIUS STRIPS	495
12.8	SUPEROSCILLATORY FIELDS	499
12.9	EXERCISES	504

REFERENCES, 505

INDEX, 533

Series Preface

This international series covers all aspects of theoretical and applied optics and optoelectronics. Active since 1986, eminent authors have long been choosing to publish with this series, and it is now established as a premier forum for high-impact monographs and textbooks. The editors are proud of the breadth and depth showcased by published works, with levels ranging from advanced undergraduate and graduate student texts to professional references. Topics addressed are both cutting edge and fundamental, basic science and applications-oriented, on subject matter that includes: lasers, photonic devices, nonlinear optics, interferometry, waves, crystals, optical materials, biomedical optics, optical tweezers, optical metrology, solid-state lighting, nanophotonics, and silicon photonics. Readers of the series are students, scientists, and engineers working in optics, optoelectronics, and related fields in the industry.

Proposals for new volumes in the series may be directed to Luna Han, senior publishing editor at CRC Press, Taylor & Francis Group (luna.han@ taylorandfrancis.com).

Preface

Visual inspection shows that there are finite regions, several wave-lengths in extent, which have, to a sufficient degree of approximation, the character of a homogeneous plane wave and also retain this character as the wave propagates. Hence, these regions indeed satisfy the above postulated conditions for regions of good approximation. Only the zero points seem to be exceptions. However, just because the amplitude vanishes there, they do not produce any stronger effect than other points of varying intensity.

ARNOLD SOMMERFELD
Optics, 1964

There is an urban legend about the infamous bank robber Willie Sutton (1901–1980) that claims that, when asked why he robbed banks, he answered, "because that's where the money is." It might similarly be said that the study of optics has traditionally, and quite reasonably, been about the regions where the intensity of light is nonzero, "because that's where the light is."

The above quotation from Arnold Sommerfeld's 1964 monograph on optics concisely highlights this view. Sommerfeld plotted the amplitude and phase of a superposition of six plane waves of different frequencies and directions of propagation. While most regions of the combined fields looked locally like plane waves, Sommerfeld noted that the phase has an unusual behavior near the neighborhoods of zeros of intensity. However, he dismissed this behavior on the grounds that there is very little light in those regions.

Some 50 years later, the attitude of researchers toward the zero-intensity regions of wavefields has changed dramatically. Those regions where the intensity of the field is zero and the phase is therefore indeterminate, or "singular" (referred to as phase singularities), have been shown to have a well-defined mathematical structure and, furthermore, this structure can strongly influence the overall behavior of the wavefield and its interactions with matter. Zeros of field intensity typically manifest as lines in three-dimensional space, around which the phase has a circulating or helical behavior; this has resulted in such objects being referred to as *optical vortices*. Such vortices possess a variety of interesting structural properties that make them potentially useful for a number of applications. Among these properties is the observation that specially prepared vortex beams can possess orbital angular momentum, and that this angular momentum can be used in manipulation of microscopic particles.

When one starts looking for singularities of wavefields, they can be found in a variety of forms. It has been shown that the field of power flow (the Poynting vector) of a light wave can possess vortex structures, and that these vortices can in turn be used to understand the transmission and radiation properties of some optical systems. We will see that the vector field of the gradient of the phase of a wavefield also circulates as a vortex. The state of polarization of a light wave possesses its own class of singularities, now known as polarization singularities, and these exhibit their own well-defined and typical mathematical behavior. Vortices have even been found in the correlation functions of partially coherent wavefields, in which the phase itself is a random function of space and time. Even more exotic singularities of wavefields can be defined, such as the so-called Riemann–Silberstein vortices.

Since the seminal papers of the 1970s, the study of singularities of wavefields has grown at an incredible rate. It now forms a subfield of optics in its own right, known as *singular optics*. There have been a number of conferences and workshops held that are dedicated to advances in the field, as well as countless special sessions at optics meetings, and there have also been several review articles written on the subject. To date, however, there has been no textbook fully dedicated to introducing the theory and applications of singular optics to a general scientific audience, and this book is an attempt to fill this void.

Why study singular optics? To my mind, the research in singular optics can be loosely divided into three separate but related areas of interest. The first of these is the study of so-called generic features of wavefields, namely, those features that appear in an optical system that has not been specially prepared. The dark interference bands produced in Young's experiment, for instance, are nongeneric, whereas the dark spots that appear in laser speckle patterns are generic. An understanding of singular optics is, therefore, an understanding of the "typical" behavior of optical waves. The second area of interest is determining the unusual properties of vortices and wavefields that are specially prepared in a pure vortex state. As we will see, there are many features of vortex beams and other singularity-containing wavefields that make them interesting topics of scientific study. The third area of interest is the use of vortices and other singularities in practical applications. Many applications are still in their nascent state, even after some 40 years of singular optics research, but others have become well established. Optical vortices have already been implemented in optical micromanipulation, and they show great promise for optical communications and remote sensing. This book will address all three facets of singular optics, and highlight their interconnections.

More generally, an understanding of singular optics provides a completely different way to look at light. Where traditional optics focuses on the shape and structure of the nonzero portions of the wavefield, singular optics provides a perspective by which an understanding of a wave's properties can come from its null regions. Through the course of this book, we will see that the behavior of a light wave is just as much defined by what happens in the empty spaces as it is by what happens in the bright spots.

I have endeavored to thread a particularly thin needle in this text, aiming to make it accessible to both advanced undergraduates in physics and optics as well as beginning graduate students in optics. A working knowledge of electromagnetic and scalar waves is somewhat essential to get the most out of this book, as is a knowledge of Fourier transforms and, ideally, Fourier optics. An understanding of complex analysis and analytic functions will also prove extremely helpful. I have attempted to include explanations of any more advanced mathematics or physics within the text itself, as needed.

It should be noted that the field of singular optics traditionally also covers the singularities that appear in ray optics, namely, the bright regions where many rays intersect, known as optical caustics. We restrict ourselves in this book to a discussion of wave singularities, as they have proven the most active field of research in recent decades. Information about the theory of ray singularities can be found in the excellent book by Nye [Nye99]; Nye's book is also the first to discuss wave singularities in detail.

In writing this text, I have been guided by the pioneering work of a number of great scientists. Names that appear quite regularly throughout the book include Professor Michael Berry, Professor John Nye, Professor Mark Dennis, Professor Isaac Freund, Professor Miles Padgett, Professor Grover Swartzlander, Jr., Professor Marat Soskin, Professor Taco Visser, and Professor Emil Wolf. I would like to thank them for their inspiring accomplishments in the field of singular optics, and I hope this book does their work credit.

I would like to acknowledge the Air Force Office of Scientific Research and Program Manager Dr. Arje Nachman, who supported much of my original research on singular optics that appears in this book.

Further thanks goes to my graduate students, Elisa Hurwitz, Matt Smith, and Charlotte Stahl for reading chapter drafts and providing useful feedback. I would also like to thank Yangyundou Wang for identifying mistakes in the draft, and Professor Michael Berry for last-minute clarifications of the history of the subject. I extend a special thanks to Professor Taco Visser for reading the entire book draft and providing me suggestions and corrections.

I would also like to thank some of my friends in helping keep me sane and happy while I was writing! This includes my personal trainer Ray Williams, my guitar instructor Toby Watson, my skating coach Tappie Dellinger, and my regular skydiving friends at Skydive Carolina, including John Solomon, Mickey Turner, Jen and Tim Reynolds, Wendy and Gene Helton, and Paul Meegan.

Additional thanks goes to Luna Han of Taylor and Francis, who provided help and encouragement during preparation of the manuscript.

Finally, I would like to thank Beth Szabo for her support, patience, and love during the long process of finishing this book.

CHAPTER 1

Introduction

Vortices in Nature

I T HAS LONG BEEN said that "Nature abhors a vacuum." This statement, the sense of which was first attributed to Aristotle, is in fact oversimplified and not really true, given that most of the universe is essentially an empty space. However, even if nature abhors a vacuum, it absolutely loves a vortex.

Vortices—circulating disturbances of an extended medium of some sort of "stuff"—are quite ubiquitous in nature. They can be found on extremely small scales, in quantum mechanical Bose–Einstein condensates, on terrestrial scales in turbulent liquids and gases, and even on cosmological scales in the swirling of spiral galaxies. We now know that they are readily found in light, as well, and an investigation into the properties of such optical vortices and related structures forms the bulk of this book. The field of study of vortices and other optical singularities is now known as *singular optics*; we will elaborate on the meaning of "singular" in this context momentarily.

What sort of behaviors do we expect a vortex of light to have? We can actually learn much about optical vortices by looking at early studies of their liquid and gas counterparts; we will learn even more from looking at their differences.

One of the earliest technical drawings of a collection of vortices was produced by the famed Leonardo da Vinci (1452–1519), in his "Studies of water passing obstacles and falling," shown in Figure 1.1. Water falling into a pond

FIGURE 1.1   Detail from Leonardo da Vinci's "Studies of water passing obstacles and falling," c. 1508.

produces an area of turbulence, and on either side of this area, we see the development of regions of opposing circulation. This already is comparable to something we will see again and again in the optical regime: the production of vortices in pairs of opposite handedness. This water example is roughly analogous to light propagation through an aperture with a size comparable to the wavelength, as will be discussed in Chapter 8. We might have predicted this pair production from conservation of angular momentum: the water entering the pond has no net circulation, so, there must also be no net circulation introduced into the water of the pond itself. In Chapter 5, we will see that vortices of light also typically have angular momentum associated with them.

Vortex production is also quite common in the atmosphere, most strikingly in the production of the so-called von Karman vortex streets. When fluid flows around a blunt rigid obstacle at a sufficient speed, it can break up into vortices of alternating handedness that are then carried downstream, as shown in Figure 1.2. At low flow rates, pairs of vortices are produced in the "shadow" of the obstacle; when the flow becomes sufficiently high, it can peel off individual vortices from this shadow, providing enough kick to release the next vortex, and thus repeating the pattern. This vortex kick actually induces a force on the obstacle, potentially making it vibrate, and

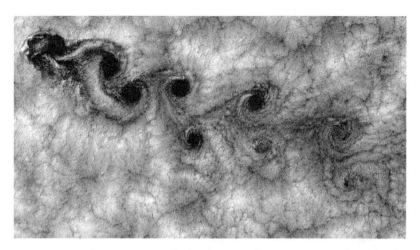

FIGURE 1.2 Landsat 7 image of clouds near the Juan Fernandez Islands on September 15, 1999. The von Karman streets are formed by the airflow around the mountains on the islands in the upper left.

in recent years, this *vortex-induced vibration* (VIV) has been proposed as a new hydroelectric energy source [BRBSG08]. The vortices persist even after being released from the obstacle that created them; we will see that optical vortices are also robust structures in optical waves, making them useful for applications such as free-space optical communications, as discussed in Chapter 6.

Another dramatic example of vortex creation in the atmosphere is the wingtip vortices created by airplanes. Roughly speaking, an airplane develops lift because the air flows faster over the top of the wing than the bottom of the wing, resulting in an upward Bernoulli force. This imbalance of air speeds also results in a net circulation of air around the wing, which manifests as an air vortex on the wingtip, as seen in Figure 1.3. Curiously, it is also possible to make a microscopic wing that can fly on a beam of light, as shown by Swartzlander et al. [SPAGR10]; we will touch upon this research in Chapter 5.

In fluids, then, it turns out that vortices are a surprisingly common phenomenon, which will often manifest without any deliberate experimental steps needed to be taken. In the study of singularities, we will refer to structures that appear "naturally" under "typical" circumstances as *generic*, and the generic properties of optical vortices will be studied in detail in Chapter 3.

FIGURE 1.3   Wingtip vortex of passing aircraft visible in smoke, from a wake vortex study on Wallops Island. (Image taken in 1990 by NASA Langley Research Center.)

The fluid vortices mentioned so far, however, involve a transport of matter. In Da Vinci's water vortices, for instance, actual water molecules are moving in circular paths around the vortex center. What we are really interested in, however, are wave vortices, in which it is a wave that circulates around a central point. Such wave vortices can also appear in water, and we can use monochromatic (single frequency) water wave vortices to better understand their analogous relatives in optical fields. In the beginning of the twentieth century, Proudman and Doodson [PD24] calculated the daily tidal elevations of the North Sea from a set of observational data, building upon earlier work by Whewell [Whe33]. Their map of the cotidal lines and corange lines are shown in Figure 1.4.

The cotidal lines represent the lines of high tide at different times of the day, while the corange lines represent the relative amplitudes of those tides. We can clearly see that all the cotidal lines converge together at several points known as *amphidromic points*, where the amplitude of the tides simultaneously goes to zero. This means that there are no tidal variations at all at the amphidromic points, which is the only possible outcome at a location where all the high-tide lines join. These points are *singularities* of the cotidal lines, the latter of which are equivalent to the *surfaces of constant phase* of an optical wavefield. As a 12-hour cycle passes, the lines of high tide circulate around the singularities, much like the hour hand of a clock moves around its central axis. This constant circulation of the tides around

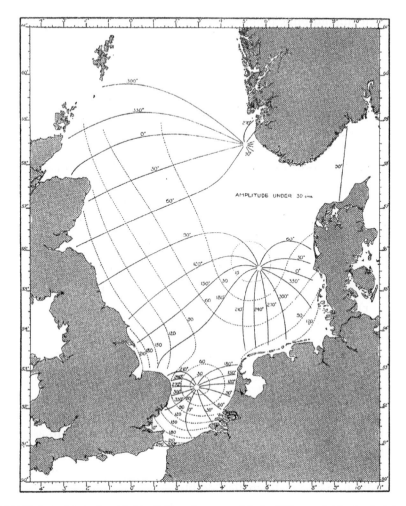

FIGURE 1.4 Map of cotidal (solid) and corange (dashed) lines of tides in the North Sea. (From J. Proudman and A.T. Doodson. *Phil. Trans. Roy. Soc. Lond. A*, 224:185–219, 1924.)

the amphidromic points justifies our use of the word "vortex" to describe such points. Unlike the earlier vortex examples, however, it should be noted that there is no physical transport of water around the tidal vortices: only energy and momentum circulate around the vortex.

It should be further noted that, following a path on the map around one of the amphidromic points, the phase of the tides increases or decreases by 360 degrees (2π radians). The periodicity of the tides (which repeat after 12 hours) and the continuity of them (there can be no discontinuous "jump" of the tide heights) implies that the phase increase or decrease around an amphidromic point must always be a multiple of 2π. From the figure, however, we note that it seems that only increases of 2π are common, or generic, to use the term introduced above. The connection between the tides and phase singularities was discussed in detail by Nye, Hajnal, and Hannay [NHH88].

But what about phase singularities in light fields? In this book, we will investigate the properties of complex scalar monochromatic wavefields $u(\mathbf{r})$ that are a function of position $\mathbf{r} = (x, y, z)$. Such complex fields may be expressed in terms of their real and imaginary parts, the real-valued functions $u_R(\mathbf{r})$ and $u_I(\mathbf{r})$, in the form

$$u(\mathbf{r}) = u_R(\mathbf{r}) + iu_I(\mathbf{r}). \tag{1.1}$$

The value of such a field at any position $\mathbf{r}$ may be represented by a point in the complex u_R, u_I plane. However, we may also write this complex function $u(\mathbf{r})$ in a polar representation in terms of a real amplitude $A(\mathbf{r})$ and real phase $\psi(\mathbf{r})$,

$$u(\mathbf{r}) = A(\mathbf{r})e^{i\psi(\mathbf{r})}. \tag{1.2}$$

The relation between the Cartesian and polar representations of a complex field is illustrated in Figure 1.5. We immediately note that a zero of the field, that is, $u(\mathbf{r}) = 0$, implies that $A(\mathbf{r}) = 0$ and that $\psi(\mathbf{r})$ is consequently undefined, that is, singular: any value of $\psi(\mathbf{r})$ is valid at such points. Regions of space where a field has zero amplitude are therefore referred to as *phase singularities*.

The singular nature of $\psi(\mathbf{r})$ at a point where $A(\mathbf{r}) = 0$ would at first glance appear to be a mathematical oddity, only significant because of our preference for working in a polar representation. If we look at the phase

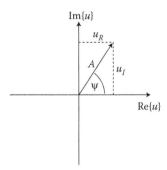

FIGURE 1.5 Representation of a complex scalar field in the complex plane in Cartesian and polar forms.

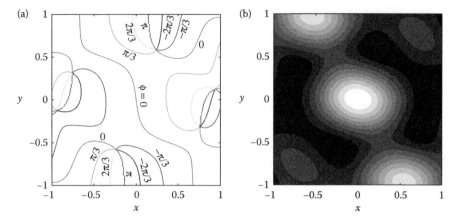

FIGURE 1.6 Cross-sectional view of the (a) phase $\psi(\mathbf{r})$ and (b) intensity $I(\mathbf{r}) = |A(\mathbf{r})|^2$ of a sum of four plane waves propagating in different directions.

structure of the field around such singular points, however, we can see that much more is going on! In Figure 1.6, we illustrate the phase and intensity contours of a sum of four arbitrarily chosen plane waves. We will not concern ourselves with the details of the calculations at this point, but note that the phase contours include multiple points where all lines of constant phase converge. These points coincide with zeros of intensity, as we would expect from the above discussion. It can be seen on comparison with Figure 1.4 that the phase structure around these singular points is directly analogous to the amphidromic points of the tides. If we include the explicit monochromatic time dependence $\exp[-i\omega t]$ in our field, we will find that the lines of constant phase circulate around the singularities, and we refer to the structure as a whole—singular point surrounded by circulating phase—as an *optical vortex*.

Though not every possible phase singularity will have a vortex structure, we will find that this is *typically* the case. We will therefore often use the terms "phase singularity" and "optical vortex" interchangeably.

Something quite profound has crept into our discussion in the transition from discussing fluid vortices to discussing vortices of monochromatic waves. The strong restriction on the behavior of the phase around a wave vortex—that it can only change by multiples of 2π—puts a strong constraint on the structure of phase singularities. We will see as we go on that singular optics may also be considered the study of the *topology* of light, that is, the study of the types of geometric structures that can appear in light waves. Many standard results from topology appear in this book in an optical context, for instance, the *hairy ball theorem* (Section 7.13), knot theory (Section 12.4), and even Möbius strips (Section 12.7). We will learn much about the structure of light in the course of this book, but don't let the mention of topology frighten you off—our discussion will be mostly free of such abstract mathematics!

The phase of a light wave strongly affects how and where it propagates through space. A singularity of phase in a wave may thus be thought of, speaking quite loosely, as a "directionless" point in a wavefield. For the vortices discussed so far, the amplitude of the wavefield is zero in the core and its phase is undefined. These are not the only types of singularities that can be found in optics. The field of power flow, or Poynting vector, of an electromagnetic wave will be seen in Chapter 8 to possess points of zero amplitude where the direction of power flow is undefined; the flow around such Poynting singularities also has a circulating structure. Even the polarization field of light can possess singularities, namely, points at which the electric field is circularly polarized, and has no definite orientation, or linearly polarized, and has no definite handedness; such polarization singularities will be discussed in Chapter 7. It is worth noting at this early stage, however, that not all optical singularities are associated with a zero of light intensity; we will see how this happens as we go on.

How did singular optics begin? The study of optical phase singularities is quite recent in comparison with the study of their fluid counterparts. The foundational paper in the field is the 1974 paper "Dislocations in wave trains" by Nye and Berry [NB74] that lays out the general theory of such objects. In hindsight, however, it was noted that circulations in light had been observed much earlier; the first observation of a vortex in light seems to be due to Wolter [Wol50], who was investigating the spatial shift of a light beam upon total internal reflection (now known as the Goos–Hänchen shift). The power flow of the beam in the medium does a sort of "loop-the-loop" before being reflected, forming a vortex of power flow. Not long after, Braunbek and Laukien [BL52] studied the power flow of an electromagnetic wave incident on a perfectly conducting half-plane. On the illuminated side of the plane, they found vortices of power flow that correspond to phase singularities of the magnetic field; all these electromagnetic examples will be discussed in Chapter 8.

We now know that optical waves have a variety of different types of singularities, each of which have their own regular and interesting structures. From the initial discussion of Nye and Berry, a vibrant field of study has grown.

CHAPTER 2

Anatomy of a Vortex Beam

A N IMPORTANT ASPECT OF singular optics is the observation that optical singularities generally have shared common features, independent of the manner in which they were created. With this in mind, we can easily observe most of these features by looking at a simple example, or arguably the simplest example, of an optical phase singularity: a highly directional beam carrying a single optical vortex.

Such a basic case allows us to make a number of simplifying approximations for the propagation of light, such as the neglect of polarization and the assumption of paraxial wavefields. Our later investigations, however, will require us to "unravel" these approximations in order to introduce more complicated singular phenomena. Anticipating this, we start our discussion with the general form of Maxwell's equations and explicitly show the assumptions and approximations made to arrive at the form of an optical vortex beam. From there, we introduce both the important characteristics of such beams as well as the notation used to describe them.

2.1 MAXWELL'S EQUATIONS AND PARAXIAL BEAMS

The most general set of macroscopic Maxwell's equations in SI units are of the form

$$\nabla \cdot \mathbf{D}(\mathbf{r}, t) = \rho(\mathbf{r}, t), \tag{2.1}$$

$$\nabla \cdot \mathbf{B}(\mathbf{r}, t) = 0, \tag{2.2}$$

$$\nabla \times \mathbf{E}(\mathbf{r}, t) = -\frac{\partial \mathbf{B}(\mathbf{r}, t)}{\partial t}, \tag{2.3}$$

$$\nabla \times \mathbf{H}(\mathbf{r}, t) = \mathbf{J}(\mathbf{r}, t) + \frac{\partial \mathbf{D}(\mathbf{r}, t)}{\partial t}. \tag{2.4}$$

These equations describe the relationship between the electric and magnetic fields and the sources of such fields. The latter consist of the free charge density $\rho(\mathbf{r}, t)$ and the free current density $\mathbf{J}(\mathbf{r}, t)$. The fields may be grouped into two classes: the vacuum electric and magnetic field $\mathbf{E}(\mathbf{r}, t)$ and $\mathbf{B}(\mathbf{r}, t)$ and the macroscopic (material) electric displacement and magnetic field $\mathbf{D}(\mathbf{r}, t)$ and $\mathbf{H}(\mathbf{r}, t)$.

The first equation, Equation 2.1, is Gauss's law. It is a mathematical expression of the observation that electric charges are sources and sinks of electric displacement field lines. The second equation, Equation 2.2, correspondingly indicates that there exist no sources or sinks of magnetic field lines. The third equation, Equation 2.3, is Faraday's law, which states that a time-varying magnetic field produces a circulating electric field. The final equation, Equation 2.4, is the Ampére–Maxwell law, which encapsulates two observations. The first of these observations is that electric currents result in circulating magnetic fields (Ampére's law), and the second is that a time-varying electric displacement produces a circulating magnetic field.

These equations cannot be solved without some knowledge of the properties of matter that the electromagnetic fields are interacting with; specifically, we need relationships between the electric field and the electric displacement, $\mathbf{E}$ and $\mathbf{D}$, and the magnetic field[*] and the "H-field," $\mathbf{B}$ and $\mathbf{H}$. Though we will find the behavior of vortices in matter to be a topic of interest later, for now we restrict ourselves to light propagating in a vacuum, so that

$$\mathbf{D}(\mathbf{r}, t) = \epsilon_0 \mathbf{E}(\mathbf{r}, t), \tag{2.5}$$

$$\mathbf{B}(\mathbf{r}, t) = \mu_0 \mathbf{H}(\mathbf{r}, t), \tag{2.6}$$

where ϵ_0 is the vacuum permittivity and μ_0 is the vacuum permeability.

We will also be primarily concerned with fields propagating in source-free regions; we therefore also restrict ourselves to the case $\rho(\mathbf{r}, t) = 0$ and

---

[*] Traditionally, $\mathbf{B}$ is known as the "magnetic induction" and $\mathbf{H}$ as the "magnetic field." We simply use the term "magnetic field," as it is almost always clear which we are referring to.

$\mathbf{J}(\mathbf{r}, t) = 0$. This leads us to the source-free Maxwell's equations in vacuum, given below:

$$\nabla \cdot \mathbf{E}(\mathbf{r}, t) = 0, \tag{2.7}$$

$$\nabla \cdot \mathbf{B}(\mathbf{r}, t) = 0, \tag{2.8}$$

$$\nabla \times \mathbf{E}(\mathbf{r}, t) = -\frac{\partial \mathbf{B}(\mathbf{r}, t)}{\partial t}, \tag{2.9}$$

$$\nabla \times \mathbf{B}(\mathbf{r}, t) = \epsilon_0 \mu_0 \frac{\partial \mathbf{E}(\mathbf{r}, t)}{\partial t}. \tag{2.10}$$

It was Maxwell himself who first noted that these equations have solutions in the form of traveling waves. This can be readily shown by taking the curl of Equation 2.9, interchanging the order of the curl and time derivative on the right, and then substituting from Equation 2.10 into it. This gives the result

$$\nabla \times [\nabla \times \mathbf{E}(\mathbf{r}, t)] = \epsilon_0 \mu_0 \frac{\partial^2 \mathbf{E}(\mathbf{r}, t)}{\partial t^2}. \tag{2.11}$$

Upon using the vector identity

$$\nabla \times [\nabla \times \mathbf{v}] = \nabla[\nabla \cdot \mathbf{v}] - \nabla^2 \mathbf{v}, \tag{2.12}$$

we may write

$$\nabla[\nabla \cdot \mathbf{E}(\mathbf{r}, t)] - \nabla^2 \mathbf{E}(\mathbf{r}, t) = \epsilon_0 \mu_0 \frac{\partial^2 \mathbf{E}(\mathbf{r}, t)}{\partial t^2}. \tag{2.13}$$

Finally, noting from Equation 2.7 that $\nabla \cdot \mathbf{E} = 0$, we have

$$\nabla^2 \mathbf{E}(\mathbf{r}, t) - \epsilon_0 \mu_0 \frac{\partial^2 \mathbf{E}(\mathbf{r}, t)}{\partial t^2} = 0. \tag{2.14}$$

This is the form of a wave equation for the vector field $\mathbf{E}(\mathbf{r}, t)$, with the speed of propagation given by $c = 1/\sqrt{\epsilon_0 \mu_0}$. A similar equation exists for the magnetic field $\mathbf{B}$.

We will begin our discussions by focusing our attention on single-frequency (monochromatic) wavefields, so that the electric field may be expressed in the form

$$\mathbf{E}(\mathbf{r}, t) = \text{Re}\left\{\mathbf{E}(\mathbf{r})e^{-i\omega t}\right\}, \tag{2.15}$$

where ω is the angular frequency of oscillation and $\mathbf{E}(\mathbf{r})$ is the complex spatially dependent part of the field. An analogous expression exists for the magnetic field. On substituting Equation 2.15 for $\mathbf{E}(\mathbf{r}, t)$ into Equation 2.14, we get a vector Helmholtz equation for the complex electric field vector, namely,

$$\nabla^2 \mathbf{E}(\mathbf{r}) + k^2 \mathbf{E}(\mathbf{r}) = 0, \tag{2.16}$$

with the wavevector $k = \omega/c$, c being the speed of light in vacuum. If we have solved this partial differential equation for $\mathbf{E}(\mathbf{r})$, the physical field can be found by substituting back into Equation 2.15.

Quantities such as the average field intensity can be determined directly from the complex field. If we consider the dot product of $\mathbf{E}$ with itself, we find that it may be written as

$$\mathbf{E}(\mathbf{r}, t) \cdot \mathbf{E}(\mathbf{r}, t) = \frac{1}{2} \mathbf{E}(\mathbf{r}) \cdot \mathbf{E}^*(\mathbf{r})$$
$$+ \frac{1}{4} \left[\mathbf{E}(\mathbf{r}) \cdot \mathbf{E}(\mathbf{r}) e^{-2i\omega t} + \mathbf{E}^*(\mathbf{r}) \cdot \mathbf{E}^*(\mathbf{r}) e^{2i\omega t} \right]. \tag{2.17}$$

At optical frequencies, $\omega \sim 10^{15}\ \text{s}^{-1}$, and such oscillations are much too fast for most detectors to resolve. We therefore only consider the cycle-averaged intensity, defined as

$$\langle I(\mathbf{r}) \rangle = 2 \langle \mathbf{E}(\mathbf{r}, t) \cdot \mathbf{E}(\mathbf{r}, t) \rangle = \frac{2}{T} \int_0^T \mathbf{E}(\mathbf{r}, t) \cdot \mathbf{E}(\mathbf{r}, t) dt, \tag{2.18}$$

where $T = 2\pi/\omega$ is the period of oscillation.[*] We readily find that

$$\langle I(\mathbf{r}) \rangle = \mathbf{E}(\mathbf{r}) \cdot \mathbf{E}^*(\mathbf{r}). \tag{2.19}$$

When talking about complex monochromatic wavefields, we will typically drop the angle brackets $\langle \cdots \rangle$ for brevity. For future reference, we note that the cycle average of the product of two monochromatic fields $\mathbf{A}$ and $\mathbf{B}$ may be written as

$$\langle \mathbf{A}(\mathbf{r}, t) \cdot \mathbf{B}(\mathbf{r}, t) \rangle = \frac{1}{T} \int_0^T \mathbf{A}(\mathbf{r}, t) \cdot \mathbf{B}(\mathbf{r}, t) dt = \frac{1}{2} \text{Re}\{\mathbf{A}^*(\mathbf{r}) \cdot \mathbf{B}(\mathbf{r})\}. \tag{2.20}$$

---

[*] A factor of 2 has been introduced into the definition of the cycle-averaged intensity for convenience.

The preceding derivations apply to fields of quite arbitrary nature, but we will be largely interested in highly directional, beamlike, fields, that is, fields that propagate primarily along a single axis. Choosing the z-axis for the direction of propagation, we note that the transverse nature of electromagnetic fields, codified in Equation 2.7, implies that the field vector $\mathbf{E}$ will lie almost entirely in the transverse xy-plane. Let us further assume for the moment that the direction of this vector is constant throughout this transverse plane, that is,

$$\mathbf{E}(\mathbf{r}) = \mathbf{E}_0 U(\mathbf{r}), \tag{2.21}$$

where $\mathbf{E}_0$ is a constant, generally complex, vector and $U(\mathbf{r})$ is a complex scalar field. We can then study the propagation of the wavefield through the function $U(\mathbf{r})$ and look for solutions to the scalar Helmholtz equation that it satisfies, namely,

$$\nabla^2 U(\mathbf{r}) + k^2 U(\mathbf{r}) = 0. \tag{2.22}$$

Furthermore, for a field propagating close to the z-axis, we expect $U(\mathbf{r})$ to be well approximated by the expression

$$U(\mathbf{r}) \approx u(\mathbf{r})e^{ikz}, \tag{2.23}$$

where $u(\mathbf{r})$ is slowly varying with respect to z. On substitution into Equation 2.22, that equation reduces to the form

$$\nabla^2 u(\mathbf{r}) + 2ik\frac{\partial u(\mathbf{r})}{\partial z} = 0. \tag{2.24}$$

We finally specify that $u(\mathbf{r})$ must be slowly varying in the z-direction on the order of a wavelength, such that

$$\lambda \left| \frac{\partial u}{\partial z} \right| \ll |u|, \tag{2.25}$$

$$\lambda \left| \frac{\partial^2 u}{\partial z^2} \right| \ll \left| \frac{\partial u}{\partial z} \right|. \tag{2.26}$$

The second derivative of $u(\mathbf{r})$ with respect to z is therefore negligible, and we are left with

$$\frac{\partial^2 u}{\partial x^2} + \frac{\partial^2 u}{\partial y^2} + 2ik\frac{\partial u}{\partial z} = 0. \tag{2.27}$$

This is known as the *paraxial wave equation* and describes a wave that propagates in a highly directional manner along the z-axis.

To summarize: to reach this point, we have restricted our attention to fields (1) in vacuum, (2) in the absence of sources, (3) that are monochromatic, (4) whose polarization effects can be neglected, and (5) that are paraxial. We are therefore looking at a restricted class of optical wavefields; nevertheless, these fields show a surprising amount of complexity, and will be used to illustrate the basic properties of phase singularities.

2.2 LAGUERRE–GAUSS PARAXIAL BEAMS

The most familiar solution to the paraxial wave equation is that of a Gaussian beam, found by first assuming that the intensity in the plane $z = 0$, called the *waist plane*, is of a Gaussian form

$$I(x, y, 0) = u(x, y, 0)u^*(x, y, 0) = I_0\, e^{-2(x^2+y^2)/w_0^2}. \qquad (2.28)$$

Here, I_0 is the on-axis intensity and w_0 is the effective width of the beam. Furthermore, we assume that the phase of the field is constant in the waist plane. With these starting conditions on the field behavior, it can be shown that the field $u_G(\mathbf{r})$ for all values of z is of the form

$$u_G(\mathbf{r}) = A_0\, e^{-i\Phi(z)} \left[\frac{1}{\sqrt{1+z^2/z_0^2}} e^{ik(x^2+y^2)/2R(z)} \right] e^{-(x^2+y^2)/w^2(z)}, \quad (2.29)$$

where

$$R(z) = z + z_0^2/z, \qquad (2.30)$$

$$w(z) = w_0\sqrt{1 + z^2/z_0^2}, \qquad (2.31)$$

$$\Phi(z) = \arctan(z/z_0), \qquad (2.32)$$

$$A_0 = \sqrt{I_0}, \qquad (2.33)$$

and

$$z_0 = \pi w_0^2/\lambda, \qquad (2.34)$$

which is known as the *Rayleigh range* of the beam. We defer the details of this calculation to the end of the chapter in order to focus for the moment

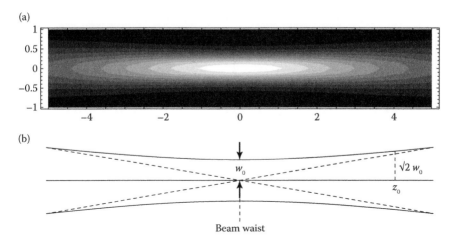

FIGURE 2.1 The cross-sectional (a) intensity $|u_G|^2$ and (b) width $w(z)$ of a Gaussian beam as a function of propagation distance z. Here, $z_0 = 4$ and $w_0 = 0.5$.

on the optical properties of the beams, rather than the math behind them. Here, $w(z)$ represents the effective width of the beam as a function of z, $R(z)$ represents the wavefront curvature of the beam, and $\Phi(z)$ is a phase term that results in the so-called *Gouy phase shift*.

The intensity $|u_G(\mathbf{r})|^2$ and the width of a Gaussian beam are illustrated in Figure 2.1. The plane $z = 0$ is the plane of minimum width and is called the waist of the beam; the Rayleigh range is the distance at which the beam width has increased to $\sqrt{2}$ times the waist width. For large values of $z \gg z_0$, the field takes on the form of an outgoing spherical wave, of the form

$$\frac{e^{ikr}}{r} = \frac{e^{ik\sqrt{x^2+y^2+z^2}}}{\sqrt{x^2+y^2+z^2}} = \frac{e^{ikz\sqrt{1+(x^2+y^2)/z^2}}}{z\sqrt{1+(x^2+y^2)/z^2}} \approx \frac{1}{z}e^{ikz}\,e^{ik(x^2+y^2)/2z}.$$

(2.35)

Relating the wavefront curvature term involving $R(z)$ for positive $z \gg z_0$ with the spherical wave, we can see the terms are comparable. For $z \ll -z_0$, the wavefield has the form of an incoming spherical wave. A focused field, therefore, acts like a perfect incoming spherical wave that then converts to an outgoing spherical wave; however, owing to the Gouy phase term $\Phi(z)$ (plotted in Figure 2.2), an actual focused Gaussian beam acquires an additional π phase shift relative to a perfect spherical wave. A number of explanations for this "mysterious" shift have been proposed, but it is only a mystery if one assumes that geometrical optics (in the form of perfect spherical waves) applies in the region of focus, which it does not.

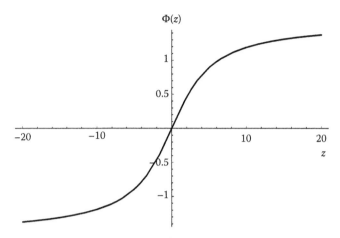

FIGURE 2.2 The Gouy phase shift as a function of z, with $z_0 = 4$.

It should be noted that a Gaussian beam is shape-invariant on propagation: it retains its Gaussian intensity profile, albeit with a different width, as it travels along the z-axis. Using this solution as a starting point, we can look for other shape-invariant beams. If we work in Cartesian coordinates, we can derive the so-called *Hermite–Gauss beams*, which will be discussed at the end of this chapter. If we work in cylindrical coordinates (ρ, ϕ, z), however, we can derive the more interesting *Laguerre-Gauss family of beams*, which inherently carry vortices within them. We therefore look for a general solution of the paraxial wave equation of the form

$$u(\mathbf{r}) = F\left[\sqrt{2}\rho/w(z)\right] G(\phi) u_G(\mathbf{r}) \exp[i\alpha(z)]. \qquad (2.36)$$

Here, F, G, and α are functions to be determined, and $u_G(\mathbf{r})$ is simply the Gaussian beam given above, with $A_0 = 1$. With some work, it can be found that the solutions to this equation are given by

$$u_{nm}^{LG}(\mathbf{r}) = \sqrt{\frac{2n!}{\pi w_0^2 (n + |m|)!}} \left(\frac{\sqrt{2}\rho}{w(z)}\right)^{|m|} L_n^{|m|}\left(\frac{2\rho^2}{w^2(z)}\right)$$

$$\times \exp[im\phi] u_G(\mathbf{r}) \exp\left[-i(2n + |m|)\Phi(z)\right], \qquad (2.37)$$

where n is a nonnegative integer, m is an integer, and $L_n^m(x)$ is an associated Laguerre function of order n and m. These beams are all defined to have an integrated cross-sectional intensity equal to unity, that is,

$$\int \left| u_{nm}^{LG}(\mathbf{r}) \right|^2 d^2 r = 1, \tag{2.38}$$

where $d^2 r$ is the area element in a plane of constant z.

For $n = 0$, $m = 0$, this expression reduces to that of an ordinary Gaussian beam, taking into account that $L_0^0(x) = 1$. For $m \neq 0$, we find that two things happen: the amplitude of the field vanishes at the center of the beam $\rho = 0$ due to the term that depends on the $|m|$-th power of ρ, and the phase changes by $2\pi m$ as one circles the origin in a counterclockwise sense in a plane of constant z, exhibited by the $\exp[im\phi]$ term. As discussed in Chapter 1, this circulation is the reason why such beams are referred to as *vortex beams*: they include a single optical vortex on their central axis. If we include the explicit time dependence $\exp[-i\omega t]$, it is found that the lines of constant phase in the xy-plane circulate around the central axis as time progresses. It circulates in a counterclockwise, or left-handed, sense for positive m, and a clockwise, or right-handed, sense for negative m.

We may make a number of other important observations from Equation 2.37. First, because m is an integer, the phase of the field can only increase or decrease by multiples of 2π as one follows a closed path around the axis of the beam. Because the argument of a complex number is invariant under changes of 2π, this implies that the field itself is completely continuous. We refer to the number m as the *topological charge* of the vortex, and we will see that any vortex in a wavefield has an integer topological charge associated with it. In the case of a Laguerre–Gauss beam, the topological charge is self-evident, but for more complicated wavefields, a rigorous mathematical formula will be helpful; we introduce this in Chapter 3.

Again inspecting Equation 2.37, we see that near the central axis of the beam, the field has the approximate functional dependence $\rho^{|m|} e^{im\phi}$; in Cartesian coordinates, this may be written as

$$u_{nm}^{LG}(\mathbf{r}) \sim (x \pm iy)^{|m|}, \tag{2.39}$$

where the $\pm$ is based on the sign of m. We again see that, though the phase is discontinuous on axis, the field itself is perfectly continuous.

As the field propagates along the z-axis, the value of m remains unchanged; the *topological charge of a vortex beam is a conserved quantity on propagation.* This is also a general property of vortices in wavefields, to be described in more detail later.

The intensity $|u|^2$ and phase $\arg[u]$ of a selection of beams in the plane $z = 0$ are illustrated in Figure 2.3. We can see the singularities of phase at the origin of the phase plots: they are the points at which all phase contours (gray levels) come together. The order n of the beam also indicates a slightly different type of singularity: n is equal to the number of dark rings

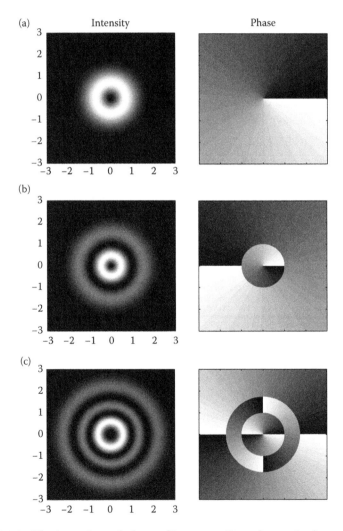

FIGURE 2.3 The intensity and phase of Laguerre–Gauss beams in the waist plane with (a) $n = 0, m = 1$, (b) $n = 1, m = -1$, and (c) $n = 2, m = 2$. The phase ranges from 0 (black) to 2π (white).

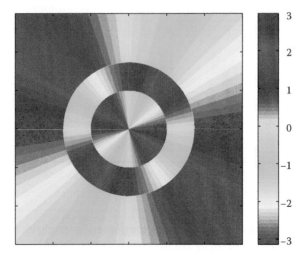

FIGURE 2.4 The phase of a Laguerre–Gauss beam of order $n = 2$, $m = 2$, displayed in color.

nested in the beam profile. At each of these dark rings, the phase jumps by a factor of π; this is a change of sign of the field as one passes the singular contour.

The gray-level plots of the phase in Figure 2.3 are potentially misleading, as the sharp jump from white to black incorrectly implies a dramatic jump in phase. It is often preferable to illustrate the phase with color plots, as in Figure 2.4. To keep the price of this book low, however, we stick to gray levels for the most part!

This is a good opportunity to note that there are a number of additional ways to indicate the phase of a complex wavefield, each of which can be useful in certain circumstances. In Figure 2.3, we have illustrated the phase as a grayscale contour plot, with black being argument zero and white being argument 2π. When there are a large number of singular points in a plot, however, it can often be difficult to see the singularities in all the shading "noise." We can also explicitly draw a selection of contour lines of constant phase, as shown in Figure 2.5a. Vortices are clearly visible as the intersection points of all the contour lines; this depiction will be useful in observing features that arise during reactions between multiple vortices. Even this diagram is overly complicated in many cases; because we are working with a complex wavefield, it is to be noted that a zero of the field is simply a point where the real and imaginary parts of the wavefield are simultaneously zero. By plotting the zeros of the real and imaginary parts (which appear in the

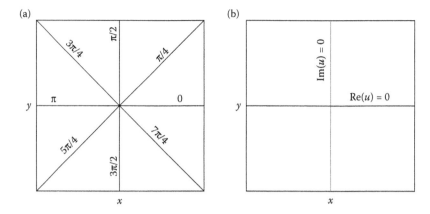

FIGURE 2.5   Other options for illustrating the phase of a wavefield, by (a) plotting the equiphase contours and (b) by plotting zeros of the real and imaginary parts of the wavefield. The examples shown are for a Laguerre–Gauss beam or order $n = 0$, $m = 1$ in the waist plane.

form of lines in the transverse plane), we can find phase singularities at their intersection; this is illustrated in Figure 2.5b. In this latter visualization, however, the handedness of the vortex is unspecified.

We use the latter of these phase descriptions to illustrate the behavior of an optical vortex as it propagates a long distance in atmospheric turbulence, based on the work of Gbur and Tyson [GT08]. Simulations of the turbulence propagation are shown in Figure 2.6 for both an $m = 1$ vortex and an $m = 3$ vortex. The phase is shown overlaid on the intensity of the beam in a plane at 3 km from the source.

In Figure 2.6a, we can see that the single $m = 1$ vortex remains, despite the fact that the field in general has been severely disrupted by the atmosphere. This is further evidence that topological charge is conserved under perturbations of the wavefield, a property important for many applications. It is to be noted, however, that the vortex is no longer centered on the original beam axis: the charge may be conserved, but the vortex position is not fixed. In Figure 2.6b, we can see that the single $m = 3$ vortex has split into 3 vortices of topological charge 1. This is illustrative again that the total topological charge is conserved, but also that a single vortex of higher-order charge will in general break into a number of single-charge vortices. We may say that the first-order vortices are stable, and that higher-order vortices are not. This is connected to the idea that first-order vortices are *generic*, and the ones most likely to appear naturally in an optical system without special preparation.

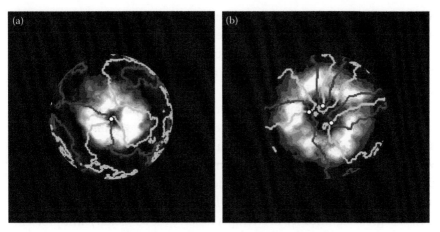

FIGURE 2.6 The intensity and zero crossings of a pair of Laguerre–Gauss beams after propagating through 3 km of atmospheric turbulence. (a) $n = 0$, $m = 1$ and (b) $n = 0$, $m = 3$. Here, the initial beam width is 1 cm, the wavelength is $\lambda = 1.055\,\mu m$, and the turbulence strength is $C_n^2 = 10^{-15}\,cm^{-2/3}$. White circles have been placed over the intersections that indicate vortex locations.

In Figure 2.7, we consider the propagation of an $n = 0$, $m = 2$ field through 3 km of atmospheric turbulence. We now see that there are evidently four intersections of the $Re(u) = 0$ and $Im(u) = 0$, implying four vortices—what has happened? If we were to add up the topological charge of

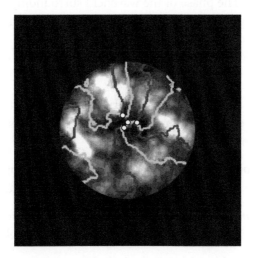

FIGURE 2.7 The intensity and zero crossings of an $m = 2$ vortex after propagating through 3 km of atmospheric turbulence. Other parameters are as in Figure 2.6. White circles have been placed over the intersections that indicate vortex locations.

all the vortices visible in the region, they would still add up to two: evidently a pair of vortices of opposite handedness was created as the field penetrated further into the turbulence. This is another general property of optical vortices: they are typically only created or annihilated in pairs of opposing circulation.

If we return to Equation 2.37 one final time, we note a curious property: the sign of m only appears in the vortex phase term, $\exp[im\phi]$. The intensities $|u(\mathbf{r})|^2$ of a pair of Laguerre–Gauss modes of order $n, |m|$ and $n, -|m|$ will be completely the same at all points in the transverse plane and for all propagation distances. It is therefore not possible to determine the phase of a vortex beam from measurements of the field intensity alone. This observation is of great significance in attempts to solve the so-called *phase problem*, to be discussed in Chapter 6.

2.3 PROPERTIES OF PHASE SINGULARITIES

From looking at the behavior of Laguerre–Gauss vortex beams, we were able to observe or deduce a number of general properties of phase singularities; these properties are listed below:

- *Topology*. Phase singularities typically manifest themselves as lines of zero intensity in three-dimensional space.

- *Circulation*. The phase of the wavefield surrounding the line of zero intensity typically has a circulating or helical behavior, increasing or decreasing continuously as one travels around the axis of the singularity. From this behavior, the structures are dubbed "optical vortices."

- *Topological charge*. These singularities have associated with them a topological charge, which must take on integer values to maintain the continuity of the wavefield.

- *Conservation*. Topological charge is conserved on propagation, and vortices are resistant to perturbations of the wavefield.

- *Phase nonuniqueness*. In the presence of optical vortices, the phase of a wavefield is not uniquely determinable from the field intensity.

- *Genericity*. Fields with topological charge $m = 1$ are stable under wavefield perturbations, while higher-order vortices will typically break up into a collection of $m = 1$ vortices.

- *Pair production.* Vortices are typically created and annihilated in pairs.

The word "typically" appears a number of times in this list, and it is an important one to acknowledge. When discussing the structure of phase singularities, it should be noted that we are usually considering the behavior that is most likely to occur, and not necessarily all of the possible behaviors. This may at first glance seem like a limitation of the theory of singular optics, but it is really its strength: when we understand the typical, or generic, behavior of phase singularities, we understand in broad terms the types of phenomena that are overwhelmingly likely to occur in wavefields. This will become more clear in Chapter 3.

2.4 DERIVATION OF HIGHER-ORDER GAUSSIAN BEAMS

In this section, we look at the explicit derivation of the Laguerre–Gauss beams and Hermite–Gauss beams used previously. As the derivations do not provide any particular insight into the properties of optical vortices, this section may be skipped on a first reading.

We begin by assuming that the field is paraxial and has the following intensity profile in the waist plane $z = 0$:

$$I(x, y, z) \sim I_0 e^{-2(x^2+y^2)/w_0^2}, \tag{2.40}$$

where I_0 is the peak intensity and w_0 is the root mean square (RMS) width in the waist plane. Knowing from experiment that the transverse intensity profile remains a Gaussian, we look for solutions of the paraxial wave equation of the form

$$u(\mathbf{r}) = A_0 e^{ik(x^2+y^2)/2q(z)} e^{ip(z)}, \tag{2.41}$$

where $q(z)$ and $p(z)$ are complex functions to be determined. On substitution from this expression into Equation 2.27, we get the equation

$$\left[\frac{k^2}{q^2}(x^2 + y^2)\left(\frac{dq}{dz} - 1\right) - 2k\left(\frac{dp}{dz} - \frac{i}{q}\right) \right] = 0. \tag{2.42}$$

Because the left term in parentheses depends on x and y and the right term is independent of these variables, this equation can only be satisfied if

$$\frac{dq}{dz} = 1, \quad \frac{dp}{dz} = \frac{i}{q}. \tag{2.43}$$

These differential equations can be readily solved by first directly integrating the q equation and then by using the results in the p equation. We find that[*]

$$q(z) = q_0 + z, \tag{2.44}$$

$$p(z) = i \log \left(\frac{q_0 + z}{q_0} \right), \tag{2.45}$$

where $q_0 = q(0)$ and we have taken $p(0) = 0$. In general it is expected that $q(z)$ will be a complex number, and we rewrite it in terms of a real and imaginary part as

$$\frac{1}{q(z)} = \frac{1}{R(z)} + \frac{i\lambda}{\pi w^2(z)}, \tag{2.46}$$

where $R(z)$ and $w(z)$ are purely real quantities. We can then write the term involving $p(z)$ as

$$\exp[ip(z)] = \frac{1}{1 + z/R_0 + i\lambda z/\pi w_0^2}, \tag{2.47}$$

where $R_0 = R(0)$ and $w_0 = w(0)$. By matching the real parts of Equations 2.44 and 2.46, we find that $R(z)$ has the general form

$$\frac{1}{R(z)} = \frac{\text{Re}(q_0) + z}{|q_0|^2 + 2z\text{Re}(q_0) + z^2}. \tag{2.48}$$

We may interpret $R(z)$ as the radius of curvature of the wavefront of the beam, and it is expected that this curvature should be infinite at $z = 0$, since the wave transitions from a converging wave to a diverging wave in that plane. Letting $R(0) = R_0 = \infty$, we find from Equation 2.46 that

$$\frac{1}{q_0} = \frac{i\lambda}{\pi w_0^2}. \tag{2.49}$$

This completely specifies the form of $q(z)$ and $p(z)$. If we now define

$$z_0 \equiv \frac{\pi w_0^2}{\lambda}, \tag{2.50}$$

---

[*] Throughout this book, we use "log" instead of "ln" to denote the natural logarithm.

then

$$R(z) = z + \frac{z_0^2}{z}, \tag{2.51}$$

$$w(z) = w_0\sqrt{1 + z^2/z_0^2}, \tag{2.52}$$

as given in the Section 2.2. Furthermore, the exponential with $p(z)$ takes on the form

$$e^{ip(z)} = \frac{1}{1 + z^2/z_0^2} \exp[-i\Phi(z)], \tag{2.53}$$

where $\Phi(z)$ is the Gouy shift. With these results, the mathematical expression for a paraxial Gaussian beam $u_G(\mathbf{r})$, given previously in Equation 2.29, has been completely derived.

Starting with the Gaussian beam solution, we can then search for other beams that are also completely shape-invariant on propagation. Working in Cartesian coordinates, we will first look for solutions to the paraxial equation of the form

$$u(\mathbf{r}) = f\left[\sqrt{2}x/w(z)\right] g\left[\sqrt{2}y/w(z)\right] u_G(\mathbf{r}) \exp[i\Psi(z)], \tag{2.54}$$

with $A_0 = 1$ in $u_G(\mathbf{r})$. Here, the functions f and g are transverse field profiles to be determined, and $\Psi(z)$ is a propagation-dependent phase shift, assumed to be more general than the Gouy shift for Gaussian beams. On substitution into the paraxial wave equation, we get the complicated expression

$$gu_G\partial_x^2 f + 2g\partial_x u_G\partial_x f + fu_G\partial_y^2 g + 2f\partial_y u_G\partial_y g + 2ik[gu_G\partial_z f + fu_G\partial_z g]$$
$$+ fg[\partial_x^2 u_G + \partial_y^2 u_G + 2ik\partial_z u_G] - 2kfgu + G\partial_z\Psi = 0. \tag{2.55}$$

The second term in square brackets satisfies the paraxial equation itself, and therefore vanishes. We simplify this expression by using new coordinates

$$\xi \equiv \sqrt{2}x/w, \tag{2.56}$$

$$\eta \equiv \sqrt{2}y/w. \tag{2.57}$$

On substitution, and dividing the result by $2fgu/w^2$, Equation 2.55 becomes

$$
\frac{\partial_\xi^2 f}{f} + \left(\frac{ikw^2}{R} - 2\right)\xi\frac{\partial_\xi f}{f} + \frac{\partial_\eta^2 g}{g} + \left(\frac{ikw^2}{R} - 2\right)\eta\frac{\partial_\eta g}{g}
$$
$$
-ikww'\left(\frac{\xi\partial_\xi f}{f} + \frac{\eta\partial_\eta g}{g}\right) - kw^2\partial_z\Psi = 0, \tag{2.58}
$$

where $w' = \partial w/\partial z$. Using Equation 2.52, it can be shown that

$$
ww' = \frac{w^2}{R}, \tag{2.59}
$$

and on substitution from this into Equation 2.58, all explicitly imaginary parts of the equation cancel. We are left with

$$
\frac{\partial_\xi^2 f}{f} - 2\xi\frac{\partial_\xi f}{f} + \frac{\partial_\eta^2 g}{g} - 2\eta\frac{\partial_\eta g}{g} - kw^2\partial_z\Psi = 0. \tag{2.60}
$$

This equation is separated in terms of its independent variables ξ, η, and z. The first two terms depend only on ξ, the second two only on η, and the last one on z. Each group must therefore be equal to a constant, and we choose for convenience that the first group equals $-2m$, the second equals $-2n$, and the final term equals C. These constants must satisfy the equation

$$
2m + 2n = C. \tag{2.61}
$$

With these choices, we get a trio of independent differential equations:

$$
\partial_\xi^2 f - 2\xi f + 2mf = 0, \tag{2.62}
$$

$$
\partial_\eta^2 g - 2\eta g + 2ng = 0, \tag{2.63}
$$

$$
\partial_z\Psi = -\frac{2m + 2n}{kw_0^2}\frac{1}{1 + z^2/z_0^2}. \tag{2.64}
$$

The third of these can be directly integrated to get

$$
\Psi(z) = -(m + n)\arctan(z/z_0). \tag{2.65}
$$

The first two equations are both in the form of the Hermite equation (described, for instance, in Chapter 18 of [Gbu11]). To achieve physically

TABLE 2.1   The First Few Hermite Polynomials

$H_0(x) = 1$
$H_1(x) = 2x$
$H_2(x) = 4x^2 - 2$
$H_3(x) = 8x^3 - 12x$
$H_4(x) = 16x^4 - 48x^2 + 12$
$H_5(x) = 32x^5 - 160x^3 + 120x$
$H_6(x) = 64x^6 - 480x^4 + 720x^2 - 120$

relevant solutions to these equations, it is necessary to have m and n take on positive integer values, which results in the solutions becoming finite polynomials. The first few of these Hermite polynomials $H_m(x)$ are shown in Table 2.1.

We then find an infinite family of solutions to the paraxial wave equation, the Hermite–Gauss beams, of the form

$$
u_{mn}^{HG}(\mathbf{r}) = \sqrt{\frac{2}{\pi}} 2^{-(m+n)/2} \frac{1}{\sqrt{n!m!w_0^2}}
$$
$$
\times H_m\left[\sqrt{2}x/w\right] H_n\left[\sqrt{2}y/w\right] u_G(\mathbf{r}) \exp[i\Psi(z)], \qquad (2.66)
$$

where normalization has been included. The intensity $|u(x, y, 0)|^2$ of a number of these modes are shown in Figure 2.8. We will often refer to these beams in shorthand as HG_{mn} beams. It can be seen from the figure that the indices m and n also indicate the number of intensity nulls along the x- and y-directions, respectively.

Though these Hermite–Gauss modes are an important class of beams, they individually do not possess optical vortices in their core. If we return to the paraxial equation and assume a shape-invariant solution in cylindrical coordinates, we can derive the Laguerre–Gauss modes. To begin, we try a solution of the form

$$
u(\mathbf{r}) = F[\sqrt{2}\rho/w(z)]G(\phi)u_G(\mathbf{r}) \exp[i\Psi(z)]. \qquad (2.67)
$$

We then attempt to find solutions to the paraxial wave equation with this form. In cylindrical coordinates, the paraxial equation may be written as

$$
\frac{\partial^2 u}{\partial \rho^2} + \frac{1}{\rho}\frac{\partial u}{\partial \rho} + \frac{1}{\rho^2}\frac{\partial^2 u}{\partial \phi^2} + 2ik\frac{\partial u}{\partial z} = 0. \qquad (2.68)
$$

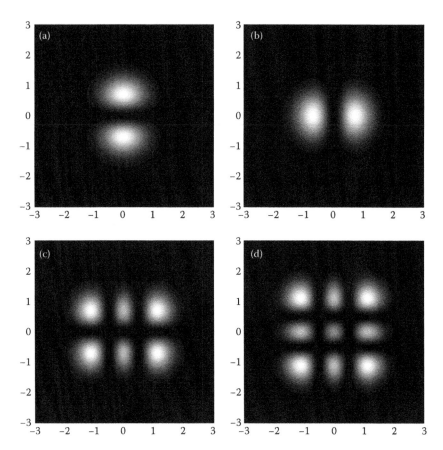

FIGURE 2.8 Intensity profiles of four Hermite–Gauss beams in the plane $z = 0$. (a) HG_{01}, (b) HG_{10}, (c) HG_{21}, and (d) HG_{22}.

On substitution from Equation 2.67 into Equation 2.68, terms related to $u_G(\mathbf{r})$ may be removed. We then introduce the new coordinate $\zeta \equiv \sqrt{2}\rho/w$ in place of ρ; with some effort, the paraxial equation becomes

$$\frac{1}{F}\frac{\partial^2 F}{\partial\zeta^2} - \frac{2\zeta}{F}\frac{\partial F}{\partial\zeta} + \frac{1}{\zeta F}\frac{\partial F}{\partial\zeta} + \frac{1}{\zeta^2}\frac{1}{G}\frac{\partial^2 G}{\partial\phi^2} - kw^2\frac{\partial\Psi}{\partial z} = 0. \qquad (2.69)$$

We set the last term, the only one now dependent on z, to be equal to a constant C; we may directly solve this equation to find that

$$\Psi(z) = -\frac{C}{2}\arctan(z/z_0). \qquad (2.70)$$

We now multiply Equation 2.69 by ζ^2; the second to last term then depends only on ϕ, and must be equal to a constant that we call $-m^2$. The function G then satisfies the harmonic oscillator equation and has a solution

$$G_m(\phi) = A_m\, e^{im\phi} + B_m\, e^{-im\phi}, \tag{2.71}$$

where m must be an integer to make the function single-valued. The expression for F is then

$$\frac{d^2F}{d\zeta^2} + \left(\frac{1}{\zeta} - 2\zeta\right)\frac{dF}{d\zeta} + \left[C - \frac{m^2}{\zeta^2}\right]F = 0. \tag{2.72}$$

At this stage, a certain amount of luck and mathematical know-how is required to proceed. We begin by introducing a new variable $v = \zeta^2$, which transforms the equation for F to

$$v\frac{d^2F}{dv^2} + (1 - v)\frac{dF}{dv} + \frac{1}{4}\left(C - \frac{m^2}{v}\right)F = 0. \tag{2.73}$$

We then introduce a new function $L(v)$ such that

$$L(v) = v^{|m|/2}F(v). \tag{2.74}$$

Writing Equation 2.73 in terms of $L(v)$ results in

$$vL'' + (|m| + 1 - v)L' + \left(\frac{C}{4} - \frac{|m|}{2}\right)L = 0. \tag{2.75}$$

This differential equation is of the form of the *associated Laguerre equation*, which may be written as

$$vL'' + (|m| + 1 - v)L' + nL = 0. \tag{2.76}$$

Like m, the quantity n must be integer-valued, which implies that we should write the constant C as

$$C = 4n + 2|m|. \tag{2.77}$$

The solutions $L_n^m(x)$ to the associated Laguerre equation are derived from the *Laguerre polynomials* $L_p(x)$ by the use of the derivative relation

$$L_n^m(x) = (-1)^m \frac{d^m}{dx^m}L_{n+m}(x). \tag{2.78}$$

TABLE 2.2   The First Few Laguerre Polynomials

$L_0(x) = 1$
$L_1(x) = -x + 1$
$2!L_2(x) = x^2 - 4x + 2$
$3!L_3(x) = -x^3 + 9x^2 - 18x + 6$
$4!L_4(x) = x^4 - 16x^3 + 72x^2 - 96x + 24$
$5!L_5(x) = -x^5 + 25x^4 - 200x^3 + 600x^2 - 600x + 120$

The first few ordinary Laguerre polynomials are shown in Table 2.2; the associated Laguerre functions can be found then by the use of Equation 2.78. One relation will be rather useful in the investigation of vortex beams, describing the behavior of $L_0^m(x)$,

$$L_0^m(x) = 1. \tag{2.79}$$

On applying the associated Laguerre function notation to the paraxial beam solution in cylindrical coordinates, we end up with the normalized solution noted earlier, namely,

$$u_{nm}^{LG}(\mathbf{r}) = \sqrt{\frac{2n!}{\pi w_0^2(n + |m|)!}} \left(\frac{\sqrt{2}\rho}{w(z)}\right)^{|m|} L_n^{|m|}\left(\frac{2\rho^2}{w^2(z)}\right)$$

$$\times \exp[im\phi] u_G(\mathbf{r}) \exp\left[-i(2n + |m|) \arctan(z/z_0)\right]. \tag{2.80}$$

We will refer to these beams in shorthand as the LG_{nm} beams.

It is somewhat striking that we have derived two infinite families of solutions for propagation-invariant beams, namely, the Hermite–Gauss and Laguerre–Gauss beams. It can be shown that each of these sets is *complete*; that is, any paraxial beam can be expressed as a coherent weighted sum of either the Hermite–Gauss beams or Laguerre–Gauss beams. For instance, an arbitrary paraxial field $v(\mathbf{r})$ may be decomposed in a series of Laguerre–Gauss beams as

$$u(\mathbf{r}) = \sum_{n=0}^{\infty} \sum_{m=-\infty}^{\infty} c_{nm} u_{nm}^{LG}(\mathbf{r}), \tag{2.81}$$

where the coefficients c_{nm} may be determined by calculating the following integrals in the plane $z = 0$:

$$c_{nm} = \int_{z=0} \left[u_{nm}^{LG}(x', y', 0) \right]^* u(x', y', 0) dx'\, dy'. \tag{2.82}$$

Equation 2.81 provides an alternative method for simulating the propagation of a paraxial field besides using, for instance, the Rayleigh–Sommerfeld diffraction formulas. If the field is known in the plane $z = 0$, its coefficients c_{nm} can be found from Equation 2.82; then the field at any z-value can be found from Equation 2.81.

It may seem curious that a beam of arbitrary width can be represented by a sum of modes with a fixed width w_0. If w_0 is taken to be significantly greater or smaller than the physical width of the beam to be decomposed, a large number of modes will be required for an accurate representation. For this reason, w_0 is typically taken to be as close to the physical width of the beam to be approximated as possible.

The completeness of the Laguerre–Gauss modes and the Hermite–Gauss modes implies that it is possible to express any member of one set as a linear combination of members of the other set. It can be shown [AV91,BAvdVW93], for instance, that a Laguerre–Gauss mode may be written as

$$u_{p,q-p}^{LG}(x, y, z) = \sum_{k=0}^{p+q} i^k b(p, q, k) u_{p+q-k,k}^{HG}(x, y, z), \tag{2.83}$$

provided $q - p \geq 0$, and

$$b(p, q, k) = (-1)^{p+k} \left(\frac{(p+q-k)!k!}{2^{p+q} p! q! k!} \right)^{1/2} \frac{d^k}{dt^k} [(1-t)^q (1+t)^p]_{t=0}. \tag{2.84}$$

As a simple example of this transformation, the field $u_{0,\pm 1}^{LG}$ may be written in terms of Hermite–Gauss beams as

$$u_{0,\pm 1}^{LG}(x, y, z) = \frac{1}{\sqrt{2}} \left[u_{10}^{HG}(x, y, z) \pm i u_{01}^{HG}(x, y, z) \right]. \tag{2.85}$$

It is worth noting that at least one more family of complete and orthogonal propagation-invariant paraxial beams may be derived, the so-called

Ince–Gaussian beams [BGV04]. These beams are derived by solving the paraxial wave equation in an elliptic cylindrical coordinate system, and form a continuous set of transition modes between the Hermite–Gaussian and Laguerre–Gaussian sets.

2.5 EXERCISES

1. Using Equation 2.85, determine the decomposition of the LG_{02} beam into Hermite–Gaussian modes.

2. Using Equation 2.85, determine the decomposition of the LG_{20} beam into Hermite–Gaussian modes.

3. Fill in the derivation from Equation 2.68, the paraxial wave equation in cylindrical coordinates, to Equation 2.69. Note that ζ depends on ρ and z, which makes things interesting.

4. Demonstrate that the Laguerre–Gauss beams of Equation 2.37 are normalized according to Equation 2.38 for any value of z. What, physically, does this independence of z imply about the beams?

5. In the source plane $z = 0$, consider the superposition of a Gaussian beam and an LG_{01} beam of the form

$$U(\mathbf{r}) = \exp[-\rho^2/w_0^2] + A(x + iy)\exp[-\rho^2/w_0^2],$$

where A is an adjustable complex amplitude. Describe what happens to the position of the vortex as A is increased from zero, and explain how it depends on the argument of A. Next, consider the sum of the vortex beam with a normally incident plane wave of amplitude A, and describe again the change in position of the vortex.

6. In the source plane $z = 0$, consider the superposition of a Gaussian beam and an LG_{02} beam of the form

$$U(\mathbf{r}) = \exp[-\rho^2/w_0^2] + A(x + iy)^2\exp[-\rho^2/w_0^2],$$

where A is an adjustable complex amplitude. Describe what happens to the second-order vortex as A is increased from zero, and explain how it depends on the argument of A. Next, consider the sum of the vortex beam with a normally incident plane wave of amplitude A, and describe again the behavior of the vortex or vortices.

CHAPTER 3

Generic Properties of Phase Singularities

T RADITIONALLY, OPTICS RESEARCHERS HAVE concentrated on designing the simplest feasible experiments for the purpose of isolating, or taking advantage of, one or more fundamental properties of light. For example, Young's double-slit experiment was conceived to demonstrate and study the wave properties of light through interference. Also, engineers and scientists have put in centuries of effort to construct optical elements that focus light with a minimum of aberrations.

These examples are concerned with the study of what is *possible* in the behavior of light; in recent years, however, much attention has been paid to studying what is *typical* in this behavior. For instance, Young's experiment is designed to interfere waves from only two sources; natural interference patterns, such as those produced by light scattering from a rough surface, involve superpositions of a large number of waves. Another example is the distinction between an image-forming lens and an arbitrary piece of glass. An imaging lens focuses light to a single, diffraction-limited spot; however, any transparent piece of glass of varying thickness will also concentrate light. Such a concentration of light, known as a *caustic*, will generally extend through a region of space. As can be seen in the photographs of Figure 3.1, there are similar features shared by all such caustics—such as the pointed edges of the bright regions, known as folds and cusps—and these common features can be analyzed theoretically. Caustics have been interpreted as singularities of the ray optics picture of light propagation, just as optical

(a)

(b)

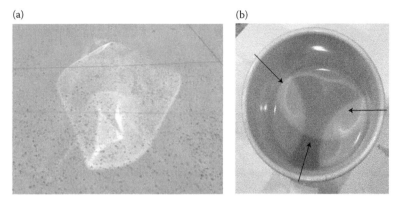

FIGURE 3.1 Some optical caustics. (a) A caustic formed on a pavement due to light transmission through an irregular piece of glass. (b) Three caustics formed on reflection on the inside of a coffee cup. The arrows indicate the directions of illumination.

vortices are singularities of the wave optics picture; a detailed discussion of the properties of caustics may be found in the book by Nye [Nye99].

Features that appear "naturally" in an optical system without any special or conscious preparation are known as *generic* features of a wavefield. Just as there are generic features in the bright spots of naturally focused light, there are generic features of dark spots (regions of complete destructive interference) as well. We will see that vortices of various types are the typical structures that arise around an intensity zero in a three-dimensional wavefield, and in this chapter, we will study all the generic forms of such phase singularities.

The problem of genericity at first seems like a purely academic one, as most optical applications involve the deliberate design of an optical system. However, the genericity of phase singularities is intimately tied to their stability under wavefield perturbations. A feature that appears naturally in an unprepared system is one that is robust under slight changes in that system. This robustness of optical vortices, and the phase singularities of a wavefield in general, is one of the most useful properties of such structures.

3.1 YOUNG'S INTERFEROMETER AND PHASE SINGULARITIES

The definitive demonstration of the wave nature of light was the interference experiment of Thomas Young, first performed in 1803 and published

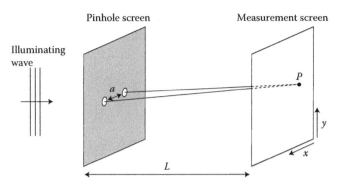

FIGURE 3.2 Illustration of Young's two-pinhole experiment.

in 1807 [You07]. Today, this experiment is referred to as "Young's double-slit experiment" or "Young's two-pinhole experiment"; we adopt the latter convention.

An illustration of the experimental geometry is shown in Figure 3.2. A pair of small circular holes, of radius comparable to the wavelength, are separated by a distance a in the screen. A monochromatic wave of wavelength λ is incident upon the screen from the left side; light transmitted through the holes arrives on an observation screen a distance L away.

We can readily determine the distribution of light on the observation screen through simple wave calculations. We will assume that each of the holes is small enough to be considered point like; furthermore, we will only consider the distribution of light near the central axis that runs between the two pinholes to the observation screen. This latter restriction allows us to neglect geometric factors that arise in diffraction. If we define the amplitude of light illuminating each of the pinholes as U_1 and U_2, and the area of the pinholes as b, the total field observed at a point P on the screen may then be written as

$$U(\mathbf{r}) = -\frac{ib}{\lambda} \left[U_1 \frac{e^{ikR_1}}{R_1} + U_2 \frac{e^{ikR_2}}{R_2} \right], \tag{3.1}$$

where

$$R_1^2 = (x - a/2)^2 + y^2 + L^2, \tag{3.2}$$

$$R_2^2 = (x + a/2)^2 + y^2 + L^2. \tag{3.3}$$

The use of the functions e^{ikR}/R implies that we are treating the light coming out of each hole as propagating as a spherical wave.

If we now assume that the observation screen is sufficiently far away, that is, L is much larger than all values of x and y of interest, we may approximate R_1 and R_2 by the first terms of their binomial expansions, for example,

$$R_1 \approx L + \frac{(x - a/2)^2 + y^2}{2L}. \tag{3.4}$$

We keep only the first term in the denominators, and both terms in the numerator; this allows us to write

$$U(x, y, L) \approx -\frac{ib}{\lambda} \frac{e^{ikL}}{L} \left[U_1 e^{+\frac{ik}{2L}[(x-a/2)^2 + y^2]} + U_2 e^{+\frac{ik}{2L}[(x+a/2)^2 + y^2]} \right]. \tag{3.5}$$

If we expand the exponents and group terms appropriately, we have

$$U(x, y, L) \approx -\frac{ib}{\lambda} \frac{e^{ikL}}{L} e^{+\frac{ika^2}{8L}} e^{+\frac{ik}{2L}(x^2 + y^2)} \left[U_1 e^{-ikxa/2L} + U_2 e^{+ikxa/2L} \right]. \tag{3.6}$$

If the illuminating wave is a normally incident plane wave, we may write $U_1 = U_2 = U_0$, and our expression for the field simplifies to

$$U(x, y, L) \approx -\frac{2ib}{\lambda} U_0 \frac{e^{ikL}}{L} e^{+\frac{ika^2}{8L}} e^{+\frac{ik}{2L}(x^2 + y^2)} \cos[kxa/2L]. \tag{3.7}$$

If we consider only the intensity of the light on the screen, we have

$$I(x, y, L) = |U(x, y, L)|^2 = \frac{4|U_0|^2 b^2}{\lambda^2 L^2} \cos^2[kxa/2L]. \tag{3.8}$$

The intensity and phase of the field on the observation screen are illustrated in Figure 3.3. The intensity pattern is the familiar alternating bands of light and darkness; the phase pattern involves the combination of two phase features. The first of these is the circular pattern produced by the light forming outgoing spherical waves. Superimposed on this are vertical lines where the phase "jumps" instantaneously; it can be seen that these jumps correspond to the zeros of intensity. These phase discontinuities result from the change of sign of the cosine term in Equation 3.7 as we cross a value of x for which it is zero. These lines are the singularities of phase of the two-pinhole experiment.

If we also consider changes in L, we find that the zeros of intensity in the experiment are surfaces in three-dimensional space. A similar

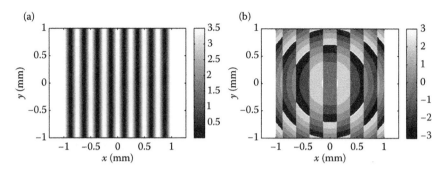

FIGURE 3.3 The (a) intensity and (b) phase for Young's two-pinhole experiment, with $a = 1$ mm, $L = 1$ m, and $\lambda = 500$ nm.

banded interference pattern will arise in any interferometric experiment in which two coherent fields are brought together, such as a Michelson, Twyman–Green, or Mach–Zender interferometer (see, for instance, Born and Wolf [BW99, Chapter 7] for a discussion of these devices). These experiments, however, are somewhat deceiving: in a complicated system involving the scattering or diffraction of light, we would naturally expect more than two fields to be brought together to interfere. What happens, then, in such cases?

We can investigate this by considering a three-pinhole interferometer, as shown in Figure 3.4. The pinholes are arranged in the form of an equilateral triangle of side a, with pinhole positions on the screen given by $(x_1, y_1) = (0, \sqrt{3}a/3)$, $(x_2, y_2) = (a/2, -\sqrt{3}a/6)$, and $(x_3, y_3) = (-a/2, -\sqrt{3}a/6)$.

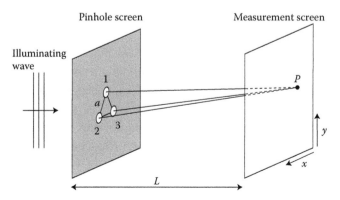

FIGURE 3.4 Illustration of "Young's three-pinhole experiment."

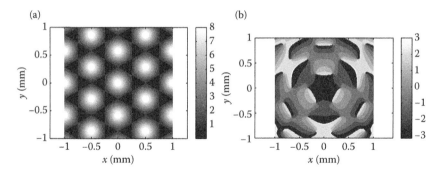

FIGURE 3.5 The (a) intensity and (b) phase for Young's three-pinhole experiment.

We may repeat our two-pinhole calculation for the case of three, and find that the field is in general given by the expression

$$U(x, y, L) \approx -\frac{ib}{\lambda}\frac{e^{ikL}}{L}\left(U_1 e^{+ik[(x-x_1)^2+(y-y_1)^2]/2L}\right.$$
$$\left.+U_2 e^{+ik[(x-x_2)^2+(y-y_2)^2]/2L} + U_3 e^{+ik[(x-x_3)^2+(y-y_3)^2]/2L}\right).$$

$$(3.9)$$

With all amplitudes U_i equal, that is, $U_i = 1$, the intensity and phase appears as in Figure 3.5. We have a hexagonal array of bright spots with a complementary array of dark spots within. We can see that those dark spots are phase singularities of the field by looking at the phase plot—in fact, the only way to be certain that we have a singularity, and not a near zero of intensity, is to look for the characteristic intersection of all the phase contours. The singularities are points in the plane, or lines in three-dimensional space, around which the phase increases or decreases continuously by 2π: they are optical vortices.

We may then continue this process and look at the interference pattern for a four-pinhole interferometer, a five-pinhole interferometer, and so on. One finds that, with limited exceptions,[*] the singularities for an interferometer with more than two pinholes are optical vortices of topological charge $m = \pm 1$; apparently these are the typical singularities that occur in a general interference experiment, and the pattern produced in the two-pinhole interferometer is a very special and unusual case.

---

[*] See the exercises for an example of one such exception.

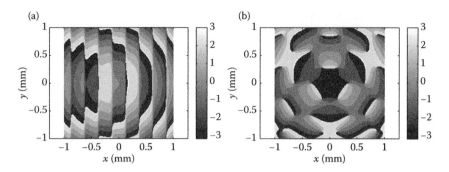

FIGURE 3.6 The phase patterns for (a) two-pinhole and (b) three-pinhole interferometers, when the amplitude of one of the pinholes is increased from $A_1 = 1$ to $A_1 = 1.2$.

We may explore the difference further by adjusting the amplitude of light coming from one of the pinholes for both the two- and three-pinhole case. If we change the amplitude of the first pinhole in each case to $U_1 = 1.2$, the intensity patterns look almost the same. However, the phase pattern of the two-pinhole case changes significantly, as shown in Figure 3.6a. The discontinuous phase jumps have turned into rapid but smooth changes in phase. We conclude from this that the new intensity pattern does not have any zeros at all; in fact, one can show that Young's two-pinhole experiment only has zeros when the amplitude of light illuminating the two holes is exactly the same.

For the three-pinhole case, illustrated in Figure 3.6b, the optical vortices are still present at or near their original locations. This is another demonstration that an optical vortex is stable under perturbations of the generating system.

We have therefore used Young-type interferometers to illustrate what we mean when we refer to "generic" singularities: such singularities must satisfy two related criteria. First, they must be the most common ones that appear in interference. We have seen that Young's two-pinhole experiment has an unusual and not typical phase structure, as compared to the infinite family of N-pinhole experiments. Second, the singularities must be stable under perturbations of the optical system. The zero surfaces of Young's two-pinhole experiment do not satisfy this condition, while the vortices of the three-pinhole experiment do.

In general, we will find that optical vortices are generic features of wavefields, regardless of how they are generated; our next step is to look more closely at their structure.

3.2 TYPICAL FORMS OF WAVE DISLOCATIONS

We have seen through various examples that phase singularities are typically lines in three-dimensional space, but have not yet explained why; in fact, this is quite easy to do in a qualitative manner. In order for a scalar wavefield to vanish at a point, the real and imaginary parts of the wave must be zero simultaneously at that point. We therefore have two homogeneous equations to satisfy, $u_R(\mathbf{r}) = 0$ and $u_I(\mathbf{r}) = 0$, with three degrees of freedom: the x, y, and z coordinates in space. We therefore have two constraints to satisfy in three-dimensional space: this leaves us with one degree of freedom, indicating that phase singularities are lines in space.

We now should have enough information to analyze the local mathematical structure of the simplest phase singularities. We apply the following observations in deriving this structure:

- A directional wave, such as an optical beam, locally has the form of a plane wave, except at regions with phase singularities.

- Singularities take the form of lines in three-dimensional space where the amplitude is zero.

- These singular lines may follow complicated curved paths, but will be locally straight.

- Along the direction of wave propagation, the field satisfies the paraxial wave equation, Equation 2.27.

As a reminder, the paraxial wave equation is given by

$$\frac{\partial^2 u}{\partial x^2} + \frac{\partial^2 u}{\partial y^2} + 2ik\frac{\partial u}{\partial z} = 0, \tag{3.10}$$

where $u(\mathbf{r})$ is the slowly varying component of the wavefield; the complete wavefield $U(\mathbf{r})$ is given by

$$U(\mathbf{r}) = u(\mathbf{r})e^{ikz}. \tag{3.11}$$

We will therefore assume that we have a paraxial wave propagating in the z-direction; clearly, this choice is arbitrary and the equation could be revised to account for propagation in any direction. With this in mind, we have three possible cases to consider.

3.2.1 Screw Dislocation

Let us begin by imagining that the phase singularity line extends along the z-axis, passing through the origin $x = y = z = 0$. In the immediate neighborhood of the axis, the field is almost zero, and because the zero line is parallel to the axis, there will be very little change in the field behavior in the z-direction. We then have, to a good approximation,

$$\frac{\partial u}{\partial z} \approx 0. \tag{3.12}$$

This means that, in the immediate neighborhood of the origin, the slow component of the wavefield satisfies the equation

$$\frac{\partial^2 u}{\partial x^2} + \frac{\partial^2 u}{\partial y^2} = 0. \tag{3.13}$$

This is simply *Laplace's equation*, which can be shown to have a general solution of the form [Gbu11, Section 9.1]

$$u(x, y) = f(x + iy) + g(x - iy), \tag{3.14}$$

where $f(v)$ and $g(v)$ are analytic functions of the complex variable v. Both of these functions can be represented by the lowest nonzero term of their Taylor series expansions, which means that we may write

$$u(x, y) \approx \alpha(x + iy) + \beta(x - iy), \tag{3.15}$$

with α and β complex constants. If we consider the case $\beta = 0$, we may write the complete field as

$$U(x, y, z) = \alpha(x + iy)e^{ikz}. \tag{3.16}$$

This is a typical example of what we will call a *screw dislocation*. (We will explain the meaning of "dislocation" soon.) We may also write this in terms of cylindrical coordinates as

$$U(\rho, \phi, z) = \alpha\rho e^{i\phi} e^{ikz}. \tag{3.17}$$

If we consider a surface of constant phase, for example, $\arg(U) = 0$, we find that it has a helical form

$$x = \rho \cos(\phi), \tag{3.18}$$

$$y = \rho \sin(\phi), \tag{3.19}$$

$$z = -\phi/k, \tag{3.20}$$

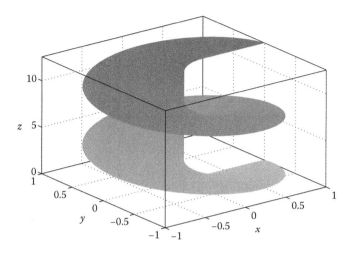

FIGURE 3.7 Illustration of a surface of constant phase for a screw dislocation.

where the pitch p of the helix is given by $p = 2\pi/k = \lambda$. This surface is illustrated in Figure 3.7, from which the origin of "screw" in "screw dislocation" should be obvious. (We will explain the meaning of "dislocation" soon.)

This is precisely the sort of vortex singularity that appears in our Laguerre–Gauss beams of order $m = 1$, as discussed in Chapter 2. From this image, we can see why a vortex with functional form $x + iy$ is called a "left-handed screw."

If we instead set $\alpha = 0$ and $\beta \neq 0$, we have a vortex of the opposite handedness and form $x - iy$. If both α and β are nonzero, the handedness depends on the relative sizes of the constants.

Our derivation also gives us an indication of why these screw dislocations are the generic type of singularities encountered. In order to encounter a higher-order singularity, both α and β must be identically zero in the Taylor expansion of the field around the central axis. In general, we do not typically expect to have both of these terms zero at the same location.

3.2.2 Edge Dislocation

The next simplest case to consider is a zero line that is parallel to the x-axis and passes through the origin. Mirroring the reasoning above, we expect that the x-derivatives near the line will be approximately zero, leaving the field $u(\mathbf{r})$ to satisfy the partial differential equation

$$\frac{\partial^2 u}{\partial y^2} + 2ik\frac{\partial u}{\partial z} = 0. \tag{3.21}$$

This is a diffusion-like equation that can in principle be solved, but as we are interested in the local form of the field near the origin, we instead consider a power series expansion of $u(\mathbf{r})$, valid up to the second order

$$u(y, z) = c_{00} + c_{10}y + c_{01}z + c_{11}yz + c_{20}y^2 + c_{02}z^2, \qquad (3.22)$$

with c_{ij} constants. On substitution of this form into the differential equation, we find that c_{10} is arbitrary, while

$$c_{20} = -ikc_{01}, \qquad (3.23)$$

and all other values are necessarily zero. This leads us to a general expression for the singularity of the form

$$U(x, y, z) = \alpha(ay + ky^2 + iz)e^{ikz}. \qquad (3.24)$$

It is to be noted immediately that we have found *two* zeros of the field, not just one: there is a zero at $(y, z) = (0, 0)$ and a second at $(y, z) = (-a/k, 0)$. Both singularities are required to find an exact solution of the paraxial wave equation, but for sufficiently small values of y, we may describe the singularity at $(0, 0)$ by the function

$$U(x, y, z) = \alpha(ay + iz)e^{ikz}. \qquad (3.25)$$

This is the simplest mathematical expression of what is called an *edge dislocation*. Equation 3.25 looks very similar to Expression 3.16 for a screw dislocation, but the phase behavior is notably different. The phase increases when following a screw dislocation along the positive z-axis, resulting in helical wavefronts, while the phase is constant when following an edge dislocation along the positive x-axis. If we look at the wavefronts in the y–z plane in the neighborhood of an edge dislocation, we find that an additional wavefront originates at the location of the dislocation. This is illustrated in Figure 3.8a for a single dislocation and in Figure 3.8b for a pair of dislocations. Following the black phase contours of the left side of the figure, for instance, we can see that an additional black contour appears to the right of $x = 0$ that is unpaired with any contour on the left.

As time passes, the surfaces of constant phase still circulate around the zero of the edge dislocation; it is therefore still a "vortex," albeit a nonhelical one.

Edge dislocations are very common in problems involving diffraction and focusing, and we will see an example of this in Section 3.5.

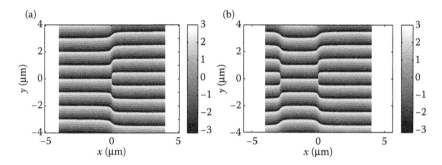

FIGURE 3.8 Illustration of (a) a single edge dislocation and (b) a pair of edge dislocations, for $\lambda = 1\,\mu\text{m}$ and $a = 18\,\mu\text{m}$.

3.2.3 Mixed Edge-Screw Dislocations

Along with dislocation lines parallel to and perpendicular to the direction of propagation, we can have lines that lie somewhere between. These are *mixed edge-screw dislocations*, and we take a few moments to consider their properties.

In this case, we assume that the zero line lies within the xz-plane, and we denote θ as the angle that the line makes with the x-axis. In keeping with our earlier discussion, we expect derivatives that act in the direction of this line to be negligible. To quantify this, we introduce new coordinates

$$x' = x\cos\theta + z\sin\theta, \tag{3.26}$$

$$z' = -x\sin\theta + z\cos\theta. \tag{3.27}$$

We have to be careful here, because the use of the paraxial equation automatically treats the derivatives along x and z on a different footing. This was not a problem when dealing with singularities that lie entirely along one of those axes, but for mixed axes, we should return to the Helmholtz equation, Equation 2.22, and rederive our paraxial expression in these terms. The Helmholtz equation is given by

$$(\nabla^2 + k^2)U(\mathbf{r}) = 0, \tag{3.28}$$

which in terms of our new coordinates becomes

$$\left(\frac{\partial^2}{\partial x'^2} + \frac{\partial^2}{\partial y^2} + \frac{\partial^2}{\partial z'^2}\right)U(\mathbf{r}) = 0. \tag{3.29}$$

Using $U(\mathbf{r}) = u(\mathbf{r})e^{ikz}$, we may write a reduced equation,

$$\left(2ik\sin\theta\frac{\partial}{\partial x'} + 2ik\cos\theta\frac{\partial}{\partial y'} + \frac{\partial^2}{\partial x'^2} + \frac{\partial^2}{\partial z'^2} + \frac{\partial^2}{\partial y^2} \right) u(\mathbf{r}) = 0. \quad (3.30)$$

We assume that our zero line is along the z'-axis; this implies that all z' derivatives are to a good approximation equal to zero. Furthermore, the second derivative with respect to x', lying only partly in the transverse plane, should be negligible compared to the second derivative with respect to y, which lies wholly in the plane. We are then left with

$$\left(\frac{\partial^2}{\partial y^2} + 2ik\sin\theta\frac{\partial}{\partial x'} \right) u(\mathbf{r}) = 0. \quad (3.31)$$

This is, in essence, Equation 3.21 for a rotated coordinate system. The simplest exact solution is therefore of the form

$$U(x, y, z) = \alpha(ay + k\sin\theta y^2 + ix')e^{ikz}, \quad (3.32)$$

and the local solution at the origin is given by

$$U(x, y, z) = \alpha(ay + ix')e^{ikz}, \quad (3.33)$$

where a again is an undetermined constant. In the limit that $\theta = 0$, we have $x' = x$ and get a screw dislocation; for $\theta = \pi/2$, we have $x' = z$ and get an edge dislocation.

What do we expect to be the behavior of the phase fronts? Let us take the simplest case, $a = -1$. For $\theta = 0$, we have a pure screw dislocation with a helical wavefront as already shown in Figure 3.7. Working in an (x', y, z')-coordinate system, we write $(x' + iy) = \rho e^{i\phi}$, which implies that the field may be written as

$$U(x', y, z') = \rho e^{i\phi} e^{ik(z'\cos\theta + x'\sin\theta)}. \quad (3.34)$$

The surface of constant phase will satisfy

$$z = -\tan\theta x - \phi/\cos\theta. \quad (3.35)$$

This is again a helical wavefront but one that is pitched in the x'-direction; this is illustrated in Figure 3.9. In the limit that $\theta \to \pi/2$, the pitch of the

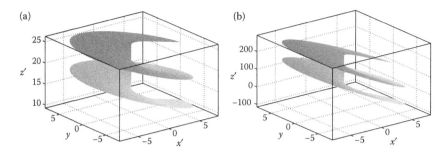

FIGURE 3.9 Illustration of (a) a mixed edge-screw dislocation for $\theta = \pi/8$ and (b) a mixed edge-screw dislocation for $\theta = 12\pi/25$.

helix becomes infinite and we end up with a pure edge dislocation. In the limit that $\theta \to 0$, the helix shape becomes independent of x and we end up with a pure screw dislocation.

This result suggests, at the very least, that dislocations with helical wavefronts are the most common and that pure edge dislocations are the exceptions to the norm. This does not, however, mean that edge dislocations are uncommon, as we will later see in the examples.

3.3 CRYSTAL DISLOCATIONS AND WAVEFRONTS

The terms "edge dislocation" and "screw dislocation" originally come from materials science, in particular the study of defects in crystal structures. There is a surprising similarity between certain classes of defects in crystals and phase singularities in waves, and it was natural to adopt the terminology of the former in explaining the latter. In fact, before J.F. Nye coauthored the first paper on singular optics [NB74], he previously had done research on crystal dislocations; see, for example, Bragg and Nye [BN47].

For our purposes, it is worth briefly discussing crystal dislocations both to aid in visualizing singularities in wavefields and to anticipate more complicated singularities to be discussed in later chapters.

The theory of crystal dislocations arose at the beginning of the twentieth century in the study of plastic (irreversible) deformations of crystals. In studying the shear stress required to create such a deformation, it was found that naive models for the process predicted critical forces between two and four orders of magnitude greater than those actually observed! One such naive model for the process is illustrated in Figure 3.10. This simple model could also not explain why a crystal has increased resistance to shearing as the amount of strain is increased.

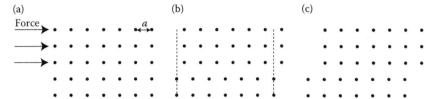

FIGURE 3.10   Illustration of an incorrect model for plastic deformation. A perfect crystal (a) has a shear force applied to it, causing the upper section of the crystal to translate to the right. When enough force is applied to displace the unit cells of the crystal by half a lattice constant *a*, it is now (b) in an unstable equilibrium; any additional force will tend to move the cells to their next stable crystal configuration (c). Dashed lines are drawn to highlight the misalignment.

A more refined picture, interpreting the plastic deformation of crystals in terms of dislocations, was introduced independently by Orowan [Oro34], Polanyi [Pol34], and Taylor [Tay34], with the latter introducing the term "dislocation." As Taylor explains it,

> It seems that the whole situation is completely changed when the slipping is considered to occur not simultaneously over all atoms in the slip plane but over a limited region, which is propagated from side to side of the crystal in a finite time.

In essence, the application of shear strain compresses a local region of the crystal, wedging an additional column of unit cells above a horizontal slip plane. The region where the additional column ends is the edge dislocation. With the strain increasing from the left, the energetically favorable response of the crystal is for the extra column to take over the position of the column to its right, effectively moving the dislocation. Eventually the dislocation results in a shifted crystal, as illustrated in Figure 3.11.

Comparing columns of unit cells with wavefronts, the similarity between Figures 3.8 and 3.11 is obvious. In a crystal, we have an additional lattice plane that originates at the dislocation line (out of the page), whereas in a wavefield, we have an additional wavefront that originates at the dislocation.

A shear strain not applied uniformly to the edge of a crystal can cause a slip over a fraction of the crystal's width, resulting in a screw dislocation. If we draw the crystal as lines that cross at the position of unit cells, edge and screw dislocations will appear as shown in Figure 3.12. In both situations, the dislocation exists as a line through the three-dimensional crystal.

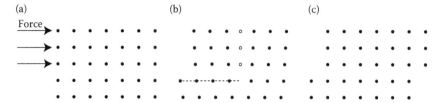

FIGURE 3.11 Formation and motion of a dislocation. (a) Shear force applied to rectangular crystal. (b) Region along the dashed line slips, leaving an extra column of unit cells out of the lattice (shown in white). (c) With shear force increasing, the extra column forces cells to the right out of position, and so on, until a stable shifted crystal results.

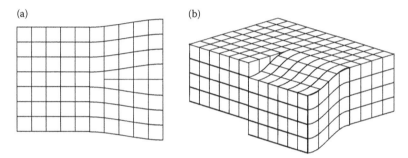

FIGURE 3.12 Illustration of (a) edge and (b) screw dislocations in crystals. The dislocation line is out of the page at the "T" junction for the edge dislocation, and is vertical in the center of the region for the screw dislocation.

We may also have mixed dislocations and, just like in the case of wavefront dislocations, the lines need only be locally straight. The increase in shear resistance can be interpreted as the formation and accumulation of multiple dislocations.

In what will be (for us) an anticipation of optical vortex properties, the structure of a crystal dislocation can be characterized by a vector known as the *Burgers vector* [Bur40]. It is defined for a given dislocation by first drawing a closed circuit around a regular region of the crystal, known as the *Burgers circuit*. If we follow an identical path around a dislocation, it will not form a closed path but will instead be separated from its starting point by a lattice vector; this is the Burgers vector.

An edge dislocation will have a Burgers vector that is perpendicular to the dislocation line, as seen in Figure 3.13a; a screw dislocation will have a Burgers vector that is parallel to the line, as seen in Figure 3.13b. A general mixed dislocation can have any lattice vector as its Burgers vector.

(a)   (b)

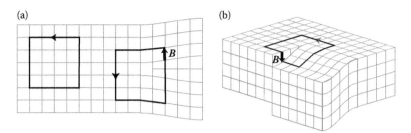

FIGURE 3.13 Illustration of the Burgers circuit and Burgers vector for (a) edge and (b) screw dislocations in crystals.

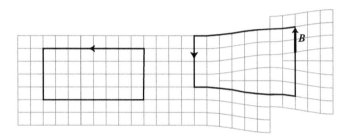

FIGURE 3.14 Burgers vector resulting from a circuit surrounding two edge dislocations of the same sense.

We typically expect the Burgers vector to be a single lattice point separation; however, the Burgers vectors for dislocations are additive, and it is possible to enclose several dislocations of the same sense in a single circuit, as illustrated in Figure 3.14.

With this in mind, two edge dislocations with a net Burgers vector of zero can come together to annihilate, leaving a perfectly regular crystal in its wake, as illustrated in Figure 3.15. In fact, one finds that Burgers vectors must be created or destroyed either in pairs or at the boundary of the crystal itself.

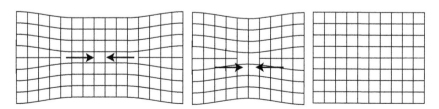

FIGURE 3.15 A pair of edge dislocations with opposite Burgers vectors coming together to annihilate.

In order to come together, as shown in the illustration, the dislocations must move from their "slip plane" (the original plane along which the crystal slipped) into a plane above or below it. This particular motion of dislocations is called *climb*; the complementary motion within the plane is called *glide*. More technically, a glide motion is any motion of a dislocation line on the surface generated by lines parallel to its Burgers vector. Because a screw dislocation has a Burgers vector parallel to the dislocation line, all motions of a screw dislocation are glide.

This discussion of dislocations has been introduced here to give a physical analog to the more abstract wave dislocations discussed in the rest of this book. However, it is important to stress that the analogy is not perfect, and the physical origin of the two types of dislocations is quite different. As stressed by Nye [Nye99, Section 5.10], wave and crystal dislocations have similar kinematics (motion), but very different *dynamics*. Crystal dislocations arise from stress forces and have energy associated with them; they move in response to these stress forces and have interaction forces between them as well. Wave dislocations have no forces associated with them and in fact can propagate at velocities faster than the vacuum speed of light, something a crystal dislocation cannot do. Crystal dislocations can provide insight into the structure of wave dislocations, but should not be assumed to perfectly mirror their behavior.

Another class of defects can arise in crystalline solids that also have analogy in wavefields. These defects are known as *disclinations*, as first coined by Frank [Fra58], and are associated with *rotations* of the crystal structure.

Illustrations of disclinations of a crystal solid are illustrated in Figure 3.16. As can be seen from the included arrows, a full circuit around the central disclination results in a rotation of the direction by 90°. When

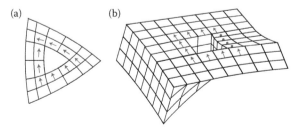

FIGURE 3.16 Illustration of two types of disclinations in cubic crystal solids. (a) A wedge disclination. (b) A twist disclination. (After Figure 9.2.18 of P.M. Chaikin and T.C. Lubensky. *Principles of Condensed Matter Physics*. Cambridge University Press, Cambridge, 1995.)

(a)   (b)

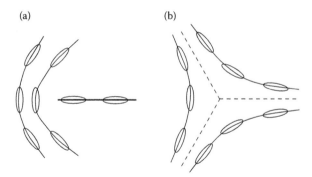

FIGURE 3.17   Illustration of two types of liquid crystal disclinations in a two-dimensional system: (a) lemon and (b) star.

the rotation is parallel to the axis of the disclination, it is called a *wedge*, and when the rotation is perpendicular to the axis of the disclination, it is called a *twist*.

Such disclinations require that the "unit cell" of the crystal be symmetric under 90° rotations, which is not generally true of optical waves and therefore not relevant for singular optics. A different set of disclinations, however, appear in liquid crystals, which are fluid-like collections of long thin molecules that can arrange themselves in crystalline form. Many such molecules (which can be simply depicted as long thin ellipsoids for our purposes) are symmetric under rotations of 180°, and a number of localized disclinations can be created that maintain this symmetry.

Illustrations of typical two-dimensional disclinations are shown in Figure 3.17. The structure on the left is referred to as a "lemon," and the structure on the right is a "star." As one follows a circular path around the central axis of the disclination, one finds that the orientation of the ellipses rotates by 180°, though in opposite senses. The ellipses rotate counterclockwise for the lemon, and clockwise for the star.

The analogous property to liquid crystal orientation in optical wavefields is the major axis of elliptical polarization. As we will see in Chapter 7, the singularities of polarization are liquid crystal-like disclinations of the major axis. The central point of these disclinations are points of circular polarization, where the major axis is undefined.

3.4 TOPOLOGICAL CHARGE AND INDEX

Returning to a discussion of optical phase singularities, we have seen that the phase of a wavefield increases or decreases by an integer multiple of 2π

as one traverses a closed path around the singularity. We have also noted that the structure of vortices is stable, in some sense, under wavefield perturbations. These properties together—discreteness and stability—suggest a "particle-like" nature for optical vortices. With this in mind, we label the integer associated with a vortex as its *topological charge*. The topological charge t of a vortex may be quantified by the following integral:

$$t \equiv \frac{1}{2\pi} \oint_C \nabla \psi(\mathbf{r}) \cdot d\mathbf{r}, \tag{3.36}$$

where $\psi(\mathbf{r})$ is the phase of the scalar wavefield $U(\mathbf{r})$, defined by

$$U(\mathbf{r}) = A(\mathbf{r})e^{i\psi(\mathbf{r})},$$

and, of course, $A(\mathbf{r})$ and $\psi(\mathbf{r})$ are real-valued functions. The contour C is taken to be a simple contour, that is, it does not intersect itself.

This definition reduces to the expected m-value for a single Laguerre–Gauss vortex beam of Chapter 2; if multiple vortices are enclosed in the contour C, their topological charges are *additive*. For example, if a charge 2 vortex and a charge -1 vortex lie within the path of integration, the total topological charge t within the contour is $t = 1$.

For simple fields such as the set of Laguerre–Gauss beams, it is clear that this charge is constant on propagation. This is self-evident because of the presence of the $\exp[im\phi]$ phase factor at all propagation distances. We will see, though, that the topological charge is also stable under small perturbations of the wavefield, such as passing the field through a weak phase screen. We will explain this stability in the next section.

Topological charge is not the only conserved property of a scalar wavefield. If we create a vector field by drawing the gradient vectors to the surfaces of constant phase, we can introduce a *topological index n* as follows. The index is the (integer) number of rotations that the vector field undergoes as one traverses a closed path around a feature of the wavefield. The topological charges and indices of typical wave features are illustrated in Figure 3.18.

The topological index of various wave features can be quantified using a formula directly analogous to that of topological charge. Labeling θ as the

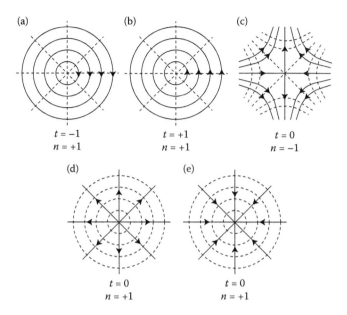

FIGURE 3.18 The topological charge t and index n values for a number of typical wave features. (a) Right-handed vortex, (b) left-handed vortex, (c) saddle, (d) source, and (e) sink. The dashed lines represent surfaces of constant phase, while the solid arrows represent the gradient of the phase.

angle that $\nabla\psi$ makes with the x-axis in the xy-plane, we may write

$$n \equiv \frac{1}{2\pi} \oint_C \nabla\theta(\mathbf{r}) \cdot d\mathbf{r}. \tag{3.37}$$

It is typically much easier to determine the index by the simple reasoning described in the previous paragraph than to explicitly calculate it using Equation 3.37. However, the formula will be quite useful in Chapter 7.

You may have noticed that we started referring to wave "features" and not "singularities." Saddle points, sources, and sinks are stationary points of the phase but not singularities, as the phase is well defined at their central points. However, it is to be seen from the values of charge and index in Figure 3.18 that all vortices have a positive topological index, while only a saddle has a negative index. Because both charge and index must be conserved under small wavefield perturbations, it is evident that the presence of saddles cannot be ignored in a discussion of vortex interactions.

3.5 CREATION AND ANNIHILATION EVENTS

The analogy between vortices and discrete fundamental "charges" goes much further than was demonstrated in the previous section. If we accept that topological charge is a conserved quantity, it follows that a vortex can only annihilate when it combines with a vortex of opposite charge, and that vortices can only be created in pairs of opposite charge. Furthermore, if we accept that topological index is also conserved, it follows that this creation/annihilation must be accompanied by the creation/annihilation of a pair of phase saddles, as well. This reaction appears to have been first elucidated by Nye, Hajnal, and Hannay [NHH88], who expressed it as the simple equation,

$$V^+ + V^- + 2S \rightleftharpoons 0. \tag{3.38}$$

In this expression, V^+ represents a left-handed vortex, V^- represents a right-handed vortex, and S represents a saddle.

This conservation law is, in essence, built into the scalar Helmholtz equation itself. We can justify it by building upon a few fundamental properties and a few appropriate visualizations.

Our starting point is the observation that a solution $U(\mathbf{r})$ to the Helmholtz equation is an *analytic* function in all three spatial variables x, y, and z. Specifically, we may state that $U(x, y, z)$ is the boundary value of an analytic function of three complex variables $\tilde{x}$, $\tilde{y}$, and $\tilde{z}$. A proof of this statement is outside of our immediate interest, but is described in Section 2.2 of Colton and Kress [CK83].

As is well known from complex analysis [Gbu11, Section 10.2], analytic functions of a single complex variable can only be zero at most at isolated points in the complex z-plane, unless it is identically zero within the domain of analyticity. This argument can be extended to analytic functions of more than one variable, with appropriate modifications. An analytic function of two variables can at most be zero on a path in the four-coordinate complex space unless it is zero everywhere; an analytic function of three variables can at most be zero on a surface in the six-coordinate complex space unless it is zero everywhere.

The same reasoning must apply to the real and imaginary parts of the wavefield separately, as the real and imaginary parts of a complex field satisfy the Helmholtz equation independently. If we therefore look at a transverse cross section of a wavefield, the real and imaginary parts of the field look like continually rolling hills, as shown in Figure 3.19.

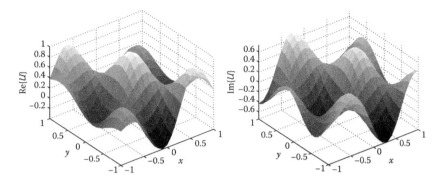

FIGURE 3.19   The real and imaginary parts of a scalar wavefield formed from four plane waves propagating in different directions.

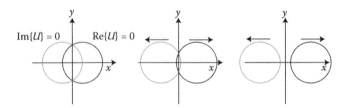

FIGURE 3.20   An idealized visualization of a vortex annihilation event.

It can be seen that the zero lines formed by the intersection of these "hills" with the $U = 0$ plane are, in general, not the same. These lines form the boundary between regions where the real and imaginary parts of the field are positive and negative. Because the field is continuous, this implies that these zero lines must either form closed paths or extend to infinity.

With this in mind, it is quite simple to see how singularities can only be created or annihilated in pairs of opposite charge. A generic singularity with topological charge $t = \pm 1$ is simply the intersection of the lines of $Re\{U\} = 0$ and $Im\{U\} = 0$; we recall the discussion of Figure 2.5. A hypothetical example of annihilation is shown in Figure 3.20, in which these zero lines are taken to be intersecting circles. The two locations where the circles intersect are a pair of vortices of opposite topological charge.[*] If the circles are made to move apart from one another by some change in system parameters (a change of focal length, aperture size, etc.), their intersection will reduce to a point and then disappear entirely. The vortices of opposite handedness have come together and annihilated. A creation event can be

---

[*] We have not specified the particular charge of the vortices, which depends on whether the interior of the circles are positive or negative values of their respective functions.

envisioned as the opposite process, in which a pair of separate circles come together and intersect, resulting in the creation of a pair of opposite-handed vortices where the circles overlap.

If we consider the full three-dimensional field, the topology in question is more complicated but the implications are the same. In three dimensions, the real and imaginary parts of the wavefield can be zero on surfaces at most; these surfaces will generally have line intersections. A simple visualization of this, building on Figure 3.20, would be the intersection of two cylinders with parallel axes along the z-direction. As the cylinders separate, the two parallel vortex lines of opposite charge will come together and annihilate.

Not all potential annihilation events require the destruction of two separate lines, however. A single vortex that bends back upon itself along the z-axis will appear, in any cross-sectional plane, as two vortices of opposite handedness. If the plane of observation is moved, these vortices appear to create or annihilate. This is illustrated in Figure 3.21. From a three-dimensional perspective, there appears to be only a single stationary vortex line. If we look at a two-dimensional cross section of the field with the z-position treated as a parameter of the system, however, we find that the in-plane vortices are "created" or "annihilated" as we change z. This example indicates that the interpretation of topological events in many circumstances depends on one's perspective. Nevertheless, the presence or absence of vortices still depends on the simple intersection of the surfaces of $\text{Re}\{U\} = 0$ and $\text{Im}\{U\} = 0$.

Nongeneric singularities, such as the zero surfaces of Young's two-pinhole interferometer, require that the surfaces $\text{Re}\{U\} = 0$ and $\text{Im}\{U\} = 0$ coincide exactly. It should be almost self-evident that this will not be a typical result.

As a simple example, we return to the three-pinhole interferometer of Figure 3.4, but multiply each of the fields emerging from the pinholes by a

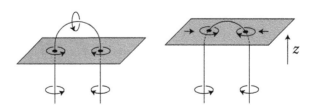

FIGURE 3.21 An idealized visualization of a vortex "annihilation" involving a single line, as the measurement plane is moved in the z-direction.

phase term as follows:

$$\phi_1 = 0, \tag{3.39}$$

$$\phi_2 = 2\pi/3, \tag{3.40}$$

$$\phi_3 = 4\pi/3. \tag{3.41}$$

This simple modification guarantees that, when all pinhole amplitudes are equal, there exists an optical vortex of topological charge $t = -1$ in the center of the observation screen. Starting with all pinhole amplitudes equal to unity, we gradually increase the amplitude U_1 of pinhole 1. In the limit of an arbitrarily large U_1, it will dominate the interference pattern and there should be no singularities present. We show the evolution in Figure 3.22.

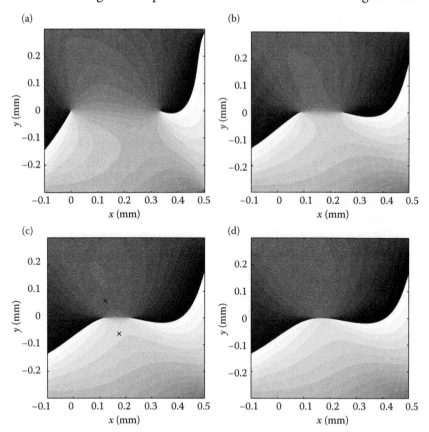

FIGURE 3.22 Annihilation of phase singularities in a three-pinhole interferometer, as (a) $U_1 = 1$, (b) $U_1 = 1.75$, (c) $U_1 = 1.9$, and (d) $U_1 = 2.0$. All other parameters are as in Figure 3.3. In (c), saddle points are marked with crosses.

In part (c), we can clearly see the presence of a pair of saddle points on either side of the line connecting the vortices: the saddles are regions where the phase is a local maximum along one direction in the plane and a local minimum along the perpendicular direction. This is in agreement with our statement that a pair of saddles must be annihilated with a pair of vortices in order to conserve the topological index. Another observation, which will be relevant in Chapter 7, is that the vortices annihilate along a fixed contour line. This implies that a vortex with local form $(x + iy)$ must annihilate with a vortex that is effectively out of phase, that is, $-(x - iy)$.

One of the most elegant demonstrations of the creation and annihilation of optical vortices was performed by Karman et al. in 1997 [KBvDW97], in the process of solving something of a minor puzzle in the theory of focusing. In Chapter 2, we derived the propagation characteristics of a paraxial Gaussian beam, and found that the beam maintains a perfect Gaussian intensity profile before, within, and beyond its waist plane. As a Gaussian function possesses no zeros, our calculation implies that a perfect Gaussian beam possesses no phase singularities at all on propagation. However, when a plane wave is focused through a hard circular aperture of radius a, it is well known [BW99, Section 8.5] that rings of zeros form in the focal plane; these are the *Airy rings*, and they are circular edge dislocations. If we focus a Gaussian beam of width w_0 through an aperture, we should see a transition from the Airy pattern to the Gaussian pattern as the ratio a/w_0 is increased, with an annihilation of all Airy ring vortices in the process. The setup is illustrated in Figure 3.23.

The experimental results of Karman et al. for the first two Airy rings (closest to the geometrical focus) are shown in Figure 3.24. As the ratio a/w_0 is increased, the first two Airy rings (seen in the cross section here as A and B) come together and annihilate, while accompanied by two other singularities C and D. The general trend, as a/w_0 increases without limit, is for all the Airy rings to shrink toward the geometrical focus, with those

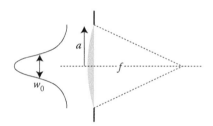

FIGURE 3.23   The focusing configuration of Karman et al. [KBvDW97].

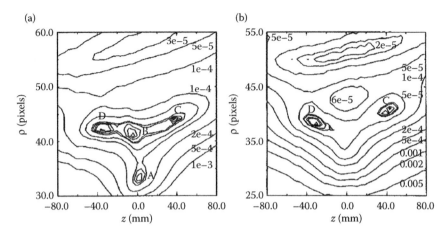

FIGURE 3.24 The intensity of the field in the neighborhood of the first two Airy rings for (a) $a/w_0 = 1.44$ ($a = 2.1$ mm) and (b) $a/w_0 = 1.64$ ($a = 2.39$ mm). Here $f = 1$ m and $w = 1.46$ mm. (Results from G.P. Karman et al. *Opt. Lett.*, 22:1503–1505, 1997.)

neighboring pairs closest to the focus annihilating in sequence. The experimental results are in excellent agreement with the researchers' theoretical predictions.

It is important to stress that, as a rule, the conservation of topological charge only holds when the system undergoes small, continuous perturbations, as we have emphasized several times. This caveat is necessary as, in the very next chapter, we will immediately see a counterexample in the form of the spiral phase plate. A Gaussian beam passing through such a plate will produce a single vortex and the net change in topological charge will be nonzero, in violation of the conservation law set out in this section. However, this seems somewhat paradoxical, as our law does not seem to have any real restrictions on its validity.

To address this contradiction, we first recall from our discussion of crystal dislocations in Section 3.3 that an isolated dislocation may disappear at the boundary of the crystal. In the optical case, this boundary is the point at infinity, and it was demonstrated by Soskin et al. [SGV+97] that superpositions of beams containing optical vortices can gain or lose vortices at infinity.

In the particular case of a spiral phase plate, we note that any vortex beam may be thought of as possessing a vortex of opposite charge, situated at the point at infinity. If we were to map the phase of the field from the xy-plane to a sphere by stereographic projection, we would find that the original vortex sits on the South pole of the sphere while the opposite vortex sits on the North pole. We will much later see, in Section 9.2, that this second vortex will move to a finite position when the spatial coherence of the wavefield is decreased.

If we were to computationally model the effect of the spiral phase plate on the illuminating wavefield, it seems reasonable that we would see a complicated system of creation and annihilation events occur as the wave propagates through the plate, culminating in one vortex on the axis and the other moving off to infinity. In practice, however, we never measure or observe such a process; so, it is most convenient to argue that our conservation law only applies to small, continuous perturbations, which accounts for a large number of situations of interest. A justification of this idea that pair production still occurs within a phase plate will be seen in Section 12.3, where fractional spiral phase plates are discussed.

3.6 EXERCISES

1. Using your preferred piece of mathematical software, calculate and plot the interference pattern produced by Young's four-pinhole interferometer, with $a = 1$ mm, $L = 1$ m, and $\lambda = 500$ nm. The pinholes should be arranged in a square, and the amplitudes and phases of the field emanating from the holes should be equal. Describe the types of singularities observed, and explain their origin. Are these generic singularities? Provide an example to illustrate your conclusion.

2. Using your preferred piece of mathematical software, calculate and plot the interference pattern of Young's three-pinhole interferometer of Figure 3.4 when it is illuminated by a plane wave shining at $10°$ to the z-axis, lying in the yz-plane, with $a = 1$ mm, $L = 1$ m, and $\lambda = 500$ nm. By varying L, determine the direction of orientation of the vortex lines.

3. Determine the topological charge of a screw dislocation that has the local mathematical form

$$u(x, y) = a(x + iy) + b(x - iy),$$

with a and b real-valued constants.

4. Determine the topological charge of a screw dislocation that has the mathematical form

$$u(x, y) = a(x + iy) + b(x + iy)^2.$$

How does the charge depend on the values of a and b?

5. Using the theory of the two-pinhole interferometer described in this chapter, prove that complete destructive interference only occurs when the amplitude of light at the two pinholes is equal, that is, $U_1 = U_2$.

CHAPTER 4

Generation and Detection of Optical Vortices

N OW THAT WE HAVE discussed the general structure and properties of optical vortices, we come to a pair of very practical questions: how do we generate pure vortex beams and how do we measure the characteristics of a vortex? As singular optics has grown as a field, a variety of different methods have been introduced for both generation and detection, each with its own advantages and disadvantages. In this chapter, we summarize a number of these, beginning with the most common and progressing to the most esoteric.

4.1 GENERATION

In the previous chapter, we noted that topological charge is conserved, and that optical vortices are only created in pairs. However, this conservation came with the caveat that it is upheld only "under small perturbations," such as scattering from a tenuous random media or refraction through a material whose properties change slowly with the wavelength. It is not difficult to violate this condition and produce a beam with a total nonzero topological charge, as the following methods illustrate.

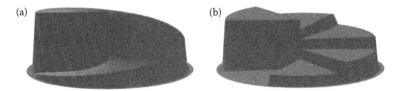

FIGURE 4.1 (a) A continuous spiral phase plate of height h. (b) A staircase spiral phase plate with six steps.

4.1.1 Spiral Phase Plate

Perhaps, the simplest method for producing a vortex beam is to pass a pure Gaussian laser mode through an optical element known as a *spiral phase plate*, as illustrated in Figure 4.1a. The element is of constant thickness in the radial direction but increases in thickness continuously in the azimuthal direction, with a height difference h between the thickest and thinnest portions.

Let us consider passing a Gaussian beam vertically through the center of this plate. If the beam is highly directional and the plate is sufficiently thin, we can evaluate its effect using a geometric model of propagation. The thickness $t(\phi)$ of the plate can be written as

$$t(\phi) = \frac{h\phi}{2\pi}. \tag{4.1}$$

We take the plate to be in a background medium of refractive index n_0. The plate's refractive index is labeled n, and the vacuum wavelength of the illuminating light is λ. As a function of azimuthal angle, the total optical thickness $O(\phi)$ of the plate (including the propagation through the background medium) is then

$$O(\phi) = \left(h - \frac{h\phi}{2\pi}\right)n_0 + \frac{h\phi}{2\pi}n. \tag{4.2}$$

Neglecting the overall constant contribution, rays passing through the plate will acquire a phase ψ given by

$$\psi(\phi) = \frac{h\phi}{\lambda}(n - n_0). \tag{4.3}$$

If the plate is fashioned so that

$$\frac{h}{\lambda}(n - n_0) = m, \tag{4.4}$$

with m as an integer, then, the field $u'(\mathbf{r})$ emerging from the plate will be related to the incident field $u(\mathbf{r})$ by

$$u'(\mathbf{r}) = e^{im\phi} u(\mathbf{r}). \tag{4.5}$$

In short, the field has been given a helical phase. On propagation, the field will evolve a dark spot in its center characteristic of a Laguerre–Gauss vortex beam.

The earliest paper[*] on using a spiral phase plate to produce vortex beams appears to be the work of Kristensen, Beijersbergen, and Woerdman [KBW94], who designed a plate to be used at microwave frequencies in a waveguide. Soon after, some of the same authors demonstrated a phase plate at optical frequencies [BCKW94] in free space. A few years later, a phase plate was fabricated for millimeter wavelengths and free space propagation [TRS+96].

By the nature of its generation, however, a vortex beam produced by a spiral phase plate will not be a pure mode. Although the phase will in principle be of the desired form, the plate does not modify the amplitude distribution of the transmitted beam, as Equation 4.5 shows. The result is that the transmitted beam will contain Laguerre–Gauss contributions of a single azimuthal order m but different radial orders n. We can demonstrate this by recalling from Section 2.4 that the Laguerre–Gauss functions are complete, and that any wavefield can be represented as a superposition of them. Restricting ourselves to the waist plane $z = 0$, we may therefore determine the weight w_{nm} of each Laguerre–Gauss mode in our transformed beam by the integrals

$$w_{nm} = \int u'(x', y') u_{nm}^*(x', y') dx' \, dy', \tag{4.6}$$

where $u_{nm}(x, y)$ are the Laguerre–Gauss modes of Section 2.4.

The values $|w_{nm}|^2$ of these weights for a Gaussian beam incident upon a spiral phase plate with $m = 1$ are shown in Table 4.1. It can be seen that only 78.5% of the beam is contained in the LG_{01} mode.

Phase plates can also be used to convert between higher-order Laguerre–Gauss modes of different helicities. For instance, a spiral phase plate with $m = 1$ can be used to convert an LG_{22} mode into an LG_{23} mode, and a plate with $m = 2$ can convert from $LG_{0,-1}$ to LG_{01}. The efficiency of the process

---

[*] It has been noted by Michael Berry that Michael Walford created an ultrasound vortex in 1974 using a spiral-cut metal sheet; Walford never published his result.

TABLE 4.1   Mode Weights $|w_{n1}|^2$ for a Gaussian Beam Incident on an $m = 1$ Spiral Phase Plate

LG_{01}	78.5%
LG_{11}	9.8%
LG_{21}	3.7%
LG_{31}	1.9%
LG_{41}	1.2%
LG_{51}	0.8%

depends upon the helicity of the plate and the mode to be converted; this is discussed by Beijersbergen et al. [BCKW94].

From Equation 4.3, it is clear that there are two parameters that control the helical phase for a fixed wavelength: the refractive index contrast ($n - n_0$) and the thickness h. If the refractive index contrast is roughly unity (such as would be for a glass plate in air), a helicity $m = 1$ requires a plate thickness on the order of the wavelength. At optical frequencies ($\lambda \approx$ 500 nm), this can be quite difficult to fabricate; in fact, the first optical spiral phase plate [BCKW94] was built on a millimeter scale and submerged in an index-matching liquid, using a small $n - n_0$ to allow for a larger h. As an alternative, the process of making an optical plate can be simplified by approximating the continuous phase ramp as a staircase structure, as illustrated in Figure 4.1b. The price of this approximation is a further loss in mode purity: now the transmitted beam will possess a mixture of modes with different values of both m and n.

We can roughly demonstrate this by assuming a staircase with six steps, equally distributed, so that

$$\psi(\phi) = 2\pi n/5, \quad 2n\pi/6 \le \phi < 2(n+1)\pi/6, \quad 0 \le n \le 5, \quad (4.7)$$

with n as an integer. We may evaluate the relative weights of pure spiral modes in this phase function by evaluating the integral

$$f(m) \equiv \frac{1}{2\pi} \int_0^{2\pi} e^{-im\phi} e^{i\psi(\phi)} \, d\phi. \quad (4.8)$$

The result is plotted in Figure 4.2. It can be seen that a majority of the transmitted field is in the $m = 1$ mode, though a certain fraction of the weight is in adjacent modes. With a larger number of steps, the azimuthal purity of the mode will improve.

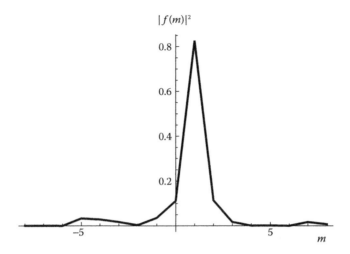

FIGURE 4.2 The quantity $|f(m)|^2$ for a staircase spiral plate with six steps.

Looking back at Equation 4.3, we may note an additional limitation of the simple spiral phase plates discussed here: they are explicitly wavelength dependent. Such a plate can only achieve the desired helicity for a single wavelength, and will have poor performance outside of a narrow bandwidth. Furthermore, the wavelength dispersion of the plate material, $n = n(\lambda)$, and possibly the background material, $n_0 = n_0(\lambda)$, introduces additional wavelength dependence.

However, this dispersion can be used to one's advantage. Just as the chromatic aberration of lenses can be ameliorated by constructing *achromatic doublets** consisting of a pair of lenses with complementary powers and dispersive properties, it is possible to construct an achromatic vortex lens [GS06] consisting of a substrate and background material of complementary dispersion. We consider such an achromat as a pair of vortex lenses of opposite handedness and different material properties; together they form a cylinder of thickness h.

We may investigate this vortex lens as follows. Over a relatively small bandwidth centered on λ_0, the refractive index of the substrate and the background medium may be written via a Taylor expansion

$$n(\lambda) \approx n(\lambda_0) + n'(\lambda_0)\delta\lambda + \frac{1}{2}n''(\lambda_0)(\delta\lambda)^2, \tag{4.9}$$

$$n_0(\lambda) \approx n_0(\lambda_0) + n_0'(\lambda_0)\delta\lambda + \frac{1}{2}n_0''(\lambda_0)(\delta\lambda)^2, \tag{4.10}$$

---

* See, for instance, Born and Wolf [BW99], Section 4.7.

where $\delta\lambda \equiv \lambda - \lambda_0$, $n'(\lambda)$ represents the derivative of n with respect to λ, and so forth. Furthermore, for sufficiently small $\delta\lambda$, we may make the Taylor series approximation

$$\frac{1}{\lambda} \approx \frac{1}{\lambda_0} \left[1 - \frac{\delta\lambda}{\lambda_0} \right]. \tag{4.11}$$

Combining these expressions into Equation 4.4 and keeping only terms linear in $\delta\lambda$, we find an effective topological charge given by

$$m(\lambda) \approx \frac{h}{\lambda_0} \left[n(\lambda_0) - n_0(\lambda_0) \right]$$

$$- \left\{ \frac{1}{\lambda_0} \left[n(\lambda_0) - n_0(\lambda_0) \right] - \left[n'(\lambda_0) - n_0'(\lambda_0) \right] \right\} \delta\lambda. \tag{4.12}$$

The first term on the right of this expression represents the topological charge at the center frequency, which can be tailored to take on an integer value. The term in the curved brackets represents the variation of the topological charge with respect to wavelength; for an appropriate choice of materials, this quantity can be adjusted to be close to zero, resulting in a vortex lens that is wavelength independent over roughly 100 nm at optical wavelengths. (Not stunningly large, but better than the chromatic lens.) There is one downside to this design: as noted in Swartzlander [GS06], for optical glasses, the refractive index difference $n(\lambda_0) - n_0(\lambda_0)$ is typically on the order of 10^{-2}, which from Equation 4.4 implies that the achromatic spiral plate is necessarily 100 wavelengths thick.

This effort to make an integer phase step dodges a potentially interesting question: what happens to a beam that passes through a spiral phase plate with a fractional step? We have seen that an optical vortex necessarily must have a topological charge of integer value, so that a fractional spiral phase plate will not result in a fractional vortex beam. A detailed discussion of the singular optics of fractional spiral phase plates and the beams generated by them will be undertaken in Chapter 12.

4.1.2 Mode Conversion

As we will see in an upcoming section, it is generally difficult to make a laser oscillate in a pure Laguerre–Gauss mode. The use of Brewster windows and other elements in the laser cavity breaks the rotational symmetry of the system, typically forcing it to oscillate in a Hermite–Gauss mode. However, it is relatively easy to select a particular Hermite–Gauss mode by placing

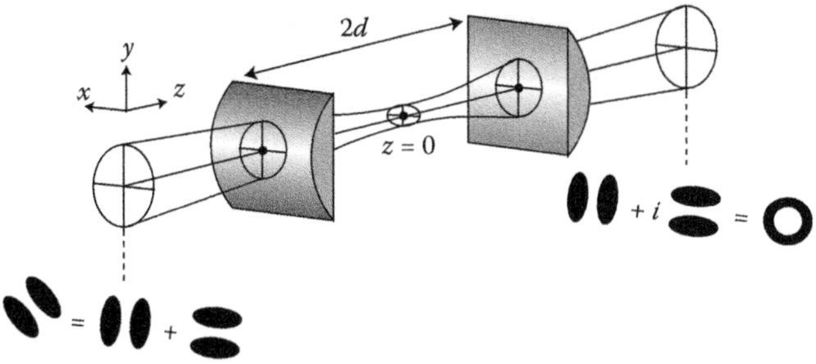

FIGURE 4.3 The conversion of a Hermite–Gauss mode into a Laguerre–Gauss mode by the use of a pair of cylindrical lenses. (Adapted from M. Padgett et al. *Am. J. Phys.*, 64:77–82, 1996.)

crossed wires in the laser cavity in positions corresponding to the nodes of the desired mode.

With this in mind, we now show that it is straightforward to convert a Hermite–Gauss mode into a Laguerre–Gauss mode via the use of a pair of cylindrical lenses. The technique is illustrated in Figure 4.3 for the simplest case; it was first demonstrated in Beijersbergen et al. [BAvdVW93] and a simplified discussion appears in Padgett et al. [PASA96].

Up to this point, we have considered beams that have a pure spherical wavefront or, to state it in a more useful way, beams that have the same wavefront curvature in the x and y transverse directions. A cylindrical lens, however, will alter the wavefront shape only along a single axis. Because the Rayleigh range (2.34) of a beam is directly related to its wavefront curvature (2.30), a cylindrical lens will result in different Rayleigh ranges for a beam along the x- and y-directions; we label these as z_{0x} and z_{0y}. This difference in turn results in two different Gouy phases (2.32) along the two transverse directions. Thanks to this asymmetric Gouy shift, a pair of Hermite–Gauss modes will acquire different phases on propagation through the system, and this difference can be tuned to produce a Laguerre–Gauss mode. A second cylindrical lens removes the asymmetry in the beam curvature.

The process is strikingly analogous to the method by which a linearly polarized beam can be converted into circular polarization by the use of a quarter-wave plate. If a linear polarization is input to a quarter-wave plate at 45° to the latter's axes, it will emerge as circularly polarized. Similarly, if a Hermite–Gauss mode is input to the mode converter at 45° to the axis of the cylindrical lens, labeled as HG_{10}^*, it will emerge as a Laguerre–Gauss mode.

Assuming that the focal length of the cylindrical lens is f and the lenses are each a distance d from the origin along the z-axis, it can be shown that, to produce a vortex, the separation of the lenses must be

$$2d = f/\sqrt{2}, \tag{4.13}$$

and the Rayleigh range of the incident beam must be taken to be

$$z_0 = (1 + 1/\sqrt{2})f. \tag{4.14}$$

This can be proven by a clever manipulation of the existing beam formulas at our disposal. We first note that the HG_{10}^{*} mode can be written as the linear sum of the HG_{10} and HG_{01} modes, namely,

$$u_{10}^{HG*}(\mathbf{r}) = \frac{1}{\sqrt{2}}\left[u_{10}^{HG}(\mathbf{r}) + u_{01}^{HG}(\mathbf{r})\right]. \tag{4.15}$$

Next, we observe that the expression for any Hermite–Gauss beam is completely separable into independent functions of x and y; this includes a factorization of the total Gouy phase into pieces that depend on the x and y orders of the beam, that is,

$$\Psi^{G}(z) = \psi_{m}^{G}(z) + \psi_{n}^{G}(z), \tag{4.16}$$

where

$$\Psi_{m}^{G}(z) = -(m + 1/2)\arctan(z/z_0), \tag{4.17}$$

$$\Psi_{n}^{G}(z) = -(n + 1/2)\arctan(z/z_0). \tag{4.18}$$

We now consider the following construction. We imagine that we have an asymmetric beam being focused to a waist at $z = 0$; the Rayleigh ranges along the x and y directions are z_{0x} and z_{0y}, respectively, and similarly the waist widths along those directions are w_{0x} and w_{0y}. The asymmetry is introduced by a cylindrical lens at position $-d$ that changes the curvature along the y-direction. The two widths along x and y must be the same at the cylindrical lens; however, using Equation 2.31, this implies that

$$\frac{z_{0x}^2 + d^2}{z_{0x}} = \frac{z_{0y}^2 + d^2}{z_{0y}}. \tag{4.19}$$

Also, the lens converts the (negative) curvature along the y-direction, which must be related to the original symmetric curvature of the beam by

$$\frac{1}{f} = \frac{1}{R_x} - \frac{1}{R_y}. \tag{4.20}$$

Using Equation 2.30, we have

$$\frac{1}{f} = \frac{d}{d^2 + z_{0x}^2} - \frac{1}{d^2 + z_{0y}^2}. \tag{4.21}$$

The pair of Equations 4.19 and 4.21 can be solved for the Rayleigh ranges, though the calculation is challenging. Following Beijersbergen et al. [BAvdVW93], the result may be written as

$$z_{0x} = dp, \tag{4.22}$$

$$z_{0y} = d/p, \tag{4.23}$$

where

$$p \equiv \sqrt{\frac{1 - d/f}{1 + d/f}}. \tag{4.24}$$

Now, we can consider the total Gouy phase shift of such a beam. Assuming that it is a Hermite–Gauss mode of order mn, its net shift may be written as

$$\Psi^G(z) = -(m + 1/2) \arctan(z/z_{0x}) - (n + 1/2) \arctan(z/z_{0y}). \tag{4.25}$$

The total shift, after traveling from $z = -d$ to $z = +d$, will be

$$\begin{aligned} \Delta\Psi_{mn} &= -2(m + 1/2) \arctan(d/z_{0x}) - 2(n + 1/2) \arctan(d/z_{0y}) \\ &= -2[(m + 1/2) \arctan(1/p) + (n + 1/2) \arctan(p)]. \end{aligned} \tag{4.26}$$

The phase between, for instance, the HG_{10} and HG_{01} modes will be

$$\Delta\Psi_{10} - \Delta\psi_{01} = -2[\arctan(1/p) - \arctan(p)]. \tag{4.27}$$

If p is taken to be $p = -1 + \sqrt{2}$, this difference will be $-\pi/2$, which implies that the output beam will be

$$u_{out}(\mathbf{r}) = \frac{1}{\sqrt{2}} \left[u_{10}^{HG}(\mathbf{r}) + i u_{01}^{HG}(\mathbf{r}) \right] = u_{01}^{LG}(\mathbf{r}). \tag{4.28}$$

This same mode converter will convert any "diagonal" Hermite–Gauss mode into a corresponding Laguerre–Gauss mode. More detail can be found in Beijersbergen et al. [BAvdVW93].

4.1.3 Computer-Generated Holograms

A hologram, which can impart a nontrivial phase structure onto an illuminating wavefield, would appear to be an ideal means of creating vortex beams from fundamental Gaussian beams. Generating the vortex hologram by computer obviates the need to record an initial hologram using a vortex beam, avoiding a potential "chicken-and-egg" problem in creation. The technique has its own special limitations, however, and we briefly review the principles of holography to highlight them. More information on holography can be found, for instance, in the book by Hariharan [Har96].

The original holographic design was the in-line hologram developed by Gabor [Gab48], and illustrated in Figure 4.4. A plane wave U_0 is normally incident upon a thin transparent scattering object. A scattered field is produced that has the form $U_s(x, y)$ on the photographic plate (assumed to be small compared to the incident field), and the total field $U_t(x, y)$ is given as

$$U_t(x, y) = U_0 + U_s(x, y). \tag{4.29}$$

The corresponding intensity on the plate is

$$I(x, y) = |U_0 + U_s(x, y)|^2 = |U_0|^2 + |U_s(x, y)|^2$$
$$+ U_0^* U_s(x, y) + U_0 U_s^*(x, y). \tag{4.30}$$

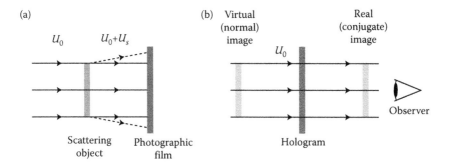

FIGURE 4.4 The (a) creation and (b) illumination of a Gabor in-line hologram. (Adapted from P. Hariharan. *Optical Holography.* Cambridge University Press, Cambridge, 2nd edition, 1996.)

For simplicity, we assume that the developed film has a transmissivity $t(x, y)$ that is linear with respect to the intensity it is exposed to, or

$$t(x, y) = t_0 + \beta I(x, y) = (t_0 + \beta |U_0|^2) + \beta |U_s(x, y)|^2$$
$$+ \beta U_0^* U_s(x, y) + \beta U_0 U_s^*(x, y). \qquad (4.31)$$

Here, t_0 is a constant representing the degree of uniform throughput, and β is a factor depending on the properties of the development process and the exposure time.

When the developed hologram is illuminated again with a uniform plane wave, there are four contributions to the transmitted field. There is a component, including t_0, that represents a directly transmitted plane wave. There is a term proportional to $|U_s(x, y)|^2$, which is assumed to be negligible compared to the others. The third term is proportional to the original scattered field $U_s(x, y)$, and produces a virtual image of the object in its original position. The fourth term, proportional to $U_s^*(x, y)$, requires a bit more explanation. It is the complex conjugate of the scattered field; because the scattered field is roughly a diverging wave, the conjugate field is roughly a converging wave. It produces a real image between the hologram and the observer.

A helical wavefront can be recorded as a hologram or, alternatively, be generated directly by computer.[*] The latter was done by Heckenberg et al. [HMSW92], who created binary spiral zone plates to generate vortex fields of topological charge 1 and 2. A simulation of such zone plates is shown in Figure 4.5. They can be produced by taking the scattered field to be the phase of a vortex beam with finite curvature, that is,

$$U_s(x, y, 0) = e^{im\phi} e^{ik(x^2 + y^2)/2R}, \qquad (4.32)$$

and then replacing the sinusoidal variation with a binary square wave, as discussed at the end of this section.

An in-line hologram is not ideal for producing vortex beams, however. The transmitted field not only includes the beam of the desired topological charge m, but also a uniform plane wave as well as a defocused conjugate beam of topological charge $-m$. The method was quickly supplanted by designing holograms of a Leith–Upatnieks type [LU62,LU63], as we now discuss.

---

[*] Computer-generated holograms have been around almost as long as holography itself; see Brown and Lohmann [BL66] as an early example.

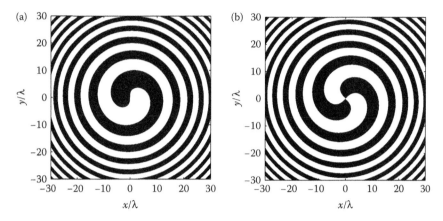

FIGURE 4.5 Binary spiral zone plates of (a) topological charge 1 and (b) topological charge 2. Here $R = 100\lambda$.

In a Leith–Upatnieks hologram, the reference wave is incident upon the photographic plate from an angle θ with respect to the normal z-axis, as illustrated in Figure 4.6a. The total field illuminating the plate is therefore

$$U_t(x, y) = U_s(x, y) + U_0 e^{-ik_x x}, \tag{4.33}$$

where $k_x^2 + k_z^2 = k^2$ and $k_x = k \sin \theta$. The intensity of the field on the plate is therefore

$$I(x, y) = |U_s(x, y)|^2 + |U_0|^2 + U_s(x, y) U_0^* e^{ik_x x} + U_s^*(x, y) U_0 e^{-ik_x x}. \tag{4.34}$$

This implies that the transmission function of the hologram is given by

$$t(x, y) = (t_0 + \beta |U_0|^2) + \beta |U_s(x, y)|^2 + \beta U_s(x, y) U_0^* e^{ik_x x}$$
$$+ \beta U_s^*(x, y) U_0 e^{-ik_x x}, \tag{4.35}$$

and the transmitted field U_{trans} produced when the developed hologram is illuminated by another incident wave at the same angle is

$$U_{trans}(x, y) = (t_0 + \beta U_0^2) U_0 e^{-ik_x x} + \beta U_0 |U_s(x, y)|^2 e^{-ik_x x}$$
$$+ \beta U_s(x, y) U_0^2 + \beta U_s^*(x, y) |U_0|^2 e^{-2ik_x x}. \tag{4.36}$$

The first term in this equation represents the portion of the plane wave directly transmitted by the hologram. The second term, dependent on

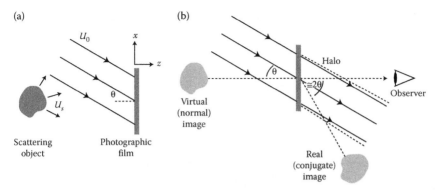

FIGURE 4.6 The (a) creation and (b) illumination of a Leith–Upatnieks hologram. (Also adapted from P. Hariharan. *Optical Holography.* Cambridge University Press, Cambridge, 2nd edition, 1996.)

$|U_s(x, y)|^2$, results in a "halo" of scattered light that spreads around the direction of the transmitted plane wave. (Note that we no longer have to assume that the scattered field is small compared to the incident wave, as it does not intersect the desired image.) The third term is directly proportional to the scattered field and therefore produces a virtual image of the original object. The fourth term is the conjugate field again, but it is now multiplied by $\exp[-2ik_x x]$. This means that the real image of the conjugate field appears at roughly an angle -2θ with respect to the z-axis. As seen in Figure 4.6b, the desired virtual image is therefore spatially separated from both the illuminating field and the conjugate image. If the "scattered field" is designed to be a vortex, we will produce a pure vortex beam propagating from the virtual image.

This method was first used by Heckenberg et al. [HMS$^+$92], who used off-axis binary computer-generated holograms to produce vortex beams of charge 1 and charge 2. After generation by computer, the holographic patterns were printed onto paper and then photoreduced to the appropriate size and developed on a slide. A hologram for a plane wave interfering with the phase of a vortex beam is illustrated in Figure 4.7; it can be seen that the dark patches, essentially representing wavefronts of the field, have an appearance analogous to an edge dislocation of a crystal, such as in Figure 3.12. Such an image is usually referred to as a *fork diagram*.

An ideal computer-generated hologram would be grayscale and have a transmission function that is a continuous function of position. However, it is often easier to produce a binary hologram, which reduces these continuous gradations into simple light and dark regions. This simplification turns

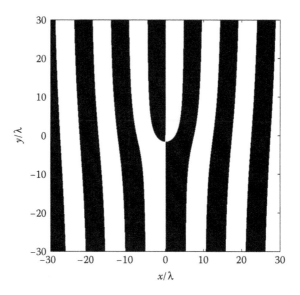

FIGURE 4.7 Simulation of the so-called "fork diagram" produced in an off-axis vortex hologram. Here $k_x = 0.1\ k$.

the hologram into a diffraction grating with many holographic orders, as can be readily shown.

Neglecting the amplitude variations of a vortex beam, we treat the scattered field in our hologram as

$$U_s(\phi) = e^{i\phi}. \tag{4.37}$$

With an appropriate choice of t_0 and β, we may write our transmission function simply as

$$t(x, \phi) = \frac{1}{2}[1 - \cos(k_x x + \phi)]. \tag{4.38}$$

We then replace the cosine function by its square wave equivalent $F(u)$, equal to -1 for $\cos(u) \leq 0$ and $+1$ for $\cos(u) > 0$. This function has a Fourier series representation equal to

$$F(u) = \sum_{n=0}^{\infty} \mathrm{sinc}(n\pi/2) \cos(nu), \tag{4.39}$$

where $\text{sinc}(x) = \sin(x)/x$. This implies that the transmission function is given by

$$t(x, \phi) = \frac{1}{2} - \sum_{n=0}^{\infty} \text{sinc}(n\pi/2) \cos[n(k_x x + \phi)], \qquad (4.40)$$

and the transmitted field $U_t(x, \phi)$ under illumination by $U_0 e^{-ik_x x}$ is

$$U_t(x, \phi) = \frac{1}{2} U_0 e^{-ik_x x} - U_0 \sum_{n=1}^{\infty} \text{sinc}(n\pi/2)$$

$$\times \left[e^{i(n-1)k_x x + in\phi} + e^{-i(n+1)k_x x - in\phi} \right]. \qquad (4.41)$$

As the summation includes an infinite number of terms with azimuthal dependence $\exp[in\phi]$, we therefore get an in-principle large collection of vortex modes of different topological charges, positive and negative, propagating with different transverse wavevector components. Because of the presence of the sinc term, the largest amplitude will be possessed by the term with the desired $m = 1$ charge, and no other beam will be propagating in the same direction. We may therefore obtain a pure vortex beam, at the cost of losing energy into the other modes.

4.1.4 Direct Laser Generation

Though many lasers are designed to operate in a single longitudinal mode and in the fundamental Gaussian transverse mode, it is relatively easy to induce such a system to lase in a higher-order Hermite–Gauss mode. It is not quite so easy to generate a Laguerre–Gauss mode in a laser cavity, however; in fact, it requires a careful manipulation of the nonlinear properties of the gain medium.

In the simplest incarnation imaginable, a laser consists of a resonator, a gain medium sandwiched within, a pumping mechanism, and an outcoupling mechanism. A schematic of the basic helium–neon laser is shown in Figure 4.8. The pumping mechanism (shown here as an electrical current) raises the atoms of the gain medium into an excited energy state. Some of these atoms release photons by the process of spontaneous emission, which then induce other atoms to release coherent photons by stimulated emission. The highly reflective mirrors of the cavity cause photons to interact

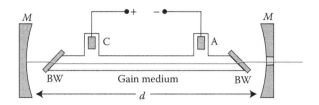

FIGURE 4.8 Illustration of a helium–neon laser. BW = Brewster window, M = mirrors, C = cathode, and A = anode.

with the gain medium multiple times, increasing the efficiency of the stimulation process. One mirror of the cavity is partially transmitting, allowing the coherent laser beam to emerge.

The resonator is essentially a Fabry–Pérot etalon, only allowing waves to oscillate at mode wavelengths $\lambda = 2d/n$, where $n = 1, 2, 3, \ldots$. If only a single etalon mode lies within the excitation wavelengths of the gain medium, the laser operates in a single temporal mode. The laser can also be excited in multiple transverse modes, usually characterized by the Hermite–Gauss or Laguerre–Gauss set of beams. These modes can "compete" with each other for the available gain medium, with often the highest-allowable mode winning out. However, the higher-order modes are increasingly wider, and a fundamental Gaussian mode can be isolated by placing a small circular aperture in the resonator, restricting the propagation of all but the lowest mode. To generate a Hermite–Gauss mode of a particular type, a larger aperture can be used along with crossed wires in the laser cavity that correspond to the nodes (zeros) of the desired HG mode.

It might seem that Laguerre–Gauss modes could be similarly selected in a circularly symmetric laser cavity. However, there are two main obstacles to this. First, Laguerre–Gauss modes of a given order are distinguished by their phase as much as by their intensity. We recall that, for instance, the LG_{01} and $LG_{0,-1}$ modes have identical intensity distributions. This means that a mode selector that depends on intensity cannot uniquely select a particular Laguerre–Gauss mode. Also, the components of a laser often break the circular symmetry of the system, making it preferentially output Hermite–Gauss modes. In the helium–neon laser illustrated above, for instance, Brewster windows are used to minimize reflections when light is transmitted out of the gain medium. These tilted windows, however, introduce an astigmatism in a finite beam, and that astigmatism introduces a frequency difference between the HG_{10} and HG_{01} modes. Even if the two can be made to oscillate simultaneously, they will produce a beat pattern at the output instead of a pure LG mode.

Nevertheless, a pure LG mode can be produced in a laser by means of a nonlinear optical process known as *cooperative frequency locking*. If the frequency difference between the HG_{01} and HG_{10} modes can be brought below a critical threshold, a nonlinear interaction in the gain medium causes the frequencies to equalize and for the phase between the two modes to equal $\pm\pi/2$, resulting in an $LG_{0,\pm1}$ mode. An early observation of this effect was made by Rigrod [Rig63], who was able to isolate some 15 axisymmetric modes in a He–Ne laser. A theoretical model of the process was later introduced by Lugiato, Oldano, and Narducci [LON88].

A detailed discussion of the nonlinear optics involved would take us too far afield from singular optics; however, a few words may be said about the process of generating an LG mode. In a situation reminiscent of the mode conversion discussed earlier, it was noted by Tamm [Tam88] that the frequency difference between an HG_{01} mode and an HG_{10} mode vanishes if those modes are aligned at 45° to the axes of the Brewster plate. By placing a movable dot of absorbing material in the laser cavity, Tamm could position it to force simultaneous oscillation in the HG_{10}^{*} and HG_{01}^{*} modes, which would at a critical threshold of frequency difference (detuning) result in an LG mode. The $\pm\pi/2$ phase difference between the HG modes represents the best use of the population inversion in the gain medium: essentially the two modes "swap" usage of the medium.

The LG modes represent a state of broken symmetry, as there is no obvious means to choose between the $+\pi/2$ and $-\pi/2$ phase difference. However, in Tamm and Weiss [TW90], it was shown that a particular mode could be selected by a system in which part of the output beam is fed back into the cavity with an appropriate phase shift. The transition between the LG_{01} and $LG_{0,-1}$ states exhibits hysteresis with respect to the phase shift.

4.1.5 Nonuniform Polarization

We have already seen in Section 4.1.2 that there is a striking analogy between circular polarization states and vortex beam states. This is not accidental; as we will elaborate upon in Chapter 5, both states represent forms of angular momentum that a light field may possess. Circular polarization is typically associated with the spin angular momentum of light and a vortex phase is associated with the orbital angular momentum of light, though in complex optical fields, the distinction between these classes may be blurred.

With this in mind, it is possible to create vortex beams by the use of a collection of polarization-sensitive and spatially varying optical elements. A summary of a number of techniques was given in Alonso, Piquero, and Serna [APS12]; here, we briefly mention two of them to highlight the possibilities.

We consider an optical beam propagating along the z-direction with polarization confined to the xy-plane; if the polarization is itself spatially varying, the field may be described by a two-component Jones vector of the form[*]

$$\mathbf{E}(\mathbf{r}) = \begin{bmatrix} E_x(\mathbf{r}) \\ E_y(\mathbf{r}) \end{bmatrix}. \tag{4.42}$$

Assuming that the paraxial approximation holds, the two components of the Jones vector propagate independently of one another in free space, and each satisfies the scalar paraxial equation.

We introduce an incident field uniformly and linearly polarized at 45° to the x-axis, which may be written as

$$\mathbf{E}(\mathbf{r}) = \frac{f(\mathbf{r})}{\sqrt{2}} \begin{bmatrix} 1 \\ 1 \end{bmatrix}, \tag{4.43}$$

where $f(\mathbf{r})$ is a function that describes the spatial variation of the field in the transverse plane, and is assumed to possess no phase vortex, that is, no phase term of the form $e^{im\phi}$.

We first use an arrangement as illustrated in Figure 4.9a. The incident field is passed through a quarter-wave plate oriented either along the x-axis or y-axis; upon transmission, the electric field now has the form

$$\mathbf{E}(\mathbf{r}) = \frac{f(\mathbf{r})}{\sqrt{2}} \begin{bmatrix} 1 \\ \pm i \end{bmatrix}, \tag{4.44}$$

where the $\pm$ represents x or y orientation, respectively. The field is now passed through the so-called "axis-finder," a commercially available device that polarizes the field along the radial direction with respect to the central axis. More technically known as a *dichroic azimuthal linear polarizer*, its

[*] For those unfamiliar with Jones vectors, we refer to Gbur [Gbu11, Section 4.1] and the first paper by Jones [Jon41].

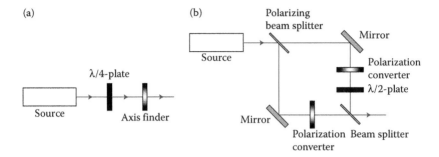

FIGURE 4.9 Polarization-based mode converters, using (a) axis finder and (b) polarization converters. (Adapted from M. Alonso, G. Piquero, and J. Serna. *Opt. Commun.*, 285:1631–1635, 2012.)

effect on the transmitted beam is represented by the Jones matrix

$$\mathbf{A}(\phi) = \begin{bmatrix} \sin^2\phi & -\cos\phi\sin\phi \\ -\cos\phi\sin\phi & \cos^2\phi \end{bmatrix}, \tag{4.45}$$

where it is to be stressed that ϕ represents the azimuthal angle of the transmitted field with respect to the axis, and not a fixed angle of orientation of the device. The field output from this axis finder is given by

$$\mathbf{E}(\mathbf{r}) = \frac{f(\mathbf{r})}{\sqrt{2}} e^{\pm i(\phi-\pi/2)} \begin{bmatrix} \sin\phi \\ -\cos\phi \end{bmatrix}. \tag{4.46}$$

The output field now carries the characteristic vortex phase $\exp[\pm i\phi]$, though the Jones vector also indicates that it is azimuthally polarized, with the direction of the polarization vector circulating around the central axis.

A significantly more complicated arrangement can produce a pure vortex beam with a uniform polarization, as illustrated in Figure 4.9b. Using the same input field as before, a polarizing beam splitter separates the horizontally and vertically polarized fields and puts them through different arms of a Mach–Zender interferometer. In the first arm of the interferometer, the field passes through a *polarization converter*, another commercial optical element that produces a radially polarized beam from a linearly polarized beam; its action can be represented by the matrix

$$\mathbf{P}(\phi) = \begin{bmatrix} \cos\phi & -\sin\phi \\ \sin\phi & \cos\phi \end{bmatrix}, \tag{4.47}$$

where again ϕ is the polar coordinate of the field. In the second arm, the field first passes through a polarization converter and then through a half-wave plate; the two arms are recombined to another beam splitter, and the output can be shown to be given by

$$\mathbf{E}(\mathbf{r}) = \frac{f(\mathbf{r})}{2} \begin{bmatrix} e^{i\phi} \\ ie^{-i\phi} \end{bmatrix}. \tag{4.48}$$

This field has vortex twists of opposite helicity in each component of the Jones vector. By the use of an ordinary polarizer, one or the other helicity, or a mixture of the two, may then be selected.

It is evident from the descriptions above that these methods of generating vortex beams will be relatively inefficient compared to many of those previously mentioned. The use of polarizers results in a significant loss of light in the generation process. Furthermore, the axis finder and polarization converter impart a spatial variation on the transmitted beam; like the spiral phase plate mentioned earlier, the output will not be a pure mode and will inevitably consist of a mixture of transverse spatial modes.

These methods nevertheless do highlight the possibilities that can arise in mixing spatial and polarization states. We will have much more to say on nonuniform polarization in Chapter 7.

4.1.6 Other Methods

Even now, we have not exhausted all the techniques that have been developed for the generation of vortex beams; in this section, we briefly mention a few others of note.

We have seen (in Section 4.1.4) that it is possible to generate vortex beams directly within a laser cavity by taking advantage of nonlinear interactions in the gain medium. In 2003, another nonlinear method was introduced by Smith and Armstrong [SA03] that can produce vortex beams in an optical parametric oscillator (OPO) by an ingenious use of a nonplanar ring cavity.

A rough schematic of the system is illustrated in Figure 4.10a. A pump beam at 532 nm is input with polarization parallel to the ordinary axis of a potassium titanyl phosphate (KTP) crystal through mirror 1. The beam undergoes nonlinear difference–frequency generation, resulting in two lower frequency output photons known as the signal and idler photons; we will discuss nonlinear processes in more detail in Chapter 10. The cavity size is chosen so that the output signal wave at 800 nm is resonant but the

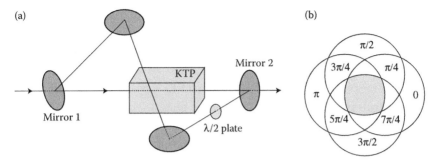

FIGURE 4.10 (a) Image-rotating nonplanar OPO. (b) Cross section of the four rotated beams passing through the nonlinear crystal, and the approximate phase in each region.

idler wave at 1588 nm is not. The cavity design is such that the entire field gets a 90° rotation as it completes a round-trip; however, the polarization is returned to its original state by the half-wave plate.

If an off-axis beam is input into the system, it will take four passes through the ring before it returns to its original state; the overall phase shift between passes can be tuned to be $\pi/2$, resulting in a collection of four beams of increasing phase sharing the resonator, as illustrated in Figure 4.10b. The output from mirror 2 is therefore a rough four-step vortex beam of uniform polarization.

Another interesting technique for generating vortices introduced by Gorodetski et al. [GDGE13] involves the coupling of surface waves known as *surface plasmons* to free-propagating vortex wavefields via the use of surface nanostructures with chirality.

Surface plasmons are electric charge density waves that can be excited on the surfaces of metals with appropriate material properties. Plasmons have become very important in recent years in a variety of applications, thanks to their short wavelength, high field intensity, and relatively long propagation range. See Sarid and Challener [SC10] and the classic text by Raether [Rae85] for more details. By etching a helical structure on a surface, it is in essence possible to make a spiral phase plate that works via plasmonic coupling, rather than by direct manipulation of the phase of an illuminating beam.

More of a means than a specific method, it should be noted that a number of different techniques have been used to generate vortex beams via the use of liquid crystals. Liquid crystal displays (LCDs) can have a spatially varying phase or amplitude pattern induced on them by an applied electric field, which allows them to be used as a *spatial light modulator* (SLM). In 1999, researchers displayed a computer-generated hologram of a set of vortices on an LCD and used them to optically trap and manipulate polystyrene particles [RHWT99]. In 2002, another group tried a different approach, creating an LCD arranged as a collection of 16 circular pie slices [GGG⁺02]. Each of these slices could be electrically induced to have its own phase delay, and the result was effectively a spiral phase plate with extremely high conversion efficiency. Even more recently, another set of researchers produced vortex beams by focusing light through micron-sized droplets of liquid crystal [BMMJ09].

4.2 DETECTION

The detection of a vortex structure, or the topological charge of a vortex, presents its own special challenges. A vortex is fundamentally a phase object, and no measurement of field intensity alone can confirm its presence. A field with extremely low, but nonzero, intensity will typically be indistinguishable from a true zero within the uncertainty of an experimental setup. Furthermore, we have already noted that Laguerre–Gauss beams of order n, $\pm m$ have identical intensity distributions and therefore cannot be distinguished by their intensity profiles. Nevertheless, a diverse selection of methods have been developed for measuring and quantifying the phase of vortex beams, as we discuss below.

4.2.1 Interference-Based Methods

Interferometry is the natural tool for the measurement of the phase structure of vortices, and can produce unambiguous representations of the topological charge. The simplest measurement technique is to interfere a vortex beam with an inclined plane wave, which results in an intensity pattern at a detector roughly of the form

$$I(x, \phi) = \left| e^{im\phi} + e^{ik_x x} \right|^2, \tag{4.49}$$

where we neglect the amplitude variations of the vortex field for simplicity. The intensity then takes on the simple form

$$I(x, \phi) = 2 + 2\cos(m\phi - k_x x). \tag{4.50}$$

Simulations of simple measured interference patterns are shown in Figure 4.11. The topological charge of the vortex beam is indicated by the

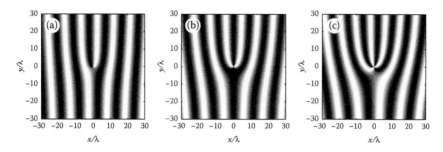

FIGURE 4.11 Fork interferograms for beams with (a) $m = 1$, (b) $m = 2$, and (c) $m = 3$. Here $k_x = 0.1k$.

number of additional wavefronts (bright lines) inserted into the pattern, making a characteristic "fork" pattern such as was already discussed in the holographic generation of vortex beams in Section 4.1.3.

This method might be used to test the quality of a vortex-generating optical element such as a spiral phase plate or computer-generated hologram in a setup such as that shown in Figure 4.12. A coherent beam is passed through a Mach–Zender interferometer, in one arm of which is placed the vortex element to be tested. The fields from each arm are brought together at the output of the system slightly misaligned (to provide the inclination of the plane wave) and then directed to the detector.

As an alternative to interference with a plane wave, it is also possible to interfere a vortex beam with its mirror image, producing a distinct azimuthal interference pattern that directly quantifies the azimuthal order. Two methods of conducting this experiment are shown in Figure 4.13. In the first [HHTV94], a beam is sent through one edge of a dual Fresnel biprism, producing a directly transmitted wave and an internally reflected wave that are mirror images of each other; the two beams are then brought

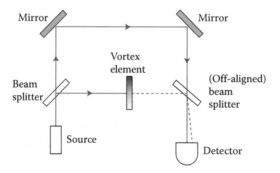

FIGURE 4.12 Mach–Zender interferometer for vortex element testing.

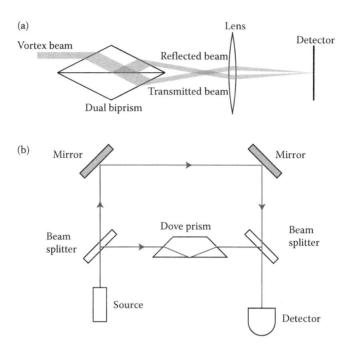

FIGURE 4.13 Vortex measurement by mirror image. (a) Via a biprism. (b) Via a Dove prism and Mach–Zender.

together at a screen. In the second, a Mach–Zender interferometer is again used, and a Dove prism is inserted into the second arm. The reflection of the beam inside the prism produces a mirror image of the beam in the first arm.

The patterns produced in the latter case may be simply deduced from the typical properties of a Laguerre–Gauss mode of order m, which has a field profile near the central axis given by $U(\rho, \phi) \sim \rho^m e^{im\phi}$. If such a beam and its mirror image, $\tilde{U}(\rho, \phi) \sim \rho^m e^{-im\phi}$, are combined along the same axis, the total intensity is

$$I(\rho, \phi) = \rho^{2m} |e^{im\phi} + e^{-im\phi}|^2 = \rho^{2m} [2 + 2\cos(2m\phi)]. \qquad (4.51)$$

Illustrations of a few typical patterns are shown in Figure 4.14. It is to be noted that there is no distinction between positive and negative m in these images, as is clear from Equation 4.51. However, if a slight inclination is introduced between the interfering beams, the result will be a fork diagram as in Figure 4.11, although one that appears to have twice the charge of the actual beam. The biprism method automatically contains this inclination.

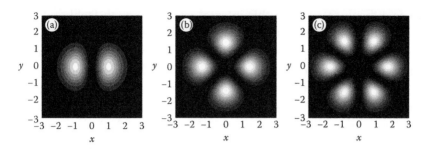

FIGURE 4.14 Interference patterns for mirror image detection, for (a) $m = 1$, (b) $m = 2$, and (c) $m = 3$.

One limitation of the simple interferometric detection of vortices discussed here is the reliance on pattern recognition: though it is easy by eye to identify distinct vortex patterns, automating the process so that it can be done rapidly (for instance, to create a vortex communication system) is a nontrivial challenge. Also, though the method allows clear discrimination between pure modes of a given topological charge, mixed modes will produce more complicated patterns that are difficult to distinguish. However, it is possible to extend the Dove prism technique to allow efficient sorting of vortex modes in a manner that could be automated, as we will see in Section 4.2.5.

4.2.2 Diffraction-Based Method

Phase also plays a significant role in diffraction theory, and one would expect that the singular behavior of a vortex beam could be deduced by an appropriately designed diffraction experiment. It has been shown, both theoretically and experimentally [HFSCC10], that light diffracting through a triangular aperture will produce a far zone radiation pattern that is related to the topological charge of the incident beam in an exceedingly simple manner. These results have been demonstrated for vortex orders up to $m = 7$ and the method works even for vortices carried in femtosecond pulses [dAA11].

We consider an equilateral triangular aperture centered on $x = y = 0$ and of side a. The field in the far zone, in a direction specified by the unit vector $\mathbf{s} = (s_x, s_y, s_z)$, is found from Fraunhofer diffraction to be proportional to

$$U(\mathbf{s}) \sim \int_A U_0(x', y') e^{-ik(s_x x' + s_y y')} dx'\, dy', \tag{4.52}$$

where A is the area of the aperture.

It is difficult to evaluate this integral exactly; though Stahl and Gbur [SGb] have shown that it is possible to do so, the calculation is quite complicated. However, a good approximation can be found by treating the diffracted field as primarily composed of a secondary wave radiated from the edges of the aperture, in accordance with the geometrical theory of diffraction [Kel62]. This reduces our calculation to an integral over the boundary of the triangular aperture. We may simplify things further by noting that the problem has threefold symmetry; we may calculate the diffracted field from one edge, and with an appropriate rotation determine the field along the other two.

Let us suppose a vortex beam of charge $\pm m$ is normally incident on a triangular aperture with geometry as shown in Figure 4.15. For a sufficiently wide beam, we may neglect the Gaussian envelope and only include the vortex core in the integral. The lower edge of the triangle contributes the following amount to the diffracted field:

$$U_m(k_x, k_y) \sim e^{-ik_y y_0} \int_{-a/2}^{a/2} (x \pm iy_0)^m e^{-ik_x x} dx. \tag{4.53}$$

This integral can be evaluated exactly in the form of a finite sum. We may use the binomial theorem

$$(x + y)^n = \sum_{k=0}^{n} \binom{n}{k} x^k y^{n-k}, \tag{4.54}$$

to write the field in the form

$$U_m(k_x, k_y) \sim e^{-ik_y y_0} \sum_{k=0}^{n} \binom{n}{k} I_k (\pm iy_0)^{n-k}, \tag{4.55}$$

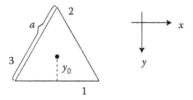

FIGURE 4.15 The geometry for the triangular aperture under consideration.

where

$$I_k \equiv a \left(\frac{ia}{2}\right)^k \left(\frac{\partial}{\partial u}\right)^k [j_0(u)]_{u=k_x a/2}, \tag{4.56}$$

with $j_0(u)$ the spherical Bessel function of the zeroth order. The contribution of the second side to the diffraction pattern may be found by rotating the vector (k_x, k_y) by $2\pi/3$ around the z-axis, with

$$\begin{bmatrix} k'_x \\ k'_y \end{bmatrix} = \begin{bmatrix} \cos(2\pi/3) & \sin(2\pi/3) \\ -\sin(2\pi/3) & \cos(2\pi/3) \end{bmatrix} \begin{bmatrix} k_x \\ k_y \end{bmatrix}. \tag{4.57}$$

A further rotation by $2\pi/3$ will give the contribution of the third edge of the triangle, and the total diffracted field in the Fraunhofer approximation will be

$$U_m^{tot}(k_x, k_y) = U_m(k_x, k_y) + e^{i2\pi m/3} U_m(k'_x, k'_y) + e^{i4\pi m/3} U_m(k''_x, k''_y). \tag{4.58}$$

The results for the Fraunhofer diffraction pattern are striking, and shown in Figure 4.16. A multilobed triangular pattern is produced on diffraction, and the number of lobes is equal to the absolute value of the topological charge *plus one*, that is, $|m| + 1$. Positive and negative charges can be distinguished by the direction that the interference pattern points.

This method of vortex measurement seems to work quite well for broadband vortices such as those generated using supercontinuum light, as demonstrated in Anderson et al. [ABdAC12].

Other apertures of broken symmetry (i.e., not square or circular) have been shown to also provide measures of the topological charge. An annular triangle aperture was shown to provide results comparable to the whole triangular aperture in Liu et al. [LTPL11] (perhaps unsurprising, considering we approximated the aperture by its boundary in the derivation above). Not long after, it was shown that the diffraction pattern of an elliptical annular aperture can also be used to measure topological charge [TLCP12]. Most recently, a diamond-shaped aperture was also found to provide information about vortex structure [LSPL13].

4.2.3 Shack–Hartmann Method

As we have discussed in Section 3.4, the topological charge of a phase singularity or collection of singularities is given as the path integral of the phase

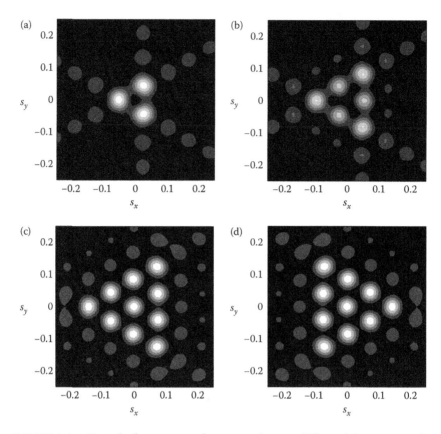

FIGURE 4.16 Fraunhofer patterns for vortex beams diffracted by a triangular aperture, for (a) $m = 1$, (b) $m = 2$, (c) $m = 3$, and (d) $m = -3$. Here, we have used $\lambda = 500$ nm and $a = 7600$ nm, and $s_x = k_x/k$, $s_y = k_y/k$.

gradient of the field around those singularities, that is,

$$t \equiv \frac{1}{2\pi} \oint_C \nabla \psi(\mathbf{r}) \cdot d\mathbf{r}. \tag{4.59}$$

With this in mind, a device that can directly measure the phase gradient of a wavefield would seem to be the ideal vortex detection tool, and the most familiar device of this form is the *Shack–Hartmann wavefront sensor*. It has been successfully used as a detector of phase singularities [CRO07], though it has limitations associated with its discrete sampling, as we will see.

A Shack–Hartmann wavefront sensor is illustrated in Figure 4.17a. A two-dimensional array of lenslets is configured to focus onto a two-dimensional detector array, as illustrated in Figure 4.17b. For a plane wave

(a)

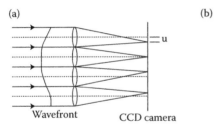

(b)

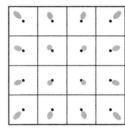

FIGURE 4.17 (a) Illustration of the cross section of a Shack–Hartmann sensor. (b) Illustration of the detector array, consisting of a collection of subapertures. Gray ellipses represent possible intensity spots from a diverging spherical wave.

normally incident on the array, each lenslet would create a focal spot directly in the middle of each subaperture. For a curved wavefront, however, the focal spot is displaced based on the local tilt of the wavefront in the corresponding lenslet. This collection of displacements gives a measure of the curvature of the incident wave.

To get a good sampling of the wavefront, lenses of sufficiently small size are required. The original sensor, developed by Roland Shack and Ben Platt to correct the aberration of telescope images induced by the atmosphere [PS01], required the fabrication of custom lenses of 1 mm diameter and 150 mm focal length.

To analyze the behavior of the system, we closely follow the arguments presented in Chen, Roux, and Olivier [CRO07]. We first consider the response of a hypothetically ideal Shack–Hartmann array, which would exactly measure the phase gradient $\nabla\psi$ of the field within the aperture, which we write as

$$\mathbf{G}(x, y) = \hat{\mathbf{x}}G_x(x, y) + \hat{\mathbf{y}}G_y(x, y) = \nabla\psi(x, y)$$
$$= \hat{\mathbf{x}}\partial_x\psi(x, y) + \hat{\mathbf{y}}\partial_y\psi(x, y). \tag{4.60}$$

The equation for topological charge then becomes

$$t = \frac{1}{2\pi}\oint_C \mathbf{G}(x, y) \cdot d\mathbf{r}. \tag{4.61}$$

If we assume that Stokes' theorem applies to the system,[*] we may rewrite the topological charge as

$$t = \frac{1}{2\pi} \int_A \nabla \times \mathbf{G}(x, y) \cdot d\mathbf{a}, \qquad (4.62)$$

where A is the area of the aperture. This result suggests that we may measure the topological charge of the wavefield by calculating the curl of the phase gradient and integrating it through the area of the aperture.

For a real Shack–Hartmann system, the phase gradient is sampled over each subaperture. From standard results of Fourier optics [Goo96, Section 5.2], we know that the amplitude of a field $U(x, y)$ in the focal plane after passing through a square aperture of width w is proportional to

$$U(x, y) \sim \int_{-w/2}^{w/2} \int_{-w/2}^{w/2} U_0(x', y') \exp[-ik(xx' + yy')/f] dx' \, dy', \qquad (4.63)$$

where $U_0(x', y')$ is the field in the aperture and f is the focal length of the lens. For a normally incident plane wave, the field will be focused to the origin with a field distribution that we call $V(x, y)$. If the field has a nontrivial phase function $\psi(x', y')$, the focal spot will be displaced due to the linear part of the phase and aberrated due to the higher terms in the phase function. Let us approximate this function by the lowest terms in its Taylor series,

$$\psi(x', y') \approx \psi(0, 0) + G_x(0, 0)x' + G_y(0, 0)y'. \qquad (4.64)$$

On substitution into Equation 4.63, we readily find that

$$U(x, y) \sim \exp[i\psi(0, 0)]V(x - fG_x/k, y - fG_y/k). \qquad (4.65)$$

The average phase gradient $\mathbf{G}^{m,n}$ in each subaperture (labeled m, n) may therefore be expressed in terms of the displacement $\mathbf{u}^{m,n}$ of each focal spot as

$$\mathbf{G}^{m,n} \approx \frac{k}{f} \mathbf{u}^{m,n}. \qquad (4.66)$$

---

[*] This is cheating somewhat, as Stokes' theorem is usually applied to *continuous* vector fields, which a phase gradient with singularities will not be. In the sense of distributions such as delta functions, however, the result is correct.

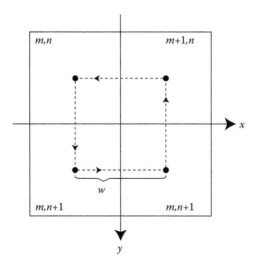

FIGURE 4.18 The circulation path through 4 subapertures of the Shack-Hartmann system.

We may estimate the curl of the phase gradient (or equivalently the topological charge in the middle of four subapertures) by calculating the circulation integral (4.61) as seen in Figure 4.18. In each subaperture, the phase gradient is assumed to be constant and of the form $\mathbf{G}^{m,n} = G_x^{m,n}\hat{\mathbf{x}} + G_y^{m,n}\hat{\mathbf{y}}$. We readily find that the result is of the form

$$t = \frac{w}{4\pi}\left[\left(G_x^{m+1,n+1} + G_x^{m,n+1} + G_y^{m+1,n} + G_y^{m+1,n+1}\right)\right.$$
$$\left. - \left(G_x^{m+1,n} + G_x^{m,n} + G_y^{m,n+1} + G_y^{m,n}\right)\right]. \tag{4.67}$$

The next natural question is to ask what this circulation looks like for a typical vortex. Because the Shack–Hartmann is essentially averaging over the curvature of the phase in each subaperture, the measured topological charge will in general deviate from the ideal value. As an example of this, we consider the case when a pure phase vortex is incident normally on the boundary between four subapertures. Taking that point to be the origin, the phase function is given by

$$\psi(x,y) = \log\left[\frac{x + iy}{\sqrt{x^2 + y^2}}\right], \tag{4.68}$$

and the phase gradient is given by

$$\nabla \phi(x, y) = \frac{x\hat{y} - y\hat{x}}{x^2 + y^2}. \tag{4.69}$$

The average value of the phase gradient in a subaperture is given by

$$\mathbf{G}^{m,n} = \frac{\int_A \nabla \phi(x, y) dx \, dy}{A}, \tag{4.70}$$

where A is the area of the particular subaperture. For the upper left one (labeled m, n), it can be shown that the integral takes on the value

$$\mathbf{G}^{m,n} = \left[\frac{\pi}{4w} + \frac{\log 2}{2w} \right]. \tag{4.71}$$

Combining this result with the other three subapertures, the total measured topological charge for a centered vortex is

$$t = \frac{1}{2} + \frac{\log 2}{\pi}, \tag{4.72}$$

a value less than unity.

This discrepancy comes from the averaging process, and it can be shown that the value of the topological charge decreases when the vortex is positioned toward the center of a subaperture. The averaging process will also tend to mask the existence of closely placed pairs of vortices with opposite charge.

4.2.4 Computer-Generated Holograms and Modans

We have already seen in Section 4.1.3 that vortex beams may be created by holographic methods, which are especially useful with the advent of computer-generated holograms. It is relatively easy to show that such holographic vortex masks may also be used to determine the structure of a vortex, and even separate out multiple vortex components into different spatial channels.

To introduce this idea, let us first imagine that we have created a Leith–Upatnieks hologram of a Laguerre–Gauss beam with field $u_{01}(\mathbf{r})$; the transmission function of the hologram, from Equation 4.35, is given by

$$t(x, y) = (t_0 + \beta U_0^2) + \beta |u_{01}(x, y)|^2 + \beta U_0^* u_{01}(x, y) e^{ik_x x}$$
$$+ \beta U_0 u_{01}^*(x, y) e^{-ik_x x}. \tag{4.73}$$

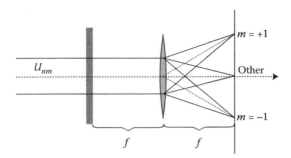

FIGURE 4.19   Using a vortex hologram as a vortex mode detector.

We now place a lens of focal length f behind this hologram and illuminate it with a *normally incident* test beam $U_{test}(\mathbf{r})$; when the hologram-lens distance is f, the field $U(x, y)$ in the rear focal plane is proportional to

$$U(x, y) \sim \int_{H} t(x', y') U_{test}(x', y') e^{-2\pi i(xx' + yy')/\lambda f} \, dx' \, dy', \qquad (4.74)$$

where H is the plane of the hologram. The geometry is illustrated in Figure 4.19. We now consider what will happen to the field in the focal plane. The integral of Equation 4.74 may be broken into three parts; we consider the third of these first,

$$U_3(x, y) \sim \beta U_0 \int_{H} u_{01}^{*}(x', y') U_{test}(x', y') e^{i(k_x - 2\pi x/\lambda f)x'} e^{-2\pi i yy'/\lambda f} \, dx' \, dy'.$$

$$(4.75)$$

At the specific point $(x_3, y_3) = (k_x \lambda f / 2\pi, 0)$, this integral reduces to the simplified form

$$U_3(x, y) \sim \beta U_0 \int_{H} u_{01}^{*}(x', y') U_{test}(x', y') \, dx' \, dy'. \qquad (4.76)$$

Recalling that the Laguerre–Gauss modes are orthonormal, we find that the amplitude of the field at (x_3, y_3) only possesses a contribution due to the $n = 0, m = 1$ component of U_{test}. Similarly, if we look at the second part of the Fraunhofer integral,

$$U_2(x,y) = \beta U_0 \int_H u_{01}(x',y') U_{test}(x',y') e^{-i(k_x + 2\pi x/\lambda f)x'}$$

$$\times e^{-2\pi i y y'/\lambda f} dx'\, dy', \tag{4.77}$$

we find that at the point $(x_2, y_2) = (-k_x \lambda f/2\pi, 0)$ it only possesses a contribution due to the $n = 0$, $m = -1$ component of U_{test}. The first part of the Fraunhofer integral

$$U_1(x,y) = \int_H \left[(t_0 + \beta U_0^2) + \beta |u_{01}(x,y)|^2 \right] U_{test}(x',y')$$

$$\times e^{-2\pi i (xx' + yy')/\lambda f} dx'\, dy', \tag{4.78}$$

will generally result in a focal spot near the center of the focal plane, due to U_{test} being normally incident and the transmission function being real valued.

We therefore find that a vortex hologram acts as a limited diffraction grating, separating out the different m-values of the test beam into different spatial locations in the focal plane. There is nothing precluding, in general, the overlap of the fields U_1, U_2, and U_3, but the system parameters can be chosen to provide sufficient spatial separation of the various spots.

This strategy—the isolation of particular transverse modes in a beam via the use of spatial filters—appears to have been pioneered in papers by Golub et al. in the 1980s [GPSS82,GKK+83,GKK+84]. The optical components designed for this task were later referred to as *modans.*[*]

The simplest modan, in theory at least, is a transparency with a *complex* transmission function $t(\mathbf{r})$. Let us assume that this transmission is of the form

$$t(\mathbf{r}) = u_{nm}^*(\mathbf{r})/u_{max}, \tag{4.79}$$

where u_{max} is the maximum value of $|u_{nm}|$. This normalization is necessary because the transmission of a passive device (without gain) is bounded by $|t(\mathbf{r})| \leq 1$. Because the Laguerre–Gauss modes form a complete set, an

---

[*] This name does not seem to have caught on, but we will use it here for convenience and because of its lovely kaiju-sounding quality.

arbitrary beam $U(\mathbf{r})$ incident upon the modan may be expanded in the form

$$U(\mathbf{r}) = \sum_{n=0}^{\infty} \sum_{m=-\infty}^{\infty} \alpha_{nm} u_{nm}(\mathbf{r}).$$

As for the hologram earlier, we pass this arbitrary beam through the modan and then focus it. The field $U_f(\mathbf{r})$ in the rear focal plane is proportional to

$$U_f(\mathbf{r}) \sim \int_D U(\mathbf{r}')t(\mathbf{r}') \exp[-2\pi i \mathbf{r} \cdot \mathbf{r}'/\lambda f]d^2 r', \qquad (4.80)$$

where D is the plane of the modan. At the geometrical focus, the field takes on the form

$$U_f(0) \sim \frac{1}{u_{max}} \int_D U(\mathbf{r}')u_{nm}^*(\mathbf{r}')d^2 r' = \alpha_{nm}/u_{max}.$$

The field amplitude at focus is therefore proportional to the weight of the (m, n)th Laguerre–Gauss mode. By placing a detector at this position, it is possible to detect the presence or absence of this mode in a field.

There are a number of limitations to this basic approach. The most obvious of them is that the modan is specific to a particular mode, and will not even detect modes with the same m (topological charge) but different n. In the early experiments, different modans were fabricated for each mode, and placed one at a time in the beam path to measure the individual weights.

It is, however, straightforward to generalize the modan concept to create a multichannel modan [AGK+90]. Let us introduce the function $w(\mathbf{r})$, where

$$w(\mathbf{r}) = \sum_{p=0}^{P} \beta_p u_p^*(\mathbf{r}) \exp[i\mathbf{k}_p \cdot \mathbf{r} + \gamma_p], \qquad (4.81)$$

where each $u_p(\mathbf{r})$ is a particular Laguerre–Gauss mode, β_p are complex weights, γ_p are phase shifts, and $\mathbf{k}_p$ are wavevectors. We next introduce the transmission function of a multichannel modan as

$$t(\mathbf{r}) = w(\mathbf{r})/w_{max}, \qquad (4.82)$$

where w_{max} is the maximum value of $|w(\mathbf{r})|$, as in the simple modan case. If again we have the modan positioned in the front focal plane of a lens and look at the image in the rear focal plane, the field in the latter plane is given by Equation 4.80. On substitution, we find that the field amplitude at the position $\mathbf{r} = \mathbf{k}_p \lambda f / 2\pi$ is given by

$$U_f(\mathbf{k}_p \lambda f / 2\pi) \sim \frac{\alpha_p \beta_p}{w_{max}}, \qquad (4.83)$$

where α_p is the weight of the "pth" Laguerre–Gauss mode in the illuminating beam. Each position in the focal plane uniquely gives the amplitude of a distinct mode of the field.

A simulation of such a modan in action, with four modes, is shown in Figure 4.20. The modan has been designed to sort the four lowest Laguerre–Gauss modes, namely, $(0, 0)$, $(0, 1)$, $(0, -1)$, and $(1, 1)$, in counterclockwise order starting from the upper left. The modes have been sorted in their waist plane, and $w_0 = 1$ mm, $\lambda = 500$ nm, and $f = 8$ cm. The modes were

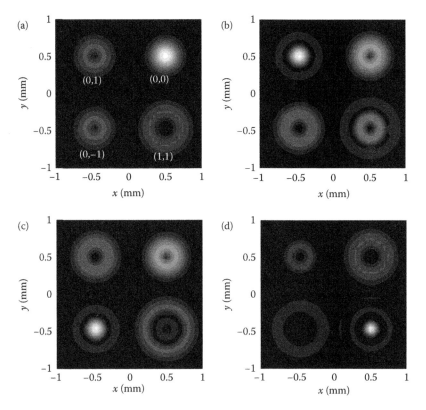

FIGURE 4.20 Simulation of a four-channel modan, described in the text. (a) Illuminated by LG_{00} mode, (b) LG_{01} mode, (c) LG_{0-1} mode, and (d) LG_{11} mode.

directed at an angle $20°$ from the propagation axis. It can be seen that, when illuminated in turn by each of the modes, a bright spot appears in the appropriate channel.

A severe limitation of the modans shown so far is the need for a complex transmission function, involving both amplitude and phase modifications, something that is in general difficult to fabricate. It is possible to adapt the idea to pure amplitude-only filters or phase-only filters, as we now show.

An amplitude-only filter may be represented by a transmission function of the form

$$t(\mathbf{r}) = t_0 + 2\frac{\Delta t}{w_{max}}\text{Re}\{w(\mathbf{r})\}, \qquad (4.84)$$

where $t_0 = A_0 + \Delta A/2$ and $\Delta t = \delta_t \Delta A/4$ dictates the amplitude transmission range of the filter, with $(A_0 + \Delta A/2)$ between 0 and 1, and $0 < \delta_t \leq 1$. Such a filter is directly analogous to the transmission hologram of Equation 4.73 introduced earlier. Because we have taken the real part of $w(\mathbf{r})$, there will be a conjugate image associated with each direction $\mathbf{k}_p$. There will also be a normally propagating "halo" associated with the baseline transmission t_0 that includes beam energy not diffracted into one of the modan channels.

An amplitude-only modan is potentially inefficient, as much of the incident light is reflected or absorbed by the filter. Alternatively, a phase-only modan [GKK+88,GSoU89] can be created by the use of the transmission function

$$t(\mathbf{r}) = \exp\{i\phi_{max}\delta_\phi Q(|w(\mathbf{r})|/w_{max})\cos[\mathbf{k}_p \cdot \mathbf{r} + \gamma_p + \arg(\beta_p)]/2\}, \qquad (4.85)$$

where ϕ_{max} is the maximum phase shift introduced by the component (typically $\phi_{max} = \pi$), $0 < \delta_\phi < 1$ is a coefficient indicating the range of variation of phase, and $Q(x)$ is a predistortion function, assumed to take the values $Q(0) = 0$ and $Q(1) = 1$, to be described momentarily.

The justification for this choice of transmission comes from a modified form of the generating function of the Bessel functions [Gbu11, Section 16.3], namely,

$$\exp[ix(t + 1/t)/2] = \sum_{n=-\infty}^{\infty} i^n J_n(x)t^n. \qquad (4.86)$$

This is found from the ordinary generating function by replacing t by it. If we let

$$x = \phi_{max}\delta_\phi Q(|w(\mathbf{r})|/w_{max})/2, \tag{4.87}$$

$$t = \exp[i(\mathbf{k}_p \cdot \mathbf{r} + \gamma_p + \arg(\beta_p))], \tag{4.88}$$

we may write

$$t(\mathbf{r}) = \sum_{n=-\infty}^{\infty} i^n J_n \left[\phi_{max}\delta_\phi Q(|w(\mathbf{r})|/w_{max})/2\right]$$

$$\times \exp[ni(\mathbf{k}_p \cdot \mathbf{r} + \gamma_p + \arg(\beta_p))]. \tag{4.89}$$

Because of the presence of n in the complex exponent, each term of this series generally has a different phase inclination; we end up with multiple diffraction orders of the transmitted beam, just as in an ordinary diffraction grating or the binary hologram considered in Equation 4.41. The first order, $n = 1$, term of this grating has a transmission function of the form

$$t_1(\mathbf{r}) = i J_1 \left[\phi_{max}\delta_\phi Q(|w(\mathbf{r})|/w_{max})/2\right] \exp[i(\mathbf{k}_p \cdot \mathbf{r} + \gamma_p + \arg(\beta_p))]. \tag{4.90}$$

On comparison of this expression with Equation 4.81, we find that the first-order diffracted field will be in agreement with a general modan if the predistortion function Q is taken to satisfy

$$J_1 \left[\phi_{max}\delta_\phi Q(|w(\mathbf{r})|/w_{max})/2\right] = C|w(\mathbf{r})|, \tag{4.91}$$

with C as a constant. This constant can be determined by evaluating the equation at the point where $w(\mathbf{r}) = w_{max}$, so that

$$C = J_1 \left[\phi_{max}\delta_\phi/2\right]/w_{max}. \tag{4.92}$$

It should be noted that, because $\phi_{max} < \pi$ and $\delta_\phi < 1$, the argument of the J_1 function will always be less than $\pi/2$. In this range, the Bessel function will be monotonically increasing and the solution for $Q(x)$ can be found computationally, or approximately by Taylor series.

Multichannel modans have been experimentally studied, both as a test of the basic principles and as a viable tool for optical demultiplexing. In the former case, the researchers Golub et al. [GKK⁺96] fabricated a six-channel modan tailored to low-order Laguerre–Gauss modes. More recently, Gibson et al. [GCP+04] used a modan in a free-space optical communications system. This latter case will be discussed in more detail in Section 6.2.

As we have noted, modans are somewhat limited in the analysis of vortex beams because they sort according to a specific (n, m) mode and not by topological charge m alone. A beam whose topological charge remains unchanged but has its radial profile distorted might be missed by a modan that is too specific. It is possible, however, to send multiple modes in the same direction; for instance, the $(0, 1)$, $(1, 1)$, and $(2, 1)$ Laguerre–Gauss modes might be directed to the same position in the focal plane.

4.2.5 Geometrical Mode Separation

The modans of the previous section are the first method described that can spatially separate the different modes of a beam. The key to this separation is the individual phase tilt—$\exp[i\mathbf{k}_p \cdot \mathbf{r}]$—imposed on each component of the transmission function. As we have seen, however, these devices are mode specific, and every mode to be detected must be deliberately encoded in the modan. It would be much more convenient to find an optical device that uses the unique phase structure of the vortex states to separate the modes automatically.

With this in mind, it should be noted that a vortex phase also increases linearly, but in the azimuthal direction. Let us suppose that we have a device that maps each point of the source field, labeled (x, y) or (r, ϕ), to the point (u, v) in the image plane, where

$$u(r, \phi) = -a\log[r/b] = -a\log(\sqrt{x^2 + y^2}/b), \qquad (4.93)$$

$$v(r, \phi) = a\phi = a\arctan(y/x), \qquad (4.94)$$

and a, b are constants with units of distance. The device, in essence, performs a coordinate transformation of the optical field; for this specific case, the transform in question is known as a "log-polar" transform. The effect of this transformation is shown in Figure 4.21 for a Laguerre–Gauss mode of order $(0, 1)$. The azimuthal phase ramp is converted into a linear phase tilt, and each pure vortex state will have a different tilt and, on focusing, end up in a different position in the focal plane. A superposition of multiple vortex states can therefore be decomposed and the individual weights can be measured. Unlike the modan, all states with a given topological charge will automatically be focused to the same position.

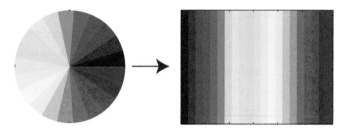

FIGURE 4.21 Illustration of the geometric transform of Section 4.2.5, showing how a radial phase is "unwrapped" into a linear phase.

This strategy was first applied to vortex measurement in 2010 by collaborators from Leiden and Glasgow [BLC+10], though the theoretical foundations go back to a 1974 paper by Bryngdahl [Bry74], who investigated geometrical transformations via optical elements. We first follow Bryngdahl's general approach, and then apply it to the vortex case.

We are interested in an optical system for which the field at a point (x, y) in the source plane is predominantly propagated to a point $[u(x, y), v(x, y)]$ in the image plane. We attempt to achieve this using the $4f$-focusing system illustrated in Figure 4.22. A phase plate in the front focal plane of the first lens imparts a phase $\varphi(x, y)$ on an illuminating field $U_0(x, y)$; in the rear focal plane of this lens, the field $U_1(u, v)$ is given by the Fourier transform of the filtered field, that is,

$$U_1(u, v) \sim \int U_0(x, y) \exp[i\varphi(x, y)] \exp[-2\pi i(xu + yv)/\lambda f]dx\, dy. \quad (4.95)$$

Let us ignore the second phase mask (PM) with phase $\varphi_1(x, y)$ for the moment and the optics that follows it. Our first objective is to determine

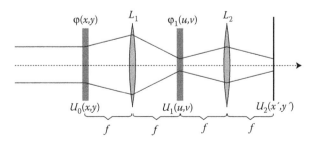

FIGURE 4.22 The basic arrangement for performing a geometrical transformation optically.

a phase function $\varphi(x, y)$ that accomplishes the desired field mapping. We note that the integral for $U_f(u, v)$ is of a standard form

$$G(u, v) \sim \int g(x, y) \exp[ikh(x, y)]dx\, dy, \qquad (4.96)$$

where we have associated $g(x, y) = U_0(x, y)$, which is typically a slowly varying function, and $h(x, y) = -(xu + yv)/f + \varphi(x, y)/k$. At optical wavelengths, λ is typically much smaller than any other length scale in the problem, and therefore $k = 2\pi/\lambda$ is extremely large. Such an integral can be approximated using the *method of stationary phase*, which will be discussed in Section 4.3, and we briefly summarize it here. In short, because of the rapid oscillation of the complex exponential, the integrand will tend to have equal weights of positive and negative contributions that will cancel in the full integral, except at points where the phase of the integral is stationary, that is, points (x, y) where

$$\frac{\partial h}{\partial x} = \frac{\partial h}{\partial y} = 0. \qquad (4.97)$$

For our particular choice of $h(x, y)$, we find that the stationary points occur at points that satisfy

$$\frac{\partial \varphi(x, y)}{\partial x} = \frac{k}{f}u, \quad \frac{\partial \varphi(x, y)}{\partial y} = \frac{k}{f}v. \qquad (4.98)$$

Our desired mapping can be achieved by substituting $[u(x, y), v(x, y)]$ into the above equations, giving us a pair of differential equations for the phase function $\varphi(x, y)$ that can be directly integrated.

Returning to Equation 4.96, we note, however that our transformation is not yet satisfactory. The method of stationary phase indicates that the contribution of the stationary point at (x_0, y_0) to the mapped point (u_0, v_0) is of the form

$$G(u_0, v_0) \sim \frac{2\pi i}{k} \frac{g(x_0, y_0)}{h_{xx}h_{yy} - h_{xy}^2} \exp[ikh(x_0, y_0)], \qquad (4.99)$$

where $h_{xx} = \partial^2 h/\partial x^2$, and so forth. It can be seen from this expression that the image "field" $G(u_0, v_0)$ is directly proportional to the object "field" $g(x_0, y_0)$. However, there is an additional nontrivial phase contribution exp[$ikh(x_0, y_0)$] that will scramble the field as it further propagates and potentially undo the transformation we desire. To cancel this additional phase, we introduce a second phase plate with phase function $\varphi_1(u, v)$ in this image plane, with $\varphi_1 = -kh$.

Before being used for vortex measurement, such geometric transformations were used to fabricate distortion-resistant optical processors. Casasent and Psaltis first developed filters that could identify scaled and rotated images [CP76], and quickly generalized their results to the detection of images subject to quite general distortions [PC77,CP78]. They applied the log-polar transformation in their work using a polar camera and scanned input SLM; sometime later, Saito, Komatsu, and Ohzu [SKO83] used a computer-generated hologram to achieve the same transformation. A few years later, researchers at King's College achieved both the forward and inverse transform optically [HDD87].

The phase function for the log-polar transformation of Equation 4.94 above can be found with some effort to be of the form

$$\varphi(x, y) = \frac{2\pi a}{\lambda f}\left[y \arctan(y/x) - x \log\left(\frac{\sqrt{x^2 + y^2}}{b}\right) + x\right], \quad (4.100)$$

and can also be confirmed by direct differentiation using Equation 4.98. By Equation 4.99, we find that this phase function must be corrected by an additional phase plate with phase

$$\varphi_1(x, y) = k(xu + yv)/f - \varphi(x, y). \quad (4.101)$$

This expression is not terribly useful in its current form, as it mixes the (x, y) and (u, v) coordinates, which are related by Equation 4.94. To simplify, we note that we may express our coordinate transforms in the alternative form

$$\exp[-u/a] = \sqrt{x^2 + y^2}/b, \quad (4.102)$$

$$\exp[iv/a] = \frac{x + iy}{\sqrt{x^2 + y^2}}. \quad (4.103)$$

We may then write

$$x = b\cos[v/a]\exp[-u/a], \quad (4.104)$$

$$y = b\sin[v/a]\exp[-u/a]. \quad (4.105)$$

Using these equations to express φ_1 entirely in terms of the coordinates (u, v), we find that it must be of the form

$$\varphi_1(u, v) = -\frac{2\pi a b}{\lambda f} \exp(-u/a) \cos(v/a). \qquad (4.106)$$

We may now look back and consider the effect of the system shown in Figure 4.22 in its entirety. The first phase screen, with phase function $\varphi(x, y)$, and the first lens produce the log-polar transformed field in the (u, v) plane, albeit with additional phase terms. These additional phase terms are negated by the use of a second phase screen with function $\varphi_1(u, v)$. What remains is a field in which every distinct vortex state has a unique phase tilt. The final lens focuses each of these tilted beams to a distinct point in the rearmost focal plane.

The impact of this system on several Laguerre–Gauss modes is shown in Figure 4.23.

This technique was first demonstrated in Berkhout et al. [BLC+10] and was applied to complex superpositions of vortex states in Berkhout et al. [BLPB11]. The effect of misalignment on the measured vortex spectrum was considered in Lavery et al. [LBCP11].

The original experimental configuration is illustrated in Figure 4.24. Three SLMs were used to perform the appropriate phase transformations. The first was used to generate a vortex mode to be measured, while the second and third SLMs create the geometric transformation and the phase correction, respectively. The use of beamsplitters in the arrangement results in a significant loss of light; later, the system was redesigned with more efficient refractive elements [LRB+12].

It can be seen in Figure 4.23 that there will be a nontrivial overlap of neighboring vortex states. It is possible, however, to shrink this overlap by modifying the geometric transformation to produce multiple copies of the tilt beam in the plane of the element ϕ_2. The angular tilt of the modes is effectively increased and the overlap reduced considerably. This modification was first discussed in O'Sullivan et al. [OMMB12] and fully studied in Mirhosseini et al. [MMSB13]. Separation efficiency of 92% was reported.

Because of its relative simplicity, stability, and versatility, this mode transformation method seems to hold the most promise for the use of vortex states in optical communications, to be discussed in Section 6.2.

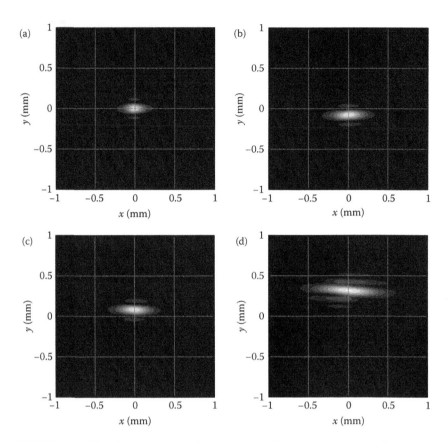

FIGURE 4.23 Simulation of a mode separator using geometric transformation, sorting (a) LG_{00}, (b) LG_{01}, (c) LG_{0-1}, and (d) LG_{0-4}. For this example, $\lambda = 500\,\text{nm}$, $f = 0.5\,\text{m}$, $a = 0.5$, $b = 1$, and $w_0 = 1\,\text{mm}$.

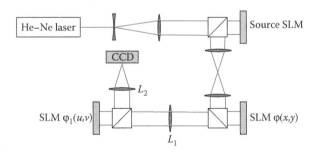

FIGURE 4.24 Illustration of an experimental configuration to separate vortex modes. (Adapted from G.C.G. Berkhout et al. *Phys. Rev. Lett.*, 105:153601, 2010.)

4.3 METHOD OF STATIONARY PHASE

We explain here in more detail the method of *stationary phase* that was applied in the use of geometric transformations for vortex detection in Section 4.2.5. This method is important in many subfields of optics but explanations of the method, especially in two variables, can be difficult to find.[*]

Let us consider an integral of the form

$$I = \int g(x, y) \exp[ikh(x, y)] dx \, dy, \tag{4.107}$$

where $g(x, y)$ is a smooth slowly varying function of x and y and $h(x, y)$ is a smooth real-valued function. We are interested in determining the approximate value of the function when k is an extremely large number. Over most of the domain of integration, the exponent will oscillate very rapidly in comparison to the slow function $g(x, y)$; these oscillations will result in a very small net contribution to the total integral.

The only place in which we expect a significant contribution is at points where the exponent is stationary, that is, where

$$\frac{\partial h}{\partial x} = \frac{\partial h}{\partial y} = 0. \tag{4.108}$$

In the neighborhood of one of these stationary points, labeled (x_0, y_0), we may approximate the slow function as a constant

$$g(x, y) \approx g(x_0, y_0), \tag{4.109}$$

and the function $h(x, y)$ by the lowest two nonzero terms of its Taylor series approximation

$$h(x, y) \approx h(x_0, y_0) + \frac{h_{xx}}{2}(x - x_0)^2 + \frac{h_{yy}}{2}(y - y_0)^2 + h_{xy}(x - x_0)(y - y_0), \tag{4.110}$$

where h_{xx} represents the second partial derivative of $h(x, y)$ with respect to x at (x_0, y_0) and so forth. Shifting the origin of our integration to (x_0, y_0), and formally extending the integral around a stationary point to infinity, we

---

[*] One place it appears is in the classical monograph on focusing by Stamnes [Sta86].

may write its contribution as

$$I_{x_0,y_0} \sim g(x_0, y_0) \exp[ikh(x_0, y_0)]$$

$$\times \int_{-\infty}^{\infty} \int_{-\infty}^{\infty} \exp\left[ik\left(\frac{h_{xx}}{2}x^2 + \frac{h_{yy}}{2}y^2 + h_{xy}xy\right)\right] dx\, dy. \quad (4.111)$$

The integrand is now of a complex Gaussian form and can be integrated via an appropriate change of coordinates. To do so, we note that the quantity in parenthesis can be written in a matrix form

$$\frac{h_{xx}x^2}{2} + \frac{h_{yy}y^2}{2} + h_{xy}xy = \begin{bmatrix} x & y \end{bmatrix}\begin{bmatrix} h_{xx}/2 & h_{xy}/2 \\ h_{xy}/2 & h_{yy}/2 \end{bmatrix}\begin{bmatrix} x \\ y \end{bmatrix}, \quad (4.112)$$

and that the matrix is real symmetric. This means that we may find a new set of variables (x', y') that diagonalize the matrix with an appropriate coordinate transformation. Using standard matrix methods, we find that the eigenvalues of the matrix are

$$\lambda_\pm = \frac{h_{xx} + h_{yy}}{4} \pm \frac{1}{4}\sqrt{(h_{xx} + h_{yy})^2 - 4(h_{xx}h_{yy} - h_{xy}^2)}, \quad (4.113)$$

and that our integral may therefore be written as

$$I_{x_0,y_0} \sim g(x_0, y_0) \exp[ikh(x_0, y_0)] \int_{-\infty}^{\infty} \int_{-\infty}^{\infty} \exp\left[ik\left(\lambda_+ x'^2 + \lambda_- y'^2\right)\right] dx'\, dy'.$$

$$(4.114)$$

The Gaussian integrals can be evaluated in the usual manner, and the result is

$$I_{x_0,y_0} \sim \frac{2\pi i}{k} \frac{g(x_0, y_0)}{h_{xx}h_{yy} - h_{xy}^2} \exp[i[kh(x_0, y_0)]. \quad (4.115)$$

The method of stationary phase is one of the key techniques in the theory of asymptotic series, which is discussed in more detail in Gbur [Gbu11, Chapter 21], Mandel and Wolf [MW95, Section 3.3], and Erdelyi [Erd55]. A few additional observations are worth mentioning. First, it is to be noted that the stationary points considered here are one of the three types that arise in two-variable stationary phase. A weaker contribution can be found from points on the boundary of a finite integration region, and an even weaker still contribution can be found from corner points on this boundary. Also, it is important to note that we have not rigorously proven that our result for a stationary point is, in fact, the most significant contribution to the integral. A proper asymptotic analysis would also include an estimate of the magnitude of the next largest contribution to the series.

4.4 EXERCISES

1. Suppose we make a "spiral" phase plate with only two steps, that is, $\phi = 0$ and $\phi = \pi$. Analyze and explain the structure of modes that come out of such a plate.

2. Neglecting dispersion, determine the effective bandwidth of a spiral phase plate made out of a material $n = 1.5$ and submerged in air, designed to produce an $m = 2$ vortex for wavelength $\lambda = 600$ nm. Define the bandwidth as those wavelengths for which the effective topological charge lies within the range $1.5 < m < 2.5$.

3. By explicit calculation, determine the type of Laguerre–Gauss beam generated by an HG_{20} beam being passed at $45°$ through a mode converter.

4. By explicit calculation, determine the type of Laguerre–Gauss beam generated by an HG_{21} beam being passed at $45°$ through a mode converter.

5. Complete the derivation of Equation 4.48 by using Jones vectors and the interference of the waves coming out of each arm of the interferometer.

6. Derive Equation 4.72 by explicit integration of the phase gradient over the four subapertures of the Shack–Hartmann interferometer. (You will need to look up some complicated integrals to do this.)

7. The fork diagrams shown in Figure 4.11 are all done for positive topological charge vortex beams. How does the pattern change when a negative topological charge beam is used? Explain qualitatively the difference in the pattern, the physics of its formation, and plot the result.

CHAPTER 5

Angular Momentum of Light

E VER SINCE LIGHT WAS discovered to be an electromagnetic wave, it has been appreciated that such a wave must carry momentum as well as energy. Indeed, James Clerk Maxwell himself discussed the pressure exerted by sunlight on the surface of Earth in his *Treatise on Electricity and Magnetism* [Max92, paragraph 793]. The momentum of light is implicit in the Lorentz force: electromagnetic fields exert forces on charged particles, and that force must be caused by an appropriate transfer of momentum from the fields themselves.

Inevitably, researchers turned to the possibility of light possessing angular momentum as well. The earliest discussion seems to be due to Poynting [Poy09], who argued for the existence of angular momentum in circularly polarized light by analogy with a mechanical system. Poynting realized that this angular momentum could be transferred to a physical object, but said that his "present experience of light forces does not give me much hope that the effect could be detected, if it has the value suggested by the mechanical model." Nevertheless, in 1936, Beth [Bet36] detected and measured the torque induced on a waveplate by circularly polarized light.

The angular momentum due to circular polarization (to be called *spin*) is not the only possible contribution to the total angular momentum; a wavefield that possesses a "twist" in it, like a vortex beam, can possess an *orbital* component of the angular momentum as well. This latter aspect was only put under serious scrutiny quite recently [ABSW92], possibly coinciding with the growth of singular optics as a field. Since then, there has been extensive research on the subject; see, for instance, the collection edited by Allen, Barnett, and Padgett [ABP03].

The study of optical angular momentum comes with certain conceptual difficulties. Light is not a rigid or even massive body, leaving significant ambiguity on how to define the "axis" of rotation for orbital angular momentum (OAM). Furthermore, there is great uncertainty in how to describe the linear momentum of light, and therefore also the angular momentum, while it is traveling in matter, resulting in the 100-year-and-counting Ambraham–Minkowski controversy.

Optical angular momentum plays an important role in both practical applications and natural phenomena. In astronomy, it has been observed that complicated light interactions with asymmetric asteroids can change the asteroids' rotation, in what is known as the YORP effect (see, for instance, [LFP+07]). In microscopy, the momentum and angular momentum of light is now regularly used to trap and rotate small objects in a technique known as *optical tweezing*, to be discussed in some detail in Chapter 6.

In this chapter, we discuss both the momentum and angular momentum of light and conceptual challenges associated with them. We start by returning to Maxwell's equations and deriving the appropriate conservation laws in terms of electric and magnetic fields.

5.1 MOMENTUM AND ANGULAR MOMENTUM IN WAVEFIELDS

We begin by considering a region of space that contains time-fluctuating electric and magnetic fields $E(r, t)$ and $B(r, t)$. Let us consider the force $F(r, t)$ on a point particle of charge q that lies within this region, which may be expressed by the Lorentz force law as

$$F(r, t) = q\left[E(r, t) + v \times B(r, t)\right], \tag{5.1}$$

where v is the velocity of the particle.

It should be noted that we are restricting ourselves to the *microscopic* form of Maxwell's equations in this derivation, working with the fields E and B and not with the macroscopic D and H. This is necessary because the "correct" form of the electromagnetic momentum law in matter, if a unique one exists, is still unclear; this will be discussed in Section 5.6.

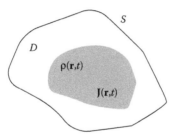

FIGURE 5.1   Illustration of the notation used in deriving the momentum conservation law. The shaded area indicates the location of sources.

We may generalize our force formula to a continuous distribution of charges and currents by replacing the charge q by the charge density $\rho(\mathbf{r}, t)$ and by replacing the quantity $q\mathbf{v}$ by the current density $\mathbf{J}(\mathbf{r}, t)$. We consider the fields and sources within a volume D, bounded by a surface S, as illustrated in Figure 5.1. Integrating the force equation over the volume, we find that the net mechanical force $\mathbf{F}_{mech}$ on the system of charges and currents is

$$\mathbf{F}_{mech}(t) = \int_D \left[\rho(\mathbf{r}', t)\mathbf{E}(\mathbf{r}', t) + \mathbf{J}(\mathbf{r}', t) \times \mathbf{B}(\mathbf{r}', t) \right] d^3 r'. \qquad (5.2)$$

We can rewrite this equation in a more helpful form through the use of the microscopic form of Maxwell's equations, which we list below for convenience:

$$\nabla \cdot \mathbf{E}(\mathbf{r}, t) = \frac{\rho(\mathbf{r}, t)}{\epsilon_0}, \qquad (5.3)$$

$$\nabla \cdot \mathbf{B}(\mathbf{r}, t) = 0, \qquad (5.4)$$

$$\nabla \times \mathbf{E}(\mathbf{r}, t) = -\frac{\partial \mathbf{B}(\mathbf{r}, t)}{\partial t}, \qquad (5.5)$$

$$\nabla \times \mathbf{B}(\mathbf{r}, t) = \mu_0 \mathbf{J}(\mathbf{r}, t) + \mu_0 \epsilon_0 \frac{\partial \mathbf{E}(\mathbf{r}, t)}{\partial t}. \qquad (5.6)$$

In these equations, ϵ_0 is the free-space permittivity and μ_0 is the free-space permeability.

We may express Equation 5.2 entirely in terms of field quantities by the use of Equations 5.3 and 5.6; the result is of the form

$$\mathbf{F}_{mech} = \int_D \left[\epsilon_0 \mathbf{E}(\nabla \cdot \mathbf{E}) - \frac{1}{\mu_0} \mathbf{B} \times (\nabla \times \mathbf{B}) - \epsilon_0 \frac{\partial \mathbf{E}}{\partial t} \times \mathbf{B} \right] d^3 r', \qquad (5.7)$$

where we now suppress the arguments of the integrand for brevity. By use of the simple derivative formula,

$$\frac{\partial}{\partial t}[\mathbf{E} \times \mathbf{B}] = \mathbf{E} \times \frac{\partial \mathbf{B}}{\partial t} + \frac{\partial \mathbf{E}}{\partial t} \times \mathbf{B}, \tag{5.8}$$

the last term of the integrand may be rewritten, resulting in the new equation

$$\mathbf{F}_{mech} = \int_D \left[\epsilon_0 \mathbf{E}(\nabla \cdot \mathbf{E}) - \frac{1}{\mu_0}\mathbf{B} \times (\nabla \times \mathbf{B}) - \epsilon_0 \frac{\partial}{\partial t}(\mathbf{E} \times \mathbf{B}) \right.$$
$$\left. + \epsilon_0 \mathbf{E} \times \frac{\partial \mathbf{B}}{\partial t} \right] d^3 r'. \tag{5.9}$$

The latter term of this new equation can also be rewritten using Equation 5.5, leaving us with

$$\mathbf{F}_{mech} = \int_D \left[\mathbf{Q} - \epsilon_0 \frac{\partial}{\partial t}(\mathbf{E} \times \mathbf{B}) \right] d^3 r', \tag{5.10}$$

where we have defined

$$\mathbf{Q} = \epsilon_0 \mathbf{E}(\nabla \cdot \mathbf{E}) - \epsilon_0 \mathbf{E} \times (\nabla \times \mathbf{E}) + \frac{1}{\mu_0}\mathbf{B}(\nabla \cdot \mathbf{B}) - \frac{1}{\mu_0}\mathbf{B} \times (\nabla \times \mathbf{B}). \tag{5.11}$$

You may notice that an additional term has mysteriously appeared in the equation, namely, $\mathbf{B}(\nabla \cdot \mathbf{B})$. Thanks to Equation 5.4, it is equal to zero but has been included to add symmetry to the expression for $\mathbf{Q}$, which now treats $\mathbf{E}$ and $\mathbf{B}$ on almost equal footing.

It is to be noted that the mechanical force on the charges may be expressed as the time rate of change of the mechanical momentum $\mathbf{P}_{mech}$, that is,

$$\mathbf{F}_{mech} = \frac{\partial \mathbf{P}_{mech}}{\partial t}. \tag{5.12}$$

With this in mind, the quantity within the integral on the right-hand side of Equation 5.10 may be considered to be the time derivative of the *momentum density of the electromagnetic field*, which suggests that the total field momentum in the volume may be defined as

$$\mathbf{P}_{field} = \int_D \epsilon_0 \mathbf{E} \times \mathbf{B} d^3 r'. \tag{5.13}$$

Equation 5.10 may then be rewritten in the suggestive form

$$\frac{\partial \mathbf{P}_{tot}}{\partial t} = \frac{\partial}{\partial t}[\mathbf{P}_{mech} + \mathbf{P}_{field}] = \int_D \mathbf{Q} d^3 r', \tag{5.14}$$

where $\mathbf{P}_{tot}$ is the total combined momentum of fields and particles within the volume.

But what does this equation mean? The physics of it is contained within the integral of the vector $\mathbf{Q}$, which we now investigate closely. It will be extremely convenient to switch to a tensor notation for vectors and derivatives, where E_j is the jth Cartesian component of $\mathbf{E}$ and ∂_j is the partial derivative with respect to the jth Cartesian coordinate. It can be shown that

$$[\mathbf{E}(\nabla \cdot \mathbf{E}) - \mathbf{E} \times (\nabla \times \mathbf{E})]_i = \partial_j \left[E_i E_j - \frac{1}{2} \delta_{ij} E_k E_k \right], \tag{5.15}$$

where δ_{ij} is the Kronecker delta and the Einstein summation convention is used.[*] If we now introduce the *Maxwell stress tensor*,[†] defined as

$$T_{ij} = -\left[\epsilon_0 E_i E_j + \frac{1}{\mu_0} B_i B_j - \frac{1}{2} \delta_{ij} \left(\epsilon_0 E_k E_k + \frac{1}{\mu_0} B_k B_k \right) \right], \tag{5.16}$$

our equation for the total rate of change of momentum may be written as

$$\left[\frac{\partial \mathbf{P}_{tot}}{\partial t} \right]_i = -\int_D \partial_j T_{ij} d^3 r'. \tag{5.17}$$

For each component i of the momentum, the right-hand side of the expression is the volume integral of the divergence of a vector with components T_{ij}. We may therefore use the divergence theorem component by

---

[*] We leave this derivation as an exercise at the end of this chapter.

[†] We define our stress tensor with the opposite sign as done, for instance, in classic texts such as Jackson [Jac75]. The sign change is made to be formally the same as in Poynting's theorem, discussed in Section 8.1.

component and express Equation 5.17 as an integral over the flux through the surface S bounding the volume as

$$\left[\frac{\partial \mathbf{P}_{tot}}{\partial t}\right]_i = -\oint_S T_{ij}\, da_j, \tag{5.18}$$

where da_j is the jth component of the infinitesimal area element $d\mathbf{a}$.

Equation 5.18 is an expression of linear momentum conservation for electromagnetic waves. It demonstrates that any change in the total momentum within the volume (left-hand side) is associated with a flux of electromagnetic momentum through the surface of the volume. For this reason, the stress tensor may also be labeled a *momentum flux density* of the field, and the quantity

$$\mathbf{p}_{field}(\mathbf{r}, t) = \epsilon_0 \mathbf{E}(\mathbf{r}, t) \times \mathbf{B}(\mathbf{r}, t) \tag{5.19}$$

may be called the *momentum density* $\mathbf{p}_{field}$ of the field.

The conservation law, Equation 5.18, must already be used with care. It is tempting to envision $T_{ij}(\mathbf{r}, t)$ and $\mathbf{p}_{field}(\mathbf{r}, t)$ as *local* quantities, that is, as the momentum flux and momentum density at the precise position $\mathbf{r}$ in space. However, this interpretation can lead to unusual and even paradoxical results. The mathematical reason for this is that the stress tensor was only defined in the context of its divergence; we may therefore add the curl of an arbitrary tensor to T_{ij} without changing the conservation law. Evidently, there are many "local" forms of the momentum flux and density functions, all of which are physically equivalent. We will see a similar issue arise when we discuss energy flow in Chapter 8.

Why does the momentum flux density need to be a tensor? Because momentum itself is a vector, we need to use a tensor to describe the flow of this vector quantity across a surface. In short, one index of T_{ij} represents the direction of flux, and the other index represents the direction of the transferred momentum.

It should be noted, and is often not mentioned, that our derivation implies that no charge is transferred across the surface S. This would result in a change in net mechanical momentum in the volume as well, but would also require a strict specification of the charge/mass ratio (or ratios) of the charge carriers.

We may do a strictly analogous calculation to describe a law of angular momentum conservation for electromagnetic fields. In this case, we begin with an expression for the mechanical *torque* on a charged particle that arises from electromagnetic fields:

$$\tau_{mech} = \mathbf{r} \times \mathbf{F} = \mathbf{r} \times \left[q\mathbf{E} + \mathbf{v} \times \mathbf{B} \right]. \tag{5.20}$$

We may again introduce the charge density and current density to convert this expression into an integral. By the use of Maxwell's equations, we may then rewrite it entirely in terms of field quantities, with the inevitable result

$$\tau_{mech} = \int_D \mathbf{r}' \times \left[\mathbf{Q} - \epsilon_0 \frac{\partial}{\partial t} (\mathbf{E} \times \mathbf{B}) \right] d^3 r', \tag{5.21}$$

where $\mathbf{Q}$ is the same vector as before. We now introduce the *angular momentum density of the field* $\mathbf{l}_{field}$ as

$$\mathbf{l}_{field}(\mathbf{r}', t) = \epsilon_0 \mathbf{r}' \times [\mathbf{E}(\mathbf{r}', t) \times \mathbf{B}(\mathbf{r}', t)], \tag{5.22}$$

and may then write the total change in angular momentum in the volume as

$$\frac{\partial \mathbf{L}_{tot}}{\partial t} = \frac{\partial}{\partial t} [\mathbf{L}_{mech} + \mathbf{L}_{field}] = \int_D \mathbf{r}' \times \mathbf{Q} d^3 r'. \tag{5.23}$$

The vector $\mathbf{Q}$ may again be written as the divergence of a tensor, as done in Equation 5.15. The integrand above may then be expressed in tensor notation as

$$[\mathbf{r} \times \mathbf{Q}]_i = -\epsilon_{ijk} r_j \partial_l T_{kl}, \tag{5.24}$$

where ϵ_{ijk} is the Levi–Civita tensor (see [Gbu11, Section 1.3.3]), r_j is the jth component of the position vector, and T_{kl} is a component of the Maxwell stress tensor. Owing to the antisymmetry of ϵ_{ijk} and the symmetry of T_{kl}, the following equation holds:

$$\epsilon_{ijk} r_j \partial_l T_{kl} = \partial_l [\epsilon_{ijk} r_j T_{kl}]. \tag{5.25}$$

If we return to a vector formalism, we may then write

$$\frac{\partial \mathbf{L}_{tot}}{\partial t} = -\int_D \nabla' \cdot (\mathbf{r}' \times \mathbf{T}) d^3 r', \tag{5.26}$$

where $\mathbf{T}$ is the tensor T_{kl}. As in the momentum case, we have the integral of a divergence, which may be rewritten using the divergence theorem in the form

$$\frac{\partial \mathbf{L}_{tot}}{\partial t} = -\oint_S (\mathbf{r}' \times \mathbf{T}) \cdot d\mathbf{a}'. \tag{5.27}$$

On comparison with Equation 5.18, we see that we have a law of angular momentum conservation that is of a similar form to the linear momentum conservation law. Any change in the *total* angular momentum $\mathbf{L}_{tot}$ within a closed volume must involve a flow of electromagnetic angular momentum through the surface. We may introduce a new tensor

$$\mathbf{M}(\mathbf{r}', t) \equiv \mathbf{r}' \times \mathbf{T}(\mathbf{r}', t), \tag{5.28}$$

which we refer to as the *angular momentum flux density* of the field, and we have already introduced the angular momentum density in Equation 5.22.

It should be noted that we are defining the torque with respect to an origin specified by the position vector $\mathbf{r}'$. Our resulting definition of electromagnetic angular momentum is therefore dependent on the choice of coordinate system. This is not necessarily a problem when looking at the interaction of light and matter: by choosing the origin, we are implicitly choosing the mechanical axis of rotation. However, we run into difficulties when we consider the angular momentum of the field in isolation: is there a unique definition of the angular momentum of an electromagnetic wave that does not depend on the choice of coordinates? This will be discussed in Section 5.4.

In studying the angular momentum of light, it is tempting to focus on the angular momentum density $\mathbf{l}_{field}$, a vector quantity, rather than the much more complicated tensor quantity $\mathbf{M}$. However, in quantifying the optical forces induced in matter, we will be more interested in the flux of angular momentum in beams, which makes $\mathbf{M}$ the more natural choice for our study. This is explained in more detail by Barnett [Bar02]. In short, the angular momentum density tells us where the angular momentum is located in a light wave, but the angular momentum flux tells us how that angular momentum is transferred across a closed boundary and therefore into matter.

5.2 ORBITAL AND SPIN ANGULAR MOMENTUM

We have already mentioned that the angular momentum of light can often be decoupled into a spin part (associated with circular polarization) and an orbital part (associated with vortex phase). This decomposition is akin to the separation of Earth's motion in the solar system into a spin part (Earth's daily rotation about its own axis) and an orbital part (Earth's yearly orbit around the Sun). Building on the results of the previous section, we now demonstrate this decoupling.

We consider an electromagnetic beam propagating paraxially in the z-direction, with electric field vector **E** lying entirely within the xy-plane. We will take the field to be monochromatic, with angular frequency ω, and consider the cycle-averaged values of field products, as done in Equation 2.19. We may therefore write the cycle-averaged Maxwell stress tensor in the form

$$\langle T_{ij} \rangle = -\frac{1}{2} \left[\epsilon_0 E_i^* E_j + \frac{1}{\mu_0} B_i^* B_j - \frac{1}{2} \delta_{ij} \left(\epsilon_0 E_k^* E_k + \frac{1}{\mu_0} B_k^* B_k \right) \right]. \quad (5.29)$$

It is important to note that only the real part of this tensor will be physically relevant. We work with the complex form for simplicity but with the understanding that we will take the real part of the resulting expression in the end to derive physical results.

It will be more convenient to write this expression entirely in terms of the electric field. To do this, we may use Faraday's law, Equation 2.3, which in monochromatic tensor form can be written as

$$\epsilon_{ijk} \partial_j E_k = i\omega B_i. \quad (5.30)$$

The resulting stress tensor is now entirely in terms of E_i, but at the cost of additional complexity:

$$\langle T_{ij} \rangle = -\frac{1}{2} \left[\epsilon_0 E_i^* E_j + \frac{1}{\mu_0 \omega^2} \epsilon_{imn} \partial_m E_n^* \epsilon_{jkl} \partial_k E_l \right.$$
$$\left. - \frac{1}{2} \delta_{ij} \left(\epsilon_0 E_k^* E_k + \frac{1}{\mu_0 \omega^2} \epsilon_{klm} \epsilon_{kpq} \partial_l E_m^* \partial_p E_q \right) \right], \quad (5.31)$$

where ϵ_{ijk} is again the Levi–Civita tensor.

To determine the angular momentum of the field, we in fact need to evaluate the tensor

$$(\mathbf{r} \times \mathbf{T})_{il} = \epsilon_{ijk} r_j T_{kl}, \quad (5.32)$$

which is the angular momentum flux density.

We now restrict ourselves to looking at the z-component of the angular momentum, that is, momentum associated with circulation about the z-axis, and furthermore look at the flow of this angular momentum across a plane of constant z, perpendicular to the direction of motion of the beam. With appropriate manipulations of the tensor identity [Gbu11, Section 5.6.5],

$$\epsilon_{ijk}\epsilon_{klm} = \delta_{il}\delta_{jm} - \delta_{im}\delta_{jl}, \tag{5.33}$$

we may write the zz-component of the angular momentum flux density as

$$(\mathbf{r} \times \mathbf{T})_{zz} = -\frac{\epsilon_0}{2}\left[r_j\epsilon_{zjk}E_k^*E_z + \frac{1}{k^2}r_m\epsilon_{zrt}\left\{\partial_z E_m^*\partial_r E_t - \partial_m E_z^*\partial_r E_t\right\}\right]. \tag{5.34}$$

This expression may be simplified further because we have assumed that the z-component of the electric field and its derivatives are zero. We then get a condensed result of the form

$$(\mathbf{r} \times \mathbf{T})_{zz} = -\frac{\epsilon_0}{2k^2}r_m\epsilon_{zrt}\partial_z E_m^*\partial_r E_t. \tag{5.35}$$

Using the properties of the Levi–Civita tensor, we may explicitly write out the sums of this expression as

$$(\mathbf{r} \times \mathbf{T})_{zz} = -\frac{\epsilon_0}{2k^2}\left[x\partial_z E_x^*\partial_x E_y - y\partial_z E_y^*\partial_y E_x + y\partial_z E_y^*\partial_x E_y - x\partial_z E_x^*\partial_y E_x\right]. \tag{5.36}$$

We now note that a paraxial wavefield has a z-dependence that is, to good approximation, simply $\exp[ikz]$. This means that we may make the replacement $\partial_z E_x^* \to -ikE_x^*$ in the above equation, with a similar substitution for E_y^*. We are left with

$$(\mathbf{r} \times \mathbf{T})_{zz} = \frac{i\epsilon_0}{2k}\left[xE_x^*\partial_x E_y - yE_y^*\partial_y E_x + yE_y^*\partial_x E_y - xE_x^*\partial_y E_x\right]. \tag{5.37}$$

To interpret this equation as a combination of spin and orbital angular momentum contributions, we rewrite the first two terms using total derivatives, for example,

$$\partial_x(E_x^*E_y) = (\partial_y E_x^*)E_y + E_x^*\partial_x E_y. \tag{5.38}$$

We may then write

$$
(\mathbf{r} \times \mathbf{T})_{zz} = \frac{i\epsilon_0}{2k} \left[x\partial_x(E_x^*E_y) - y\partial_y(E_y^*E_x) + yE_y^*\partial_xE_y - xE_x^*\partial_yE_x \right.
$$
$$
\left. -x(\partial_xE_x^*)E_y + y(\partial_yE_y^*)E_x \right]. \tag{5.39}
$$

There is one final step to perform. As noted earlier, this flux density only appears in the context of an integral over a surface—in this case, a plane of constant z—which means that we need to integrate over x and y to get a physically meaningful result. Assuming that the field is finite in extent (and vanishes far from the origin), we may perform an integration by parts on the first two terms of Equation 5.39, leaving us with the final expression

$$
(\mathbf{r} \times \mathbf{T})_{zz} = \frac{i\epsilon_0}{2k} \left[-E_x^*E_y + E_y^*E_x + yE_y^*\partial_xE_y - xE_x^*\partial_yE_x \right.
$$
$$
\left. -x(\partial_xE_x^*)E_y + y(\partial_yE_y^*)E_x \right]. \tag{5.40}
$$

The real part of this equation represents the angular momentum flux density of the field. We may separate this into two parts; the first of these is found by combining the first two terms of Equation 5.40, and will be denoted the *spin contribution* to the angular momentum

$$
M_{spin} = \frac{\epsilon_0}{2k}\text{Im}\left\{ E_x^*E_y - E_y^*E_x \right\}. \tag{5.41}
$$

The latter four terms will be denoted the *orbital contribution* to the angular momentum, and may be written as

$$
M_{orbit} = -\frac{\epsilon_0}{2k}\text{Im}\left\{ yE_y^*\partial_xE_y - xE_x^*\partial_yE_x - x(\partial_xE_x^*)E_y + y(\partial_yE_y^*)E_x \right\}. \tag{5.42}
$$

The connection between this expression and the OAM can be made clearer by applying Gauss's law in the paraxial approximation, which has the form

$$
\partial_xE_x + \partial_yE_y = 0. \tag{5.43}
$$

We then have

$$
M_{orbit} = -\frac{\epsilon_0}{2k}\text{Im}\left\{ yE_y^*\partial_xE_y - xE_x^*\partial_yE_x + x(\partial_yE_y^*)E_y - y(\partial_xE_x^*)E_x \right\}. \tag{5.44}
$$

This expression can be converted into polar coordinates, with the result

$$M_{orbit} = \frac{\epsilon_0}{2k}\text{Im}\left\{(\partial_\phi E_x)E_x^* + (\partial_\phi E_y)E_y^*\right\}. \tag{5.45}$$

It is now not difficult to demonstrate by example that M_{spin} and M_{orbit} are associated with the appropriate types of angular momentum. For instance, let us consider a field that passes through a waveplate that performs the transformation

$$E_x \rightarrow E_x e^{i\psi_x}, \quad E_y \rightarrow E_y e^{i\psi_y}. \tag{5.46}$$

These overall constant phase changes to E_x and E_y will not have any effect on the orbital contribution to the angular momentum, Equation 5.45, but will affect the spin contribution, Equation 5.41. Next, let us consider a field that passes through a spiral phase plate that makes the transformation

$$E_x \rightarrow E_x e^{i\phi}, \quad E_y \rightarrow E_y e^{i\phi}, \tag{5.47}$$

with ϕ being the azimuthal angle in cylindrical coordinates. This transformation will not change the spin angular momentum but, thanks to the presence of the derivatives in Equation 5.45, the OAM will change.

It should be noted that our decomposition of the angular momentum into spin and orbit parts can also be done using the angular momentum density, namely,

$$\mathbf{l}_{field}(\mathbf{r}',t) = \epsilon_0 \mathbf{r}' \times [\mathbf{E}(\mathbf{r}',t) \times \mathbf{B}(\mathbf{r}',t)]. \tag{5.48}$$

This was done, for example, in [Ber98], and is suggested as an exercise. It was shown by Barnett [Bar02], however, that the separation of the angular momentum density is only valid in the paraxial approximation, while the separation of the angular momentum *flux* density holds even in the nonparaxial case.

5.3 ANGULAR MOMENTUM OF LAGUERRE–GAUSSIAN BEAMS

Now that we have general formulas for the spin and orbital angular momentum of paraxial beams of light, it is worth applying them to the specific case of Laguerre–Gaussian beams, which we will see are, in a sense, the "purest" beams of OAM. For simplicity, we will investigate beams of the form LG_{0m}, and only consider them in the $z = 0$ waist plane. (We can get away with this because we used the paraxial approximation to remove all z-derivatives of the field in our derivation of the spin and orbital angular momentum.) The spatial part of the beam will therefore be of the form[*]

$$u_{0m}(\mathbf{r}) = \sqrt{\frac{2}{\pi m!}} \left(\frac{\sqrt{2}}{w_0}\right)^{|m|} \rho^{|m|} \exp[im\phi] \exp[-\rho^2/w_0^2], \qquad (5.49)$$

and the electric field polarization will be taken to be constant across the beam's cross section, so that we may write

$$\mathbf{E}(\mathbf{r}) = \boldsymbol{\alpha} E_0 u_{0m}(\mathbf{r}), \qquad (5.50)$$

with $\boldsymbol{\alpha}$ a constant unit transverse electric field vector.

We begin by calculating the spin angular momentum for such a beam. Using Equation 5.41, we may readily find that

$$M_{spin} = \frac{\epsilon_0 E_0^2}{2k} \mathrm{Im} \left\{ \alpha_x^* \alpha_y - \alpha_y^* \alpha_x \right\} |u_{0m}|^2. \qquad (5.51)$$

If the field is linearly polarized, for instance, if $\boldsymbol{\alpha} = \hat{\mathbf{x}}$, then $M_{spin} = 0$. If the field is circularly polarized, however, with

$$\boldsymbol{\alpha} = \frac{\hat{\mathbf{x}} \pm i\hat{\mathbf{y}}}{\sqrt{2}}, \qquad (5.52)$$

we find that

$$M_{spin} = \pm \frac{\epsilon_0 E_0^2}{2k} |u_{0m}|^2. \qquad (5.53)$$

We can find an even more intuitive result if we normalize this result by the flux of photons across the $z = 0$ plane. We first calculate the power flow using the Poynting vector

$$\mathbf{S} = \frac{1}{2} \mathbf{E} \times \mathbf{H}^* = \frac{1}{2\mu_0} \mathbf{E} \times \mathbf{B}^*. \qquad (5.54)$$

The Poynting vector and its significance will be discussed in detail in Chapter 8; for now, we note that it represents the electromagnetic power flux per

[*] We have taken a factor of w_0 out of the usual LG normalization in order that Equation 5.50 has the proper units of electric field.

unit area. Via the use of Faraday's law, the ith component of the Poynting vector may be written as

$$S_i = \frac{i}{2\mu_0\omega}\left[E_j\partial_i E_j^* - E_j\partial_j E_i^*\right].$$

(5.55)

If we use the paraxial approximations of the previous section, namely, $E_z \approx 0$, $\partial_z E_j^* \approx -ikE_j^*$, we find that the z-component of the Poynting vector, representing power flow across the $z = 0$ plane, has the simple form

$$S_z(\rho) = \frac{E_0^2 C^2 k}{2\mu_0\omega}\rho^{2|m|}\exp[-2\rho^2/w_0^2],$$

(5.56)

where we have defined

$$C \equiv \sqrt{\frac{2}{\pi m!}}\left(\frac{\sqrt{2}}{w_0}\right)^{|m|}$$

(5.57)

for brevity. On integrating over the $z = 0$ plane, we find the total power flow across this plane is

$$P_z = \frac{w_0^2}{2\mu_0 c}E_0^2.$$

(5.58)

On a quantum level, every photon possesses an energy $E = \hbar\omega$. The flux of photons N, or photons per second, crossing the $z = 0$ plane is then simply given by

$$N = \frac{P_z}{\hbar\omega} = \frac{w_0^2 E_0^2}{2\mu_0 c\hbar\omega}.$$

(5.59)

We may also calculate the total angular momentum flux L_{spin} crossing the $z = 0$ plane by integrating Equation 5.53 over this plane; the result is

$$L_{spin} = \pm\frac{\epsilon_0 w_0^2 E_0^2}{2k}.$$

(5.60)

If we divide the total angular momentum flux of our beam by the photon flux, we arrive at the spin angular momentum per photon, l_{spin}, namely,

$$l_{spin} = \pm\hbar.$$

(5.61)

The Laguerre–Gauss beams in pure circular polarization states therefore possess a single positive or negative quantum of angular momentum in the z-direction.

We take a similar approach to investigate the OAM, initially working only with nonnegative m. The calculation can be quite involved, but can be simplified dramatically with a few tricks. We first note that we can change variables and express the field in terms of $\zeta = x + iy$ and $\zeta^* = x - iy$ as

$$u_{0m}(\zeta, \zeta^*) = C\zeta^m \exp[-\zeta\zeta^*/w_0^2], \qquad (5.62)$$

with C defined as before. Next, we apply the antisymmetric property of the imaginary part of a complex number, namely,

$$\text{Im}\{AB^*\} = -\text{Im}\{BA^*\}, \qquad (5.63)$$

to rewrite Equation 5.44 in the slightly different form

$$M_{orbit} = -\frac{\epsilon_0}{2k}\text{Im}\left\{yE_y^*\partial_x E_y - xE_x^*\partial_y E_x - xE_y^*\partial_y E_y + yE_x^*\partial_x E_x\right\}. \qquad (5.64)$$

From here, we may transform the x- and y-derivatives into derivatives with respect to ζ and ζ^*, using

$$\partial_x = \partial_\zeta + \partial_{\zeta^*}, \qquad (5.65)$$

$$\partial_y = i\partial_\zeta - i\partial_{\zeta^*}. \qquad (5.66)$$

With a certain amount of rearranging (left as an exercise), our original long expression reduces to the relatively simple form

$$M_{orbit} = -E_0^2\frac{\epsilon_0}{2k}\text{Re}\left\{-\zeta u_{0m}^*\partial_\zeta u_{0m} + \zeta^* u_{0m}^*\partial_{\zeta^*} u_{0m}\right\}. \qquad (5.67)$$

The appropriate derivatives can now be taken. We readily find that

$$-\zeta u^*\partial_\zeta u + \zeta^* u^*\partial_{\zeta^*} u = -C^2 m(\zeta^*)^m\zeta^m e^{-2\zeta\zeta^*/w_0^2}. \qquad (5.68)$$

Returning to polar coordinates, we may write

$$-\zeta u^*\partial_\zeta u + \zeta^* u^*\partial_{\zeta^*} u = -C^2 m\rho^{2m} e^{-2\rho^2/w_0^2}, \qquad (5.69)$$

and therefore that

$$M_{orbit}(\rho) = C^2 m E_0^2 \frac{\epsilon_0}{2k} \rho^{2m} e^{-2\rho^2/w_0^2}. \tag{5.70}$$

The total OAM flux across the $z = 0$ plane can then be found by integration; introducing a new variable $v = 2\rho^2/w_0^2$, this integral may be written as

$$L_{orbit} = C^2 m E_0^2 \frac{\epsilon_0}{2k} \frac{w_0^{2m+2}}{2^{m+2}} \int_0^{2\pi} \int_0^{\infty} v^m e^{-v} \, d\phi \, dv. \tag{5.71}$$

The integral over ϕ simply gives 2π; the integral over v is the familiar integral for the factorial function

$$n! = \int_0^{\infty} v^n e^{-v} \, dv. \tag{5.72}$$

Performing the integrals, and using the value of C, we find that

$$L_{orbit} = \frac{\epsilon_0 w_0^2}{2k} m E_0^2. \tag{5.73}$$

We immediately see that the OAM is proportional to m. If we divide the total angular momentum flux of our beam by the photon flux, Equation 5.59, we arrive at the OAM per photon, l_{orbit}. This is readily found to be

$$l_{orbit} = m\hbar. \tag{5.74}$$

This result was done for a positive value of m, but it is clear that the result must also hold for negative m. We therefore find that the OAM of a photon in a pure Laguerre–Gauss mode is proportional to its topological charge and is quantized in units of $\hbar$.

After this derivation, it is tempting to draw a strong quantitative connection between OAM and topological charge. However, Berry [Ber98] introduced a clever and elegant counterexample to this hypothesis.

We consider a field that has the form

$$v(\mathbf{r}) = \exp[-\rho^2/w_0^2] f(\zeta), \tag{5.75}$$

where $f(\zeta)$ is an Mth-order polynomial of the form

$$f(\zeta) = \sum_{m=0}^{M} f_m \zeta^m,$$ (5.76)

with $M > 0$. This field is a superposition of pure LG_{0m} states with weights f_m. We leave off the normalization of the individual modes in this case, as it is irrelevant to the final result. By following the same approach as done for a pure Laguerre–Gauss mode, we find that the OAM per photon is given by

$$l_{orbit} = \hbar \frac{\sum_{m=1}^{M} m\, m! |f_m|^2 w_0^{2m}}{\sum_{m=0}^{M} m! |f_m|^2 w_0^{2m}}.$$ (5.77)

This expression is mathematically equivalent to the average value of m, with unnormalized "probability" weights $m! |f_m|^2 w_0^{2m}$. Only in the case of a pure Laguerre–Gauss mode LG_{0M} will this take on the value $M\hbar$; it is clear that in other cases the result will be less than $M\hbar$.

The total topological charge of the beam, however, can be readily found to be M. By the fundamental theorem of algebra, the polynomial can always be factorized into the product of M terms

$$f(\zeta) = (\zeta - \zeta_1)(\zeta - \zeta_2) \cdots (\zeta - \zeta_M),$$ (5.78)

where the ζ_m are the zeros of the polynomial, and one or more may be identical. All the vortices circulate in the same direction, and their combined topological charge is M. We therefore find that OAM and topological charge are generally different quantities.

5.4 INTRINSIC AND EXTRINSIC ANGULAR MOMENTUM

Though we have demonstrated that the OAM of a pure Laguerre–Gauss mode is proportional to its topological charge, we have avoided an important question: is this truly an inherent property of the mode? The problem arises because, unlike the spin angular momentum density given by Equation 5.41, the OAM density, given by Equation 5.44, explicitly depends upon the transverse position of the z-axis, that is, the x- and y-coordinates. A coordinate-dependent quantity is said to be *extrinsic*, while a coordinate-independent quantity is said to be *intrinsic*. We can understand this distinction by considering the Earth–Sun system again: we *define* the separation of angular momentum into spin and orbit by declaring the spin to be the angular momentum of Earth around its own center of mass, while the OAM is associated with the motion around an external axis roughly centered on the Sun. Similarly, the spin of light is associated with the inherent angular momentum associated with a photon, while the OAM is associated with an extended circulation of the wavefield.

It is a little unnerving to think of a fundamental conserved quantity having an explicit dependence on the choice of coordinates. As first shown by Berry [Ber98], however, and further elaborated upon by Glasgow researchers [OMAP02], under many circumstances, the OAM of light will be an intrinsic quantity.

For simplicity, we restrict ourselves to considering the angular momentum density of a paraxial monochromatic field, and work with the combined spin and orbital angular momentum, namely,

$$l_{field}(\mathbf{r}) = \frac{\epsilon_0}{2}\mathbf{r} \times [\mathbf{E}(\mathbf{r}) \times \mathbf{B}^*(\mathbf{r})]. \tag{5.79}$$

If we consider the total angular momentum in the z-direction, we have

$$L_{field} = \frac{\epsilon_0}{2}\hat{\mathbf{z}} \cdot \int \mathbf{r} \times [\mathbf{E}(\mathbf{r}) \times \mathbf{B}^*(\mathbf{r})]d^2r. \tag{5.80}$$

Let us now consider the angular momentum as measured from a different origin. Introducing $\mathbf{r}' = \mathbf{r} - \mathbf{r}_0$, we may write

$$L'_{field} = \frac{\epsilon_0}{2}\hat{\mathbf{z}} \cdot \int (\mathbf{r}' + \mathbf{r}_0) \times [\mathbf{E}(\mathbf{r}' + \mathbf{r}_0) \times \mathbf{B}^*(\mathbf{r}' + \mathbf{r}_0)]d^2r'. \tag{5.81}$$

In terms of the original angular momentum, we have

$$L'_{field} = L_{field} + \frac{\epsilon_0}{2}\hat{\mathbf{z}} \cdot \int \mathbf{r}_0 \times [\mathbf{E}(\mathbf{r}) \times \mathbf{B}^*(\mathbf{r})]d^2r = L_{field} + \Delta L, \tag{5.82}$$

where

$$\Delta L = \hat{\mathbf{z}} \cdot \int \mathbf{r}_0 \times \mathbf{p}_{field} \, d^2r. \tag{5.83}$$

Since an intrinsic quantity must be completely independent of the choice of x_0 and y_0, we find that the angular momentum is intrinsic if and only if the transverse momenta p_x, p_y of the field are zero, namely,

$$p_x = \epsilon_0 \hat{\mathbf{x}} \cdot \int \mathbf{E} \times \mathbf{B}^* d^2 r = 0, \qquad (5.84)$$

$$p_y = \epsilon_0 \hat{\mathbf{y}} \cdot \int \mathbf{E} \times \mathbf{B}^* d^2 r = 0. \qquad (5.85)$$

If the total angular momentum is intrinsic, and the spin is intrinsic, this implies that the OAM will also be intrinsic.

For the paraxial Laguerre–Gauss beams of the previous section, this condition holds to a good approximation. This is yet another reason why such beams may be considered pure angular momentum states. What is not so clear, however, is what Equations 5.84 and 5.85 actually mean, physically.

It can be helpful for our intuition, and later discussions, to consider the analogous formulas in classical mechanics. Let us imagine that we have a system of N particles, with a specific particle labeled by index α, with mass m_α and velocity $\mathbf{v}_\alpha$. The net momentum of the system can be written as

$$\mathbf{P} = \sum_{\alpha=1}^{N} m_\alpha \mathbf{v}_\alpha, \qquad (5.86)$$

while the net angular momentum with respect to a particular coordinate system can be written as

$$\mathbf{L} = \sum_{\alpha=1}^{N} \mathbf{r}_\alpha \times m_\alpha \mathbf{v}_\alpha. \qquad (5.87)$$

If we calculate the angular momentum with respect to a different coordinate system, in which the coordinate axes are parallel but the origin of coordinates has shifted, the velocities $\mathbf{v}_\alpha$ are unchanged but the positions of the particles are now represented by $\mathbf{r}'_\alpha$. The angular momentum in this new coordinate system is

$$\mathbf{L}' = \sum_{\alpha=1}^{N} \mathbf{r}'_\alpha \times m_\alpha \mathbf{v}_\alpha. \qquad (5.88)$$

Let us define the relation between the coordinate systems as

$$\mathbf{r}_\alpha = \mathbf{r}'_\alpha - \mathbf{r}_0. \qquad (5.89)$$

Because the position vectors are different in the two systems, we expect the value of the angular momentum to be in general different. On substitution from Equation 5.89 into Equation 5.88, we readily find the relationship between the values of the angular momentum with respect to the two origins:

$$\mathbf{L}' = \sum_{\alpha=1}^{N} \mathbf{r}_\alpha \times m_\alpha \mathbf{v}_\alpha - \sum_{\alpha=1}^{N} \mathbf{r}_0 \times m_\alpha \mathbf{v}_\alpha = \mathbf{L} - \sum_{\alpha=1}^{N} \mathbf{r}_0 \times m_\alpha \mathbf{v}_\alpha. \quad (5.90)$$

Clearly, the two values will be the same if the latter term on the right side of the preceding equation vanishes, that is, if

$$\sum_{\alpha=1}^{N} \mathbf{r}_0 \times m_\alpha \mathbf{v}_\alpha = 0. \quad (5.91)$$

This expression is directly analogous to Equations 5.84 and 5.85. One situation in which Equation 5.91 is true is immediately apparent: if every mass has its "mirror image" on the opposite side of the rotation axis with equal and opposite transverse velocity, the momentum of each pair will cancel out. The simplest example of such a system is a rotationally symmetric distribution of masses with a radial-dependent velocity. Two special cases of this are a rigid body, for which the azimuthal velocity is proportional to the axial distance ρ, that is, $v(\rho) = \omega\rho$, and an ideal fluid, for which the azimuthal velocity is inversely proportional to the axial distance, $v(\rho) = A/\rho$, with A a constant. If we consider the angular momentum l per unit mass, we find it is of the form $l = \omega\rho^2$ for the rigid body and $l = A$ for an ideal fluid.

For a vortex beam, the analogous quantity is the angular momentum flux density per photon, that is, $\hbar\omega M_{orbit}(\rho)/S_z(\rho)$. Using our previous expressions, Equations 5.56 and 5.70, we get the exceedingly simple result

$$\hbar\omega M_{orbit}(\rho)/S_z(\rho) = m\hbar. \quad (5.92)$$

The angular momentum flux density per photon is a constant with respect to transverse position ρ, suggesting that the circulation in a Laguerre–Gauss vortex beam is "fluid-like." Later, in Chapter 9, we will see that certain partially coherent vortex beams act like rigid rotators, and others as a mix of rigid and fluid.

The analogy between mechanical and optical angular momentum suggests one interpretation of the possible intrinsic nature of OAM. In mechanical systems, spin angular momentum is *defined* as that angular momentum associated with rotations around the center of mass. Though we consider circular polarization to be associated with the spin of a light beam, it is evident that, for beams that satisfy Equations 5.84 and 5.85, the OAM acts very much like an effective mechanical spin.

5.5 TRAPPING FORCES

The momentum and angular momentum of light, though small on a macroscopic scale, can be put to use on the microscopic scale in the trapping and manipulation of particles. The first work on this subject was performed by Ashkin [Ash70] in the 1970s, who demonstrated the acceleration of micron-sized particles in a single laser beam and the trapping of such particles between a pair of beams. Not long after, Ashkin also demonstrated [Ash78] the trapping and cooling of atoms using resonance radiation pressure, a technique that has developed into a fundamental tool for studying quantum phenomena [MvdS99].

A quite different breakthrough was made in 1986, however, when it was demonstrated [ADBC86] that large dielectric particles (tens of wavelengths in diameter) can be trapped within a *single beam* of light, pinned at the geometrical focus of the beam by a gradient force. A schematic of such a system is illustrated in Figure 5.2a. From a geometric optics perspective, this gradient force can be understood as resulting from the refraction of light within the particle, resulting in a net force toward the region of highest intensity; this is illustrated in Figure 5.2b.

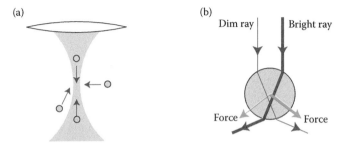

FIGURE 5.2 (a) Illustration of a gradient beam trap; arrows roughly indicate the direction of the gradient force. (b) Illustration of the gradient force for a pair of parallel rays of unequal intensity refracted by a spherical particle.

Detailed calculations of the relevant forces in the geometrical optics limit have been done by Ashkin [Ash92]. However, the same forces appear even when working with subwavelength-size particles, in the so-called Rayleigh regime; we explore this model in some detail. Angular momentum does not play a direct role in the interaction of light with such particles: we will nevertheless see that it has an influence.

To begin, we treat a single particle as a small but finite-sized dipole with a positive charge and negative charge separated by vector distance **d**, as illustrated in Figure 5.3a. From the Lorentz force law, Equation 5.1, the electric force on the dipole may be written as

$$\mathbf{F} = q\mathbf{E}(\mathbf{r} + \mathbf{d}) - q\mathbf{E}(\mathbf{r}). \tag{5.93}$$

With a suitable rearrangement, this can be written as

$$\mathbf{F} = qd \left\{ \frac{\mathbf{E}(\mathbf{r} + \mathbf{d}) - \mathbf{E}(\mathbf{r})}{d} \right\}. \tag{5.94}$$

The term in brackets looks like the derivative along the $\hat{\mathbf{d}}$-direction, and may be written as $\hat{\mathbf{d}} \cdot \nabla\mathbf{E}(\mathbf{r})$ in the limit of vanishing d. If, at the same time, we increase the magnitude of the charge so that the quantity $\mathbf{p} = q\mathbf{d}$ is constant, we find that the electric force on a point dipole is given by

$$\mathbf{F} = (\mathbf{p} \cdot \nabla)\mathbf{E}(\mathbf{r}). \tag{5.95}$$

Next, we consider the magnetic component of the Lorentz force law. With the notation of Figure 5.3b, the magnetic force on the dipole may be written as

$$\mathbf{F} = q\mathbf{v}_+ \times \mathbf{B}(\mathbf{r} + \mathbf{d}) - q\mathbf{v}_- \times \mathbf{B}(\mathbf{r}), \tag{5.96}$$

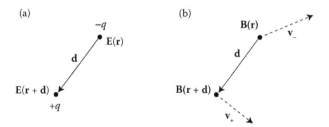

FIGURE 5.3 Illustration of the (a) electric force and (b) the magnetic force on an electric dipole.

where

$$\mathbf{v}_+ = \frac{d(\mathbf{r} + \mathbf{d})}{dt}, \quad \mathbf{v}_- = \frac{d\mathbf{r}}{dt}. \tag{5.97}$$

Assuming that the magnetic field varies little over the region of the dipole, that is, $\mathbf{B}(\mathbf{r} + \mathbf{d}) \approx \mathbf{B}(\mathbf{r})$, we may simplify our expression for the magnetic force to

$$\mathbf{F} = \frac{d(q\mathbf{d})}{dt} \times \mathbf{B}(\mathbf{r}) = \frac{d\mathbf{p}}{dt} \times \mathbf{B}(\mathbf{r}).$$

The total force on a point dipole is found to be

$$\mathbf{F} = (\mathbf{p} \cdot \nabla)\mathbf{E}(\mathbf{r}) + \frac{d\mathbf{p}}{dt} \times \mathbf{B}(\mathbf{r}) \tag{5.98}$$

or, if we look at a cycle-averaged time-harmonic force,

$$\mathbf{F} = \mathrm{Re}\left\{\frac{1}{2}(\mathbf{p} \cdot \nabla)\mathbf{E}^*(\mathbf{r}) - \frac{1}{2}i\omega\mathbf{p} \times \mathbf{B}^*(\mathbf{r})\right\}. \tag{5.99}$$

We are interested in the specific case of a small dielectric particle, for which the induced polarization is proportional to the electric field, that is,

$$\mathbf{p} = \alpha\mathbf{E}, \tag{5.100}$$

and α is the complex polarizability of the particle. On substitution, we find that the (complex) force on our particle due to an electromagnetic field is given by

$$\mathbf{F} = \frac{\alpha}{2}(\mathbf{E} \cdot \nabla)\mathbf{E}^* - \frac{i\omega}{2}\alpha\mathbf{E} \times \mathbf{B}^*. \tag{5.101}$$

We now apply the monochromatic form of Faraday's law, Equation 5.5, to eliminate the magnetic field entirely from the expression. We arrive at the equation

$$\mathbf{F} = \frac{\alpha}{2}\left[(\mathbf{E} \cdot \nabla)\mathbf{E}^* + \mathbf{E} \times (\nabla \times \mathbf{E}^*)\right]. \tag{5.102}$$

A little vector calculus will help write the second bracketed term in a more useful form. We note that

$$\mathbf{E} \times (\nabla \times \mathbf{E}^*) = \nabla_2(\mathbf{E} \cdot \mathbf{E}_2^*) - (\mathbf{E} \cdot \nabla)\mathbf{E}^*, \tag{5.103}$$

where the subscript 2 indicates that the gradient only acts on the second term in the expression. Then, in summation form, the total (real-valued) force may be written as

$$\mathbf{F} = \mathrm{Re}\left\{\frac{\alpha}{2}\nabla_2(\mathbf{E}\cdot\mathbf{E}_2^*)\right\} = \mathrm{Re}\left\{\frac{\alpha}{2}\sum E_i\nabla E_i^*\right\}. \tag{5.104}$$

The polarizability α is in general complex, with the imaginary part characterizing the absorption properties of the particle. If we write $\alpha = \alpha_R + i\alpha_I$, we may express the force as

$$\mathbf{F} = \frac{\alpha_R}{2}\mathrm{Re}\left\{\sum E_i\nabla E_i^*\right\} - \frac{\alpha_I}{2}\mathrm{Im}\left\{\sum E_i\nabla E_i^*\right\}. \tag{5.105}$$

The real-valued quantity may be written explicitly as

$$\mathrm{Re}\left\{\sum E_i\nabla E_i^*\right\} = \frac{\sum E_i\nabla E_i^* + \sum E_i^*\nabla E_i}{2} = \frac{1}{2}\nabla|\mathbf{E}|^2, \tag{5.106}$$

so that we have

$$\mathbf{F} = \frac{\alpha_R}{4}\nabla|\mathbf{E}|^2 - \frac{\alpha_I}{2}\mathrm{Im}\left\{\sum E_i\nabla E_i^*\right\}. \tag{5.107}$$

We now need to simplify the rightmost term. We use Equation 5.103 in reverse to write

$$\mathrm{Im}\left\{\sum E_i\nabla E_i^*\right\} = \mathrm{Im}\left\{(\mathbf{E}\cdot\nabla)\mathbf{E}^*\right\} + \mathrm{Im}\left\{\mathbf{E}\times(\nabla\times\mathbf{E}^*)\right\}. \tag{5.108}$$

Applying Faraday's law again to the latter term, we may write the total force on our dipole as

$$\mathbf{F} = \frac{1}{4}\alpha_R\nabla|\mathbf{E}|^2 + \frac{1}{2}\alpha_I\omega\mathrm{Re}\left\{\mathbf{E}\times\mathbf{B}^*\right\} - \frac{1}{2}\alpha_I\mathrm{Im}\left\{(\mathbf{E}\cdot\nabla)\mathbf{E}^*\right\}. \tag{5.109}$$

The first of these terms is what is known as the *gradient force*. It is the effect illustrated in Figure 5.2b, in which the light refracted/scattered by the particle results in a net force toward regions of high intensity. The second term is the *radiation pressure force* or *scattering force*, which can be seen to be proportional to the momentum density, Equation 5.19. It depends on the imaginary part of the polarizability, and represents the net momentum imparted to the particle due to light absorption. The derivation of these two forces was first evidently done by Gordon [Gor73], and later elaborated upon by Chaumet and Nieto-Vesperinas [CNV00].

The third term of Equation 5.109, which we will refer to as the *spin force*, has traditionally been considered negligible as compared to the other two. However, in 2009, researchers from Spain and France noted [AMLS09] that this term can be quite large if there is a nonuniform circular polarization in the field.

By expanding out the imaginary part of the spin force, we may write it as

$$\mathbf{F}_{spin} = -\frac{1}{4i}\alpha_I \left[\mathbf{E}\cdot\nabla\mathbf{E}^* - \mathbf{E}^*\cdot\nabla\mathbf{E}\right]. \tag{5.110}$$

We may simplify this by using the vector identity

$$\nabla\times(\mathbf{E}\times\mathbf{E}^*) = \mathbf{E}^*\cdot\nabla\mathbf{E} - \mathbf{E}^*(\nabla\cdot\mathbf{E}) + \mathbf{E}(\nabla\cdot\mathbf{E}^*) - \mathbf{E}\cdot\nabla\mathbf{E}^*, \tag{5.111}$$

and noting that $\nabla\cdot\mathbf{E} = 0$ in the absence of sources. From what remains, we may express the spin force in the form

$$\mathbf{F}_{spin} = \frac{\alpha_I}{4i}\nabla\times(\mathbf{E}\times\mathbf{E}^*). \tag{5.112}$$

If we introduce a new vector

$$\mathbf{L}_S \equiv \frac{\alpha_I}{4i}\mathbf{E}\times\mathbf{E}^*, \tag{5.113}$$

we may write the spin force as the curl of $\mathbf{L}_S$:

$$\mathbf{F}_{spin} = \nabla\times\mathbf{L}_S. \tag{5.114}$$

The quantity $\mathbf{L}_S$ may be considered the time-averaged spin density of the electromagnetic field. It will only be nonzero if the polarization is of circular or elliptic type, and will vanish for linear polarization. This can readily be shown by considering the fields $\mathbf{E}(\mathbf{r}) = \hat{\mathbf{x}}E(\mathbf{r})$ and $\mathbf{E}(\mathbf{r}) = (\hat{\mathbf{x}} + i\hat{\mathbf{y}})E(\mathbf{r})$. The cross-product of $\mathbf{E}\times\mathbf{E}^*$ will be zero for the linear case, and will be $-2i\hat{\mathbf{z}}|E(\mathbf{r})|^2$ for the circular case. Locally, we can imagine a field having the form

$$E(\mathbf{r}) = E_0[1 - x/x_0], \tag{5.115}$$

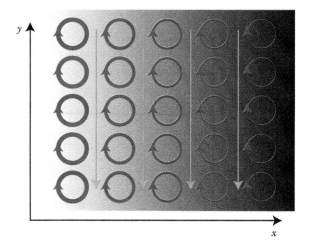

FIGURE 5.4 Illustration of the spin force. The imbalance in spin strengths (thickness of circles) from right to left results in a net downward force throughout the region.

and in this region the spin force will have the value

$$\mathbf{F}_{spin} = -\frac{\alpha_I}{2}\hat{\mathbf{y}}\frac{|E_0|^2}{x_0}. \tag{5.116}$$

The interpretation of such a spin force is not difficult to come by as well, and is illustrated in Figure 5.4. The circles represent circular polarization rotating in a clockwise sense, and thicker arrows represent a higher intensity of light (also represented by the gradient background). Each spin region produces a stronger downward force relative to the upward force produced by its right-side neighbor, resulting in a net downward force.

Such a picture may look familiar in the context of *shear forces* in fluids. When there is an transverse imbalance in the flow of a fluid, the difference of speed introduces a vorticity. Here, we have the converse: an imbalance in vorticity produces a shear force in the electromagnetic field.

5.6 MOMENTUM IN MATTER AND THE ABRAHAM–MINKOWSKI CONTROVERSY

Up to this point, we have worked entirely with the microscopic form of Maxwell's equations, and have avoided looking at the momentum of light *within* a macroscopic material. We have good reason to be cautious: the proper theoretical form for the momentum in matter has been a hotly-contested problem for over a century, and is now known as the *Abraham–Minkowski controversy*, after Hermann Minkowski [Min08] and Max Abraham [Abr09], whose conflicting results in 1908 and 1909 sparked the debate.

The basic issue may be described, in a perhaps oversimplified way, as a debate over whether the wave or particle nature of light determines its momentum in matter. For a monochromatic wave, the phase velocity of light is inversely proportional to the refractive index n in matter, that is, $v_p = c/n$. If we consider light to be particle-like, then we would expect that its momentum is proportional to its velocity, and that it decreases when it enters matter, i.e.

$$p_A = p_0/n, \tag{5.117}$$

where p_0 is the momentum in vacuum. This is the Abraham result for the momentum of light in matter. However, we also know that the wavelength of light λ decreases by a factor n in matter, which implies that the wavenumber k increases. For waves, the momentum is proportional to the wavenumber, so it should increase in matter as

$$p_M = np_0. \tag{5.118}$$

This is the Minkowski result for the momentum of light in matter.

The difference between the two results could not be more stark. We can illustrate this with a thought experiment, as shown in Figure 5.5. We imagine that a pulse of light is incident upon a block of transparent material with $n > 1$. In the Abraham formulation, the momentum of the pulse decreases as it enters the block; because momentum is conserved, the block must acquire a net momentum to the right while the pulse is present.

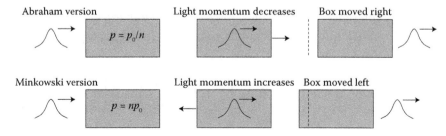

FIGURE 5.5 Simple illustration of the difference between the Abraham and Minkowski formulations of the momentum of light in matter.

When the pulse leaves, the block has shifted right. In the Minkowski formulation, the momentum of the pulse increases, which by conservation of momentum implies that the block must shift left.

It is to be noted that, in both cases, the difference only occurs during the time that the pulse is within the block of material. For a monochromatic wave, there would be no observable difference, as any increase/decrease of momentum from light entering the material would be balanced by a decrease/increase from light leaving. A little thought suggests that the discernible effect is typically quite small, and we will see that this is part of the challenge in experimentally resolving the problem.

How is it possible for there to be a controversy at all, considering that the answer should in principle follow directly from Maxwell's equations, as in the vacuum case? Let us consider the macroscopic Maxwell's equations, combined with the Lorentz force law, and see what develops.

From Section 2.1, we have the macroscopic Maxwell's equations

$$\nabla \cdot \mathbf{D}(\mathbf{r}, t) = \rho_f(\mathbf{r}, t), \tag{5.119}$$

$$\nabla \cdot \mathbf{B}(\mathbf{r}, t) = 0, \tag{5.120}$$

$$\nabla \times \mathbf{E}(\mathbf{r}, t) = -\frac{\partial \mathbf{B}(\mathbf{r}, t)}{\partial t}, \tag{5.121}$$

$$\nabla \times \mathbf{H}(\mathbf{r}, t) = \mathbf{J}_f(\mathbf{r}, t) + \frac{\partial \mathbf{D}(\mathbf{r}, t)}{\partial t}. \tag{5.122}$$

To avoid confusion, we have explicitly labeled ρ_f and $\mathbf{J}_f$ as the free charge and current densities, which will be important in our derivation. We again consider the application of these to the integral form of the Lorentz force law

$$\mathbf{F}_f = \int_D [\rho_f \mathbf{E} + \mathbf{J}_f \times \mathbf{B}] d^3 r', \tag{5.123}$$

where $\mathbf{F}_f$ represents the mechanical force on the free charges of the system alone. On substituting from Maxwell's equations, this free charge force law becomes

$$\mathbf{F}_f = \int_D \left[\mathbf{E}(\nabla \cdot \mathbf{D}) + \mathbf{H}(\nabla \cdot \mathbf{B}) + \left(\nabla \times \mathbf{H} - \frac{\partial \mathbf{D}}{\partial t} \right) \times \mathbf{B} \right] d^3 r', \tag{5.124}$$

where we have sneakily added the term $\mathbf{H}(\nabla \cdot \mathbf{B})$, which is zero according to Equation 5.120. By rewriting the time derivative related to $\mathbf{D}$ as a total derivative, we may further write

$$\mathbf{F}_f = \int_D \left[\mathbf{E}(\nabla \cdot \mathbf{D}) + \mathbf{H}(\nabla \cdot \mathbf{B}) - \mathbf{B} \times (\nabla \times \mathbf{H}) \right.$$

$$\left. -\mathbf{D} \times (\nabla \times \mathbf{E}) - \frac{\partial}{\partial t}(\mathbf{D} \times \mathbf{B}) \right] d^3 r'. \tag{5.125}$$

Then in analogy with Section 5.1, we interpret the last term as the time derivative of the electromagnetic momentum, and define the total change in electromagnetic momentum as

$$\mathbf{F}_M \equiv \frac{\partial}{\partial t} \int_D \mathbf{D} \times \mathbf{B} d^3 r', \tag{5.126}$$

where "M" in this case refers to Minkowski.

We now switch to a tensor notation; again in analogy with Section 5.1, we may write the force as

$$[\mathbf{F}_f + \mathbf{F}_M]_i = \int_D \left[\partial_j(E_i D_j) - \partial_i(E_m D_m) + \partial_j(H_i B_j) - \partial_i(H_m B_m) \right.$$

$$\left. + E_m \partial_i D_m + H_m \partial_i B_m \right] d^3 r'. \tag{5.127}$$

The first four terms on the left of the integrand look reminiscent of the Maxwell stress tensor. In order to have agreement with the vacuum result, we define

$$T_{ij} \equiv - \left[E_i D_j + H_i B_j - \frac{1}{2} \delta_{ij}(E_m D_m + H_m B_m) \right], \tag{5.128}$$

and after an application of Gauss's theorem and some rearranging, we have

$$[\mathbf{F}_f + \mathbf{F}_M]_i = -\oint_S T_{ij} \, da_j + \int_D f_i^m d^3 r', \tag{5.129}$$

where we have introduced a "matter force density" as

$$f_i^m \equiv \frac{1}{2} \left[E_m \partial_i D_m - D_m \partial_i E_m + H_m \partial_i B_m - B_m \partial_i H_m \right]. \tag{5.130}$$

If we assume the highly unphysical result* that $\mathbf{D} = \epsilon E$ and $\mathbf{B} = \mu \mathbf{H}$, with ϵ and μ constant in space and time, this matter force density vanishes identically. We then find, in what turns out to be the Minkowski formulation, that we have a momentum conservation law strictly analogous to the vacuum case.

Let us investigate the matter force density further. Using the standard relations

$$\mathbf{D} = \epsilon_0 \mathbf{E} + \mathbf{P}, \tag{5.131}$$

$$\mathbf{B} = \mu_0 (\mathbf{H} + \mathbf{M}), \tag{5.132}$$

we may rewrite the matter force density as

$$f_i^m \equiv \frac{1}{2} [E_m \partial_i P_m - P_m \partial_i E_m + B_m \partial_i M_m - M_m \partial_i B_m]. \tag{5.133}$$

This looks rather mysterious in its current form, but we try and evaluate it by guessing that it has a relation with

$$[\mathbf{P} \times (\nabla \times \mathbf{E})]_i = - \mathbf{P} \times \frac{\partial \mathbf{B}}{\partial t}\bigg|_i = P_m \partial_i E_m - P_j \partial_j E_i. \tag{5.134}$$

We now use derivative product rules, one in time and two in space, to rewrite all three terms on the left of the above expression. We get

$$\left[-\frac{\partial}{\partial t}(\mathbf{P} \times \mathbf{B}) + \frac{\partial \mathbf{P}}{\partial t} \times \mathbf{B} \right]_i = \partial_i (E_m P_m) - \partial_j (E_i P_j) + E_i \partial_j P_j - E_m \partial_i P_m. \tag{5.135}$$

The terms on the right-hand side look suspiciously like a stress tensor again; we may write

$$\left[-\frac{\partial}{\partial t}(\mathbf{P} \times \mathbf{B}) + \frac{\partial \mathbf{P}}{\partial t} \times \mathbf{B} \right]_i = -\partial_j \left[E_i P_j - \frac{1}{2} \delta_{ij}(E_m P_m) \right]$$

$$+ E_i \partial_j P_j + \frac{1}{2} P_m \partial_i E_m - \frac{1}{2} E_m \partial_i P_m. \tag{5.136}$$

---

* Perhaps a greater mystery than the Abraham–Minkowski controversy is why anyone would make this assumption, since momentum effects in matter generally require short pulses and therefore explicitly dispersive constitutive relations, unlike the nondispersive relations assumed here.

The rightmost two terms are exactly what we need for our matter force density. Similarly, we may write

$$[\mathbf{B} \times (\nabla \times \mathbf{M})]_i = -\partial_j \left[M_i B_j - \frac{1}{2} \delta_{ij}(B_m M_m) \right]$$

$$+ M_i \partial_j B_j + \frac{1}{2} B_m \partial_i M_m - \frac{1}{2} M_m \partial_i B_m. \qquad (5.137)$$

Again, the rightmost two terms are needed for the matter force density. Also, $\partial_j B_j = \nabla \cdot \mathbf{B} = 0$. On substitution, we may finally write

$$f_i^m = \frac{\partial}{\partial t}(\mathbf{P} \times \mathbf{B})_i - \partial_j T_{ij}^m - \frac{\partial \mathbf{P}}{\partial t} \times \mathbf{B} \Big|_i + E_i \partial_j P_j - (\nabla \times \mathbf{M}) \times \mathbf{B}|_i, \qquad (5.138)$$

where we have defined a matter stress tensor of the form

$$T_{ij}^m = E_i P_j - M_i B_j - \frac{1}{2} \delta_{ij} \left[E_m P_m - B_m M_m \right]. \qquad (5.139)$$

Let us finally write our entire conservation law, which with rearrangement has the form

$$\mathbf{F}_f + \mathbf{F}_M + \int_D \left[-\mathbf{E}(\nabla \cdot \mathbf{P}) + \left(\frac{\partial \mathbf{P}}{\partial t} + \nabla \times \mathbf{M} \right) \times \mathbf{B} \right] d^3 r' - \frac{\partial}{\partial t} \int_D \mathbf{P} \times \mathbf{B} d^3 r'$$

$$= -\oint_S (T_{ij} + T_{ij}^m) \, da_j. \qquad (5.140)$$

The complicated term involving polarization and magnetization simplifies immensely if we note that the bound charge and current densities are

$$\rho_b = -\nabla \cdot \mathbf{P}, \qquad (5.141)$$

$$\mathbf{J}_b = \nabla \times \mathbf{M} + \frac{\partial \mathbf{P}}{\partial t}, \qquad (5.142)$$

which means that the expression is the Lorentz force of the fields on the bound charges:

$$\mathbf{F}_L = \int_D \left[-\mathbf{E}(\nabla \cdot \mathbf{P}) + \left(\frac{\partial \mathbf{P}}{\partial t} + \nabla \times \mathbf{M} \right) \times \mathbf{B} \right] d^3 r'. \qquad (5.143)$$

The final unspecified term, which related the polarization to the magnetic field, is evidently an interaction term between the polarization and magnetization fields; we write it as

$$\mathbf{F}_{I,M} \equiv -\frac{\partial}{\partial t} \int_D \mathbf{P} \times \mathbf{B} d^3 r', \qquad (5.144)$$

where "I, M" indicates the interaction in the Minkowski interpretation. Our final conservation law has the quite elaborate form

$$\mathbf{F}_f + \mathbf{F}_M + \mathbf{F}_L + \mathbf{F}_{I,M} = -\oint_S (T_{ij} + T_{ij}^m)\, da_j. \qquad (5.145)$$

Let us simplify things a little by assuming that we have no free charges, only a dielectric, so that

$$\mathbf{F}_M + \mathbf{F}_L + \mathbf{F}_{I,M} = -\oint_S (T_{ij} + T_{ij}^m)\, da_j. \qquad (5.146)$$

This expression is somewhat similar to Equation 5.18 of Section 5.1, in that it suggests that the net change in momentum in a volume is the result of momentum flowing across the surface. However, the momentum in the volume now includes a field component and two matter components, and the flow across the surface includes both a field and matter component. It is to be noted that there is evidently no flow of matter across the surface, but rather energy and momentum carried by disturbances in the polarization and magnetization densities.

We are already potentially in some trouble here because we have a force acting on macroscopic matter but no way to account for any physical motion of the matter itself. Such a *quasi-static* approximation to the momentum problem is commonly used, but it is an example of how we have implicitly built in assumptions to our calculation that make its validity suspect.

One significant problem with the results given is that the Minkowski field momentum density, $\mathbf{p}_M$, given by

$$\mathbf{p}_M = \mathbf{D} \times \mathbf{B}, \qquad (5.147)$$

leads to an asymmetric energy momentum tensor for the electromagnetic field, which is not invariant under relativistic frame transformations. This was noted by Abraham, who suggested that the proper field momentum density is of the form

$$\mathbf{p}_A = \frac{1}{c^2}\mathbf{E} \times \mathbf{H}. \tag{5.148}$$

We can arrive at this result directly from the Minkowski one by expanding out the Minkowski formula

$$\mathbf{p}_M = \mu_0(\epsilon_0\mathbf{E} + \mathbf{P}) \times (\mathbf{H} + \mathbf{M}) = \frac{1}{c^2}\mathbf{E} \times \mathbf{H} + \mathbf{P} \times \mathbf{B} + \frac{1}{c^2}\mathbf{E} \times \mathbf{M}. \tag{5.149}$$

On substitution, we may write our conservation law in an Abraham form

$$\mathbf{F}_A + \mathbf{F}_L + \mathbf{F}_{I,A} = -\oint_S (T_{ij} + T_{ij}^m)\, da_j, \tag{5.150}$$

where

$$\mathbf{F}_{I,A} = \frac{\partial}{\partial t} \int_D \frac{1}{c^2}\mathbf{E} \times \mathbf{M} d^3 r', \tag{5.151}$$

and all other terms remain the same.

We therefore find that Abraham and Minkowski predict different force densities for the light and matter within a medium. Abraham predicts a field density $\mathbf{F}_A$ and a matter density $\mathbf{F}_{I,A}$, while Minkowski predicts a field density $\mathbf{F}_M$ and a matter density $\mathbf{F}_{I,M}$. At least in the context of Maxwell's equations alone, both appear to be self-consistent momentum conservation laws.

Looking at the field momentum density in isolation, however, we find that they predict quite different results. We consider a monochromatic plane wave propagating in the z-direction in a homogeneous medium with (real) permittivity ϵ and (real) permeability μ, with electric field

$$\mathbf{E} = \hat{\mathbf{x}}E_0\, e^{i(kz-\omega t)}, \tag{5.152}$$

where $k = nk_0 = n\omega/c$. From Faraday's law, we have

$$\mathbf{B} = \hat{\mathbf{y}}\frac{n\omega}{c}E_0\, e^{i(kz-\omega t)}. \tag{5.153}$$

The cycle-averaged form of the Minkowski momentum density is

$$|\mathbf{p}_M| = \frac{1}{2}|\mathbf{D} \times \mathbf{B}^*| = \frac{1}{2}\frac{\epsilon n}{c}|E_0|^2, \tag{5.154}$$

while the cycle-averaged form of the Abraham momentum density is

$$|\mathbf{p}_A| = \frac{1}{2c^2}|\mathbf{E} \times \mathbf{H}^*| = \frac{1}{2c^2}\frac{n\omega}{\mu c}|E_0|^2. \tag{5.155}$$

We would now like to calculate the average momentum per photon. Borrowing ahead from Section 8.1, the energy density U of a monochromatic electromagnetic wave is

$$U = \frac{1}{4}(\mathbf{D}^* \cdot \mathbf{E} + \mathbf{H}^* \cdot \mathbf{B}) = \frac{1}{2}\epsilon|E_0|^2. \tag{5.156}$$

Considering that the quantum energy of a photon is $\hbar\omega$, we may write the density N of photons as

$$N = \frac{U}{\hbar\omega} = \frac{1}{2\hbar\omega}\epsilon|E_0|^2. \tag{5.157}$$

Finally, we may determine the momentum per photon p_M or p_A by dividing each momentum density by the photon density to get

$$p_M = n\hbar k_0, \tag{5.158}$$

$$p_A = \hbar\frac{k_0}{n}, \tag{5.159}$$

where $k_0 = \omega/c$. These results match those found by simple arguments in Equations 5.117 and 5.118.

With such a large discrepancy, one might expect that experimental work is the best way to resolve the controversy. However, directly measuring the momentum of an electromagnetic wave in matter is a difficult proposition; this is illustrated nicely by the block example of Figure 5.5. The momentum of light only has an effect on the block when pulses are used, and only while the pulse is interacting with the block. The pulse must therefore be quite short, but also the block itself must be very small due to the small magnitude of electromagnetic momentum. A short pulse, however, will be subject to significant dispersion effects, which can be difficult to account for. Absorption of light in the material will furthermore reduce the difference between the Abraham and Minkowski cases.

A number of experiments have nevertheless been done, and results have been found that support both the Abraham and the Minkowski views; we only mention two illustrative examples. In 2008, She, Yu, and Feng [SYF08] performed a clever experiment very similar in spirit to the block example. They fired a short pulse out of a free-hanging vertical fiber. Under the Abraham interpretation, the pulse should increase its momentum on leaving the fiber, giving the fiber an upward "kick." According to Minkowski, the pulse should decrease its momentum on leaving the fiber, and the fiber will be pulled downward, with no observable kick. In practice, though, due to imperfections at the fiber tip, it will also be pushed slightly sideways in either case. The researchers observed an upward push, suggesting the Abraham theory is correct.

A quite different experiment was performed by Ashkin and Dziedzic much earlier [AD73]. A pulsed focused beam of light was used to illuminate a transparent liquid from above. The radiation pressure is then expected to either cause the liquid to bulge outward (Minkowski) or inward (Abraham), and this in turn will affect the focusing properties of the light, as illustrated in Figure 5.6. The transmitted field was imaged by microscope, and the results appeared to be consistent with the Minkowski result; however, a new version of the experiment by Zhang et al. [ZSPL15] was consistent with the Abraham version.

It is not clear what to make of all the conflicting results, both theoretical and experimental. Intuitively, one suspects that the wrong question is being asked if a physics question is unresolved for over 100 years. This seems to be partly the take of some recent theoretical work on the subject, which essentially answer the question with: "It depends." Leonhardt [Leo14], in analyzing the momentum effects on fluids, concluded that the proper momentum formula depends on whether the fluid is brought into motion by the electromagnetic forces or not. In the rather boldly titled paper "Resolution of the Abraham–Minkowski dilemma," Barnett [Bar10] suggests that

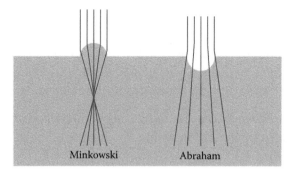

FIGURE 5.6 Simple idea behind the Ashkin and Dziedzic experiment, showing the different predictions for Abraham and Minkowski.

both forms are correct, but represent different types of momenta. In particular, he argues that the Abraham momentum is the kinetic momentum and
the Minkowski momentum is the canonical momentum. A review article by
Pfeifer et al. [PNHRD07] are not entirely in disagreement when they argue
that there is, in fact, no controversy, and any differences in experimental
predictions comes from not properly taking into account the material contributions to the momentum (recall the footnote after Equation 5.130). In
fact, this same article notes that there are many possible mathematically
sound decompositions of the momentum of light in matter beyond those
of Abraham and Minkowski, such as one by Einstein and Laub [EL08].

We conclude by noting that there are connections between this discussion of linear momentum and angular momentum. Leach et al. [LWG+08]
have studied the effect of a rotating medium on the propagation of light
through that medium, and have observed the rotation of interference patterns. They suggested that such experiments might shed additional light on
the Abraham–Minkowski dilemma.

Though much work has been done, it seems that discussions and arguments about the proper interpretation of electromagnetic momentum in
matter will go on. For more information, see also the review by Baxter and
Loudon [BL10] and Kemp [Kem15].

5.7 EXERCISES

1. Prove the expression given by Equation 5.15. You will need to use the
   relation

   $$\epsilon_{ijk}\epsilon_{klm} = \delta_{il}\delta_{jm} - \delta_{im}\delta_{jl}.$$

2. Demonstrate that Equation 5.16 has the proper sign convention for
   momentum flow by considering the momentum flow of a linearly
   polarized plane wave, propagating in the z-direction, across a surface
   of constant z.

3. Doing things the hard way: use the Maxwell stress tensor to calculate the force between a pair of point charges $+q$ and $-q$ located
   symmetrically a distance d away from either side of the plane $z = 0$.

4. A Poincaré beam (to be discussed in Section 7.8) is one that possesses
   every state of polarization within its cross section. An example is a
   beam of the form

   $$\mathbf{E}(\mathbf{r}) = \frac{1}{\sqrt{2}}[\hat{\mathbf{x}}u_{00}(\mathbf{r}) + \hat{\mathbf{y}}u_{01}(\mathbf{r})],$$

where u_{nm} is a Laguerre–Gauss mode. In the $z = 0$ plane, calculate the spin force acting on such a beam as a function of transverse position (x, y).

5. Calculate the OAM of a Hermite–Gauss HG_{01} beam using Equation 2.85 to write the Hermite–Gauss beam as a sum of Laguerre–Gauss beams.

6. Show that, under the paraxial approximation, Equation 5.48 for the angular momentum density separates into a spin and orbital contribution.

7. Derive Equation 5.45 for the OAM density in terms of ϕ-derivatives from the Cartesian form, Equation 5.44.

CHAPTER 6

Applications of Optical Vortices

A N IMPRESSIVELY DIVERSE COLLECTION of applications has arisen from research into singular optics. Some of these come directly from the unique phase and angular momentum properties of vortex beams, while others take advantage of the generic properties of vortices. In this chapter, we review some of the most significant applications to date; it should be noted that additional possibilities are considered in later chapters.

6.1 MICROMANIPULATION, SPANNING, AND TRAPPING

In Section 5.5, we considered the rather complicated forces that can be exerted on a microscopic particle in a nonuniform optical field. These forces cause some types of particles to become trapped in regions of high intensity, leading to the concept of optical tweezing.

Though the technique has already become extremely important and widespread as described, the use of vortex beams in an optical tweezer can increase the applicability of the technique to a broader variety of particles. Furthermore, in many cases, the spin or orbital angular momentum of the vortex beams will rotate particles, allowing advanced micromanipulations of particles as well as the development of light-powered micromachines.

We begin by looking again at Equation 5.109, which describes the total force of an electromagnetic wave on a Rayleigh particle:

$$\mathbf{F} = \frac{1}{4}\alpha_R\nabla|\mathbf{E}|^2 + \frac{1}{2}\alpha_I\omega\mathrm{Re}\left\{\mathbf{E}\times\mathbf{B}^*\right\} - \frac{1}{2}\alpha_I\mathrm{Im}\left\{(\mathbf{E}\cdot\nabla)\mathbf{E}^*\right\}. \quad (6.1)$$

Speaking qualitatively, it is to be noted that optical tweezing requires the terms containing α_I to be relatively small compared to the gradient term containing α_R. This is readily understood: if the particle is highly absorbing (with a large α_I), then it will absorb much of the light incident upon it and transmit very little. This simultaneously results in a large push downstream from radiation pressure and a small gradient force. We therefore do not expect conventional tweezers to work well with absorptive particles. Furthermore, an absorbing particle can be overheated and destroyed at even low beam powers, putting a limit on the strength with which particles can be trapped.

Another limitation of conventional tweezing can be found by investigating the physical origins of α_R and α_I. These quantities are readily derivable from the theory of Rayleigh scattering. We outline the calculation here, and refer readers to Section 5.2 of [BH83] for more details.

Let us consider, as shown in Figure 6.1, a spherical particle of radius d, with a (generally complex) permittivity ϵ_i embedded in a background medium of permittivity ϵ_o, illuminated by an electromagnetic wave with electric field pointing in the $\hat{\mathbf{z}}$-direction. If the particle is much smaller than the wavelength λ of the illuminating field, we may treat the system using an electrostatics approximation, where the particle is illuminated by a uniform electric field $\mathbf{E} = E_0\hat{\mathbf{z}}$.

Following traditional boundary value problems with dielectrics, we work with the scalar potentials $\phi_i(\mathbf{r})$ inside the sphere and $\phi_o(\mathbf{r})$ outside the

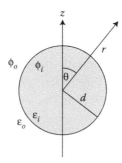

FIGURE 6.1 Notation related to Rayleigh scattering.

sphere. The potentials must satisfy three boundary conditions, namely,

$$\phi_i(d) = \phi_o(d), \tag{6.2}$$

$$\epsilon_i \frac{\partial \phi_i(d)}{\partial r} = \epsilon_o \frac{\partial \phi_o(d)}{\partial r}, \tag{6.3}$$

$$\phi_o(\mathbf{r}) \to -E_0 r \cos \theta \text{ as } r \to \infty. \tag{6.4}$$

The third condition is the assumption that the potential created in the presence of the external field is local, and that the electric field approaches $\mathbf{E} = E_0 \hat{\mathbf{z}}$ far away from the scatterer.

In the electrostatics approximation, $\phi_i(\mathbf{r})$ and $\phi_o(\mathbf{r})$ satisfy Laplace's equation, and the general solutions are of the form

$$\phi_i(\mathbf{r}) = \sum_{l=0}^{\infty} A_l r^l P_l(\cos \theta), \tag{6.5}$$

$$\phi_o(\mathbf{r}) = \sum_{l=0}^{\infty} \frac{B_l}{r^{l+1}} P_l(\cos \theta), \tag{6.6}$$

where $P_l(\cos \theta)$ are the Legendre polynomials and A_l and B_l are determined by the boundary conditions. Applying the conditions of Equations 6.2, 6.3, and 6.4, we find that

$$\phi_o(\mathbf{r}) = \frac{\epsilon_i - \epsilon_o}{\epsilon_i + 2\epsilon_o} E_0 \frac{\cos \theta}{r^2} d^3 - E_0 r \cos \theta. \tag{6.7}$$

The first term on the right-hand side is the potential of the scattered field. On comparing it with the potential of an ideal electric dipole in the same background medium,

$$\phi(\mathbf{r}) = \frac{p \cos \theta}{4\pi \epsilon_o r^2}, \tag{6.8}$$

we find that the dipole moment p of the sphere may be written as

$$p = 4\pi \epsilon_o \frac{\epsilon_i - \epsilon_o}{\epsilon_i + 2\epsilon_o} E_0 d^3. \tag{6.9}$$

Finally, we note that the polarizability of a particle is defined by the relation $p = \alpha E_0$; this leads us to a general expression for the polarizability as

$$\alpha = 4\pi \epsilon_o \frac{\epsilon_i - \epsilon_o}{\epsilon_i + 2\epsilon_o} d^3. \tag{6.10}$$

Let us consider nonabsorbing materials, with $\text{Im}\{\epsilon_\alpha\} = 0$. In general, we expect $\epsilon_i > 0$ and $\epsilon_o > 0$, but α depends on the difference $\epsilon_i - \epsilon_o$. A high-index particle in a lower-index background medium will therefore have a positive α, but a low-index particle in a high-index medium will have a negative α. The gradient force for such particles will push away from regions of high intensity, a problem for a trap based on an ordinary Gaussian beam.

Both absorptive and low-index particles can, however, be caught in the center of a focused vortex beam. In both cases, the low-intensity core of the vortex provides a natural location for the particle to settle. For absorptive particles, it can be pushed there by scattering forces, while for low-index particles, it can be pulled there by gradient forces. The first demonstration of absorptive particle trapping was done by He, Heckenberg, and Rubinsztein-Dunlop [HHRD95], who used a focused beam of topological charge 3 for the trap. The dark core of a vortex beam also reduces the amount of light absorbed by the particle and allows for higher beam intensities.

The trapping of low-index particles was demonstrated by Gahagan and Swartzlander [GS96,GS98], who trapped 20-µm diameter hollow glass spheres surrounded by water in an argon-ion laser beam at 514 nm. The researchers also trapped density-matched water droplets in acetophenone. They also noted that high-index polystyrene spheres of 10-µm diameter could be trapped in a ring in the bright donut of the vortex beam.

It is to be noted that even high-index particles can be trapped in a vortex core, if they are of sufficiently small size. The particles end up being pushed into the dark core due to scattering forces arising from the bright donut of the vortex. In fact, the trapping efficiency can be higher than that of an ordinary Gaussian beam, due to the absence of the strong direct scattering force of the beam's core. This effect was demonstrated early by Sato, Ishigure, and Inaba [SII91], and detailed calculations of the forces involved were done by Ashkin [Ash92]. Vortex beams, therefore, have been shown to enhance the applicability and efficiency of optical tweezing in many circumstances.

Upon the discovery of optical OAM, it was more or less immediately recognized that it could be used to rotate microscopic particles. In their aforementioned work on trapping [HHRD95], He, Heckenberg, and Rubinsztein-Dunlop also noted that their vortex-trapped particles could be set into rotation; they followed up these initial observations with a detailed study soon after [HFHRD95]. Using an LG_{03} beam produced from a

15-mW He–Ne laser operating at 632.8 nm, they trapped and rotated particles on the order of 1–2 μm in size at a speed ranging from 1 to 10 Hz, depending on the shape and size of the particle.

Further experiments demonstrated that spin and orbital angular momentum could be used in combination to produce more or less rotation. Frieese et al. [FERDH96] demonstrated this using absorbing CuO particles with sizes ranging from 1 to 5 μm and a He–Ne beam as the light source. The trap used, however, was not fully three-dimensional: the particles were pushed downstream by the scattering force and held in place against a microscope slide. The first fully three-dimensional vortex rotator was made by Simpson et al. [SDAP97], and they trapped a variety of particle types in a Nd:YLF laser beam of wavelength 1047 nm. These researchers also demonstrated that the total angular momentum is a linear sum of the spin and orbit contributions. Experimental images of particle rotation from their paper are shown in Figure 6.2. Typical rotation speeds were on the order of 1 Hz.

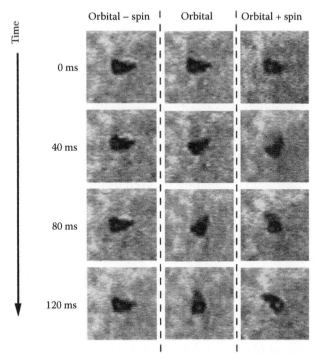

FIGURE 6.2 The rotation of a 2-μm diameter Teflon particle in an $m = 1$ vortex beam in the presence and absence of spin angular momentum. (After N.B. Simpson et al. *Opt. Lett.*, 22:52–54, 1997.)

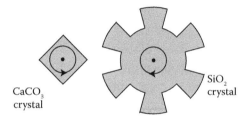

FIGURE 6.3 The optical gear arrangement of Reference [FRDG⁺01].

Simpson et al. referred to their system as an "optical spanner," recognizing that it represented a new tool in noncontact micromanipulation. One limitation, however, is the need for absorbing particles, which limits the amount of light intensity, and consequently torque, that can be applied to the system. An alternative was presented by Friese et al. [FNHRD98], who noted that small birefringent but transparent particles would experience a torque from the change of polarization on transmission, directly inspired by Beth's 1936 experiment [Bet36] mentioned at the beginning of Chapter 5. With no restrictions on illumination power, they were able to rotate a 1-µm thick calcite crystal at 357 Hz in a 300-mW laser beam.

Not long after, Friese et al. used this alternative method of producing optical torque to drive an optical gear [FRDG⁺01]. A birefringent calcite crystal fragment roughly 1 µm in size was trapped and rotated, and a small gear made of nonbirefringent and transparent silicon dioxide was trapped adjacent to it, as illustrated in Figure 6.3. The calcite, rotating somewhere between 100 and 200 Hz, imparted a rotation of roughly 0.2 Hz to the silicon dioxide gear. The gear and calcite fragment did not directly touch, but motion was imparted via fluid flow induced by the rotating calcite.

The use of particle rotation to generate fluid flow introduces other intriguing possibilities. Ladavac and Grier [LG04] constructed a "microoptomechanical pump" using a 2×3 array of optical vortices. The upper row of 3 possessed a total charge $m = +21$, while the lower row possessed $m = -21$. The vortices trapped collections of silica spheres in their bright rings, and the circulating motion of these spheres funneled fluid through the device, making it act as a simple pump. The flow of fluid was confirmed by following single free spheres as they were pushed along between the vortex rows, as illustrated in Figure 6.4. This pump takes advantage of the observation that off-axis particles trapped in a vortex beam will actually orbit the center of the beam. This was noted independently by Curtis and Grier [CG03] as well as Garcés-Chávez et al. [GCMP⁺03].

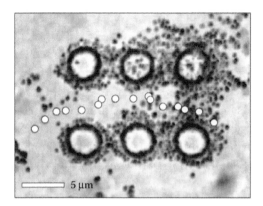

FIGURE 6.4 Time-lapse composite of 16 images at half-second intervals, showing the trajectory of a single sphere as it moved to the left through the pump system. (After K. Ladavac and D.G. Grier. *Opt. Exp.*, 12:1144, 2004.)

If particles have an asymmetric shape, angular momentum is not necessarily required to set them in rotation: an imbalance in scattering forces can have the same effect. This was acknowledged in early experiments, where it was found that rough-shaped particles would continue to rotate even when the angular momentum was "switched off." Another means of designing microrotators, then, is to fabricate particles with a particular handedness. Galajdá and Ormos [GO01] designed a variety of such light-driven rotors, of "helix," "sprinkler," and "propeller" shapes, and made a complex micromachine driven by one of these rotors, as illustrated in Figure 6.5.

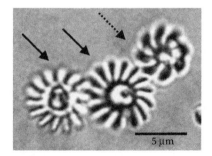

FIGURE 6.5 Top view of two cogwheels (solid arrows) being turned by a light-driven rotor (dashed arrow). The cogwheels rotate on axes fixed to the glass surface below, while the rotor is held and rotated by optical tweezers. (After P. Galajdá and P. Ormos. *Appl. Phys. Lett.*, 78:249–251, 2001.)

It is to be noted that complex optical forces do not need to be used simply for rotation; in 2010, Swartzlander et al. [SPAGR10] demonstrated that a semicylindrical rod could be "flown" on a beam of light, the scattering forces serving in rough analogy to the aerodynamic forces of airplanes.

In summary, then, it has been shown that optical rotation can be imparted to microscopic particles through at least three distinct mechanisms: OAM acting on absorbing particles, spin angular momentum acting on birefringent particles, and scattering forces acting on particles with distinct handedness. These options provide a lot of possibilities for future optical micromachines.

6.2 OPTICAL COMMUNICATIONS

As Internet and mobile data usage continues to grow, it is of interest to ask if additional unexploited strategies exist that can increase the amount of data transferred simultaneously along a single communications channel. The standard technique for encoding multiple signals is, of course, the use of different frequency bands. Even in fiber optics, however, the spectral carrying capacity of fibers is reaching its limit. A simple strategy to double the number of signals that can be sent at the same time is the use of orthogonal polarization states; however, this is a small increase and the two polarization states typically mix on propagation, resulting in significant crosstalk between independent signals.

A related but distinct problem is the development of reliable free-space *optical* communication. The use of visible light in free-space communications affords a great increase in bandwidth over low-frequency radio waves, and potentially affords a great degree of security: a point-to-point laser-encoded message is extremely hard to intercept, due to the directionality of the laser. Unfortunately, visible light is highly susceptible to atmospheric turbulence, which manifests as random spatial and temporal variations in the refractive index of the atmosphere. This turbulence can cause the intensity of light to fluctuate dramatically at the detector; as systems usually encode their data in intensity variations, turbulence can introduce unacceptable levels of error even over a few kilometers.

Optical vortex beams, or OAM states, have been proposed as solutions to both problems. In principle, it is possible to transmit information simultaneously in orthogonal beams with different topological charges; these beams can be separated at the detector by using the modal method of Section 4.2.5. It would seem, then, that angular momentum could be used to encode an arbitrary number of signals on the same communications channel. Furthermore, because the topological charge of an optical beam is stable under weak perturbations of a wavefield, it may serve as an alternate way to encode information in a beam of light, one that is resistant to atmospheric effects.

For fiber optics, experimental tests have already yielded promising results. Bozinovic et al. [BYR+13] have experimentally demonstrated 1.6 terabit per second data transmission through a 1.1-km fiber, using two vortex modes over 10 wavelengths.

In the case of free-space optical communications, however, the challenges are much more formidable, even controversial. One of the earliest demonstrations of free-space communication using OAM was a laboratory demonstration performed by Gibson et al. [GCP+04], who used modans to encode and decode information in beams of one of eight OAM states. Notably, they used nonadjacent states, with topological charges $t = \pm4, \pm8, \pm12, \pm16$; this can reduce crosstalk between independent states, as we will see. This experiment, however, only tested a propagation distance of 15 m.

A computational study of topological charge conservation over long propagation distances in atmospheric turbulence was undertaken by Gbur and Tyson [GT08]. They investigated the propagation of Laguerre–Gauss beams of different orders through weak-to-moderate turbulence, and calculated the average topological charge within the detector by numerically implementing Equation 3.36. Some of their results are shown in Figure 6.6.

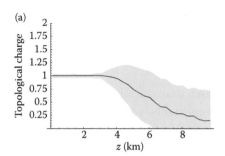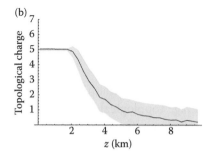

FIGURE 6.6 The average topological charge and its variance measured at the detector in moderate $C_n^2 = 10^{-15}\,\mathrm{m}^{-2/3}$ atmospheric turbulence, for (a) $n = 1$, $m = 1$ and (b) $n = 5$, $m = 5$. Here, $w_0 = 2\,\mathrm{cm}$, $\lambda = 1.55\,\mu\mathrm{m}$, and the detector radius is 4 cm. (Taken from G. Gbur and R.K. Tyson. *J. Opt. Soc. Am. A*, 25:225–230, 2008.)

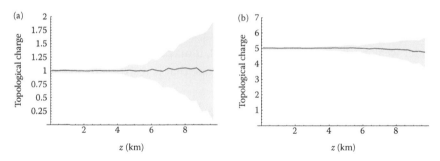

FIGURE 6.7 The average topological charge and its variance measured at a variable detector in moderate $C_n^2 = 10^{-15}\,\mathrm{m}^{-2/3}$ atmospheric turbulence, for (a) $n = 1$, $m = 1$ and (b) $n = 5$, $m = 5$. Here, $w_0 = 2\,\mathrm{cm}$, $\lambda = 1.55\,\mu\mathrm{m}$, and the starting detector radius is 4 cm. (Taken from G. Gbur and R.K. Tyson. *J. Opt. Soc. Am. A*, 25:225–230, 2008.)

It can be seen that the transmission of the topological charge is nearly perfect over a finite propagation distance, and then drops dramatically, with a corresponding increase in the variance. This drop can be attributed to the vortex "wandering" out of the detector region; the small distortions of atmospheric turbulence amount to a random walk of the vortex away from the center axis. For a larger topological charge, the loss can be slowed somewhat, as seen in Figure 6.6b.

If the aperture is increased, the range over which charge is reliably transmitted can be extended, but even here there are fundamental limits. In Figure 6.7, the aperture size is increased in the simulations at the same rate as which a free-space beam would diffract, that is,

$$r(z) = r_0 - w_0 + w_0\sqrt{1 + 4z^2/(k^2 w_0^4)}. \qquad (6.11)$$

Still, we find that the variance increases dramatically, even though the average topological charge is effectively constant—so, what is happening? As the beam becomes more distorted, pairs of oppositely charged vortices are created in the field due to self-interference effects. Though these additional charges should average to zero, they will often be produced at the edge of the aperture, and one of the pairs will end up outside the aperture, the other within, resulting in a net change in the detected charge. The loss of charge from vortex wander and pair production is illustrated in Figure 6.8.

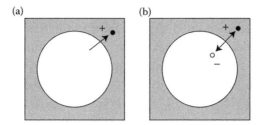

FIGURE 6.8 Two ways in which positive topological charge can be lost from a detector: (a) charge wander and (b) pair production at the aperture boundary.

It appears that loss of topological charge is inevitable, once propagation distances and turbulence strength become large enough. In principle, this loss can be offset by choosing states of extremely large order. If multiple signals are to be sent simultaneously, they must also be sufficiently separated in order. However, creation of high-order beams is generally difficult. The turbulence-induced crosstalk in an OAM-based optical communications system has been studied numerically by Anguita, Neifeld, and Vasic [ANV08] as well as experimentally by Rodenburg et al. [RLM$^+$12].

These limitations have not stopped researchers from conducting long-range tests of OAM beams for communication, though these tests have resulted in significant pushback from radio engineers, who have argued that there is, in fact, nothing new in the approach. An illustrative example of this controversy centered around a radio experiment performed in Venice in 2012 [TMs$^+$12a]. In this experiment, crudely illustrated in Figure 6.9a, a pair of transmitters was used to broadcast an untwisted ($l = 0$) beam and

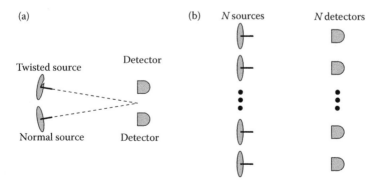

FIGURE 6.9 (a) Illustration of the Venice experiment and (b) illustration of an N-channel MIMO. Horizontal distance greatly compressed for the figure.

a twisted beam ($l = 1$) to a pair of detectors. With both transmitters aimed at the midpoint between the detectors, the untwisted beam produces signals completely in phase at the detectors, while the twisted beam produced signals completely out of phase. Taking the sum of or difference between the detector outputs reproduces the untwisted or twisted signals, respectively.

It was soon argued [TCPC12] that the original experiment is not technically different from a standard MIMO (multiple-input-multiple-output) technique in radio, as illustrated in Figure 6.9b. In a general N-channel MIMO system, N spatially separated transmitters produce signals that overlap at N separated detectors. Because each transmitter produces a different phase pattern at the detector array, broadcasting can be treated as a linear transformation, with the source and detector signals being N-element vectors and the transmission treated as an $N \times N$ matrix. Since the matrix is in general invertible, it is in principle possible to determine the signals from the N independent transmitters from measurements at the detector. It can be seen that the Venice experiment uses a very similar strategy, although they precondition the phase at the twisted transmitter.

The Venice group disputes that their experiment is quite so simple [TMs$^+$12b]. So, who is correct? It is arguable that they are both partially correct in their assertions. The specific experiment, as shown, is more or less equivalent to a MIMO system; however, in general, it appears that vortex modes may provide certain advantages over MIMO. A MIMO system may be said to encode multiple signals through the use of linear momentum, by using fields propagating in different directions. At longer communication ranges, the sources and detectors must be separated by larger transverse distances in order to clearly distinguish the different channels. Vortex beams, in contrast, can all be broadcast along the same direction and in principle be separated at the detector regardless of distance (though, as we have seen, turbulence is a strong limiting factor at optical frequencies).

Other researchers have correctly noted [EJ12,ABB15] that OAM does not represent a new set of hitherto undiscovered modes for use in radio communication. This means, for example, that Hermite–Gauss beams can be used to form different communications channels just as Laguerre–Gauss beams can. The novelty, though, seems to lie in the stability and discreteness of the vortex core of OAM modes, and the ability to separate and measure the vortex twist by a variety of methods.

Additional experiments have been done. Wang et al. [WYF$^+$12] demonstrated a 1.37 terabit per second free-space data transmission rate using visible light in the laboratory. More recently, Krenn et al. [KFF$^+$14] demonstrated transmission of information using 16 different OAM modes in visible light across a 3-km distance in Vienna.[*]

At this point, it seems much more work will need to be done to determine the practical value of vortex beams in atmospheric propagation. Those results that have been achieved present a cautiously optimistic view for the future.

6.3 PHASE RETRIEVAL

In optics, much effort has been devoted to the determination of the phase of a wavefield from measurements of its intensity only, a problem broadly referred to as *phase retrieval*. At optical frequencies, the field oscillates much too rapidly to be directly measured and must typically be found through sensitive interferometric techniques. The phase of the field plays a fundamental role in its propagation, and consequently knowledge of the field is required in a variety of imaging applications.

However, it is not difficult to see that the presence of phase singularities makes the problem of phase retrieval nonunique. Let us consider again Equation 2.37 for a Laguerre–Gauss mode of order n, m:

$$u_{nm}^{LG}(\mathbf{r}) = \sqrt{\frac{2n!}{\pi w_0^2 (n + |m|)!}} \left(\frac{\sqrt{2}\rho}{w(z)} \right)^{|m|} L_n^{|m|} \left(\frac{2\rho^2}{w^2(z)} \right)$$

$$\times \exp[im\phi] u_G(\mathbf{r}) \exp\left[-i(2n + |m|)\Phi(z) \right].\qquad (6.12)$$

In this expression, the sign of the azimuthal order m only appears in the azimuthal phase term, $\exp[im\phi]$. This term cancels out of the intensity expression, $|u_{nm}^{LG}(\mathbf{r})|^2$, which implies that Laguerre–Gauss modes of order n, m and $n, -m$ have exactly the same intensity profile, even on propagation. We therefore conclude that the phase retrieval problem is generally nonunique in the presence of vortices.

All is not lost, however, as we have already seen that only first-order vortices are generic and that they have a relatively simple structure. By making educated guesses about the nature of the phase around zero points, we should be able to remove the phase ambiguity associated with the vortices and correctly deduce the complete phase profile. We discuss strategies to do so in this section.

---

[*] Cities whose names start with "V" seem to be popular for optical communication experiments. Next stop: Vancouver?

There is a long history of phase retrieval in optics, and a number of different techniques that can be used to determine the phase. We first consider a method based on the so-called *transport-of-intensity equation* (TIE), introduced by Teague [Tea82,Tea83]. The basis of this strategy is the observation that the phase of a wavefield plays a major role in its propagation, and consequently the evolution of the intensity of the field. By measuring the intensity in several planes along the direction of propagation, one can in principle deduce the phase from the changes in intensity.

We start with the paraxial wave equation, Equation 2.27, namely,

$$\nabla_\perp^2 u + 2ik\frac{\partial u}{\partial z} = 0, \tag{6.13}$$

where $\nabla_\perp$ is the Laplacian with respect to x and y and again $U(\mathbf{r}) \approx u(\mathbf{r}) \exp[ikz]$.

We now separate $u(\mathbf{r})$ into an amplitude and phase in the form

$$u(\mathbf{r}) = A(\mathbf{r})e^{i\psi(\mathbf{r})}. \tag{6.14}$$

With some effort (left as an exercise), we may manipulate the paraxial wave equation into two evolution equations for the intensity $I(\mathbf{r}) = A^2(\mathbf{r})$ and phase $\psi(\mathbf{r})$, determined from the real and imaginary parts of the paraxial equation. The expression for the phase is of the form

$$k\frac{\partial I}{\partial z} = -\nabla_\perp \cdot (I\nabla_\perp\psi), \tag{6.15}$$

which is the TIE. If the intensity of the field and its derivative along the z-direction are known, this equation becomes a linear second-order partial differential equation for the phase $\psi(\mathbf{r})$.

Teague's original solution to this problem involved introducing an auxiliary function $\psi(\mathbf{r})$, defined as

$$\nabla\Psi = I\nabla\psi, \tag{6.16}$$

so that the TIE equation takes the form

$$\nabla^2\Psi = -k\frac{\partial I}{\partial z}. \tag{6.17}$$

The function $\Psi(\mathbf{r})$ therefore satisfies a two-dimensional Poisson equation. This equation can be solved by the use of Green's function techniques, as

Teague did, or by orthogonal function decomposition [GRN95b], and then the solution for $\psi(\mathbf{r})$ can be found from Equation 6.16 by direct integration. In both cases, it is assumed that the phase goes to zero at large distances from the origin, that is, it satisfies a Dirichlet boundary condition.

We can immediately see a limitation of the technique, however, from Equation 6.16. To solve for $\psi(\mathbf{r})$, we must divide by $I(\mathbf{r})$; at points where $I(\mathbf{r}) = 0$, the expression for $\nabla\psi$ is therefore undefined, that is, a phase singularity. It is reasonable to expect that the TIE phase reconstruction method will run into difficulties when such zero points exist in the field.

In fact, as we have noted, the phase is nonunique in the presence of vortices. This was rigorously demonstrated by Gureyev, Roberts, and Nugent [GRN95a], and can be summarized as follows. Let us assume that there exist two phases, $\psi_1(\mathbf{r})$ and $\psi_2(\mathbf{r})$, that satisfy the TIE, Equation 6.15, with the same intensity $I(\mathbf{r})$ and z-derivative. Then, we may take the difference of the two TIEs, leaving an expression of the form

$$\nabla_\perp \cdot \{I(\mathbf{r}) [\nabla_\perp \psi_1(\mathbf{r}) - \nabla_\perp \psi_2(\mathbf{r})]\} = 0. \tag{6.18}$$

Because the divergence of a curl is zero, we may add the curl of a vector $\mathbf{A}(\mathbf{r})$ to the expression in the $\{\cdots\}$ brackets without changing the expression. We therefore have

$$I(\mathbf{r})\nabla_\perp \psi_1(\mathbf{r}) = I(\mathbf{r})\nabla_\perp \psi_2(\mathbf{r}) + \nabla \times \mathbf{A}(\mathbf{r}). \tag{6.19}$$

However, since the divergence of Equation 6.18 is only transverse, we must choose a function $\mathbf{A}(\mathbf{r})$ of the form

$$\mathbf{A}(\mathbf{r}) = A(\mathbf{r})\hat{z}. \tag{6.20}$$

As the curl represents the local circulation of a vector field, we expect $\nabla \times \mathbf{A}$ to form closed loops in the xy-plane. Its contribution to the phase, in other words, is a vortex circulation. In the presence of vortices, we expect a curl contribution to the phase that cannot be determined directly from the TIE.

We may readily determine the local form of $\mathbf{A}(\mathbf{r})$ in the presence of any mixed edge-screw dislocation of the form

$$U(x, y) = \frac{U_0}{w_0}(x \pm i\alpha y). \tag{6.21}$$

If we calculate $I\nabla\psi$, we have

$$I\nabla\psi = \pm\frac{\alpha U_0^2}{w_0^2}(-y\hat{x} + x\hat{y}), \tag{6.22}$$

and this can be represented by a vector function $\mathbf{A}(\mathbf{r})$ as

$$\mathbf{A}(\mathbf{r}) = \pm \frac{\alpha U_0^2}{2w_0^2}(x^2 + y^2)\hat{\mathbf{z}}. \tag{6.23}$$

Curiously, all first-order vortices have the same rotationally symmetric functional form for $\mathbf{A}(\mathbf{r})$, regardless of their angle with respect to the z-axis.

The preceding discussion diagnoses the problem in phase retrieval when vortices are present, but does not describe how to solve it. It was noted by Nugent and Paganin [NP00], however, that Equation 6.19 is very close to the result of Helmholtz's theorem [AW01, Section 1.16], which states that a general vector field $\mathbf{F}(\mathbf{r})$ may be decomposed into divergence-free and curl-free terms, in the form

$$\mathbf{F}(\mathbf{r}) = \nabla \psi_S(\mathbf{r}) + \nabla \times \psi_V(\mathbf{r}), \tag{6.24}$$

where $\psi_S(\mathbf{r})$ and $\psi_V(\mathbf{r})$ are scalar and vector contributions to the decomposition. Nugent and Paganin assumed that $\nabla \psi(\mathbf{r})$, when vortex effects are included, may be generally decomposed in the Helmholtz manner. In this way, the TIE equation is modified to

$$\nabla_\perp \cdot [I(\nabla_\perp \psi_S + \nabla_\perp \times \psi_V)] = -k \frac{\partial I}{\partial z}. \tag{6.25}$$

In their model, which they then tested on images containing first-order vortices [AFN+01], they found that this modifies the TIE with a vortex-specific contribution of the form

$$\nabla_\perp \cdot [I \nabla_\perp \psi_S] = -\sum_i \frac{m_i}{r_i} \frac{\partial I}{\partial \theta_i} - k \frac{\partial I}{\partial z}, \tag{6.26}$$

where the sum is over the vortices in the field, m_i is the topological charge of the ith vortex, r_i is the distance from the ith vortex, and θ_i is the azimuthal angle around the vortex.

This modified TIE is quite compelling, as it seems to incorporate vortex effects directly into the algorithm with minimal change. Though the vortex charges are not known from experiment, we know that they will typically be of first order; one can therefore try different combinations of first-order vortices in the reconstruction until a suitable solution is found. Furthermore, as we will see in Section 11.4, there are strong correlations between the signs of adjacent vortices; taking advantage of this can further reduce the guesswork and the time required to make a phase reconstruction.

However, it should be noted that the vortex terms in Equation 6.26 are zero for pure screw dislocations, as the intensity in such a case will be rotationally symmetric. It seems reasonable to conclude, then, that these additional terms still do not present the complete solution, but only separate out a contribution to the gradient phase due to the tilt of the vortex.

A more comprehensive approach for the problem of phase retrieval with vortices was recently presented by Lubk et al. [LGBV13]. The vortices are properly treated as "holes" in the simply connected domain over which the phase must be reconstructed, and Neumann boundary conditions are applied at the boundaries of these holes as well as at infinity. One additional complication presents itself, however, because the phase around a series of vortices is a multivalued function, one must introduce a series of branch cuts connecting pairs of oppositely charged vortices, or running from a vortex to infinity. An additional set of boundary conditions must be imposed at these branch cuts, where the phase jumps by an integer multiple of 2π crossing them; it is here that the charges of vortices are incorporated, rather than on the boundary surrounding the vortex itself. A rough illustration of this is shown in Figure 6.10. In this method, as discussed in the incomplete method earlier, one must guess at the values of the vortex charges and consequently the branch cut jumps on these boundaries.

Though the use of TIE for phase retrieval in the presence of vortices has apparently not yet been used extensively, knowledge gained from its study

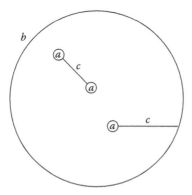

FIGURE 6.10 Boundary conditions to be used in a vortex field phase retrieval problem. (a) Neumann boundary conditions around phase singularities. (b) Neumann boundary condition at infinity. (c) Branch cut boundary conditions.

has led to practical results. An understanding of the curl effects in the TIE has proven beneficial in the imaging of lithographic photomasks [STS⁺14].

It should be noted that there has been some success in directly reconstructing vortex phase from iterative phase retrieval methods [MA07]. Apparently, there is enough information present in the intensity profile of a field to at least indicate the presence of a vortex, even if there is not enough to determine the precise charge of the said vortex.

6.4 CORONOGRAPH

One of the most exciting areas of astronomy in modern times is the search for planets beyond our solar system. As of this writing, more than 1900 exoplanets have been found in some 1200 planetary systems, an impressive number that has grown rapidly since the first confirmed discoveries in 1988 [CWY88]. The earliest observations of these planets used indirect methods, such as detecting the slight wobble of the star due to the planet's motion or measuring the slight dimming of a star as a planet makes a transit across its face [Per00]. However, the search for extraterrestrial life requires direct observations of planets, from which spectral data can be taken to look for the signatures of biological material.

The difficulty in such direct searches is the extreme brightness of stars compared to planets; starlight is typically over 10 million times brighter than the reflected light from the planet, and the planet is typically no more than a few arc-seconds distant from the star. The result of these two factors is that the planet signal is lost in the noise from the star.

One strategy to overcome this is the use of a *coronagraph*, so named because such a device was first introduced by Lyot [Lyo39] to study the *corona*, the aura of plasma that extends millions of miles into space from the Sun's surface. An illustration of a Lyot-style coronagraph is shown in Figure 6.11. The system is aimed at the center of the star, and light from the

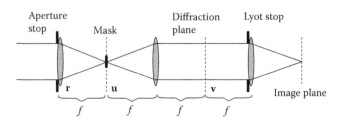

FIGURE 6.11 Illustration of a coronagraph. The transverse vector coordinate for each plane is labeled.

planet (or corona) is assumed to come from a slightly off-axis angle. This light is imaged via the first lens into the mask plane; for a Lyot coronagraph, the mask is a simple occulting mask that blocks the central image of the star. A fraction of the starlight will be diffracted by the mask, and this—and the planet light—is collimated by the second lens and propagated to a second aperture, the Lyot stop, which will obstruct most of the diffracted light. The third lens then images the remaining light onto the detector plane.

Diffraction effects make it impossible to completely remove starlight from the image, and diffraction is therefore a fundamental limitation in the effectiveness of the device. One difficulty that arises from diffraction is that it irretrievably mixes some of the starlight with the planet light, making a certain fraction of the planet light unusable.

An alternate strategy, introduced by Foo, Palacios, and Swartzlander [FPGS05], is to replace the occulting mask with a pure vortex mask. The light from the star, centered on the mask, will be divided into regions of opposing phase; these regions will result in destructive interference in the central part of the image plane, forcing the starlight to diffract to wide angles. The light from the planet will locally experience only a linear phase shift, causing its position to shift in the image plane but without any loss of brightness.

We may demonstrate the effect by explicitly calculating the diffraction of light from the aperture plane to the diffraction plane in the figure. Assuming a normally incident plane wave of unit amplitude on the aperture stop, the field $U(\mathbf{u})$ in the mask plane is proportional to the Fourier transform of the field in the aperture A:

$$U(\mathbf{u}) = \frac{1}{i\lambda f} \iint\limits_{A} e^{-ik\mathbf{r}\cdot\mathbf{u}/f} d^2 r, \tag{6.27}$$

where f is the focal length of the lens and $k = 2\pi/\lambda$ is the wavenumber of light. In polar coordinates, this may be written as

$$U(\mathbf{u}) = \frac{1}{i\lambda f} \int\limits_{0}^{2\pi}\int\limits_{0}^{R} e^{-ikru\cos\phi/f} r\, dr\, d\phi, \tag{6.28}$$

where R is the radius of the aperture stop. These integrals can readily be evaluated in terms of Bessel functions (see, for instance, [Gbu11],

Section 16.5) in the form

$$U(\mathbf{u}) = -iR\frac{J_1(kRu/f)}{u}. \tag{6.29}$$

Next, we assume that the field is transmitted through a vortex phase mask with transmission function

$$t(\phi) = e^{im\phi}, \tag{6.30}$$

with m the integer order of the mask. The field $U(\mathbf{v})$ in the diffraction plane is proportional to the Fourier transform of $U(\mathbf{u})$:

$$U(\mathbf{v}) = -\frac{R}{\lambda f} \iint \frac{J_1(kRu/f)}{u} e^{im\phi} e^{ikuv \cos \phi/f} u \, du \, d\phi. \tag{6.31}$$

The azimuthal angle can again be evaluated in terms of Bessel functions, and the result is

$$U(\mathbf{v}) = -\frac{i^m kR}{f} \int_0^\infty J_1(kRu/f) J_m(kuv/f) du. \tag{6.32}$$

This integral of a product of Bessel functions is in general quite complicated, except in special cases; we will focus on a pair of those special cases that are particularly illustrative, namely, $m = 0$ and $m = 2$. The $m = 0$ case is the result in the absence of a phase mask.

It can be shown[*] that the following is true:

$$\int_0^\infty J_\mu(at) J_{\mu-1}(bt) dt = \begin{cases} b^{\mu-1}/a^\mu, & 0 < b < a, \\ 1/2b, & 0 < b = a, \\ 0, & b > a > 0. \end{cases} \tag{6.33}$$

Applying this to the $m = 0$ case, we find that

$$U(\mathbf{v}) = \begin{cases} -1, & 0 < v < R, \\ 0 & v > R. \end{cases} \tag{6.34}$$

---

[*] This result is derived in Watson [Wat44], Section 13.4.

This is simply the image of the field from the aperture plane, as one would expect.

For the $m = 2$ case, we get the result

$$U(\mathbf{v}) = \begin{cases} 0, & 0 < v < R, \\ -(R/v)^2 & v > R. \end{cases} \tag{6.35}$$

This result is quite surprising. For an $m = 2$ vortex mask, an on-axis plane wave will be diffracted *completely* outside a circle of radius R. With an appropriate choice of Lyot stop, this light can be completely removed from the image plane.

This diffraction pattern is illustrated in Figure 6.12. It has been dubbed the "ring of fire" by Swartzlander [Swa09] in a paper in which he demonstrates that a similar exclusion of light arises for any even-charged vortex mask.

This result may seem paradoxical, especially in light of our discussion in previous chapters. We know that owing to the analytic properties of light (recall Section 3.5), a monochromatic wave can only vanish at most on a surface in three-dimensional space, yet this result implies that the field is completely zero over at least some three-dimensional volume. Looking back at the $m = 0$ result, however, is should be clear that this result is itself only approximate: no conventional imaging system can produce a perfect image of a hard-edged aperture. Approximations built into the Fourier optics used

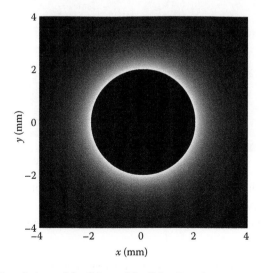

FIGURE 6.12   Simulation of the "ring of fire," for $R = 2$ mm.

here produces this idealized result; for practical purposes, however, the theory matches the experiment exceptionally well.[*]

The first experimental test of such a coronagraph was performed [LFJS06] in 2006. In this experiment, the light emitted from two closely spaced optical fibers was sent through the device and the red output of an off-axis fiber was successfully isolated from the green output of an on-axis fiber. In this first test, an intensity contrast of 95% was achieved, with lens aberrations, multiple reflections, and vortex mask fabrication errors limiting the effectiveness.

The first astronomical demonstration of a vortex coronagraph was achieved not long after [JFAM+08]. An 8-inch refractive finder telescope at the Steward Observatory at the University of Arizona was used as the primary image system. The target of the investigation was the binary star system Cor Caroli, and the angular separation of the stars is 19.3 arcseconds. The stars only have a relative magnitude difference of 2.7, making them both relatively visible and a good test of a coronagraph in practice.

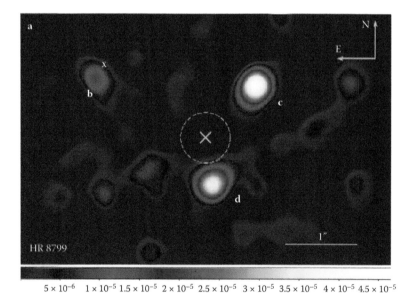

$$5 \times 10^{-6} \quad 1 \times 10^{-5} \quad 1.5 \times 10^{-5} \quad 2 \times 10^{-5} \quad 2.5 \times 10^{-5} \quad 3 \times 10^{-5} \quad 3.5 \times 10^{-5} \quad 4 \times 10^{-5} \quad 4.5 \times 10^{-5}$$

FIGURE 6.13 Calibrated image of the HR 8799 system obtained using a vector vortex coronagraph, showing the three exoplanets around the (eliminated) star. (Taken from E. Serabyn, D. Mawet, and R. Burruss. *Nature*, 464:1018–1020, 2010.)

---

[*] It is worth noting that there has been some interest in *nodal areas* in wavefields, however, in which the field is designed to be effectively zero over an extended area [RSS+15].

In this case, the system demonstrated a 97% suppression of the primary star intensity.

In 2010, a vortex coronagraph was used for the first time to directly image an exoplanet [SMB10]. A 1.5-m diameter off-axis subaperture of the Hale telescope was used to image three exoplanets around the star HR 8799; one such image is shown in Figure 6.13. In this case, however, the vortex mask used was a so-called vector phase mask [MRAS05], which uses a pattern of subwavelength grooves to make a spatially varying artificial birefringence.

The full potential of vortex coronagraphs has not yet been realized, and there are many practical limitations to be overcome. However, there is good reason for optimism: a theoretical study [GPK+06] indicated that an $m = 6$ vortex coronagraph can come very close to the ideal terrestrial limits on coronagraph performance.

6.5 SIGNAL PROCESSING AND EDGE DETECTION

We have seen in Section 6.3 that the presence of optical vortices significantly complicates problems of phase retrieval, presenting a significant challenge to coherent optical imaging. However, we now show that deliberately adding a phase twist to a field in the Fourier plane of an imaging system can in fact be useful for image processing.

We consider the problem of spatial filtering using a $4f$-focusing system, as illustrated in Figure 6.14; a classic discussion of spatial filtering can be found by O'Neill [O'N56]. A field $U_0(x, y)$ is input in what we refer to as the source plane 0; the field that arises in the Fourier plane 1, between the lenses, is the Fourier-transformed version of that in plane 0, and is labeled $U_1(u, v)$. The second lens performs another Fourier transform operation, resulting in a field $U_2(x', y')$ at the output plane 2.

In the Fourier plane, we may input a phase or amplitude screen with transmission function $t(u, v)$ in order to selectively filter out certain Fourier

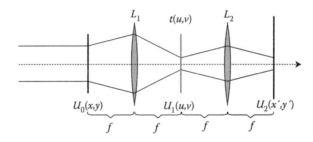

FIGURE 6.14   Illustration of a $4f$-focusing system.

components of the field. One common choice is a *low-pass filter*: a small (usually circular) aperture that blocks high-frequency components of the field and produces a blurred image, with the amount of blur inversely related to the radius of the aperture. Such low-pass filtering can be used to screen out high-frequency noise in an image. Another common choice is a *high-pass filter*, a small circular obstruction that blocks the low-frequency components of the field, only allowing high-frequency components to pass. Such high-pass filters will emphasize sharp gradients in an image, and are often used for edge detection. However, this process is extremely lossy, as most of the light intensity is contained in the low-frequency components of the field.

To understand more quantitatively the effect of a given spatial filter on the output wavefield, we turn to some basic Fourier analysis. For brevity, we take the field in the Fourier plane as the Fourier transform of the field in the source plane, and neglect additional propagation factors that are constant within the plane, that is,

$$U_1(u, v) = \frac{1}{(2\pi)^2} \int U_0(x, y) e^{-i(ux+vy)} dx\, dy. \tag{6.36}$$

After transmission through the filter, the field is given by $U_1(u, v) t(u, v)$; the field in the output plane is then the inverse Fourier transform of this*:

$$U_2(x', y') = \int U_1(u, v) t(u, v) e^{i(ux'+vy')} du\, dv. \tag{6.37}$$

This expression can be evaluated using the convolution theorem, to be of the form

$$U_2(x', y') = \frac{1}{(2\pi)^2} \int U_0(x, y) \tilde{t}(x' - x, y' - y) dx\, dy, \tag{6.38}$$

where $\tilde{t}(X, Y)$ is the inverse Fourier transform of $t(u, v)$, that is,

$$\tilde{t}(X, Y) = \int t(u, v) e^{i(uX+vY)} du\, dv. \tag{6.39}$$

From Equation 6.38, we see that the filtering operation represents a "smearing" of sorts of the field U_0, which in the output plane is integrated in a

---

* Actually, it should be another direct Fourier transform, resulting in an inverted image at the output; for simplicity, we use the inverse transform, which just corresponds to flipping the output coordinates.

region around the point (x', y'). If there is no filter in the Fourier plane, then $\tilde{t}(X, Y) = (2\pi)^2 \delta(X)\delta(Y)$, and the output field is equal to the input field. If the filter is a low-pass Gaussian filter, than the output field is smeared out (blurred) to an extent inversely related to the width of the filter.

But what happens if we use a spiral phase plate as a filter? If we let $t(u, v) = \exp[i\phi_u]$, we can in fact determine $\tilde{t}(X, Y)$ directly in polar coordinates to be

$$\tilde{t}(\rho, \phi) = \frac{\pi i}{\rho^2} e^{i\phi}. \tag{6.40}$$

The field in the output plane, given by Equation 6.38, is therefore convolved with a decaying spiral phase function. What effect does this have on the overall image? Because a spiral phase function has opposing values on either side of the central singularity, an integral of a smooth function convolved with it will tend to vanish. Only in regions where there is a rapid change in the image amplitude will one find an imperfect cancellation of these opposing phases, resulting in a nonzero value. In short, the spiral phase filter acts as an all-directional edge enhancer. Unlike the high-pass filter mentioned earlier, however, the spiral phase filter does not, in principle, obstruct the propagation of light at all, making it a *lossless* edge enhancer.

We illustrate the effect in Figure 6.15. Focusing on the latter two images, one can see that the discontinuities in the image are highlighted in both cases. In both cases, gamma correction is required to lighten low brightness regions; however, overall, the spiral filtered image produces much brighter edges.

The use of spiral phase filtering in image processing was apparently first considered by Davis et al. [DMCC00], where they considered it a generalization of the Hilbert transform in image processing and referred to it as a "radial Hilbert transform." This work was studied in more detail by Crabtree, Davis, and Moreno [CDM04], who elaborated upon the theory and considered the case where the vortex mask is placed directly against a lens in an imaging geometry.

Spiral phase filtering was first applied directly to microscopy by Fürhapter et al. [FJBRM05]. A sample was illuminated by a laser diode and collected by a microscope objective; after exiting the objective, the light passed through a $4f$-system with a spatial-light modulator providing the appropriate vortex phase in the Fourier plane.

In an interesting twist, the same group pointed out [JFBRM05] that a spiral phase function with an interior "cut-out" can produce shadowing effects in images. A comparison of a traditional spiral phase filter and a

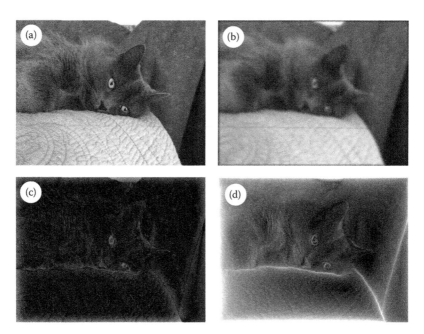

FIGURE 6.15 Simulation of optical image processing with feline Rascal. (a) Original image, 3264 × 2448 resolution. (b) Low-pass filtered version, using a Gaussian filter of width 25 pixels. (c) High-pass filtered version, using a Gaussian filter with inner width 5 pixels, and gamma correction 0.5. (d) Spiral filtered version, with gamma correction 0.5.

"cut-out" filter is shown in Figure 6.16. In an ordinary spiral phase filter (a), the central pixel is typically made perfectly absorbing, which makes it work somewhat like a high-pass filter. By replacing a central portion of the filter with a perfectly transmissive region, we allow the central frequencies to transmit but maintain some of the high-frequency shadowing performed by the spiral phase. Because there is now a definite phase relation between the

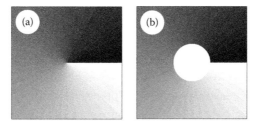

FIGURE 6.16 Illustration of the phase of (a) a standard spiral phase filter and (b) a cut-out filter.

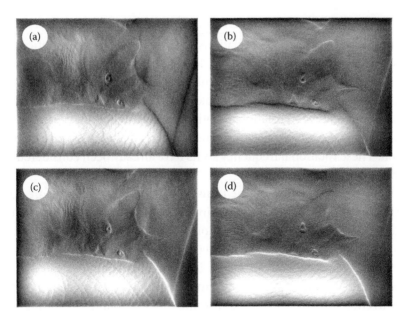

FIGURE 6.17 Shadowing effects with a cut-out with phase (a) $\psi = 0$, (b) $\psi = \pi/2$, (c) $\psi = \pi$, and (d) $\psi = 3\pi/2$. Here, the central cut-out is 5 pixels in radius, and gamma correction 0.5 is again used.

central region and the spiral phase, different choices of the central phase ψ will produce different shadowing effects, as illustrated in Figure 6.17. One can see that different horizontal and vertical features are emphasized in the processed image, depending on the choice of ψ.

Work with spiral phase filters has continued. Guo et al. [GHXD06] introduced a Laguerre–Gaussian spatial filter with amplitude and phase transmission equal to the field profile of a Laguerre–Gauss beam. The smooth amplitude profile of the Laguerre–Gauss filter was found to reduce suboscillations that typically appear with a hard-edged aperture. Situ, Pedrini, and Osten [SPO09] introduced two new methods for shadowing/edge enhancement, by using a fractional spiral filter (to be discussed much later in Section 12) or by lateral shifting of a standard spiral filter. These strategies were tested in a microscopy system not long after [SWPO10].

Though perhaps the greatest appeal of such filtering is being able to implement it directly in optical systems, it should be noted that spiral phase filtering may also be done computationally, as was done for the figures of this section, with similar benefits.

6.6 ROTATIONAL DOPPLER SHIFTS

One of the most surprising revelations in the study of singular optics was the discovery of a new type of Doppler effect that applies exclusively to optical fields with a "twist," either circular polarization, a helical wavefront, or both. Unlike the traditional Doppler effect, in which frequency shifts appear due to relative translational motion between source and detector, the *rotational Doppler effect* involves frequency shifts due to relative rotational motion. These rotational frequency shifts have already been considered for a number of applications.

It will be helpful to review the translational Doppler effect before considering the rotational version. For simplicity, we restrict ourselves to the nonrelativistic limit, $v \ll c$; a discussion of the full relativistic equation can be found, for instance, in French [Fre68, Chapter 5] and Rindler [Rin91, Chapter 3]. We follow French's discussion of the nonrelativistic effect.

We consider a source and detector moving at speeds $v_s \ll c$ and $v_d \ll c$, respectively, along the z-axis, and imagine that the source is emitting planar wavefronts with a frequency $\nu = 1/\tau$. The situation is illustrated in Figure 6.18.

After emitting the first wavefront at $t = 0$, the source moves a distance $v_s \tau$, while the wavefront propagates a distance $c\tau$. The next wavefront is therefore emitted at a distance $\lambda' = (c - v_s)\tau$, which is also the effective wavelength of the wave with respect to a stationary observer. Because the detector is also moving, however, the effective speed of the wave is given by $c - v_d$, and the detector sees a time interval between the wavefronts of

$$\tau' = \frac{\lambda'}{c - v_d} = \frac{c - v_s}{\nu(c - v_d)}. \tag{6.41}$$

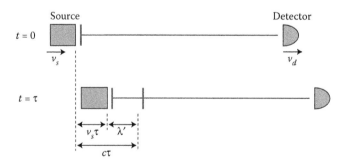

FIGURE 6.18 Illustration of the arrangement used to illustrate the translational Doppler effect. (Adapted from A.P. French. *Special Relativity*. W.W. Norton, New York, 1968.)

The detected frequency of the wave is therefore

$$\nu' = \nu \frac{1 - v_d/c}{1 - v_s/c}. \tag{6.42}$$

Because this calculation was done with respect to an absolute frame of reference[*] in which the speed of the pulses is c, the source and detector motion have asymmetric roles in Equation 6.42. For very low speeds, however, we may approximate $1/(1 - v_s/c) \approx 1 + v_s/c$ and, neglecting terms quadratic or higher in speed, get the result

$$\nu' \approx \nu \left[1 - (v_d - v_s)/c\right], \tag{6.43}$$

which depends only on the relative motion of source and detector. If the source and detector are moving away from each other, that is, $v_d - v_s > 0$, the measured frequency is lower. If the source and detector are moving toward one another, that is, $v_d - v_s < 0$, the measured frequency is higher.

The translational Doppler effect can therefore be summarized as a change in the observed spacing of planar wavefronts due to relative motion. We recall, however, that a vortex beam has screw-like wavefronts, as was illustrated in Figure 3.7. As time evolves, the screw rotates with the same period $\tau = 1/\nu$ of the wave itself. If we now imagine, however, that we move to a frame of reference which is rotating around the same axis as the vortex beam itself, the wavefront screw will appear to move slower/faster when we rotate in the same/opposite direction to its own motion, respectively. Because the period of screw rotation is the same as the wave frequency, this change in observed rotational speed will correspond to a change in observed frequency. This is what is generally known as a rotational Doppler shift. A state of circular polarization will also exhibit such a shift, however, as its electric field vector traces out a screw-like path in three-dimensional space.

We may make our observations quantitative via a transformation of a vortex beam to a rotating coordinate system. For simplicity, we consider a pure vortex beam with either left-handed $\hat{\mathbf{e}}_+$ or right-handed $\hat{\mathbf{e}}_-$ circular polarization, and with radial order n and azimuthal order m. Considering both polarization options simultaneously, we write the electric field in a

---

[*] The relativistic Doppler equation is of the form $\nu' = \nu\sqrt{(1 - v/c)/(1 + v/c)}$, where v is the relative velocity between source and detector.

stationary reference frame as

$$\mathbf{E}(\mathbf{r}, t) = \hat{\mathbf{e}}_{\pm} u_{nm}(\mathbf{r}) e^{-i\omega t}, \tag{6.44}$$

where $u_{nm}(\mathbf{r})$ is a Laguerre–Gauss mode as defined in Equation 2.37. It is important to note that, even though we have a monochromatic field, we need to keep the time dependence explicit in our formula, as it will change in the rotating frame. It is to be noted, however, that we may also write

$$u_{nm}(\mathbf{r}) = f_{nm}(r) e^{im\phi} = 2^{|m|/2} \frac{f_{nm}(r)}{r^{|m|}} z_{\pm}^{|m|}, \tag{6.45}$$

where $z_{\pm} \equiv (x \pm iy)/\sqrt{2}$, and the sign depends on the sign of m. Everything in the expression $u_{nm}(\mathbf{r})$, with the exception of the $z_{\pm}$ term, is therefore rotationally invariant. This observation will make the transformation to a rotating reference frame almost trivial, after a small amount of additional work.

In terms of the Cartesian coordinates x and y, the matrix defining the change to the rotating reference frame is given by

$$\mathbf{R}_{\Omega}(t) = \begin{bmatrix} \cos(\Omega t) & \sin(\Omega t) \\ -\sin(\Omega t) & \cos(\Omega t) \end{bmatrix}, \tag{6.46}$$

where Ω is the angular frequency of rotation. This transformation corresponds to a rotation of the coordinate axes, and the reference frame, in the *positive-ϕ* direction. Formally, we may write the electric field $\mathbf{E}'$ in this new frame in terms of two applications of this matrix, both on the polarization of the field and the position vector $\mathbf{r}$:

$$\mathbf{E}'(\mathbf{r}', t) = \mathbf{R}_{\Omega}(t) \cdot \mathbf{E}\left[\mathbf{R}_{\Omega}(t) \cdot \mathbf{r}, t\right]. \tag{6.47}$$

However, both the polarization and the position variables are written in terms of complex quantities, namely, $\hat{\mathbf{e}}_{\pm} = (\hat{\mathbf{x}} \pm i\hat{\mathbf{y}})/\sqrt{2}$ and $z_{\pm} = x \pm iy$. It is therefore convenient to introduce the transformation matrix $\mathbf{T}$, which converts from a Cartesian system to an orthogonal complex system of coordinates:

$$\mathbf{T} \equiv \frac{1}{\sqrt{2}} \begin{bmatrix} 1 & -i \\ 1 & i \end{bmatrix}. \tag{6.48}$$

This same matrix will perform the transformation from $\hat{\mathbf{x}}, \hat{\mathbf{y}}$ to $\hat{\mathbf{e}}_+, \hat{\mathbf{e}}_-$ as well as the transformation from x, y to z_+, z_-. If we make a similarity transform of $\mathbf{R}_\Omega(t)$ with $\mathbf{T}$, that is,

$$\mathbf{R}'_\Omega(t) = \mathbf{T}\mathbf{R}_\Omega(t)\mathbf{T}^\dagger, \tag{6.49}$$

we quickly find that

$$\mathbf{R}'_\Omega(t) = \begin{bmatrix} e^{i\Omega t} & 0 \\ 0 & e^{-i\Omega t} \end{bmatrix}. \tag{6.50}$$

With this matrix, our transformation into a reference frame rotating at angular frequency Ω becomes trivial. We find that $\hat{\mathbf{e}}_\pm \to \hat{\mathbf{e}}_\pm e^{\pm i\Omega t}$ and $z_\pm \to z_\pm e^{\pm i\Omega t}$. On substitution, we readily find that the electric field in the transformed coordinate system is given by

$$\mathbf{E}'(\mathbf{r}, t) = \hat{\mathbf{e}}_\pm u_{nm}(\mathbf{r}) e^{\pm i\Omega t} e^{im\Omega t} e^{-i\omega t}, \tag{6.51}$$

where we have replaced $\mathbf{r}'$ by $\mathbf{r}$ for brevity. Combining the temporal terms, we find that the frequency of the electric field in the rotating frame is given by

$$\omega' = \omega \mp \Omega - m\Omega. \tag{6.52}$$

This result combines the rotational Doppler shift from spin and orbital contributions to the angular momentum; it is readily understood by considering particular cases. If $\Omega > 0$, the reference frame is rotating in the $+\phi$-direction. Left-handed circular polarization, with vector $\hat{\mathbf{e}}_+$, then rotates in the same direction as the frame and the motion of the field vector appears slower, corresponding to a lower observed frequency. Also, a left-handed vortex, with positive m, rotates in the same direction as the reference frame and also results in a lower observed frequency. Of course, a reference frame rotating in the opposite sense as the polarization or orbital phase will produce an increase in frequency.

It is important to note a number of differences, both in principle and in practice, between the rotational Doppler effect and its translational counterpart. Unlike the ordinary Doppler effect, the rotational effect only occurs for fields lacking rotational invariance: it will not be seen, for instance, in a fundamental Gaussian mode. Furthermore, the effect will only exhibit the simple shift of Equation 6.52 for pure angular momentum states, that is, states that have a pure exp[$im\phi$] dependence. As other fields can always be expressed as a coherent superposition of pure states, and each pure state will generally have a different spectral shift, one expects a quite complicated rotational spectrum.

In practice, it is simply not possible to rotate sources/detectors at speeds greater than, at most, 10^3 Hz, due to the extreme centrifugal forces involved. Typically, experiments will use much lower rotation speeds on the order of 1 Hz, which means that the bandwidth of the source must be of the same scale, in order for the shift to be detectable. This is obviously not a typical situation—an ordinary helium–neon laser operating at 633 nm has a bandwidth of 1.5×10^9 Hz—but can be achieved with well-stabilized lasers.

The rotational Doppler shift here should not be confused with the rotation-induced version of the translational Doppler effect, as is observed in astronomical objects such as galaxies. The difference is illustrated in Figure 6.19, for a rotating ring source. In the plane of rotation, one sees red and blue translational Doppler shifts due to the parts of the ring moving away from/toward the observer, respectively. The rotational Doppler shift would be seen along the axis of rotation.

The existence of an "angular" Doppler effect was first introduced by Garetz [Gar81], who noted the possibility for circular polarization and proposed applications in Raman scattering and the measurement of rotation of microscopic particles. Garetz's introduction of this effect was motivated by an earlier experiment by Garetz and Arnold [GA79] in which frequency shifts were observed when circularly polarized light was passed through a spinning half-wave plate. The orbital analog of this effect was not introduced until 1996, when Nienhuis [Nie96] suggested that a rotating cylindrical lens system could introduce a frequency shift in a Laguerre–Gauss

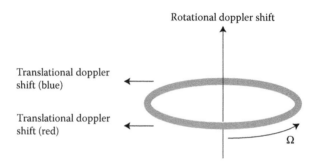

FIGURE 6.19 The distinction between the rotational Doppler effect and the rotation-induced translational Doppler effect.

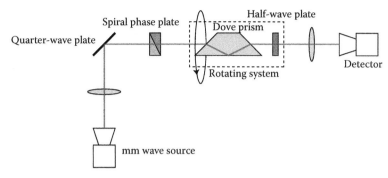

FIGURE 6.20 Illustration of the experiment of Courtial et al. [CRD⁺98], used to measure rotational frequency shifts.

mode. The term "rotational frequency shift" was introduced by Bialynicki-Birula and Bialynicka-Birula [BBBB97], who considered the effect of such frequency shifts on atomic emissions. Allen, Babiker, and Power [ABP94] theoretically considered how an atom moving in a vortex beam experiences shifts in its resonance frequency.

The first experimental observations of a rotational frequency shift due to OAM were performed by researchers at the University of St. Andrews [CDR⁺98,CRD⁺98]. The first of these papers studied the effect of OAM alone, while the second considered both orbital and spin angular momentum simultaneously. We take a few moments to describe the latter experiment, which is illustrative of the general principles involved.

A diagram of the configuration is shown in Figure 6.20. The source is a diode source operating in the millimeter wave regime, with a center frequency of 94×10^9 Hz. It is phase-locked to a crystal oscillator, giving a short-term bandwidth of roughly 1 Hz. The wave is reflected from a quarter-wave plate, resulting in circular polarization, and then transmitted through a spiral phase plate, resulting in a vortex mode.

Though up to this point we have considered the rotation of the source or detector, it is in general difficult to do so without introducing additional off-axis motion that can mask the rotational shift. Instead, a Dove prism and half-wave plate are rotated together in the system; the prism performs image rotation and the half-wave plate rotates the "image" of the polarization, both at 2Ω, where $\Omega = 1.25$ Hz is the frequency of rotation. The observed frequency shift was in agreement with Equation 6.52, the sum of the spin and orbital contributions.

The rotational Doppler shift has been considered for a number of applications. In 2002, researchers used such shifts to create precise and continuous variations in interference patterns [AMP+02]; this ability allows for very fine control of the motion and rotation of optically trapped particles. Much more recently, researchers in Glasgow applied the rotational Doppler shift to detect the rotation of a spinning object [LSBP13]. A spinning object illuminated by a vortex beam will produce a scattered field, which is rotationally shifted, and this shift can then be used to deduce the object's rotational speed. At about the same time, other researchers were able to use the rotational Doppler shift to observe the synchronous spinning of gas molecules, using both deuterium and nitrogen gas [KSG+13].

Fields that are not spatially coherent will also in general show rotational shifts, though the shifted spectrum can be quite complicated, owing to the variety of OAM modes in the partially coherent field. Hautakorpi et al. considered such shifts for scalar fields [HLSK06], while Agarwal and Gbur [AG06] looked at the shifts associated with both spin and orbital angular momentum.

6.7 EXERCISES

1. We consider a simple MIMO system, as shown below, consisting of two point sources A and B separated by distance $2a$ and two detectors 1 and 2 separated by distance $2d$, and a distance L between the source and detector arrays. Assume that the fields radiated by the sources are effectively plane waves of the form

$$U_A = Ae^{-iks \cdot a\hat{x}},$$

with $\mathbf{s}$ the unit vector pointing from the origin to the detector plane, and a similar expression for U_B. Write a matrix relation between the amplitudes A and B of the sources and the fields U_1 and U_2 at the detectors in terms of d, a, and L. Find the eigenvalues of this matrix and demonstrate from these that there must be a minimum separation of sources and detectors to be able to separate out the signals.

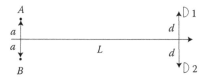

2. Starting from the paraxial wave equation, express the field as an intensity and a phase and, from this, derive the TIE, Equation 6.15, as well as the complementary equation relating to the evolution of phase.

3. Using the local form of the vector $\mathbf{A}(\mathbf{r})$ for a first-order vortex, Equation 6.23, derive the modified form of the TIE equation, Equation 6.26.

4. The "ring of fire" for a vortex coronagraph is effective for any even-order vortex mask. Show this using the calculations of Section 6.4 and the following integral:

$$\int_0^\infty J_{\nu+2n+1}(ax)J_\nu(bx)dx$$
$$= \begin{cases} b^\nu a^{-\nu-1}P_n^{(\nu,0)}(1 - 2b^2/a^2), & \operatorname{Re}(\nu) > -1 - n, \ 0 < b < a, \\ 0, & \operatorname{Re}(\nu) > -1 - n, \ 0 < a < b, \end{cases}$$

where $P_n^{(\alpha,\beta)}(x)$ are Jacobi's polynomials. Plot the intensity of the diffraction pattern for an $m = 4$ and an $m = 6$ coronagraph. (You will have to look up the details of Jacobi's polynomials.)

5. A question that requires significant computational work and image processing—What happens if a higher-order spiral phase plate, say $m = 2$, is used for edge detection on an image? Using your own image (or one provided by your instructor), compare the spiral filtering using an $m = 1$, $m = 2$, and $m = 3$ spiral plate.

CHAPTER 7

Polarization Singularities

S O FAR, WE HAVE for the most part ignored effects related to the polarization of light. Even when we have explicitly taken it into account, we have usually considered the case of *uniform polarization*, in which the state of polarization is exactly the same throughout the wavefield. For paraxial waves, we were then able to treat the propagation of light through the use of the scalar wave equation. As a result, we demonstrated that the generic phase singularities of light are optical vortices, and we have devoted many pages to their properties and potential applications.

We must now, confess, however, that these optical vortices are themselves not generic! Even a purely transverse electromagnetic field will in general have two independent complex electric field components, say E_x and E_y. In order for there to be a true zero of intensity in a plane of constant z, we must satisfy four equations—$\mathrm{Re}\{E_x\} = 0$, $\mathrm{Re}\{E_y\} = 0$, $\mathrm{Im}\{E_x\} = 0$, and $\mathrm{Im}\{E_y\} = 0$—at the same (x, y)-point in the plane. This overspecified system of equations in general will not have a solution, and therefore the optical vortices we have previously discussed will not typically be observed in electromagnetic waves.

The electric field itself, however, is a complex vector, and this vector field of polarization may have its own singularities of direction. In the cross section of a monochromatic electromagnetic wave where the most general polarization is elliptical, two types of polarization singularities are found in the transverse electric field: C-points, points of circular polarization at which the orientation of the polarization ellipse is undefined, and L-lines, lines on which the handedness of the polarization ellipse is undefined. These singularities can be shown to be generic and have their own distinct features and topological properties, as well as a simple relationship to scalar optical vortices. Furthermore, just as nongeneric higher-order vortex beams have found use in applications, nongeneric polarization singularities also turn out to have useful properties.

In this chapter, we discuss the theory and applications of polarization singularities, beginning with a general discussion of the different ways to characterize the polarization of light. We then discuss generic polarization singularities and finally consider a number of important special cases.

7.1 BASICS OF POLARIZATION IN OPTICAL WAVEFIELDS

We begin by deriving the most general state of polarization of a monochromatic electromagnetic field. Regardless of whether the field is paraxial or not, we will find that the electric field vector generally follows an elliptical path. Our initial discussion uses the notation of that in Born and Wolf [BW99, Section 1.4.3] and Nye [Nye83a].

Though an electromagnetic plane wave will always have an electric field transverse to the direction of propagation due to Gauss' law, $\nabla \cdot \mathbf{E}(\mathbf{r}) = 0$, the same is not true of most fields.[*] To study the most general electromagnetic wave, then, we should begin by assuming that all three components of the field are nonzero; we write them in Cartesian coordinates (x, y, z) in the form

$$E_x(\mathbf{r}, t) = a_x(\mathbf{r}) \cos[\omega t + \delta_x(\mathbf{r})], \tag{7.1}$$

$$E_y(\mathbf{r}, t) = a_y(\mathbf{r}) \cos[\omega t + \delta_y(\mathbf{r})], \tag{7.2}$$

$$E_z(\mathbf{r}, t) = a_z(\mathbf{r}) \cos[\omega t + \delta_z(\mathbf{r})], \tag{7.3}$$

where $a_i(\mathbf{r})$ is the amplitude of the ith component of the field and $\delta_i(\mathbf{r})$ is the phase of the ith component, with $i = x, y, z$. We may expand the cosine of each component using the relation

$$\cos[\omega t + \delta_i] = \cos(\omega t) \cos(\delta_i) - \sin(\omega t) \sin(\delta_i), \tag{7.4}$$

---

[*] As discussed in Section 2.1, for paraxial waves, we can usually consider the field to be transverse to a good approximation.

and then write the total vector form of the field as

$$\mathbf{E}(\mathbf{r}, t) = \mathbf{p}(\mathbf{r}) \cos(\omega t) + \mathbf{q}(\mathbf{r}) \sin(\omega t), \tag{7.5}$$

where

$$p_i(\mathbf{r}) = a_i(\mathbf{r}) \cos[\delta_i(\mathbf{r})], \quad q_i = -a_i(\mathbf{r}) \sin[\delta_i(\mathbf{r})]. \tag{7.6}$$

The real-valued vectors $\mathbf{p}$ and $\mathbf{q}$ in general define a plane, and because the field is periodic in time, we may immediately state that the electric field vector at any point in space traces out a closed path within the plane.

To find the specific shape of this path, let us now assume, without loss of generality, that the electric field vector is confined to the xy-plane. This includes the important case of a paraxial electromagnetic wave with electric field transverse to the direction of propagation z, but even in the general case, we may always define a new Cartesian coordinate system for which the electric field vector lies entirely along two coordinate axes. We therefore write

$$E_x = a_x \cos[\omega t + \delta_x],$$
$$E_y = a_y \cos[\omega t + \delta_y], \tag{7.7}$$

and again may introduce the vectors $\mathbf{p}$ and $\mathbf{q}$ from Equations 7.6. We may then express the electric field in a vector–matrix form

$$|\mathbf{E}\rangle = \mathbf{P} |\mathbf{C}\rangle, \tag{7.8}$$

where

$$|\mathbf{E}\rangle = \begin{bmatrix} E_x \\ E_y \end{bmatrix}, \quad |\mathbf{C}\rangle = \begin{bmatrix} \cos(\omega t) \\ \sin(\omega t) \end{bmatrix}, \tag{7.9}$$

and

$$\mathbf{P} = \begin{bmatrix} p_x & q_x \\ p_y & q_y \end{bmatrix}. \tag{7.10}$$

The simple matrix Equation 7.8 may be inverted, to write

$$|\mathbf{C}\rangle = \mathbf{P}^{-1} |\mathbf{E}\rangle, \tag{7.11}$$

and $\mathbf{P}^{-1}$ is readily found from standard linear algebra techniques to be

$$\mathbf{P}^{-1} = \frac{1}{\det[\mathbf{P}]} \begin{bmatrix} q_y & -q_x \\ -p_y & p_x \end{bmatrix}, \tag{7.12}$$

where $\det[\mathbf{P}] = p_x q_y - q_x p_y$ is the determinant of $\mathbf{P}$, which may be written in terms of the field parameters as

$$\det[\mathbf{P}] = a_x a_y \sin \delta, \tag{7.13}$$

and $\delta \equiv \delta_y - \delta_x$.

We would like to remove the explicit time dependence from our description of the electric field; the result should be an equation that describes the complete path of the real electric field vector. This can be done by considering the matrix product

$$\langle \mathbf{C} | \mathbf{C} \rangle = (\mathbf{P}^{-1} | \mathbf{E} \rangle)^T (\mathbf{P}^{-1} | \mathbf{E} \rangle), \tag{7.14}$$

where T labels the transpose of a matrix and $\langle \mathbf{C} | = (|\mathbf{C}\rangle)^T$. This product is simply equal to unity, and time has been eliminated from the expression; in terms of the electric field vector, we may write

$$\langle \mathbf{E} | (\mathbf{P}^{-1})^T \mathbf{P}^{-1} | \mathbf{E} \rangle = \langle \mathbf{E} | \mathbf{Q} | \mathbf{E} \rangle = 1, \tag{7.15}$$

where

$$\mathbf{Q} \equiv (\mathbf{P}^{-1})^T \mathbf{P}^{-1} = \frac{1}{[\det(\mathbf{P})]^2} \begin{bmatrix} p_y^2 + q_y^2 & -(p_x p_y + q_x q_y) \\ -(p_x p_y + q_x q_y) & p_x^2 + q_x^2 \end{bmatrix}. \tag{7.16}$$

This matrix may also be written in terms of field parameters as

$$\mathbf{Q} = \frac{1}{(a_x a_y \sin \delta)^2} \begin{bmatrix} a_y^2 & -a_x a_y \cos \delta \\ -a_x a_y \cos \delta & a_x^2 \end{bmatrix}. \tag{7.17}$$

Equation 7.15 is in a quadratic form in terms of the electric field components E_x and E_y, implying that the electric field vector must trace out a hyperbolic, parabolic, or elliptical path. Only an elliptical path is periodic, as our electric field vector must be. We have therefore demonstrated that the electric field vector traces out a path in the form of an ellipse.

It is important to note that the polarization ellipse is an ellipse only in terms of **E**-space; that is, the electric field vector does not physically trace

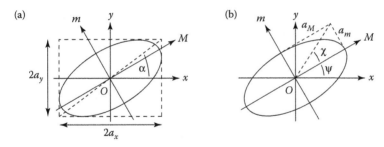

FIGURE 7.1 Illustration of the polarization ellipse in terms of (a) field properties and (b) geometric properties. The angle α is the angle of the diagonal of the box of side $2a_x$, $2a_y$, which is generally different than the orientation angle ψ.

out an ellipse of finite size in x and y. It is also worth noting that our result has been derived without applying Maxwell's equations at all—this result therefore applies to any monochromatic vector field in three dimensions.

We now want to specify in an explicit manner how the geometry of the polarization ellipse, namely, its major and minor semi-axes, a_M and a_m, and the angle of orientation of the major axis ψ, are related to the field amplitudes a_x, a_y and the phase difference δ. The relevant geometry is illustrated in Figure 7.1.

We can readily find these parameters by diagonalizing the matrix $\mathbf{Q}$, which is real symmetric and therefore always diagonalizable by rotation; we introduce a matrix Θ that rotates the coordinate system by the angle ψ,

$$\Theta = \begin{bmatrix} \cos\psi & \sin\psi \\ -\sin\psi & \cos\psi \end{bmatrix}. \tag{7.18}$$

The form of $\mathbf{Q}$ in this new coordinate system is found by similarity transformation,

$$
\begin{aligned}
\mathbf{Q}' &= \Theta\mathbf{Q}\Theta^T \\
&= \begin{bmatrix} Q_{xx}\cos^2(\psi) + Q_{xy}\sin(2\psi) + Q_{yy}\sin^2(\psi) \\ (Q_{yy} - Q_{xx})\cos(\psi)\sin(\psi) + Q_{xy}\cos(2\psi) \end{bmatrix}
\end{aligned}
$$

$$
\begin{matrix}
(Q_{yy} - Q_{xx})\cos(\psi)\sin(\psi) + Q_{xy}\cos(2\psi) \\
Q_{yy}\cos^2(\psi) - Q_{xy}\sin(2\psi) + Q_{xx}\sin^2(\psi)
\end{matrix}\Bigg], \tag{7.19}
$$

where Q_{ij} is the i,jth component of **Q**. The matrix will be diagonal when the off-diagonal components are zero, that is, when

$$\tan(2\psi) = \frac{2Q_{xy}}{Q_{xx} - Q_{yy}} = \frac{2a_x a_y \cos\delta}{a_x^2 - a_y^2}. \qquad (7.20)$$

This seemingly simple equation can cause quite a bit of mischief if we are not careful, so, we should spend a little time exploring it. Ideally, we would like the angle ψ to represent the angle that the major axis of the ellipse makes with the x-axis. However, because the standard tangent function is π-periodic, and we have a double angle in its argument, there are *four* angles that satisfy Equation 7.20. These four choices coincide with rotating the system to align the new x-axis along the two orientations of the major or minor axes of the ellipse. This is a problem for our future work, as we would like to plot the major axis of the ellipse as a function of position, and the angle ψ as defined above can "jump" from representing a major axis to representing a minor axis.

The difficulty may be identified by noting that, for ψ, the ordinary arctangent is only defined over the range $-\pi/4 \le \psi < \pi/4$, which will not always include the major axis within it. We need to use instead the four-quadrant arctangent atan2(y, x), which takes as arguments a pair of x, y values in the xy-plane and returns the polar angle $-\pi \le \theta < \pi$ associated with them.

For our case, we define

$$\sin(2\psi) \equiv \frac{-2Q_{xy}}{\sqrt{4Q_{xy}^2 + (Q_{xx} - Q_{yy})^2}}, \qquad (7.21)$$

$$\cos(2\psi) \equiv \frac{Q_{yy} - Q_{xx}}{\sqrt{4Q_{xy}^2 + (Q_{xx} - Q_{yy})^2}}, \qquad (7.22)$$

and then introduce angle ψ as

$$\psi = \text{atan2}[\sin(2\psi), \cos(2\psi)]/2, \qquad (7.23)$$

which is defined over the range $-\pi/2 \le \psi < \pi/2$. It should be noted that, if we flip the signs of our definitions of the sine and cosine functions, the tangent is unchanged. This choice of sign amounts to choosing ψ to align with either the major or minor axes, though we will need to try a choice to see which is which.

In the diagonal representation at angle ψ, the equation of the ellipse may be written in the form

$$\frac{|E_M|^2}{a_M^2} + \frac{|E_m|^2}{a_m^2} = 1, \tag{7.24}$$

where we make what amounts to a lucky guess and define

$$\frac{1}{a_M^2} = Q'_{xx} = Q_{yy}\sin^2(\psi) + Q_{xy}\sin(2\psi) + Q_{xx}\cos^2(\psi), \tag{7.25}$$

$$\frac{1}{a_m^2} = Q'_{yy} = Q_{yy}\cos^2(\psi) - Q_{xy}\sin(2\psi) + Q_{xx}\sin^2(\psi). \tag{7.26}$$

In terms of the field variables, these may be written as

$$\frac{1}{a_M^2} = \frac{a_y^2\cos^2(\psi) - a_x a_y \cos\delta \sin(2\psi) + a_x^2\sin^2(\psi)}{a_x^2 a_y^2 \sin^2\delta}, \tag{7.27}$$

$$\frac{1}{a_m^2} = \frac{a_y^2\sin^2(\psi) + a_x a_y \cos\delta \sin(2\psi) + a_x^2\cos^2(\psi)}{a_x^2 a_y^2 \sin^2\delta}. \tag{7.28}$$

A significant amount of effort must be made to write the major and minor semi-axes in their final form; the details are left as an exercise. On making the appropriate calculations, we end up with the relations

$$a_M^2 = a_y^2 \sin^2\psi + a_x^2\cos^2\psi + a_x a_y \sin(2\psi)\cos\delta, \tag{7.29}$$

$$a_m^2 = a_y^2 \cos^2\psi + a_x^2\sin^2\psi - a_x a_y \sin(2\psi)\cos\delta. \tag{7.30}$$

These may be combined with the definition of ψ, in terms of a_x, a_y,

$$\tan(2\psi) = \frac{2a_x a_y \cos\delta}{a_x^2 - a_y^2}. \tag{7.31}$$

If a_M and a_m represent the major and minor axes, respectively, we should find that

$$\Delta \equiv a_M^2 - a_m^2 \geq 0. \tag{7.32}$$

The quantity Δ is easy to evaluate, and we find that

$$\Delta = (a_x^2 - a_y^2)\cos(2\psi) - 2a_x a_y \sin(2\psi)\cos\delta. \tag{7.33}$$

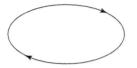
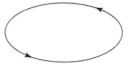

Right-handed polarization   Left-handed polarization

FIGURE 7.2   The direction of circulation for left-handed and right-handed ellipti-cal polarization.

If we use Equations 7.21 and 7.22, we readily find that $\Delta \geq 0$, and our definitions of a_m and a_M are consistent.

One more quantity is useful in describing the ellipse; we introduce the *ellipticity* as the angle χ satisfying

$$\tan\chi = \mp a_m/a_M. \qquad (7.34)$$

The two signs are introduced to distinguish the two paths the electric field vector may traverse around the ellipse, and may formally be considered the sign of a_m. Because $0 \leq |a_m/a_M| \leq 1$, this gives the ellipticity of the total range $-\pi/4 \leq \chi \leq \pi/4$. The polarization is *left handed* or *right handed* if it appears to be circulating counterclockwise or clockwise with respect to an observer looking directly into the oncoming field, respectively. This is illustrated in Figure 7.2. A negative ellipticity is right handed, and a positive ellipticity is left handed. The handedness of the polarization ellipse is also referred to as its *helicity*.

There are two special cases of the polarization ellipse, which we now consider in terms of the field parameters.

7.1.1 Linear Polarization

Returning to Equation 7.7, the ellipse will reduce to a straight line if

$$\delta = \delta_y - \delta_x = m\pi, \quad (m = 0, \pm 1, \pm 2, \ldots). \qquad (7.35)$$

In such a case, we find that

$$\frac{E_y}{E_x} = (-1)^m \frac{a_y}{a_x}, \qquad (7.36)$$

or in another form,

$$E_y = (-1)^m \frac{a_y}{a_x} E_x. \qquad (7.37)$$

This is the equation of a straight line; the light is said to be *linearly polarized*. From Equations 7.29 and 7.30, we find with some effort that the geometric parameters of the ellipse are given by

$$a_M^2 = a_x^2 + a_y^2, \tag{7.38}$$

$$a_m^2 = 0, \tag{7.39}$$

$$\tan(2\psi) = (-1)^m \frac{2a_x a_y}{a_x^2 - a_y^2}. \tag{7.40}$$

It is to be noted that the handedness, represented by the sign of $\tan\chi$, is undefined for linear polarization, as $\tan\chi = 0$.

7.1.2 Circular Polarization

Returning to the general equation for an elliptically polarized field, it might seem at first that letting

$$a_x = a_y \equiv a \tag{7.41}$$

would define a purely *circularly polarized* field; however, it is possible to inscribe an ellipse in a square such that $a_x = a_y$ but $a_M \neq a_m$. An example is shown in Figure 7.3.

To make the path of the electric field vector truly circular, we also require

$$\delta = \delta_y - \delta_x = (2m + 1)\pi/2, \quad (m = 0, \pm1, \pm2, \ldots). \tag{7.42}$$

With this requirement, the electric field satisfies the equation

$$\left(\frac{E_x}{a}\right)^2 + \left(\frac{E_y}{a}\right)^2 = 1, \tag{7.43}$$

which is the equation for a circle.

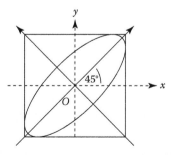

FIGURE 7.3   An ellipse inscribed within a square, such that $a_x = a_y$ but $a_M \neq a_m$.

It is to be noted that the angle of orientation ψ is undefined for circular polarization, as Equation 7.31 reduces to $\tan(2\psi) = 0/0$.

7.2 STOKES PARAMETERS AND THE POINCARÉ SPHERE

We now have two complete descriptions of the state of polarization of an electromagnetic wave at a point, via the field parameters a_x, a_y, and δ and the geometric parameters a_M, a_m, and ψ. However, neither of these descriptions is readily measurable via an experiment, at least in the optical domain, where the field is oscillating much too rapidly.[*]

In 1852, G.G. Stokes [Sto52] provided a set of *four* parameters that can be measured experimentally and can be directly related to both the field and geometric parameters. In terms of the field properties, these four parameters are

$$S_0 = a_x^2 + a_y^2, \tag{7.44}$$

$$S_1 = a_x^2 - a_y^2, \tag{7.45}$$

$$S_2 = 2a_x a_y \cos \delta, \tag{7.46}$$

$$S_3 = 2a_x a_y \sin \delta. \tag{7.47}$$

The fact that there are four quantities, where three uniquely describe the state of polarization, suggests that the Stokes parameters are not independent. In fact, it can readily be shown that

$$S_0^2 = S_1^2 + S_2^2 + S_3^2. \tag{7.48}$$

For future reference, we note that the matrix $\mathbf{Q}$ can be written entirely in terms of the Stokes parameters, in the form

$$\mathbf{Q} = \frac{2}{S_3^2} \begin{bmatrix} S_0 - S_1 & -S_2 \\ -S_2 & S_0 + S_1 \end{bmatrix}, \tag{7.49}$$

which already demonstrates that the Stokes parameters provide the same information as the geometric and field parameters. However, this relationship can be made clearer, as we will see.

It will be convenient at this point to describe the field using a complex notation, as we can then readily describe the average properties of products

---

[*] It should be noted that both of these descriptions have lost the *absolute* phase of the field, which is characterized by both δ_x and δ_y.

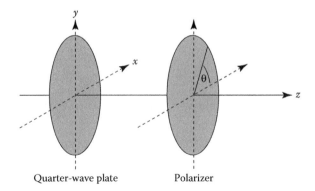

FIGURE 7.4 The experimental arrangement for measuring Stokes parameters.

of fields using the strategy outlined between Equations 2.15 and 2.20. The complex form of the electric field is

$$\mathbf{E}(\mathbf{r}, t) = \mathrm{Re}\left\{\mathbf{E}(\mathbf{r})e^{-i\omega t}\right\}, \tag{7.50}$$

where we have

$$\mathbf{E}(\mathbf{r}) = \hat{\mathbf{x}}a_x e^{i\delta_x} + \hat{\mathbf{y}}a_y e^{i\delta_y}. \tag{7.51}$$

We now consider a system of *six* time-averaged field measurements, using a polarizer and a quarter-wave plate to project an arbitrary state of the field onto the Stokes parameters. The simple system is illustrated in Figure 7.4.

The complex field emerging from this system will be labeled $\mathbf{E}(\theta, q)$, where θ is the angle of acceptance of the linear polarizer and $q = \pi/2, 0$ represents the presence or absence of the wave plate, which imparts a $\pi/2$ phase shift to the x-component of the field relative to the y-component. The six measurements, and the complex fields that emerge, are listed below.

$$\mathbf{E}(0, 0) = \hat{\mathbf{x}}a_x e^{i\delta_x}, \tag{7.52}$$

$$\mathbf{E}(\pi/2, 0) = \hat{\mathbf{y}}a_y e^{i\delta_y}, \tag{7.53}$$

$$\mathbf{E}(\pi/4, 0) = \hat{\mathbf{x}}\frac{a_x}{\sqrt{2}}e^{i\delta_x} + \hat{\mathbf{y}}\frac{a_y}{\sqrt{2}}e^{i\delta_y}, \tag{7.54}$$

$$\mathbf{E}(-\pi/4, 0) = \hat{\mathbf{x}}\frac{a_x}{\sqrt{2}}e^{i\delta_x} - \hat{\mathbf{y}}\frac{a_y}{\sqrt{2}}e^{i\delta_y}, \tag{7.55}$$

$$\mathbf{E}(\pi/4, \pi/2) = \hat{\mathbf{x}}\frac{a_x}{\sqrt{2}}e^{i(\delta_x+\pi/2)} + \hat{\mathbf{y}}\frac{a_y}{\sqrt{2}}e^{i\delta_y}, \tag{7.56}$$

$$\mathbf{E}(-\pi/4, \pi/2) = \hat{\mathbf{x}}\frac{a_x}{\sqrt{2}}e^{i(\delta_x+\pi/2)} - \hat{\mathbf{y}}\frac{a_y}{\sqrt{2}}e^{i\delta_y}. \tag{7.57}$$

We may calculate the cycle-averaged intensity of each of these cases, using Equation 2.19 in the form

$$\langle I(\theta, q)\rangle = \mathbf{E}^*(\theta, q) \cdot \mathbf{E}(\theta, q). \tag{7.58}$$

The results are as follows:

$$\langle I(0, 0)\rangle = a_x^2, \tag{7.59}$$

$$\langle I(\pi/2, 0)\rangle = a_y^2, \tag{7.60}$$

$$\langle I(\pi/4, 0)\rangle = \frac{1}{4}\left[2a_x^2 + 2a_y^2 + 4a_x a_y \cos\delta\right], \tag{7.61}$$

$$\langle I(-\pi/4, 0)\rangle = \frac{1}{4}\left[2a_x^2 + 2a_y^2 - 4a_x a_y \cos\delta\right], \tag{7.62}$$

$$\langle I(\pi/4, \pi/2)\rangle = \frac{1}{4}\left[2a_x^2 + 2a_y^2 + 4a_x a_y \sin\delta\right], \tag{7.63}$$

$$\langle I(-\pi/4, \pi/2)\rangle = \frac{1}{4}\left[2a_x^2 + 2a_y^2 - 4a_x a_y \sin\delta\right]. \tag{7.64}$$

The Stokes parameters can then be easily found from the measured intensities

$$S_0 = \langle I(0, 0)\rangle + \langle I(\pi/2, 0)\rangle, \tag{7.65}$$

$$S_1 = \langle I(0, 0)\rangle - \langle I(\pi/2, 0)\rangle, \tag{7.66}$$

$$S_2 = \langle I(\pi/4, 0)\rangle - \langle I(-\pi/4, 0)\rangle, \tag{7.67}$$

$$S_3 = \langle I(\pi/4, \pi/2)\rangle - \langle I(-\pi/4, \pi/2)\rangle. \tag{7.68}$$

This set of Stokes measurements is clearly inefficient—six measurements are used to determine three field parameters—but this arrangement weighs each measurement equally in the derivation of the parameters. Other arrangements using a smaller number of parameters will be considered in the exercises.

From Equation 7.65, it is clear that S_0 is equal to the total intensity of the electromagnetic wave at that point, and that intensity does not play a role in the orientation and ellipticity of the polarization ellipse. Also, we have seen from Equations 7.47 and 7.48 that S_0 is determined by the other Stokes parameters. It is convenient, therefore, to introduce a smaller set of normalized Stokes parameters of the form

$$s_1 = S_1/S_0, \qquad (7.69)$$

$$s_2 = S_2/S_0, \qquad (7.70)$$

$$s_3 = S_3/S_0. \qquad (7.71)$$

These normalized parameters satisfy the relation

$$s_1^2 + s_2^2 + s_3^2 = 1, \qquad (7.72)$$

and we will find that they have a special significance.

We now have *three* equivalent sets of parameters that can be used to characterize the state of polarization of an electromagnetic wave: the field parameters, the geometric parameters, and the Stokes parameters. It can be seen from Equations 7.29 and 7.30 that there is not a simple algebraic relation between these sets of parameters. However, it is possible to use a geometric construction due to Poincaré to unify all three sets in a particularly elegant manner; this construction, the *Poincaré sphere*,[*] also leads to physical implications to be discussed in Section 7.11.

We begin with the complex form of the electric field components, namely,

$$E_x = a_x e^{i\delta_x} e^{-i\omega t}, \qquad (7.73)$$

$$E_y = a_y e^{i\delta_y} e^{-i\omega t}. \qquad (7.74)$$

We now introduce a complex number defined as the ratio of these components as

$$\zeta = u + iv \equiv \frac{E_y}{E_x} = \frac{a_y}{a_x} e^{i\delta}, \qquad (7.75)$$

where u and v are real numbers. It can be seen that each point in the complex plane represents a distinct state of polarization; in polar coordinates,

---

[*] See, for instance, the paper by Jerrard [Jer54].

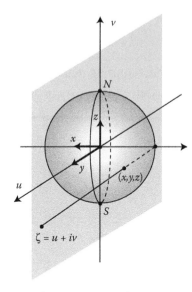

FIGURE 7.5 The stereographic projection from the complex plane onto the Poincaré sphere.

the ratio $\rho \equiv a_y/a_x$ determines the distance from the origin and $\phi \equiv \delta$ represents the azimuthal angle in the complex plane. We can map these points from the plane onto a sphere of unit radius by the use of a *stereographic projection*; the particular projection to be used here is illustrated in Figure 7.5.

The real and imaginary parts of ζ are taken to lie along the y- and z-axes in three-dimensional space. The mapping is done as follows: a line is drawn from the point where the $-x$-axis intersects the sphere to the point $u + iv$ in the complex plane. The point (x, y, z) at which the sphere intersects this line is the mapping from the plane to the sphere.

With some inspection, we may note a few special cases. Points on the equator of the sphere are defined by $v = 0$, which implies $\delta = 0$ or $\delta = \pi$ and corresponds to all possible states of linear polarization. The North and South poles, for which $u = 0$, are given by $a_x = a_y$ and $\delta = \pm\pi/2$, corresponding to left- and right-circular polarization, respectively. All other locations on the sphere represent elliptical polarization, with left-handed elliptical above the equator and right-handed elliptical below.

The spatial variables x, y, and z can be determined generally in terms of the field parameters by solving for the intersection point between the sphere, $x^2 + y^2 + z^2 = 1$, and the line from the $-x$-axis. This line satisfies the equations

$$x = \frac{y}{u} - 1, \quad z = \frac{v}{u}y. \tag{7.76}$$

Ignoring the fixed intersection point at $(-1, 0, 0)$, the other intersection point has the coordinates

$$x = \frac{a_x^2 - a_y^2}{a_x^2 + a_y^2} = s_1, \tag{7.77}$$

$$y = \frac{2a_x a_y \cos \delta}{a_x^2 + a_y^2} = s_2, \tag{7.78}$$

$$z = \frac{2a_x a_y \sin \delta}{a_x^2 + a_y^2} = s_3. \tag{7.79}$$

We therefore have the extremely useful result that the coordinates of the state of polarization on the sphere correspond to the normalized Stokes parameters of Equations 7.69 through 7.71.

The most striking part of the Poincaré sphere, however, is how locations on the sphere map to the geometrical properties of the polarization ellipse. To derive this, it is convenient to apply the following trigonometric formulas to the equations for a_M and a_m:

$$\cos^2 \psi = \frac{1}{2}[1 + \cos(2\psi)], \tag{7.80}$$

$$\sin^2 \psi = \frac{1}{2}[1 - \cos(2\psi)], \tag{7.81}$$

which after some effort allows us to write

$$2a_M^2 = S_0 + S_1 \cos(2\psi) + S_2 \sin(2\psi), \tag{7.82}$$

$$2a_m^2 = S_0 - S_1 \cos(2\psi) - S_2 \sin(2\psi), \tag{7.83}$$

and also to write

$$\tan(2\psi) = \frac{S_2}{S_1}. \tag{7.84}$$

We may write the tangent of twice the ellipticity angle using the tangent formula

$$\tan(2\chi) = \frac{2\tan(\chi)}{1 - \tan^2(\chi)} = \frac{2a_m a_M}{a_M^2 - a_m^2}. \tag{7.85}$$

Solving for $\sin(2\chi)$ and $\cos(2\chi)$, it can be shown that the normalized Stokes parameters, and hence the position on the Poincaré sphere, are given by

$$s_1 = x = \cos(2\chi)\cos(2\psi), \tag{7.86}$$

$$s_2 = y = \cos(2\chi)\sin(2\psi), \tag{7.87}$$

$$s_3 = z = \sin(2\chi), \tag{7.88}$$

where ψ is again the angle of orientation of the ellipse and χ is the ellipticity. These expressions indicate that 2χ and 2ψ represent the latitude and longitude angles on the sphere, with χ measured from the equator.

A point on the Poincaré sphere therefore simultaneously determines the Stokes parameters and the geometric parameters of the polarization ellipse. The field parameters can be determined from these. The relationship between the Stokes parameters, the state of polarization, and the Poincaré sphere is illustrated in Figure 7.6.

The Poincaré sphere also highlights the somewhat special nature of linear and circular polarization, which occupy only a small part of the total sphere. Linear polarization states cover only a line on the sphere, while circular polarization states are only two isolated points. This specialness will be elaborated upon in the following sections.

The Poincaré sphere, and most of the mathematics of polarization described in the previous two sections, is typically used to describe paraxial electromagnetic waves whose electric fields are, to a good approximation,

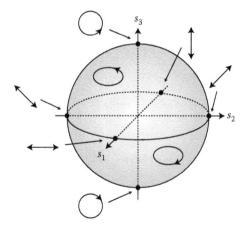

FIGURE 7.6 The relationship between the state of polarization, the Poincaré sphere, and the Stokes parameters.

completely transverse. However, this amounts to treating the electric field as having only two complex components, E_x and E_y, which is analogous to the *spinor representation* of spin-1/2 particles in quantum mechanics; the Poincaré sphere is then the electromagnetic analog of the Bloch sphere for spinors, as described in Chapter 12 of Miller [Mil08]. However, the quantum carrier of electromagnetic waves, the photon, is a spin-1 particle, which has significantly different properties. Though in many cases our transverse approximation will not cause any problems, it is worthwhile to note that there is an exact geometric representation of spin-1 states, originally introduced by Majorana [Maj32] in the context of quantum mechanics.

We give a brief discussion of the Majorana representation as used by Hannay [Han98] to describe light polarization. To do so, it is first worthwhile to recall what is being neglected when we consider a purely transverse electric field. Returning to the discussion at the beginning of Section 7.1, we noted that the electric field at any point in a general monochromatic electromagnetic wave is confined to a plane defined by the vectors **p** and **q**. The cross-product of these two vectors defines a *direction*, and in a transverse field, this is trivially taken to be the direction of propagation. We need two angles to define this direction; combined with the orientation angle ψ and the ellipticity χ of the polarization ellipse, we in general need four parameters to define the state of polarization of a nontransverse wave.

These parameters can be defined by *two* points on a sphere instead of one as in the Poincaré representation. We consider two unit vectors **u** and **v** that lie on the surface of the sphere, and we take the direction of the bisector of these two vectors as the normal to the polarization ellipse **n**, that is,

$$\mathbf{n} \equiv \frac{\mathbf{u} + \mathbf{v}}{\sqrt{2 + 2\mathbf{u} \cdot \mathbf{v}}}. \tag{7.89}$$

The state of polarization is then found by projecting the two vectors down into the plane normal to **n**. These two points in the plane represent the two foci of the polarization ellipse, which is assumed to have a semimajor axis of unit length. The general case and special cases are illustrated in Figure 7.7. Circular polarization is specified by the two vectors **u** and **v** being coincident, while linear polarization is specified by the two vectors being at antipodal points on the sphere.

In the linear case, of course, $\mathbf{u} + \mathbf{v} = 0$ and there is no normal to the polarization ellipse. In the circular polarization case, we no longer need to specify "left-handed" and "right-handed" circular polarization, because these two orthogonal states are represented by opposing **n** vectors. It

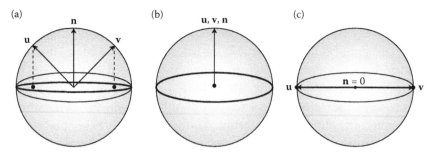

FIGURE 7.7 (a) A general polarization state on the Majorana sphere, and the special cases of (b) circular and (c) linear polarization.

should be clear from this simple discussion that all states and directions of polarization are represented on the Majorana sphere.

7.3 POLARIZATION SINGULARITIES: C-POINTS AND L-LINES

With a general description of the polarization properties of light behind us, we may now investigate the types of singularities possible in a transverse electromagnetic wavefield. We have already noted that the zeros of a complex monochromatic field are nongeneric, and as such, a zero would require the simultaneous vanishing of four components—the real and imaginary parts of E_x and E_y—at a point specified by the two transverse coordinates x, y.

A paraxial wave, however, may be considered to be the superposition of two independent, orthogonally polarized fields. The E_x and E_y components of the total field will each possess their own optical vortex lines, which can be isolated experimentally by the use of a polarizer. Such a decomposition is dependent on the choice of the x- and y-axes, though, and therefore is not a measure of something universal about the field.

An alternative is to instead decompose the field in terms of left and right circularly polarized light, with components E_+ and E_-, respectively, and consider phase singularities of these two components, which will also generically be optical vortices. These singularities can even be measured directly from the total state of polarization, as a zero line of left circularly polarized light will be a line of purely right circularly polarized light in the total field, and vice versa. These lines will typically intersect a plane at points, which are labeled as *C-points*. We may therefore say that, far from there being no singularities at all in a general electromagnetic wave, there are two families, namely, the left- and right-handed C-points.

It is a bit parochial, however, to simply regard C-points as the phase singularities of particular wavefield components, similar to our discussion in previous chapters: as we will see, C-points possess their own unique topological properties and reactions that are associated with their complex vector nature. The state of polarization can be described at every point in a monochromatic wavefield by an ellipse, and this ellipse has a direction specified by the orientation angle ψ of its major axis. At points of circular polarization, this orientation angle is undefined, making a C-point a singularity of direction in the field of polarization ellipses.

In terms of the vectors $\mathbf{p}$ and $\mathbf{q}$ defined in Section 7.1, a C-point will appear when

$$\mathbf{p}^2 - \mathbf{q}^2 = 0, \quad \mathbf{p} \cdot \mathbf{q} = 0. \tag{7.90}$$

This is equivalent to saying that the denominator and numerator, respectively, of Equation 7.31 must simultaneously vanish.

Linear polarization may also be considered a singularity of polarization, albeit of a type distinct from C-points. Looking at the Poincaré sphere, circular polarization exists only at isolated points, while linear polarization encircles the equator, dividing the sphere into left-handed and right-handed polarization.[*] In a transverse plane, we therefore expect that linear polarization appears as lines, either closed or stretching off to infinity, that separate the plane into regions of left- and right-handedness. These lines, known as *L-lines*, are singularities of the handedness of the polarization ellipse, which is undefined on the line and flips across it.

C-points and L-lines were first discussed by Nye [Nye83a], and further elaborated upon by Hajnal and Nye [NH87] and Hajnal [Haj87a]. Hajnal [Haj87b] also tested the theoretical predictions using sources of three and four interfering microwave beams with a wavelength of 3.5 cm, and employed a crossed dipole probe to measure the transverse electric field at points within a three-dimensional region.

If we restrict our attention to only the transverse component of the electric field, this description of polarization singularities can be readily extended to all of three-dimensional space. In a volume of space, the C-points are generally lines, C-lines, that intersect a transverse plane at a point. The L-lines are generally surfaces, L-surfaces, in such a volume.

---

[*] It can be helpful to imagine the Poincaré sphere as engraved with all the states of polarization. Covering it with ink and rolling and spinning it on a rubber sheet, followed by stretching of the sheet, we can roughly create diagrams of any possible transverse polarization field.

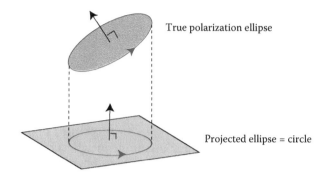

FIGURE 7.8 Illustrating how C-points of the transverse electric field are not generally C^T-points of the true polarization ellipse. An ellipse inclined to the transverse plane can project into a circle, and vice versa.

It is important to note, however, that the transversality of the electric field is almost always only approximate, and that the true geometry of the polarization field can be somewhat different than that of an idealized transverse paraxial wave. To account for this, we distinguish two classes of singularities, the C^T-points and L^T-lines of the *true* polarization ellipse, which can have a general orientation in three-dimensional space, and the transverse C-points and L-lines of the projection of the polarization ellipse onto transverse coordinates. These two sets of singularities do not, in general, coincide; this is illustrated in Figure 7.8. To describe the complete state of polarization in the true field, the handedness must be replaced by the vector area of the ellipse, with a magnitude given by the area of the ellipse and the direction defined in accordance with the usual right-hand rule.

As pointed out by Berry and Dennis [BD01c], the topology of the true polarization field has nontrivial differences from the transverse field. Though the transverse field has L-surfaces in three-dimensional space, the true field has L^T-lines. We can see this directly by returning to the three-dimensional definitions of $\mathbf{p}$ and $\mathbf{q}$, in Equation 7.6. Linear polarization will only exist when $\mathbf{p}$ is parallel to $\mathbf{q}$, or

$$\mathbf{p} \times \mathbf{q} = 0. \tag{7.91}$$

This vector condition is equivalent to the pair of scalar conditions

$$\frac{p_x}{q_x} = \frac{p_y}{q_y}, \quad \frac{p_y}{q_y} = \frac{p_z}{q_z}. \tag{7.92}$$

With two constraints in three-dimensional space, the regions of true linear polarization must be lines.

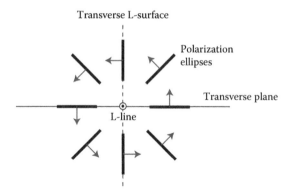

FIGURE 7.9 Relationship between an L-surface in the transverse polarization and an L^T-line in the true polarization.

An explanation for this difference in topology between transverse and true polarization fields is illustrated in Figure 7.9. A cross section of a field possessing an L^T-line is shown, and the polarization ellipses and their handedness are also illustrated. Around the L^T-line in three-dimensional space, the ellipse can make a full 360-degree rotation. Measurements in the transverse plane, however, will show a discontinuous jump of the handedness across the line; furthermore, the projection of the true ellipse into transverse coordinates will show an apparent L-surface in the vertical direction. With this picture, the L^T-lines may properly be thought of as singularities in the *vector area* of the polarization ellipse, which circulates around the line very much like a vortex and has zero magnitude on the L^T-line itself.

For simplicity, we will focus mostly on the transverse C-points and L-lines for the rest of this chapter. The differences in their behavior as compared to the true polarization singularities should be nevertheless be kept in mind.

Transverse fields possess their own surprises. It is not difficult to see that circular and linear singularities must also exist for the elliptical path of the magnetic field **H**. However, as discussed by Hajnal [Haj87a], the singularities of the **H**-field do not generally coincide with the singularities of the **E**-field. This seems surprising at first glance, as a plane wave with a circular **E**-field will also have a circular **H**-field of the same handedness. For a general elliptically polarized plane wave, however, the ellipse of the magnetic field is distinct from that of the electric field. When multiple plane waves are combined, then, the sums of the electric and magnetic fields at a point will generally not be proportional.

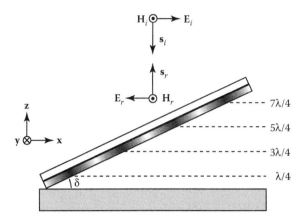

FIGURE 7.10   Illustration of Wiener's 1890 experiment.

This effect is even evident in simple interference experiments. We demonstrate it using the classic 1890 experiment by Otto Wiener [Wie90], originally performed to specify the role of the electric field in optics.

Wiener's experiment is illustrated in Figure 7.10. Of interest at the time was determining whether the electric field or the magnetic field is the "active ingredient" in chemical processes such as photo development. Wiener realized that a plane wave normally reflected from a perfect mirror will produce spatially displaced standing waves, with the electric field having antinodes at positions

$$z^E_{max} = (2m + 1)\lambda/4, \quad m = 1, 2, 3, \ldots, \tag{7.93}$$

and the magnetic field having antinodes at positions

$$z^H_{max} = m\lambda/2, \quad m = 1, 2, 3, \ldots. \tag{7.94}$$

With a photographic plate oriented at a small angle δ to the mirror, Wiener was able to determine that the photographic emulsion was sensitive to the E-field, and not the H-field. This was a major contributor to the decision to define the electric field vector as the polarization of light.

The corresponding nodes of **E** and **H** will also clearly not be the same. The singularities of the electric and magnetic fields are therefore in different spatial locations. Furthermore, because the sign of a complex field flips across a node, there will exist regions where the electric and magnetic fields are in phase, and regions where they are 180 degrees out of phase. Even for a simple case such as this, the relationship between the oscillations of **E** and **H** can be nontrivial.

Though this experiment involves nongeneric interference patterns, it indicates that the singularities of the electric and magnetic fields can be spatially separated from each other. This independence of the electric and magnetic singularities can be shown to extend to polarization singularities. Even more, because the L-lines of the electric and magnetic fields do not have to coincide, there generally will exist regions where the electric and magnetic field states have opposite handedness.

It is also possible to characterize the C-points of polarized light via the Stokes parameters. As a point of circular polarization is characterized by $a_x = a_y$ and $\delta = \pm\pi/2$, C-points are defined by

$$S_1 = 0, \quad S_2 = 0. \tag{7.95}$$

An L-line, on the other hand, is a point on the equator of the Poincaré sphere, for which

$$S_3 = 0. \tag{7.96}$$

We have therefore found that the natural generalization of singular optics to vector fields is the identification of C-points and L-lines as singularities of the field of polarization ellipses. Furthermore, we have seen that these singularities are closely connected to the phase singularities of the circularly polarized components of the field. This latter observation will be important in Section 7.9.

7.4 GENERIC FEATURES OF POLARIZATION SINGULARITIES

As in the case of phase singularities, we expect that the structure of the field around singularities of polarization will be topologically nontrivial and possess a small number of standard, generic forms. As the major axes of the polarization ellipses form a set of orientations, we might expect the structures of polarization singularities to be identical to those of a vector field such as in Figure 3.18, with integer values of the topological index. However, this is in fact not the case—as we will see, the topological index of a generic polarization singularity is a half-integer, and the structures of such polarization singularities look quite different from those of a vector field.

Qualitatively, the difference exists because the vectors in a vector field not only have an orientation—an angle with respect to a given coordinate axis—but a definite direction—the direction the arrow points, while the major axis of an ellipse is only defined by its orientation, and the ellipse is invariant with respect to a 180-degree rotation. Mathematically, the difference arises because the polarization ellipse is only defined through the quadratic form of Equation 7.15, and the symmetric matrix $\mathbf{Q}$ that contains all the properties of the polarization ellipse. The eigenvectors of this matrix, let us refer to them as $\mathbf{M}$ and $\mathbf{m}$, represent the orientations of the major and minor axes of the ellipse, respectively. However, they are only unique up to a sign, as we can readily see that, if $\mathbf{M}$ represents an eigenvector of $\mathbf{Q}$, then $-\mathbf{M}$ does, as well.

The subtle distinction between a polarization ellipse and a vector field, therefore, is that the ellipse is defined by the *eigenvectors* of $\mathbf{Q}$, and these do not possess a unique direction. In studying the singularities of polarization, we must therefore investigate the singularities of a two-dimensional, second-order *tensor field*.

The mathematics of tensor fields is elegant and well established, though it is significantly more involved than that of vector fields. We leave a detailed discussion for Section 7.12, and here borrow the main results and talk about polarization singularities in an intuitive manner.

We have already noted that monochromatic electromagnetic fields are analytic functions in all three spatial variables.[*] Because of this, they are necessarily continuous, and therefore the matrix $\mathbf{Q}$ derived from them is also continuous. In any closed circuit around a singularity, the orientation of the polarization ellipse can change, but field continuity requires that it returns to its original orientation upon completion of the circuit. Because the ellipse looks identical when rotated by 180 degrees, however, we expect that the simplest, and presumably generic, singularities will have the major axis rotate by 180 degrees over the circuit. In terms of the orientation angle ψ, the topological index of such a singularity should therefore be of the form

$$n \equiv \frac{1}{2\pi} \oint_L \nabla \psi(\mathbf{r}) \cdot d\mathbf{r} = \pm 1/2, \qquad (7.97)$$

where L is a closed path of integration around the singularity. Such a topological index suggests that the orientation of the polarization ellipse, which is a continuous function of position almost everywhere, will jump discontinuously across a point of circular polarization. To get a better feeling for the behavior of the ellipses near such a point, we consider the streamlines that join together the major axis lines; these lines can only intersect at a

---

[*] This was discussed in Section 3.5.

singularity. A general singularity can then possess three distinct types of streamlines in an angular region:

- Hyperbolic. The streamlines do not begin or end at the singularity.
- Elliptic. The streamlines begin and end at the singularity.
- Parabolic. The streamlines end at the singularity.

Regions of these classes of streamlines cannot mix, and pairs of distinct regions will be locally separated from each other by a straight line called a *separatrix*. An illustration of a hypothetical singularity containing all three types of regions is shown in Figure 7.11.

From the figure, we can deduce how the orientation angle changes in each region as a function of its angular width. In a hyperbolic region with an opening angle of α, it can readily be determined that the angle changes by an amount $\alpha - \pi$. It is to be noted that the sense of rotation in a hyperbolic region changes according to whether $\alpha < \pi$ or $\alpha > \pi$. In an elliptic region with an opening angle of γ, the angle changes by an amount $\gamma + \pi$. Both of these conditions can most readily be seen by considering the limiting cases $\alpha, \gamma \to 0$ and $\alpha, \gamma \to \pi$. In a parabolic region with opening angle β, the angle changes by β.

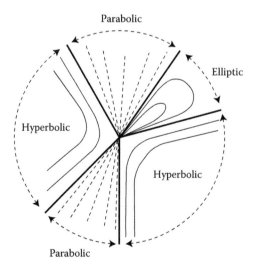

FIGURE 7.11 Illustration of the streamlines around a hypothetical polarization singularity, with the thick lines between regions representing the separatrices.

With this simple observation, we can prove a powerful and elegant theorem by Bendixson [Ben01]. He showed that the index n of a two-dimensional second-order tensor singularity is simply related to the number N_h of hyperbolic regions and number N_e of elliptic regions by the relation

$$n = 1 + \frac{N_e - N_h}{2}. \tag{7.98}$$

The proof is surprisingly straightforward. Let us assume that the ith hyperbolic, parabolic, and elliptic regions have opening angles α_i, β_i, and γ_i, respectively. As the index n of a singularity is the number of times that the orientation of the ellipse changes by 2π, we may write

$$2\pi n = \sum_{i=1}^{N_h}(\alpha_i - \pi) + \sum_{i=1}^{N_p}\beta_i + \sum_{i=1}^{N_e}(\gamma_i + \pi). \tag{7.99}$$

However, the total opening angles of all regions must add to 2π, that is,

$$\sum_{i=1}^{N_h}\alpha_i + \sum_{i=1}^{N_p}\beta_i + \sum_{i=1}^{N_e}\gamma_i = 2\pi. \tag{7.100}$$

On substitution, we find that

$$2\pi n = 2\pi - N_h\pi + N_e\pi. \tag{7.101}$$

Dividing by 2π, Equation 7.98 follows.

Next, we argue that there cannot, in fact, be any elliptic regions for the simplest singularities. For now, we base this on the plausible assumption (to be proven in Section 7.12) that the "simplest" singularities will be those for which the major axes always rotate in the same direction and never "double-back" on themselves. We then consider the implications of Equation 7.98 under this assumption.

If the index $n = -1/2$, we immediately find that there cannot be an elliptic region. Such a region would make a positive contribution to the index, automatically implying that the major axis orientation doubles back at least once during a path around the singularity. For $n = +1/2$, let us assume that there is a single elliptical region. We then find that $N_h = 2$; however, for a hyperbolic region to produce a positive rotation of the major axis, and not double back, it must have an opening angle greater than 180 degrees. It is clearly not possible to have two hyperbolic regions each with such an angle; we therefore find that no elliptic regions exist for the simplest singularities with half-integer index.

Equation 7.98 then suggests the following. The simplest, presumably generic, polarization singularities come in a small number of distinct flavors. For an index $n = -1/2$, the singularity will have three hyperbolic regions. It cannot have any parabolic regions, as they would produce a positive rotation. For an index $n = +1/2$, the singularity will have a single hyperbolic region. It will either possess no parabolic regions, and a single separatrix, or a single parabolic region, with two separatrices. In the latter case, one can show that an additional straight-line trajectory exists within the parabolic region, separating it into two regions of differing curvature; considering this as an additional separatrix, we can say that three exist.

The symmetric forms of the three generic polarization singularities are illustrated in Figure 7.12. They are named the *lemon*, with index $n = +1/2$ and a single separatrix, the *star*, with index $n = -1/2$ and three separatrices, and the *monstar*, with index $n = -1/2$ and three separatrices. The (le)monstar is considered to be a hybrid between the lemon and the star, with the index of the lemon and the separatrices of the star.

It should be noted that the arguments presented here assume that the C-line intersects the transverse plane at a point, and does not lie entirely within the plane. In the latter case, the polarization singularities will look quite different; this nongeneric case was discussed by Nye [Nye83a].

Figure 7.12 can be compared with Figure 3.17 of Section 3.3, showing some typical disclinations in liquid crystals. The lemon and star patterns appear in both cases, as does the monstar (though not shown in the crystal case). Clearly, we may consider polarization singularities as disclinations

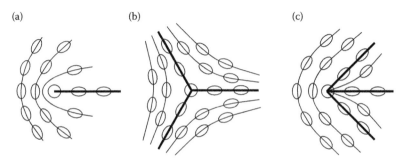

FIGURE 7.12 The generic ellipse fields around polarization singularities, named (a) lemon, (b) star, and (c) monstar.

of an electromagnetic wave, just as we considered phase singularities as dislocations of a scalar wave.

Another type of disclination in the polarization of light was introduced by Nye [Nye83b], which we briefly discuss. We have noted that zeros of the transverse complex electric field $\mathbf{E}(x, y)$ are nongeneric, as the four components of the field must vanish simultaneously at a single (x, y)-point in the transverse plane. However, the real-valued instantaneous transverse field $\mathbf{E}(x, y, t)$ has only two components E_x and E_y that need to vanish simultaneously to make a singularity, and therefore point zeros of this field are generic. The zero of the electric field is a singularity of direction, and typical patterns of the field around the singularity will be the same as shown in Figure 3.18, where the solid arrows represent the instantaneous electric field vector.

These disclinations depart significantly from the singularities described previously in that they are explicitly time dependent and will typically move in the plane as time evolves, even for monochromatic fields. For this reason, perhaps, such singularities have not been as extensively studied as the C-points and L-lines, though Hajnal [Haj87b] also measured them in his 1987 microwave experiments.

7.5 TOPOLOGICAL REACTIONS OF POLARIZATION SINGULARITIES

Though we have now determined the generic types of polarization singularities, we have not connected them to actual wavefields. In other words: under what field conditions do lemons, stars, and monstars arise? This question must also be addressed in order to consider topological reactions of polarization singularities.

Let us consider an idealized polarization singularity, given by the expression

$$\mathbf{E}(x, y) = \gamma(\hat{\mathbf{x}} + i\alpha\hat{\mathbf{y}})(x + \beta iy) + (\hat{\mathbf{x}} - i\alpha\hat{\mathbf{y}}). \qquad (7.102)$$

Here α and β are independent quantities each allowed to be ± 1. This electric field represents a vortex of left (right) circular polarization combined with a plane wave of right (left) circular polarization. For instance: with $\alpha = 1$, $\beta = -1$, we have a right-handed vortex in left-handed circularly polarized light. The quantity γ is generally complex and represents the relative amplitude and phase of the vortex compared to the plane wave.

We would like to determine the type of polarization singularity that exists at the origin, as a function of α, β, and γ. To do so, we will use results from Section 7.12. We first note that, in the neighborhood of a polarization singularity, we may make a linear approximation to the tensor **Q**; from Equation 7.194, we may specifically write

$$\frac{Q_{xx} - Q_{yy}}{2} = 2(ax + by), \tag{7.103}$$

$$Q_{xy} = 2(cx + dy), \tag{7.104}$$

where all coefficients are real-valued and constant and the factor of 2 has been added for notational convenience. From Equation 7.49, this is equivalent to saying that

$$\frac{-S_1}{S_3^2} = ax + by, \tag{7.105}$$

$$\frac{-S_2}{S_3^2} = cx + dy. \tag{7.106}$$

In the neighborhood of a C-point $S_3 \approx \pm 1$. We may thus neglect the denominator of the preceding equations and simply write

$$-S_1 = ax + by, \quad -S_2 = cx + dy. \tag{7.107}$$

In such a linear approximation, we may write

$$a = -S_{1x}, \quad b = -S_{1y}, \quad c = -S_{2x}, \quad d = -S_{2y}, \tag{7.108}$$

where S_{1x} represents the partial derivative of S_1 with respect to x evaluated at the origin, and so forth. We may explicitly calculate the Stokes parameters from Equation 7.102, keeping only the linear terms. The result is

$$S_1 = 4\gamma_R x - 4\gamma_I \beta y, \tag{7.109}$$

$$S_2 = -4\alpha\gamma_I x - 4\alpha\beta\gamma_R y, \tag{7.110}$$

where γ_R and γ_I are the real and imaginary parts of γ, respectively.
From these, the partial derivatives give the coefficients

$$a = -4\gamma_R, \quad b = 4\gamma_I \beta, \quad c = 4\gamma_I \alpha, \quad d = 4\gamma_R \alpha\beta. \tag{7.111}$$

Now, we may use the results of Section 7.12, in which it is shown that the topological index of a tensor singularity is given by

$$n = \frac{1}{2}\text{sign}(D_I), \tag{7.112}$$

where

$$D_I = ad - bc = S_{1x}S_{2y} - S_{1y}S_{2x}. \tag{7.113}$$

For our example, we have

$$D_I = -16\alpha\beta[\gamma_R^2 + \gamma_I^2]. \tag{7.114}$$

We get the following simple, and striking, result: the topological index is negative if the handedness of the vortex and the polarization of the vortex is the same, and the index is positive if they are opposite. The relative phase between components does not play a role at all.

Also from Section 7.12, we note that the number of separatrices is determined by the discriminant D_L of a cubic equation

$$D_L = -18d(2b + c)(2a - d)c + 4(2b + c)^3 c + (2b + c)^2(2a - d)^2$$
$$- 4d(2a - d)^3 - 27d^2 c^2, \tag{7.115}$$

and that there are three for $D_L > 0$ and one for $D_L < 0$. This discriminant may be written in terms of Stokes parameters as

$$D_L = -18S_{2y}(2S_{1y} + S_{2x})(2S_{1x} - S_{2y})S_{2x} + 4(2S_{1y} + S_{2x})^3 S_{2x}$$
$$+ (2S_{1y} + S_{2x})^2(2S_{1x} - S_{2y})^2$$
$$- 4S_{2y}(2S_{1x} - S_{2y})^3 - 27S_{2y}^2 S_{2x}^2. \tag{7.116}$$

With some effort, it can be shown that $D_L > 0$ for $\alpha\beta = +1$ and $D_L < 0$ for $\alpha\beta = -1$. We therefore have the following two cases:

- $\alpha\beta = +1$. We have $n = -1/2$ and three separatrices. The singularity is a star.

- $\alpha\beta = -1$. We have $n = +1/2$ and one separatrix. The singularity is a lemon.

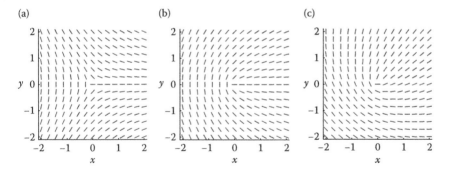

FIGURE 7.13 Simple model polarization singularities, for (a) $\alpha = 1$, $\beta = 1$, $\gamma = 1$, (b) $\alpha = -1$, $\beta = 1$, $\gamma = 1$, and (c) $\alpha = -1$, $\beta = 1$, $\gamma = \exp[i\pi/4]$.

We test these results by plotting the major axis of the polarization ellipse as a function of position for several values of α, β, and γ in Figure 7.13. The plots confirm our calculations, and demonstrate that the only effect of the phase difference γ between the orthogonal polarization states is the orientation of the singularity.

At this point, the reader might wonder what happened to the $n = +1/2$, three separatrix monstar, which did not make an appearance at all in our limited model. The monstar is a transition singularity that only appears just before creation/annihilation events, which we now consider.

As topological index is conserved, we expect that creation and annihilation always involves a star and lemon. It is evident that both of these singularities must appear in the same component of circular polarization, as the underlying vortices in that component must also be created or destroyed. We can investigate this by considering Young's three-pinhole vortex annihilation event discussed in Section 3.5; we make the pinhole field left-hand circular and interfere it with a normally incident plane wave that is right-hand circular.

The results are shown in Figure 7.14. We have a lemon and star to begin with, in part (a). In part (b), the horizontal line connecting the two singularities appears to be a new separatrix, suggesting that the lemon has evolved into a monstar, with two new separatrices splitting off from the original single lemon line. In part (c), the singularities have annihilated, leaving a curved, but smooth, set of streamlines.

What happens to the local separatrices? The central one is "eaten up" between the two singularities as they come together, while the next two closest below come together and annihilate themselves. The remaining

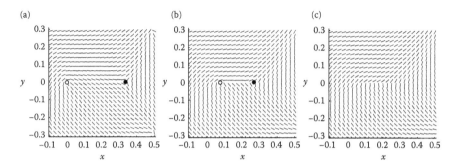

FIGURE 7.14 Annihilation of polarization singularities in electromagnetic Young's three-pinhole system plus a plane wave, with the amplitude of hole 1 given by (a) $A_1 = 1.0$, (b) $A_1 = 1.7$, and (c) $A_1 = 2.1$. The white circle represents a lemon/monstar, while the black circle represents a star.

separatrix of each singularity becomes a single locally straight region of the polarization field.

There are potentially many different ways for generic polarization singularities to annihilate, with different separatrices playing different roles; the process, however, involves a lemon evolving into a monstar and annihilating with a star.

7.6 HIGHER-ORDER POLARIZATION SINGULARITIES

With a discussion of the generic polarization singularities behind us, we briefly turn to the discussion of higher-order polarization singularities, which are nongeneric but turn out to be of some importance. Just as a vortex of topological charge $|t| > 1$ can be considered as a sum of first-order vortices, a higher-order polarization singularity can be considered as an overlap of the generic ones. We stick to a qualitative description here, and note that a bit more information can be found in Delmarcelle and Hesselink [DH94].

The merging of two stars, each with topological index $n = -1/2$, must produce a total singularity of index $n = -1$. As for vector fields, we expect that this singularity must be a saddle point. The stars may join along a separatrix line or pairs of separatrices can merge together; these possibilities are illustrated in Figure 7.15.

Similarly, the merging of two lemons, each with topological index $n = +1/2$, must result in a singularity of total index $n = +1$. There are a number of possibilities in such a case; the simplest two, the node and the center, are illustrated in Figure 7.16.

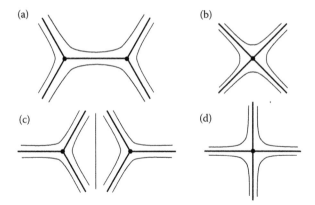

FIGURE 7.15 Illustration of how two stars along (a) the same separatrix can join to form a (b) saddle, or two stars can merge (c) pairs of separatrices to form a (d) saddle.

These singularities are analogous to those seen in vector fields, as in Figure 3.18. In a vector field, however, the saddles, nodes, and centers are topologically stable, whereas in a tensor field they are unstable. Nevertheless, we will see in the next section that they turn out to have their own uses.

It is worth noting that there are two distinct ways for a higher-order polarization singularity to occur. The first is for a second-order vortex to appear in a single handedness of polarization, which results in a net index $n = \pm 1$, depending on the relative sign of the vortices and handedness. The second is for first-order vortices of opposite handedness, and opposite

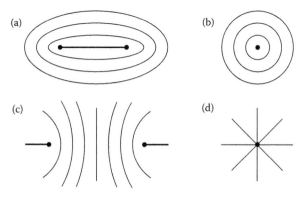

FIGURE 7.16 Illustration of how two lemons along (a) the same separatrix can join to form a (b) center singularity, two lemons on (c) different separatrices can join to form a (d) node singularity.

polarization handedness, to come together. The two cases are distinguishable, in that the electric field in the core of the singularity will be nonzero in the case of a second-order vortex combined with a local plane wave, and it will be zero in the case of two first-order vortices combined.

7.7 NONUNIFORMLY POLARIZED BEAMS

We have seen in Chapter 6 that nongeneric phase singularities have found use in optics applications; in this section, we shall see that this is also true of nongeneric polarization singularities. The simplest examples of such singularities are Gaussian-like beams with radial or azimuthal polarization; these have also been referred to as *vector beams* or *cylindrical vector beams*. In general, any beam whose state of polarization is a function of transverse position is referred to as a *nonuniformly polarized beam*. We first introduce the simplest incarnations of such beams, discuss their relationships to generic singularities, and then consider their application.

Simulations of the intensity and polarization in cross sections of radial and azimuthal beams are illustrated in Figure 7.17. The field is taken to be linearly polarized at every point in space for each beam, which means that we may draw the actual direction of the complex electric field vector instead of just the major axis of the polarization ellipse. Both beams have identical intensity profiles and, in fact, have profiles identical to the first-order Laguerre–Gauss beams $LG_{0,\pm 1}$.

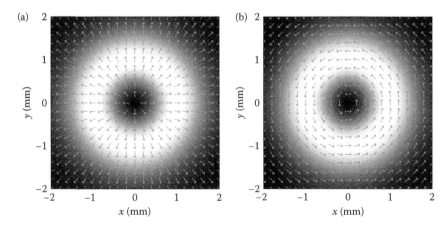

FIGURE 7.17 The construction of (a) a radially polarized beam and (b) an azimuthally polarized beam from Hermite–Gauss modes of orthogonal polarization.

A radially polarized beam may be constructed by combining an $\hat{\mathbf{x}}$-polarized Hermite–Gauss HG_{10} mode with a $\hat{\mathbf{y}}$-polarized HG_{01} mode with equal amplitudes and phases. Referring back to Section 2.4, and neglecting normalization factors of the Hermite–Gauss modes, the total electric field in the source plane may be written in the form

$$\mathbf{E}(x, y) = E_0 \frac{x\hat{\mathbf{x}} + y\hat{\mathbf{y}}}{w_0} \exp[-(x^2 + y^2)/2w_0^2], \qquad (7.117)$$

where w_0 is the width of the modes in the waist plane. It can be seen immediately from the resulting expression that the field is linearly polarized at every point in space, and that the polarization vector points in the radial direction. Similarly, an azimuthally polarized beam may be constructed by combining a $\hat{\mathbf{y}}$-polarized Hermite–Gauss HG_{10} mode with an $\hat{\mathbf{x}}$-polarized HG_{01} mode with equal amplitudes and a 180-degree phase shift between them. Now the field in the source plane may be written as

$$\mathbf{E}(x, y) = E_0 \frac{-y\hat{\mathbf{x}} + x\hat{\mathbf{y}}}{w_0} \exp[-(x^2 + y^2)/2w_0^2]. \qquad (7.118)$$

The field is again linearly polarized, but the electric field vector is orthogonal to that of the radial field at every point in space: the field is polarized in the azimuthal direction.

The construction of these fields is illustrated schematically in Figure 7.18. Again, we may draw arrows to represent the direction of the real-valued electric field vectors at each point. Because the lobes of the Hermite–Gauss modes are π out of phase, the vectors point in opposite directions in each lobe.

The radial and azimuthally polarized beams may be thought of as vector analogs of the Laguerre–Gauss modes LG_{01} and $LG_{0,-1}$. Just as the two Laguerre–Gauss modes are formed from out-of-phase sums of the HG_{01} and HG_{10} modes, as noted in Equation 2.85, so the vector beams are formed from orthogonally polarized sums of the same Hermite–Gauss modes.

The polarization properties of these vector beams persist on propagation, as the HG_{01} and HG_{10} have the same phase and amplitude evolution on propagation. A radially (azimuthally) polarized beam therefore remains radially (azimuthally) polarized at every transverse plane, though the phase curvature and width of the beam evolves.

The topological index of both radial and azimuthal beams can readily be seen to be $n = 1$. As seen in the previous section, they can be considered

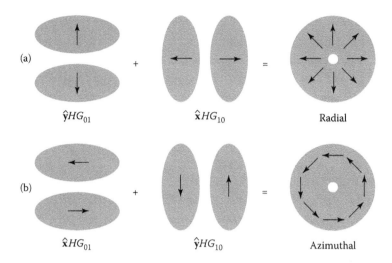

FIGURE 7.18 The construction of (a) a radially polarized beam and (b) an azimuthally polarized beam from Hermite–Gauss modes of orthogonal polarization.

to be the combination of two $n = 1/2$ polarization singularities, namely, lemons, superimposed on each other. As beams of pure linear polarization, however, the radial and azimuthal beams can be considered to be "special cases among the special cases," as nodes and centers may have more general elliptical polarization in their neighborhood.

The observation of such vector beams goes back quite some time, though the recent revival in interest seems to have started with the development of a quantum well semiconductor laser that could emit azimuthally polarized light [EKH+92]. Not long after, an analytic model of an azimuthal beam of a Bessel–Gauss form was developed and extensively studied [JH94, Hal96,GH96]. (Properties of ordinary, nonvector, Bessel–Gauss beams will be discussed in Section 12.1.)

Vector beams have now been demonstrated to have novel properties that are potentially useful for applications. Youngworth and Brown [YB00a] performed a thorough theoretical and experimental study of the tight focusing of radial and azimuthal beams, noting that the electric field in the region of focus has a very strong longitudinal component, as seen in Figure 7.19. This longitudinal component can be understood quite readily, and is explained by Figure 7.20. The radial electric field vectors, all pointing away from the axis of propagation, are tilted toward the axis on focusing, and in the region of focus combine in phase to produce a longitudinal

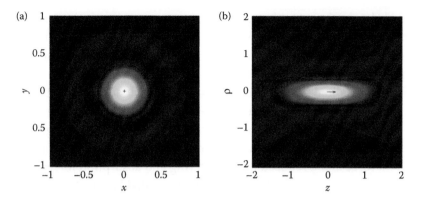

FIGURE 7.19 The normalized intensity of the longitudinal component of a focused radially polarized beam ($NA = 1.32$) (a) in the focal plane and (b) in a radial and longitudinal cross section. The spatial units are given in wavelengths. (From K.S. Youngworth and T.G. Brown. *Opt. Exp.*, 7:77–87, 2000.)

field. Longitudinal fields have been suggested as a technique for electron acceleration, among other uses. It was also noted that an azimuthal electric field remains purely transverse on focusing, and possesses no longitudinal component at all.

Not long after, Dorn, Quabis, and Leuchs [DQL03] experimentally demonstrated that the spot size of a radially polarized field can be significantly smaller than that of a linearly polarized field. Observing that the longitudinal component of the field has a smaller spot than the transverse component, the researchers focused a radial beam through an annular aperture, effectively keeping only the longitudinal piece of the beam. Using a helium–neon beam in a system with a numerical aperture $NA = 0.9$, they

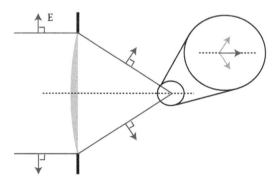

FIGURE 7.20 Illustration of the formation of a strong longitudinal field in radially polarized light.

produced a focal spot with area $0.16\lambda^2$, as compared with that produced with linear polarization, $0.26\lambda^2$. This arrangement was studied in more detail, and with an eye toward lithography and optical data storage, by Sheppard and Choudhury [SC04].

The longitudinal field provides a number of other advantages. Trapping metal particles with optical tweezers can be difficult due to strong scattering and absorption along the line of propagation. The strong longitudinal field in radial polarization, however, simultaneously provides a strong axial gradient force and a reduced axial scattering and absorption forces. It was shown computationally by Zhan [Zha04] that radial beams should be able to trap metal particles of Raleigh size.

The longitudinal component can also be applied to sensing applications. Novotny et al. [NBYB01] used the longitudinal field as an excitation source to probe the orientation of single molecules. Conversely, they used such molecules to study the longitudinal field structure.

A number of techniques may be used to efficiently generate radially or azimuthally polarized light, and many of these are directly analogous to the techniques used to create vortex beams as discussed in Section 4.1. A commercial device known as an "axis-finder" or, more technically, a dichroic azimuthal linear polarizer, will directly convert uniform linear polarization into radial polarization, and might be thought of as a vector analogy to a spiral phase plate. Certain lasers, such as the quantum well semiconductor laser discussed earlier [EKH$^+$92], can directly produce a cylindrical vector beam, just as lasers can be induced to directly output Laguerre–Gauss modes. The combination of Hermite–Gauss modes, as discussed in the beginning of this section, is also experimentally viable and analogous to combining such modes to make Laguerre–Gauss modes. Finally, we note that a Mach–Zender interferometer may be combined with polarizing beamsplitters to generate radial or azimuthal beams [YB00b].

More information about fields with nonuniform polarization states can be found in the review article by Brown [Bro08].

7.8 POINCARÉ BEAMS

Radial and azimuthal beams may be considered the simplest forms of nonuniformly polarized beams possible, in that they have a cylindrical symmetry and the polarization state is linear at every point in space. One may naturally wonder if it is possible to go to another extreme and create a beam that features *every* state of polarization on the Poincaré sphere in every cross section. This was proven to be possible by Beckley, Brown, and Alonso [BBA10], and such beams are now referred to as *full Poincaré beams*.

The simplest Poincaré beam can be constructed from a pair of low-order Laguerre–Gauss modes, namely, u_{00}^{LG} and u_{01}^{LG}, given from Section 2.2 by

$$u_{00}(\mathbf{r}) = \sqrt{\frac{2}{\pi w_0^2}} \frac{1}{1 + z^2/z_0^2} e^{-i\Phi(z)} e^{-ik\rho^2/2R(z)} e^{-\rho^2/w^2(z)}, \tag{7.119}$$

$$u_{01}(\mathbf{r}) = \sqrt{\frac{2}{\pi w_0^2}} \frac{1}{1 + z^2/z_0^2} \frac{\sqrt{2}\rho}{w(z)} e^{i\phi} e^{-2i\Phi(z)} e^{-ik\rho^2/2R(z)} e^{-\rho^2/w^2(z)}, \tag{7.120}$$

where $\Phi(z)$ is the Gouy shift, $w(z)$ is the propagation-dependent beam width, $R(z)$ is the propagation-dependent wavefront curvature, and $z_0 = \pi w_0^2/\lambda$ is the Rayleigh range. Here w_0 is again the beam width at $z = 0$. We add these two modes together to construct a transverse electric field vector as follows:

$$\mathbf{E}(\mathbf{r}, \gamma) = \cos \gamma \hat{\mathbf{e}}_1 u_{00}(\mathbf{r}) + \sin \gamma \hat{\mathbf{e}}_2 u_{01}(\mathbf{r}), \tag{7.121}$$

where $\mathbf{e}_1$ and $\mathbf{e}_2$ are orthogonal unit vectors and γ is a real-valued free parameter. Because of the orthogonality of these unit vectors, it is easy to show that the intensity of the combined beam is cylindrically symmetric

$$|\mathbf{E}(\mathbf{r}, \gamma)|^2 = \cos^2 \gamma |u_{00}(\mathbf{r})|^2 + \sin^2 \gamma |u_{01}|^2. \tag{7.122}$$

As both the fundamental Gaussian mode and the first-order Laguerre–Gaussian mode have rotationally symmetric intensities, so does the combined mode. We may write the electric field itself in the explicit form

$$\mathbf{E}(\mathbf{r}, \gamma) = \cos \gamma \sqrt{\frac{2}{\pi w_0^2}} \frac{1}{1 + z^2/z_0^2} e^{-i\Phi(z)} e^{-ik\rho^2/2R(z)} e^{-\rho^2/w^2(z)}$$
$$\times \left[\hat{\mathbf{e}}_1 + e^{i[\phi - \Phi(z)]} \frac{\sqrt{2}\rho}{w(z)} \hat{\mathbf{e}}_2 \tan \gamma \right]. \tag{7.123}$$

Everything outside of the square brackets in the preceding equation represents an overall scaling of amplitude and phase that does not influence the state of polarization of the field at a point $\mathbf{r}$. We therefore introduce the state vector $\mathbf{e}(\mathbf{r})$ of our combined beam as

$$\mathbf{e}(\mathbf{r}) \equiv \hat{\mathbf{e}}_1 + e^{i[\phi - \Phi(z)]} \overline{\rho} \hat{\mathbf{e}}_2, \tag{7.124}$$

where we introduce the dimensionless parameter $\overline{\rho} \equiv \sqrt{2} \tan \gamma \rho/w$.

Because of the azimuthal ϕ dependence, every possible phase relation between $\hat{e}_1$ and $\hat{e}_2$ is expressed in a plane of constant z; because of the radial $\bar{\rho}$ dependence, every possible amplitude ratio between the two polarizations is also expressed. Furthermore, each combination appears exactly one time, demonstrating that the polarization in the plane covers the Poincaré sphere exactly one time. The polarization pattern will be invariant on propagation except for an overall rotation based on the value of the Gouy phase $\Phi(z)$.

The specific nature of the pattern will depend on the choice of polarization states $\hat{e}_1$ and $\hat{e}_2$. Let us first take them to be the circular states, namely, $\hat{e}_1 = \hat{e}_+$ and $\hat{e}_2 = \hat{e}_-$. The position on the Poincaré sphere can then be readily found by writing the normalized Stokes parameters in terms of $\bar{\rho}$ and $\phi' \equiv \phi - \Phi(z)$. We have

$$s_1 = \frac{2\bar{\rho}\cos(\phi')}{1 + \bar{\rho}^2}, \tag{7.125}$$

$$s_2 = \frac{2\bar{\rho}\sin(\phi')}{1 + \bar{\rho}^2}, \tag{7.126}$$

$$s_3 = \frac{1 - \bar{\rho}^2}{1 + \bar{\rho}^2}. \tag{7.127}$$

When $\bar{\rho} = 0$, we have $s_1 = 0$, $s_2 = 0$, and $s_3 = 1$, which indicates that we have left-hand circular polarization. As $\bar{\rho} \to \infty$, we approach $s_1 = 0$, $s_2 = 0$, and $s_3 = -1$, which is right-hand circular polarization. When $\bar{\rho} = 1$, we have $s_3 = 0$ and a pure state of linear polarization.

A pair of Poincaré beams, using different basis vectors $\hat{e}_1$ and $\hat{e}_2$, are shown in Figure 7.21. The mapping of the entire Poincaré sphere can be seen, and the change of handedness across lines of linear polarization is also evident.

Poincaré beams were first experimentally produced in Beckley, Brown, and Alonso [BBA10] by transmitting a circularly polarized beam through a stressed optical window. The window used had a birefringence that depended linearly on radius and a fast axis that precessed azimuthally. The results were in good agreement with the theoretical predictions.

The Poincaré beam may find use in polarimetry as a source whose state of polarization can be adjusted simply through a lateral shift in the beam position. Another use was discovered by Gu, Korotkova, and Gbur [GKG09], who noted that such a nonuniformly polarized beam will have lower-intensity fluctuations in atmospheric turbulence than an ordinary

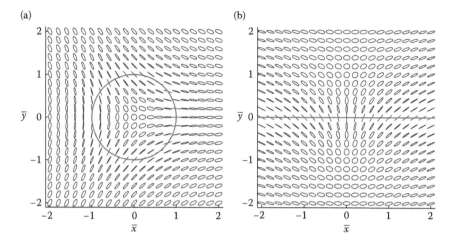

FIGURE 7.21 Illustration of the polarization ellipses of Poincaré beams, with (a) $\hat{e}_1 = \hat{e}_-$, $\hat{e}_2 = \hat{e}_+$ and (b) $\hat{e}_1 = \hat{y}$, $\hat{e}_2 = \hat{x}$. The thick lines indicate L-lines.

Gaussian beam or an equivalent uniformly polarized beam. At the time, the researchers arrived at this choice of beam without realizing its Poincaré significance. Later work demonstrated that this scintillation reduction can be improved further by using an incoherent array of Poincaré beams [GG13].

An even broader class of Poincaré beams can be constructed that span not only the surface of the Poincaré sphere but also its interior, the interior representing states of partial polarization [BBA12].

7.9 POINCARÉ VORTICES

We have seen that there is a close relationship between the phase singularities of scalar wavefields and the polarization singularities of a field with general polarization properties. In a somewhat self-referential application, it was shown by Freund [Fre01] that this connection can be used to detect optical vortices using polarization measurements.

The technique is based on the Stokes parameters of Section 7.2. We consider an interferometric system as illustrated in Figure 7.22, in which a sample beam containing optical vortices $E_y(\mathbf{r})$ is directly combined via a beamsplitter with a reference beam of orthogonal polarization, $E_x(\mathbf{r})$. The combination is then filtered, as described in Section 7.2, to measure the normalized Stokes parameters s_2 and s_3 as a function of transverse position.

We next define a new complex field

$$P \equiv s_2 + is_3, \tag{7.128}$$

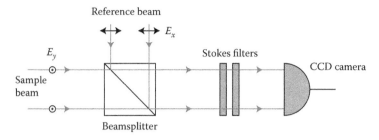

FIGURE 7.22 Simple experimental apparatus for Poincaré vortex measurement. (After I. Freund. *Opt. Lett.*, 26:1996–1998, 2001.)

which can be readily shown from Equations 7.1 and 7.2 to have the form

$$P = \frac{a_x a_y}{a_x^2 + a_y^2} \exp[i(\delta_y - \delta_x)]. \tag{7.129}$$

Let us now suppose that the phase δ_x of the reference beam is constant. Then, the phase of P is simply a shifted version of the phase of the sample field $E_y(\mathbf{r})$, and vortices of the field P are identical in position and topological charge to the vortices of the sample field. As a version of the complex field P was originally described by Poincaré, Freund has dubbed the singularities of this field as *Poincaré vortices*.

Even if the field E_x is not uniform, the Poincaré vortices will be reasonable representatives of the singularities of E_x, provided there are no singularities in E_x itself. Variations in δ_x will distort the fine structure of the vortex, for example, making a screw vortex in E_y look like a mixed edge-screw vortex in P, but such details are usually not important in measurements.

The technique is still, at heart, an interferometric one, such as discussed at the beginning of Section 4.2, and in its basic operation will suffer from the same limitations of all interferometers. However, the technique can be extended to the observation of C-points in a straightforward way, something standard interferometers cannot do. Instead of measuring P, consider the measurement of C instead, defined as

$$C \equiv s_1 + is_2, \tag{7.130}$$

with s_1 and s_2 again the normalized Stokes parameters. A point of circular polarization is defined as a point where $S_1 = a_x^2 - a_y^2 = 0$ and $S_2 = 2a_x a_y \cos \delta = 0$, which corresponds with a zero of C. A point of circular polarization will therefore be a vortex of the constructed field C. Further-more, since the argument of C is given by $\Phi_C = \arctan(s_2/s_1)$ and the major axis of the ellipse in a polarized field is given by $\psi = \frac{1}{2}\arctan(s_2/s_1)$, it follows that the index n of the polarization singularity and the topological charge t of the vortex of C are related by

$$t = 2n. \tag{7.131}$$

In measurements of the properties of C-points, interference with a refer-ence field is no longer necessary. The sample field can be directly filtered to determine the Stokes parameters and the C field constructed from these.

7.10 PARTIALLY POLARIZED LIGHT AND SINGULARITIES OF THE CLEAR SKY

Up to now, we have focused entirely on fully polarized light, which can be characterized at every point in space by a complex electric field vector. How-ever, many sources of light—and most natural sources of light—produce light that is partially polarized or completely unpolarized. A thermal source, for instance, consists of a large number of atoms independently radiating. At any instant of time, the sum of electromagnetic fields emitted from all atoms defines a definite state of polarization; however, the phase relations between the different atomic fields evolve in time, and the overall state of polarization will also evolve. Restricting ourselves to transverse electric fields, we may therefore generalize the components of the electric field from Equations 7.7 to the form

$$E_x = a_x \cos[\omega t + \delta_x(t)],$$
$$E_y = a_y \cos[\omega t + \delta_y(t)], \tag{7.132}$$

where $\delta_x(t)$ and $\delta_y(t)$ are now fluctuating in time. If we assume that the field is quasi-monochromatic, this implies that the fluctuations occur on a timescale that is much slower than $1/\omega$.

To characterize the polarization properties of the field, we must now introduce the polarization matrix $\mathbf{J}$ of the electric field

$$\mathbf{J} \equiv \begin{bmatrix} \langle E_x^*(t)E_x(t) \rangle & \langle E_x^*(t)E_y(t) \rangle \\ \langle E_y^*(t)E_x(t) \rangle & \langle E_y^*(t)E_y(t) \rangle \end{bmatrix}, \tag{7.133}$$

where the angle brackets represent a long time average

$$\langle E_x^*(t)E_x(t)\rangle = \lim_{T\to\infty} \frac{1}{2T} \int\limits_{-T}^{T} E_x^*(t)E_x(t)dt. \tag{7.134}$$

It is important to note immediately that this matrix is completely different from the matrix **P** defined earlier, which described the polarization ellipse in terms of the field components.

The matrix **J** is Hermitian; that is, $\mathbf{J}^\dagger = \mathbf{J}$, and it is also nonnegative definite, which means that for an arbitrary complex vector $|c\rangle$, we have

$$\langle c| \mathbf{J} |c\rangle \geq 0. \tag{7.135}$$

The Hermitian property is obvious from the definition of the matrix; the nonnegative definiteness property follows from expanding the product in the inequality explicitly

$$|c_x|^2 \langle E_x^*E_x\rangle + c_x^*c_y\langle E_x^*E_y\rangle + c_y^*c_x\langle E_y^*E_x\rangle + |c_y|^2 \langle E_y^*E_y\rangle$$
$$= \langle |c_xE_x + c_yE_y|^2\rangle \geq 0. \tag{7.136}$$

We can determine the form of the polarization matrix in the extreme cases of an unpolarized field and a completely polarized field by simple physical reasoning without extensive calculation. In an unpolarized field, there should be no long-term correlation between the x and y components of the electric field, and therefore $J_{xy} = J_{yx} = 0$. Furthermore, there should be no distinction between the intensities of the x and y components, and so that $J_{xx} = J_{yy} = I_0/2$, where I_0 is the average intensity of the field. We therefore find that an unpolarized field should have a polarization matrix proportional to the identity matrix

$$\mathbf{J}_{unpol} = \frac{I_0}{2} \begin{bmatrix} 1 & 0 \\ 0 & 1 \end{bmatrix}. \tag{7.137}$$

For a completely polarized field, the phase relationships between the various components of the fields are independent of time. The averages of the field products have the trivial forms

$$\langle E_i^*(t)E_j(t)\rangle = E_i^*E_j, \quad i,j = x,y, \tag{7.138}$$

where E_i and E_j are the effective monochromatic components of the field as we have considered in the previous sections. Writing the complex polarization vector as $|\mathbf{E}\rangle$, we may then write

$$\mathbf{J}_{pol} = |\mathbf{E}\rangle\,\langle\mathbf{E}| = \begin{bmatrix} E_x^* E_x & E_x^* E_y \\ E_y^* E_x & E_y^* E_y \end{bmatrix}. \qquad (7.139)$$

We have also expressed the polarization matrix for a completely polarized field above as the direct product of the complex polarization vector with itself for future reference. One can readily show that the determinant of this matrix always vanishes, that is, $\text{Det}(\mathbf{J}_{pol}) = 0$.

It is also possible for an electromagnetic field to be partially polarized. In a result that dates back to Stokes [Sto52], we can demonstrate that the polarization matrix of the field at any point in space may be uniquely written as the sum of an unpolarized part and a polarized part, that is,

$$\mathbf{J} = \mathbf{J}_{unpol} + \mathbf{J}_{pol}. \qquad (7.140)$$

Stokes implied that this decomposition could be made globally, that is, that a partially polarized electromagnetic beam could be uniquely written as the sum of a fully polarized beam and an unpolarized beam, each of which independently satisfies a paraxial wave equation. Over 150 years later, however, Wolf [Wol08] demonstrated that this is generally not true, and this decomposition can only be applied point by point in a field. Nevertheless, it is an important decomposition, especially for the discussion of polarization singularities, as we will see.

We follow the proof as given by Wolf [Wol07, Section 8.2.3], and imagine that we want to decompose a general polarization matrix $\mathbf{J}$ into a polarized part of the field given by

$$\mathbf{J}_{pol} = \begin{bmatrix} B & D \\ D^* & C \end{bmatrix}, \qquad (7.141)$$

and an unpolarized part given by

$$\mathbf{J}_{unpol} = \begin{bmatrix} A & 0 \\ 0 & A \end{bmatrix}. \qquad (7.142)$$

For these to individually represent physical polarization matrices, we require the determinant of $\mathbf{J}_{pol}$ to vanish, that is,

$$BC - |D|^2 = 0, \qquad (7.143)$$

and further require $A, B, C \geq 0$.

In order for the sum $\mathbf{J}_{pol} + \mathbf{J}_{unpol}$ to equal a given polarization matrix $\mathbf{J}$, we need to determine if coefficients A, B, C, and D exist that satisfy the above constraints and the equations

$$A + B = J_{xx}, \quad A + C = J_{yy}, \quad D = J_{xy}. \tag{7.144}$$

We now attempt to solve for these coefficients. We may solve for B and C in the above equations and therefore write the determinant condition entirely in terms of A,

$$(J_{xx} - A)(J_{yy} - A) - |J_{xy}|^2 = 0, \tag{7.145}$$

which results in a quadratic formula for A,

$$A^2 - (J_{xx} + J_{yy})A + (J_{xx}J_{yy} - |J_{xy}|^2) = 0. \tag{7.146}$$

The first term in parenthesis is simply $\mathrm{Tr}(\mathbf{J})$, while the second term in parenthesis is $\mathrm{Det}(\mathbf{J})$. This expression is also the equation for the eigenvalues of the polarization matrix $\mathbf{J}$; as it is nonnegative definite, we know that both solutions for A must be nonnegative, satisfying one of our required inequalities.

Now solving explicitly for A, we have

$$A = \frac{\mathrm{Tr}(\mathbf{J}) \pm \sqrt{[\mathrm{Tr}(\mathbf{J})]^2 - 4\mathrm{Det}(\mathbf{J})}}{2}. \tag{7.147}$$

There are two possible values for A, but only one can lead to valid polarization decomposition if the decomposition is unique. Looking at the negative root first, it leads to the values

$$B = \frac{1}{2}(J_{xx} - J_{yy}) + \frac{1}{2}\sqrt{[\mathrm{Tr}(\mathbf{J})]^2 - 4\mathrm{Det}(\mathbf{J})}, \tag{7.148}$$

$$C = \frac{1}{2}(J_{yy} - J_{xx}) + \frac{1}{2}\sqrt{[\mathrm{Tr}(\mathbf{J})]^2 - 4\mathrm{Det}(\mathbf{J})}. \tag{7.149}$$

The discriminant, however, is of the form

$$\sqrt{[\mathrm{Tr}(\mathbf{J})]^2 - 4\mathrm{Det}(\mathbf{J})} = \sqrt{(J_{xx}^2 - J_{yy})^2 + 4|J_{xy}|^2} \geq |J_{xx} - J_{yy}|, \tag{7.150}$$

which indicates that $B, C \geq 0$, as required. The positive root for A will conversely result in negative values of B and C, which is not a valid solution.

Now we are, in fact, done: we have found a complete, unique set of values of A, B, C, and D for our decomposition. Furthermore, we note that the intensity of the polarized part of the field is $\text{Tr}(\mathbf{J}_{pol}) = B + C$, while the total intensity of the field $\text{Tr}(\mathbf{J}) = 2A + B + C$; we may define a *degree of polarization P* of the form

$$P = \frac{\text{Tr}(\mathbf{J}_{pol})}{\text{Tr}(\mathbf{J})} = \sqrt{1 - \frac{4\text{Det}(\mathbf{J})}{[\text{Tr}(\mathbf{J})]^2}}. \qquad (7.151)$$

In the case that the field is completely polarized, $\text{Det}(\mathbf{J}) = 0$ and $P = 1$; in the case that the field is completely unpolarized, $(\text{Tr}(\mathbf{J}))^2 = 4\text{Det}(\mathbf{J})$ and $P = 0$. Any other case is considered partially polarized.

At first glance, the decomposition of the field into polarized and unpolarized parts may seem to be purely academic; however, we can easily show that the components of the matrix $\mathbf{J}$, and therefore the polarized and unpolarized parts, may be directly determined from the Stokes parameters.

We generalize the Stokes parameters from Equations 7.44 through 7.47 to averaged quantities of the form

$$S_0 = \langle |E_x|^2 + |E_y|^2 \rangle, \qquad (7.152)$$

$$S_1 = \langle |E_x|^2 - |E_y|^2 \rangle, \qquad (7.153)$$

$$S_2 = \langle E_y^* E_x + E_x^* E_y \rangle, \qquad (7.154)$$

$$S_3 = \langle i(E_y^* E_x - E_x^* E_y) \rangle, \qquad (7.155)$$

or in terms of the polarization matrix

$$S_0 = J_{xx} + J_{yy}, \qquad (7.156)$$

$$S_1 = J_{xx} - J_{yy}, \qquad (7.157)$$

$$S_2 = J_{yx} + J_{xy}, \qquad (7.158)$$

$$S_3 = i(J_{yx} - J_{xy}). \qquad (7.159)$$

Now, let us consider what the Stokes parameters look like in terms of the polarized and unpolarized decomposition. We have

$$S_0 = 2A + B + C, \tag{7.160}$$

$$S_1 = B - C, \tag{7.161}$$

$$S_2 = D^* + D, \tag{7.162}$$

$$S_3 = i(D^* - D). \tag{7.163}$$

What we have found is quite remarkable: the unnormalized Stokes vectors S_1, S_2, and S_3 depend only on the polarized part of the field in the parameters B, C, and D. The unpolarized part only appears in S_0. Measurement of the Stokes parameters allows us to directly measure the polarized and unpolarized parts of the field.

Furthermore, we can generalize our original description of the Poincaré sphere to take into account a field with a general degree of polarization. From Equation 7.159, we can readily show that

$$\text{Det}(\mathbf{J}) = \frac{1}{4}\left[S_0^2 - S_1^2 - S_2^2 - S_3^2\right], \tag{7.164}$$

$$\text{Tr}(\mathbf{J}) = S_0. \tag{7.165}$$

On substitution from these expressions into Equation 7.151, we find that

$$P = \frac{(S_1^2 + S_2^2 + S_3^2)^{1/2}}{S_0} = \sqrt{s_1^2 + s_2^2 + s_3^2}, \tag{7.166}$$

where we have used the normalized Stokes parameters, Equations 7.69 through 7.71, in the last step. Interpreting the normalized Stokes parameters as positions on the Poincaré sphere, we find that the degree of polarization determines the radial distance from the origin of the sphere, and the latitude and longitude on or within the sphere determines the polarization ellipse of the polarized part of the field. A fully polarized field will be characterized by a point on the surface of the Poincaré sphere, while an unpolarized field is characterized by the center of the sphere.

This entire discussion leads us back to a discussion of the polarization singularities of light. Even when a field is partially polarized, we can still measure the polarized part of the field and study any singularities that appear in it. All the previous arguments related to the singularities of a polarized wavefield directly apply to the polarized *part* of a general wavefield.

An excellent illustration of this can be found in the polarization properties of the clear sky, which have been known for a long time and extensively studied. In fact, over 1000 years ago, Viking sailors may have used dichroic "sunstones" to detect the polarization of the sky, allowing them to navigate even when the Sun was obscured by clouds [RGF$^+$12]. Much more recently, the first scientific observations of the polarization of skylight were performed by Arago [Ara11], and though it was generally appreciated that the blue sky must arise from the scattering of sunlight by small particles in the atmosphere, a satisfactory explanation of the polarization effects did not arise until the work of Lord Rayleigh much later [Ray71]. This theory itself, however, does not explain all features of the polarization of the sky, and we will see that our understanding of polarization singularities will complete the picture. We will treat the problem in a very qualitative manner here; a detailed mathematical derivation can be found in the paper by Berry, Dennis, and Lee [BDL04].

Let us, to begin, imagine the Earth to be completely transparent. We may therefore observe the light coming from all directions, and treat the atmosphere as a spherical surface surrounding us. In the simplest approximation, the observed skylight comes from sunlight scattered off of small particles in the atmosphere. The (unpolarized) light from the Sun excites electric dipoles in the particles; however, dipoles do not radiate parallel to their direction of oscillation, which results in partial polarization at nearly all points in the sky. This is illustrated in Figure 7.23.

Light that scatters off of the top of the atmosphere, as our figure shows, produces mostly horizontally polarized light, while light that scatters off of the side of the atmosphere produces mostly vertically polarized light. From this simple picture, we also predict that there will be two points where the field is completely unpolarized: the direction of the Sun, and the point directly opposite on the sphere, the *anti-Sun*. These locations on the sky sphere must be polarization singularities, as the polarized part of the field vanishes at these points. Further consideration of the simple Rayleigh scattering picture given above indicates that the polarization lines must form circles around these two singularities; in other words, the Sun and anti-Sun points are center singularities, as discussed in Section 7.6, each with topological index $n = +1$.

That the total topological index on the sky sphere is $n = +2$ could have been readily predicted. In topology, the delightfully named *hairy ball*

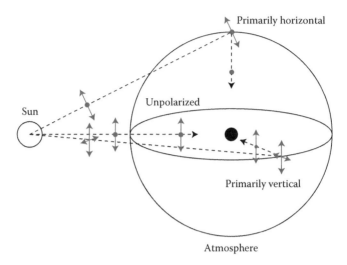

FIGURE 7.23 Illustration of Rayleigh scattering in the atmosphere. Light that scatters off of the "top" of the atmosphere produces mostly horizontally polarized light, while light that scatters off of the "side" of the atmosphere produces mostly vertically polarized light.

theorem demonstrates that a tangent vector or tensor field on a sphere always has a net index of 2. The name for the theorem comes from its colloquial statement: "It is impossible to comb a completely hairy ball flat without leaving at least one cowlick on the surface." A simple proof of this theorem is given in Section 7.13 at the end of this chapter.

However, the simple prediction of two center singularities in the clear sky was known to be inaccurate even when Lord Rayleigh published his results; there are in fact two seemingly unpolarized points near the Sun. One that appears above it, and is known as the Babinet point, and one below, known as the Brewster point. Furthermore, the anti-Sun also has two singularities in its vicinity, the Arago point (above the anti-Sun) and the second Brewster point (below the anti-Sun).

We can qualitatively understand the existence of these singularities from the topological properties of polarization singularities. The $n = +1$ centers at the Sun and anti-Sun are topologically unstable and, as noted in Section 7.6, will tend to break into a pair of lemons on the same separatrix. In this case, the instability comes from multiple scattering effects in the atmosphere. Symmetry requires that this separation either occurs along the line connecting the Sun to the zenith or perpendicularly to it; it can be shown that multiple scattering causes it to occur along the zenith line.

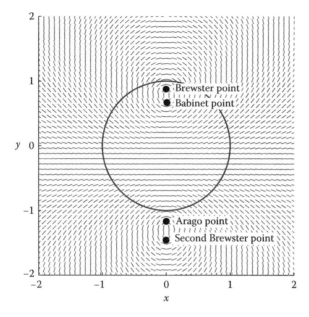

FIGURE 7.24 The polarization lines and singularities of the sky, using a stereographic projection of the sky sphere into the plane. The interior of the circle represents the visible sky, using the same parameters $\alpha = 15°$, $\delta = 4\arctan A = 12°$. (Adapted from M.V. Berry, M.R. Dennis, and R.L. Lee Jr. *New J. Phys.*, 6:162, 2004.)

An illustration of the lines of polarization in the sky is shown in Figure 7.24. At any time of day, only two of the four singularities will be unobscured by the Earth. The second Brewster point is generally not visible, as it appears only after sunset, though it has been observed from a hot air balloon [HBS⁺02].

It is to be noted that a center of index $n = +1$ is a true point of unpolarized light, whereas the lemons of index $n = +1/2$ are points of circular polarization. From the perspective of a polarization analyzer, however, both give the same result, as circular polarization also does not have a definite orientation.

7.11 THE PANCHARATNAM PHASE

The Poincaré sphere has played a significant role in the description of polarization throughout this chapter. With this in mind, it would be remiss to not spend some time discussing a surprising and unusual effect that is intimately related to the geometry of the sphere. Though this effect—known as the *Pancharatnam phase*—is not directly connected to the polarization singularities discussed previously, it may be considered somewhat "singular" in its nonintuitive nature.

This effect was first introduced by Shivaramakrishnan Pancharatnam in a seminal 1956 research paper [Pan56] on a generalized theory of interference, and may be described as follows. We imagine an optical system in which the state of polarization of a light beam is changed in a nontrivial cyclic manner. For instance, we may use optical elements to change the state from left-hand circular polarization to horizontal linear polarization, then to vertical linear polarization, and then back to left-hand circular polarization; on the Poincaré sphere, the path forms a spherical triangle. Though in this process the beam will naturally acquire a phase shift on propagation—to be referred to as a *dynamic phase*—Pancharatnam demonstrated that it will acquire an additional phase shift that may be attributed purely to geometrical effects on the Poincaré sphere. Since this discovery, such *geometric phases* have been found to appear in a remarkably diverse set of physical systems beyond optics, as we will discuss.

We begin by considering a pair of general transversely polarized electric fields $\mathbf{E}_A$ and $\mathbf{E}_B$ described in a general orthonormal polarization basis $\mathbf{e}_1$ and $\mathbf{e}_2$. The combined intensity of these two waves is proportional to the absolute square of the total electric field, that is,

$$I = (\mathbf{E}_A + \mathbf{E}_B)^* \cdot (\mathbf{E}_A + \mathbf{E}_B) = |\mathbf{E}_A|^2 + |\mathbf{E}_B|^2 + 2\mathrm{Re}(\mathbf{E}_A^* \cdot \mathbf{E}_B). \quad (7.167)$$

Talking about the overall "phase" of an electromagnetic wave is somewhat problematic, as there are *two* phases of a transverse electromagnetic field, namely, the phases of the x and y components of the field. Pancharatnam resolved this by defining two fields as being in phase when the intensity of their interference pattern takes on the maximum possible value. This condition, labeled *Pancharatnam's connection* by Berry [Ber87], involves varying the overall relative phase between $\mathbf{E}_A$ and $\mathbf{E}_B$ until they satisfy the conditions

$$\mathrm{Re}(\mathbf{E}_A^* \cdot \mathbf{E}_B) > 0, \quad (7.168)$$

$$\mathrm{Im}(\mathbf{E}_A^* \cdot \mathbf{E}_B) = 0. \quad (7.169)$$

With this definition, we may now introduce a simple proof and demonstration of Pancharatnam's theorem; this particular proof is due to Aravind [Ara92]. We will show, given a closed path formed by three points on the

Poincaré sphere connected by segments of great circles, that there must be a phase difference between the initial and final states, although they are the same point on the Poincaré sphere.

We first note that a general electric field of unit magnitude may be written as

$$\mathbf{E} = \mathbf{e}_1 \cos(\theta/2)e^{-i\phi/2} + \mathbf{e}_2 \sin(\theta/2)e^{i\phi/2}. \tag{7.170}$$

When $\mathbf{e}_1$ and $\mathbf{e}_2$ represent the points of right- and left-circular polarization, respectively, then, θ and ϕ represent the usual angles on the sphere. As a state is defined only up to an overall phase, however, it will be more convenient to write

$$\mathbf{E} = \mathbf{e}_1 \cos(\theta/2) + \mathbf{e}_2 \sin(\theta/2)e^{i\phi}, \tag{7.171}$$

where we have simply shifted the overall phase by $e^{i\phi/2}$. We will leave it as an exercise to demonstrate that the final result of our calculation does not change. We now consider three points on the Poincaré sphere. Point A is taken to lie along $\mathbf{e}_1$, point B is taken along the prime meridian of the sphere, and C is taken to lie somewhere between $\phi = 0$ and $\phi = \pi$. The evolution of polarization is taken along the geodesics of the sphere from A to B to C, then back to A, and we label the final state as $\tilde{A}$. The path is illustrated in Figure 7.25.

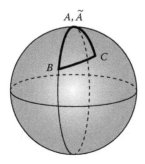

FIGURE 7.25 The circuit taken on the Poincaré sphere in illustrating the Pancharatnam phase.

We now define the electric fields at these points as follows:

$$\mathbf{E}_A = \mathbf{e}_1, \tag{7.172}$$

$$\mathbf{E}_B = e^{-i\chi_B} \left[\cos(\theta_B/2)\mathbf{e}_1 + \sin(\theta_B/2)\mathbf{e}_2\right], \tag{7.173}$$

$$\mathbf{E}_C = e^{-i(\chi_B+\chi_C)} \left[\cos(\theta_C/2)\mathbf{e}_1 + \sin(\theta_C/2)e^{i\phi_C}\mathbf{e}_2\right], \tag{7.174}$$

$$\mathbf{E}_{\tilde{A}} = e^{-i\chi_{\tilde{A}}}\mathbf{e}_1. \tag{7.175}$$

The phase χ_B represents the phase delay of state B with respect to A, while χ_C represents the phase delay of C with respect to B. The phase $\chi_{\tilde{A}}$ represents the delay of $\tilde{A}$ with respect to A, and this is the quantity that is of most interest to us.

We now consider the possibility of bringing all states into phase with respect to their neighbors as defined by the Pancharatnam connection. Clearly, the requirement that A and B be in phase reduces to the conditions $\cos\chi_B > 0$ and $\sin\chi_B = 0$, which has the simple solution $\chi_B = 0$. Similarly, the requirement that C and $\tilde{A}$ be in phase has the simple solution $\chi_{\tilde{A}} = \chi_C$.

We now attempt to bring B and C into phase. The condition given by Equation 7.169 becomes

$$-\sin(\chi_{\tilde{A}})\cos(\theta_B/2)\cos(\theta_C/2) + \sin(\phi_C - \chi_{\tilde{A}})\sin(\theta_B/2)\sin(\theta_C/2) = 0. \tag{7.176}$$

The sine term containing ϕ_C may be expanded out using the familiar trigonometric formula. With some rearrangement, we find that

$$\tan(\chi_{\tilde{A}}) = \frac{\sin(\phi_C)\sin(\theta_B/2)\sin(\theta_C/2)}{\cos(\theta_B/2)\cos(\theta_C/2) + \sin(\theta_B/2)\sin(\theta_C/2)\cos(\phi_C)}. \tag{7.177}$$

We now multiply the numerator and denominator of the right-hand side of this equation by $4\cos(\theta_B/2)\cos(\theta_C/2)$; on applying appropriate trigonometric formulas to remove the half-angles, the final result is

$$\tan(\chi_{\tilde{A}}) = \frac{\sin(\phi_C)\sin(\theta_B)\sin(\theta_C)}{[1 + \cos(\theta_B)][1 + \cos(\theta_C)] + \sin(\theta_B)\sin(\theta_C)\cos(\phi_C)}. \tag{7.178}$$

In bringing fields B and C into phase, we find that the phase of $\tilde{A}$ must be different than the field of A. This phase $\chi_{\tilde{A}}$ is the Pancharatnam phase, and we have indirectly demonstrated that a field taken through a continuous cycle must accrue this additional phase.

The phase of Equation 7.178 has a simple geometric interpretation. Our path along the geodesics forms a spherical triangle on the surface of the sphere. It can be shown [Eri90] that the spherical triangle defined by three unit vectors **a**, **b**, **c** has a solid angle Ω that satisfies

$$\tan(\Omega/2) = \frac{\mathbf{a} \cdot (\mathbf{b} \times \mathbf{c})}{1 + \mathbf{b} \cdot \mathbf{c} + \mathbf{c} \cdot \mathbf{a} + \mathbf{a} \cdot \mathbf{b}}. \tag{7.179}$$

Using $\mathbf{a} = (0, 0, 1)$, $\mathbf{b} = (\sin\theta_B, 0, \cos\theta_B)$, and $\mathbf{c} = (\sin\theta_C \cos\phi_C, \sin\theta_C \sin\phi_C, \cos\theta_C)$, one readily finds that

$$\tan(\Omega/2) = \frac{\sin(\phi_C)\sin(\theta_B)\sin(\theta_C)}{[1 + \cos(\theta_B)][1 + \cos(\theta_C)] + \sin(\theta_B)\sin(\theta_C)\cos(\phi_C)}, \tag{7.180}$$

which agrees with Equation 7.178. We have therefore shown that a field taken through a closed cycle of three polarization states accrues a geometric phase that is equal to half of the solid angle Ω of the spherical triangle defined by those states.[*]

It is to be noted that this result is surprisingly general. Though we have implicitly treated $\mathbf{e}_1$ and $\mathbf{e}_2$ as the states of circular polarization, our calculation is valid for any pair of antipodal points on the sphere, and therefore the triangle may be in principle positioned anywhere on the sphere itself. This result can also be generalized to arbitrary closed circuits on the Poincaré sphere, though the proof is more involved [Ber87]. The total geometric phase is simply half the total solid angle of the path taken on the sphere. We can see how to do this, roughly, by creating a complicated path of overlapping multiple spherical triangles of different sizes and handedness, with a negative area assigned to right-handed triangles.

The Pancharatnam phase is relatively easy to introduce, but there are a number of challenges associated with it. The first of these is simply attempting to explain what the result *means*, physically. The second challenge, not unrelated to the first, is the difficulty in unambiguously measuring the phase.

We may approach the first challenge by noting that, since Pancharatnam's original work, many physical systems have been recognized as possessing a geometric phase. The first result along this line was discovered by

---

[*] It should be noted that, throughout this calculation, we have made the reasonable but unproven assumption that the phase differences calculated between points are those along a great circle connecting the points.

Berry [Ber84], who demonstrated that certain quantum systems taken adiabatically around a circuit will acquire a purely geometric phase factor. The connection to polarization and Pancharatnam was noted not long after by Berry [Ber87], and can be immediately seen by considering our earlier observation that a transverse polarization state is mathematically equivalent to the spinor of a spin-1/2 particle.

Berry's discovery led to a reassessment of other early experimental observations in terms of geometric phases. The most famous of these is Foucault's pendulum [Opr95], first introduced in 1851, that indirectly demonstrates the rotation of the Earth. The pendulum is freely hanging and may swing in any direction. After being set into motion, it slowly changes its direction of oscillation during the course of a day as the Earth in essence rotates underneath it.

For simplicity, we first consider two extreme positions of the pendulum. For a pendulum situated at the North Pole, its direction of oscillation[*] will complete a full rotation in the course of a single day. For a pendulum situated at the equator, its path of oscillation will not change at all. For any intermediate latitude, assuming the behavior of the pendulum continuously varies with position, the pendulum must therefore complete somewhere between zero and one rotation in 24 hours, and generally will be oscillating in a different direction than it did the day before. Regarding the angle of orientation as a phase, this change is a manifestation of a geometric phase. The actual difference in orientation $\Delta\varphi$ is given by

$$\Delta\varphi = -2\pi \sin\theta, \tag{7.181}$$

where θ is the latitude, and not the usual angle in spherical coordinates. The situation is illustrated in Figure 7.26. The pendulum follows a circular path on the sphere during a 24-hour day, and it can be shown from Equation 7.181 that the phase change is equal, modulo 2π, to the solid angle defined by the path.

The Foucault pendulum illustrates the geometric phase as a phenomenon purely related to the *parallel transport* of a vector along a surface. The precise definition of parallel transport takes us well outside of our current interests, but in the case of the pendulum implies that the pendulum is moved slowly, and allowed to freely oscillate, so that any change in its

---

[*] This is cheating a bit, as a pendulum generally has an elliptical path such as polarization and should properly be characterized by a tensor, and not a vector. Since we are only considering "linear polarization" of the pendulum, however, we can get away with it.

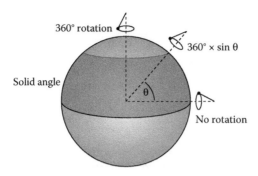

FIGURE 7.26 Pendulums at different latitudes on the Earth and the geometric phases accumulated by them.

direction is entirely due to gravity and the overall change in position on the sphere. It is to be noted that the "phase" in this case—the direction of the pendulum—is a distinct parameter independent of the angles that define position on the sphere.

Another example of a geometric phase, and a somewhat surprising one, is given by the ability of a cat to (almost) always land on its feet when dropped, even when released upside down and at rest.[*] The strangeness of this ability was really driven home when Étienne-Jules Marey released the first high-speed photographs of a cat performing it in 1894 [Mar94]. These images are shown in Figure 7.27.

At the time, Marey's photos, and conclusions, caused great consternation. An article in the *New York Herald* about Marey's presentation at the Paris Academy gives a sense of the response he received:

> When M. Marey laid the results of his investigations before the Academy of Sciences, a lively discussion resulted. The difficulty was to explain how the cat could turn itself round without a fulcrum to assist it in the operation. One member declared that M. Marey had presented them with a scientific paradox in direct contradiction with the most elementary mechanical principles.

The seeming paradox was, in fact, a symptom of oversimplified physical thinking. In elementary mechanics courses, students are taught that there is a linear relationship between the angular momentum of a rigid rotating body and its angular velocity. An object dropped at rest, so the

---

[*] The connection between falling cats and the geometric phase was first brought to my attention by a paper by Batterman [Bat03].

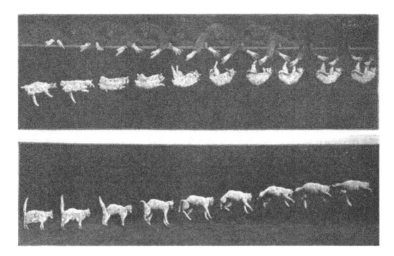

FIGURE 7.27 Étienne-Jules Marey's photographs of a falling cat, 1894. The photographs should be read from right to left and top to bottom.

reasoning goes, cannot spontaneously produce a net angular momentum and therefore cannot turn over.

The problem is in the assumption that a cat behaves as a rigid body, which most cat owners will immediately know to be untrue. We can make a simplified model of a cat as being two joined cylinders, and its body position is defined by two angles: θ, which represents the bend in the body, and φ, which represents the relative rotation of the segments about their axes. It is to be noted that φ plays no role when $\theta = 0$. The overall orientation of the cat with respect to the vertical is indicated by another angle, ϕ.

This model is shown in Figure 7.28a. If the cat's body is straight, that is, $\theta = 0$, it has no ability to turn itself over. It could, in principle, counter-rotate its upper and lower body sections by $\varphi = \pm\pi$ to make its front paws face the ground, but any untwisting will reverse this process—and it can be seen in Figure 7.27 that a cat lands with its body untwisted.

If the cat completely folds its body, however ($\theta = \pi/2$), it can rotate its upper and lower body sections in opposite directions. The angular momenta of the two sections are equal and opposite and, once the cat's paws are facing the ground, it can unbend to $\theta = 0$. The cat has turned over without any net angular momentum.

The beginning of this motion can be seen in the sixth photograph from the right on the top row. In reality, this motion is only part of the cat's maneuvering; a detailed model was introduced by Kane and Scher [KS69]. For our purposes, we note that the angles of the cat's body orientation,

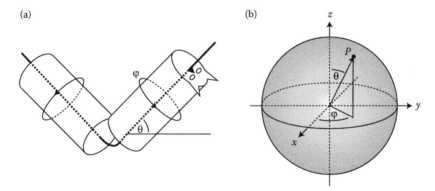

FIGURE 7.28 (a) Two cylinder models of a falling cat and (b) the spherical representation of it.

θ and φ, are analogous to the angles used in spherical coordinates, as in Figure 7.28b. It was demonstrated by Gbur [Gbu] that in at least one special case, the overall orientation angle of the cat after changes in θ and φ can be related to the solid angle of the path on the sphere. Again, we have an overall phase—the orientation of the cat with respect to the vertical—modified by a change in the internal variables of the system, θ and φ.

This reasoning gives some intuition as how a geometric phase arises on the Poincaré sphere, as well. A cycle through the polarization state variables, in this case the ellipticity and orientation angle, produces a change in the overall phase of the field.

Experimental measurement of the Pancharatnam phase is challenging because the geometric phase produces the same observable effect as a dynamic phase, and can be masked or mimicked by it. The simplest experiment imaginable to test the geometric phase would be to divide a beam into two halves, one of which passes through free space and the other of which passes through an optical system that produces a polarization cycle. In the end, the two beams are combined again and their relative phase difference is measured through interferometry. The easiest way to change locations on the Poincaré sphere, however, is to use a waveplate to modify the relative phases of the x and y components of the field. This change of phase and polarization state presumably includes both dynamic and geometric phase variations, but it is not clear how to distinguish between the two. A more robust experiment should change the polarization state in both arms of the interferometer in complementary ways, so that the dynamic phase is the same for both and the only difference is the geometric phase.

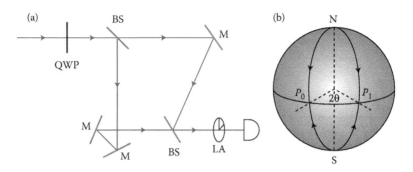

FIGURE 7.29 (a) Interferometer for measuring Pancharatnam phase and (b) the equivalent path on the Poincaré sphere.

One such experiment was conducted by Hariharan, Mujumdar, and Ramachandran [HMR99]. A schematic of their apparatus is shown in Figure 7.29a. After passing through a quarter-wave plate, a beam of circularly polarized light is divided and subjected to three reflections in one arm of an interferometer, and two reflections in the other arm. On reflection, the handedness of the polarization flips, so that the overall field that comes out of the second beamsplitter consists of an equal mix of left- and right-circularly polarized light.

The combined beam then passes through a linear analyzer. This projects the circular states to the same linear polarization state at P_0. When the analyzer is rotated, however, the beams follow a different path on the Poincaré sphere to the position P_1, as illustrated in Figure 7.29b. The combined paths from the North and South poles to the pair of linear states on the equator form a region of finite solid angle, and this solid angle must coincide with a change in the geometric phase of the fields. A rotation of the linear analyzer must produce a shift in the interference pattern detected. When the analyzer is rotated by $\pi/2$, the solid angle is π and the maxima in the interference pattern will become minima, and vice versa; this is exactly what was observed. It is worth noting that the geometric phase for stacks of wave plates has also been calculated and experimentally tested by Berry and Klein [BK96]. In this experiment, they compared the interference fringes produced by a twisted stack of wave plates to a uniform stack; excellent agreement with theory was found.

7.12 SECOND-ORDER TENSOR FIELDS AND THEIR SINGULARITIES

In this section, we would like to elaborate on the properties of second-order tensor fields, specifically the nature of the singularities, and rigorously prove some important results related to them. Though not the earliest work on the subject, the PhD thesis of Delmarcelle [Del94] is an extremely useful reference, and it can be freely read on the Internet.

We consider the structure of a real symmetric second-order, two-dimensional tensor field $\mathbf{Q(r)}$ that can be written in matrix form as

$$\mathbf{Q(r)} = \begin{bmatrix} Q_{xx} & Q_{xy} \\ Q_{xy} & Q_{yy} \end{bmatrix}, \tag{7.182}$$

where all the Q_{ij} are real-valued and functions of position $\mathbf{r}$. "Second-order," in this case, refers to the number of indices used to label the components of the tensor (two), and "two-dimensional" refers to the space in which this tensor is defined, namely, the xy-plane.

Because the tensor is symmetric and real-valued, it generally possesses two distinct eigenvalues λ_x and λ_y and two orthogonal eigenvectors $|\mathbf{v}_x\rangle$ and $|\mathbf{v}_y\rangle$. These eigenvectors can be normalized, but are only defined to within a sign. In terms of polarization ellipses, the smaller eigenvalue represents the inverse square of the major semi-axis of the ellipse, while the larger represents the inverse square of the minor semi-axis. We may draw streamlines following the trajectories of the major and minor axes, and the streamlines will form an orthogonal mesh almost everywhere in the plane, with the exception of singular points.

Singularities may be formally defined as those regions in the plane where the eigenvalues are equal, and $\mathbf{Q}$ is therefore proportional to the identity matrix. From our general expression (7.182) for the matrix, we find that this requires two conditions, namely,

$$Q_{xx}(\mathbf{r}) - Q_{yy}(\mathbf{r}) = 0, \tag{7.183}$$

$$Q_{xy}(\mathbf{r}) = 0. \tag{7.184}$$

We have two equations and two degrees of freedom, x and y, and therefore generally expect our singularities to be isolated points in the plane.

In the context of polarization, we have seen from Equation 7.20 that these two conditions represent a point where the orientation angle of the major axis is undefined; this equation, again, is

$$\tan(2\psi) = \frac{2Q_{xy}}{Q_{xx} - Q_{yy}}, \tag{7.185}$$

which is undefined at a tensor singularity.

The main quantity used for classifying such singularities is the topological index n, which we define as

$$n \equiv \frac{1}{2\pi} \oint_L \nabla\psi(\mathbf{r}) \cdot d\mathbf{r}, \tag{7.186}$$

where L is a simple (i.e., nonself-intersecting) closed path taken counterclockwise in the plane. Again, ψ is the orientation angle, which may be written as

$$\psi = \frac{1}{2} \arctan\left[\frac{2Q_{xy}}{Q_{xx} - Q_{yy}}\right]. \tag{7.187}$$

We now prove several fundamental theorems relating to the topological index, which will seem quite familiar as we progress. First, we consider the topological index over a path L that contains no singularities. We have noted before that the electromagnetic field, and therefore the matrix elements Q_{ij}, are analytic with respect to all spatial variables, and therefore have continuous partial derivatives with respect to those variables everywhere. The function $\psi(\mathbf{r})$ will consequently be analytic and differentiable everywhere except at singular points where Equations 7.183 and 7.184 are satisfied. When integrating $\nabla\psi$ over a region containing no singularities, then, Stokes' theorem will apply, namely,

$$\oint_L \mathbf{A} \cdot d\mathbf{r} = \int_S \nabla \times \mathbf{A} \cdot d\mathbf{a}, \tag{7.188}$$

where S is the surface bounded by the path L and $d\mathbf{a}$ is the vector infinitesimal area element. For the specific choice of $\mathbf{A} = \nabla\psi$, we have

$$\oint_L \nabla\psi \cdot d\mathbf{r} = \int_S \nabla \times \nabla\psi \cdot d\mathbf{a} = 0, \tag{7.189}$$

since the curl of a gradient is zero.

Those readers familiar with complex analysis will immediately recognize Equation 7.189 as the tensor field equivalent of Cauchy's integral theorem, as described for instance in Gbur [Gbu11, Section 9.5]. The results that follow also have direct analogies in complex analytic fields.

Building on Equation 7.189, it is straightforward to show that the index of two nonintersecting paths that enclose the same set of singularities is

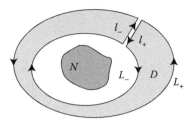

FIGURE 7.30 A closed path that encircles a region with singularities, but does not enclose them. The light-gray region D is that enclosed by the path, while the dark-gray region N is not enclosed.

the same. We use a closed contour as shown in Figure 7.30, in which the path has four segments, L_+, L_-, l_+, and l_-, and the path encircles a region with singularities N but does not include them in its domain D. From Equation 7.189, we have

$$\int_{L_+} \nabla\psi \cdot d\mathbf{r} + \int_{L_-} \nabla\psi \cdot d\mathbf{r} + \int_{l_+} \nabla\psi \cdot d\mathbf{r} + \int_{l_-} \nabla\psi \cdot d\mathbf{r} = 0. \qquad (7.190)$$

Because the field is analytic everywhere outside of N, we may take the limit as the paths l_+ and l_- come together and the integrations over them cancel out, leaving the path integral over two closed regions L_+ and L_-. As we want to take all paths to be counterclockwise by convention, we introduce a new path $L'_- = -L_-$, meaning that we may write

$$\oint_{L_+} \nabla\psi \cdot d\mathbf{r} = \oint_{L'_-} \nabla\psi \cdot d\mathbf{r}. \qquad (7.191)$$

We therefore find that the index with respect to two nonintersecting paths enclosing the same singular region is the same or, in other words, that the index of a collection of singularities is path independent.

The specific value of the index around any group of singularities is necessarily some multiple of a half-integer. This is self-evident due to the requirement that the continuity of ψ holds everywhere except on a singularity and that the polarization ellipse is symmetric with respect to 180-degree rotations in the plane.

We now would like to focus on the simplest possible types of tensor singularities; such singularities will, presumably, be generic. It is reasonable to expect that the index of these simple singularities will be $n = \pm 1/2$, but would like to rigorously prove this.

We have already noted that, at a singularity, we expect the tensor $\mathbf{Q}$ to be proportional to the identity matrix. If we consider the behavior of the tensor in the immediate vicinity of a singularity, the simplest possible case is for the coefficients of the tensor to vary linearly with position

$$\mathbf{Q} = \begin{bmatrix} \lambda + \alpha x + \beta y & \alpha_1 x + \beta_1 y \\ \alpha_1 x + \beta_1 y & \lambda + \gamma x + \delta y \end{bmatrix}, \tag{7.192}$$

where all coefficients are real-valued and independent of x and y. Because the major ellipse orientation depends entirely upon $Q_{xx} - Q_{yy}$ and Q_{xy}, it will be more convenient to define the linear approximation to the tensor in the form of the equations

$$\frac{Q_{xx} - Q_{yy}}{2} = ax + by, \tag{7.193}$$

$$Q_{xy} = cx + dy, \tag{7.194}$$

where a, b, c, and d are real constants.

A key to the derivation of the next result is the observation that the determinant-like quantity

$$D_I \equiv ad - bc, \tag{7.195}$$

is independent of the choice of the coordinate system. This observation, though plausible, is rather tedious to derive explicitly, as it effectively involves the transformation of a third-order tensor. Naturally, we leave the proof as an exercise, though details may also be found in Chapter 4 of Delmarcelle [Del94].

The simplest singularities will be points where $D_I \neq 0$. Under this assumption, we can show that the index n of a simple singularity is of the form

$$n = \frac{1}{2}\text{sign}(D_I). \tag{7.196}$$

We prove this by explicitly evaluating Equation 7.186, using the definition of ψ from Equation 7.187, and substituting the linear approximations from

Equations 7.193 and 7.194. The integral takes the form

$$n \equiv \frac{1}{4\pi} \oint_L \frac{(ax+by)d(cx+dy) - (cx+dy)d(ax+by)}{(ax+by)^2 + (cx+dy)^2}, \tag{7.197}$$

where $d(ax+by)$ is the differential element of $(ax+by)$, and so forth. We have used the standard formula for the derivative of an arctangent here, namely,

$$\frac{d\arctan(x)}{dx} = \frac{1}{1+x^2}. \tag{7.198}$$

Because, from Equation 7.191, we know that the integral around a singularity is independent of path, we may take as our path the ellipse

$$(ax+by)^2 + (cx+dy)^2 = 1. \tag{7.199}$$

The denominator of Equation 7.197 is therefore unity, and we have

$$n \equiv \frac{1}{4\pi} \oint_L [(ax+by)d(cx+dy) - (cx+dy)d(ax+by)]. \tag{7.200}$$

We now choose new variables of integration u and v, defined as

$$u \equiv ax + by, \tag{7.201}$$

$$v \equiv cx + dy, \tag{7.202}$$

and in terms of u and v our path of integration is a circle C of unit radius. The Jacobian of the coordinate transformation is simply D_I; the sign of D_I indicates whether our new path of integration is clockwise or counterclockwise, which implies that we may write

$$n = \frac{\text{sign}(D_I)}{4\pi} \oint_C [u\,dv - v\,du], \tag{7.203}$$

where C by definition is a counterclockwise path. Finally, we make the usual transformation for a circular path, $u = \cos\phi$ and $v = \sin\phi$, which leads to the simple result

$$n = \frac{\text{sign}(D_I)}{2}. \tag{7.204}$$

As promised, a simple singularity—which in this case is one with a nonzero D_I—must have a topological index of $\pm 1/2$. With this proven, we then note that the result of Section 7.4 follows immediately, namely, that there are no elliptic regions and either three hyperbolic regions, for $n = -1/2$, or one hyperbolic region, for $n = +1/2$. In the latter case, however, there may or may not be a parabolic region, which determines whether or not we have a monstar or a lemon.

The remaining question is to determine the conditions under which one or three separatrices appear at a singularity. These can be found by considering the form of the matrix $\mathbf{Q}$ as a function of azimuthal angle ϕ from the x-axis. Borrowing from Equation 7.19 for a rotated coordinate system, as the mathematics is the same, we find that the elements defined with respect to axes oriented at an angle ϕ may be written in the form

$$Q_{rr} = \frac{1}{2}(Q_{xx} + Q_{yy}) + \frac{1}{2}(Q_{xx} - Q_{yy})\cos(2\phi) + Q_{xy}\sin(2\phi), \quad (7.205)$$

$$Q_{r\phi} = Q_{\phi r} = -\frac{1}{2}(Q_{xx} - Q_{yy})\sin(2\phi) + Q_{xy}\cos(2\phi), \quad (7.206)$$

$$Q_{\phi\phi} = \frac{1}{2}(Q_{xx} + Q_{yy}) - \frac{1}{2}(Q_{xx} - Q_{yy})\cos(2\phi) - Q_{xy}\sin(2\phi). \quad (7.207)$$

The major axis of the polarization ellipse will align with a line from the origin when $Q_{r\phi} = 0$, or

$$\tan(2\phi) = \frac{2Q_{xy}}{Q_{xx} - Q_{yy}}. \quad (7.208)$$

Finally, using Equations 7.193 and 7.194 for the components of $\mathbf{Q}$ and using $x = \cos\phi, y = \sin\phi$, we find that separatrices will appear at angles ϕ such that

$$\tan(2\phi) = \frac{2\tan(\phi)}{1 - \tan^2(\phi)} = \frac{c + d\tan(\phi)}{a + b\tan(\phi)}. \quad (7.209)$$

This ends up being a cubic equation in $\tan(\phi)$, with at most three possible roots and therefore a maximum of three separatrices; letting $\alpha = \tan(\phi)$, this cubic may be written as

$$d\alpha^3 + (2b + c)\alpha^2 + (2a - d)\alpha - c = 0. \quad (7.210)$$

It is known that a cubic equation with real parameters always has at least one real root. It will typically have either one real root or three real roots, though

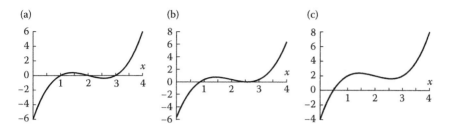

FIGURE 7.31   Illustration of the roots of a cubic function, for (a) $f(x) = (x − 1)$ $(x − 2)(x − 3)$, (b) $f(x) = (x − 1)(x − 2)(x − 3) + 0.39$, and (c) $f(x) = (x − 1)$ $(x − 2)(x − 3) + 2$.

it is possible, though rare (nongeneric), to have repeated roots. This is illustrated in Figure 7.31 for a few simple cubic functions. Typically there are either one or three intersections with the x-axis, though two intersections may appear when one of the inversion points just touches the axis.

The number of roots is well known to be determined by the sign of the discriminant of the cubic equation. For a cubic of the form

$$Ax^3 + Bx^2 + Cx + D = 0,\qquad(7.211)$$

the discriminant may be written as

$$D_L = 18ABCD − 4B^3D + B^2C^2 − 4AC^3 − 27A^2D^2.\qquad(7.212)$$

In terms of our singularity parameters, this becomes

$$D_L = −18cd(2b + c)(2a − d) + 4c(2b + c)^3 + (2b + c)^2(2a − d)^2$$
$$− 4d(2a − d)^3 − 27c^2d^2.\qquad(7.213)$$

It should be noted that there is one more property associated with a singularity of a tensor field, characterized by a quantity D_C. Referred to as the *contour property*, it describes the shape of the contour lines of constant eigenvalues of the tensor. It depends on the sign of D_C, given by

$$D_C = 4(\alpha\gamma − \alpha_1^2)(\beta\delta − \beta_1^2) − (\alpha\delta + \beta\gamma − 2\alpha_1\beta_1)^2.\qquad(7.214)$$

For positive D_C, the contours are elliptic and for negative D_C, the contours are hyperbolic. This property does not seem to play a significant role in the topological behavior of tensor field singularities, and we therefore do not discuss it further.

We have therefore seen the topological origins of the major features of a polarization singularity, namely, the number of separatrices and the topological index. The discussion is still rather abstract, and does not readily connect the polarization properties of the field with the topological properties of the singularity. Such connections are not easy to come by, but several papers by Dennis [Den02,Den08] have addressed the relationship between the Stokes parameters and the polarization topology.

One important lesson to take from this section is that the nature of polarization singularities is not due to the structure of Maxwell's equations but rather from the fact that the polarization ellipses must be described in the context of two-dimensional, second-order tensor fields. It is also important to note that the structure of polarization singularities may also be described using *catastrophe theory*, and in fact this is where the original research on such singularities began, for example, in the work of Berry and Hannay [BH77] and Thorndike, Cooley, and Nye [TCN78]. We have used the tensor field description instead because it involves the least amount of additional mathematics.

7.13 THE HAIRY BALL THEOREM

In Section 7.10, we referenced the "hairy ball theorem" in discussing the polarization properties of sunlight. This important theorem of topology, first proven by Luitzen Egbertus Jan Brouwer in 1912, is popularly stated as either "it is impossible to comb a hairy ball flat without leaving a cowlick" or "you can't comb a coconut." More rigorously, it may be stated that a continuous vector or tensor field tangent to the surface of a sphere must possess a topological index of $n = +2$.

There exist numerous proofs of this theorem, some quite technical; see, for instance Milnor [Mil78] and Eisenberg and Guy [EG79]. We consider here a fun and elegant proof due to Jarvis and Tanton [JT04] that is almost entirely geometrical but rigorous and readily understandable without advanced mathematics.

The proof is built upon a result involving triangulated polygons known as *Sperner's lemma*. This lemma is given below.

Lemma 7.1   Sperner's lemma. With the boundary vertices of an arbitrary triangulated polygon labeled A, B, or C, with all the $A - B$ edges oriented in the same sense (let us say clockwise), any interior labeling of the polygon

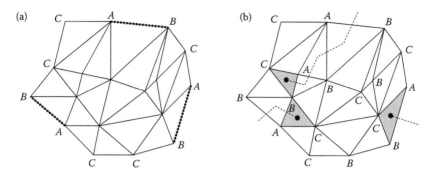

FIGURE 7.32 The labeling of a triangulated polygon, with (a) boundary vertices labeled with $A - B$ in a clockwise orientation and (b) with an arbitrary interior labeling.

will produce at least d subtriangles $A - B - C$, where d is the number of exterior $A - B$ edges.

An example of such a polygon, and its boundary labeling, is shown in Figure 7.32a. It can be seen that there are three $A - B$ sections on the boundary of this particular polygon; in Figure 7.32b, it can be seen that there are exactly three triangles with $A - B - C$ labeling.

The proof of this lemma is also hinted at in Figure 7.32b. Let us imagine our polygon as the walls and doorways of a very large palace, and let $A - B$ sides be doorways, and all other sides be walls. A path starting from an exterior doorway will continue through any $A - B$ side of a triangle, and any triangle has at most two, until it reaches an $A - B - C$ triangle and a dead end. There may be additional $A - B - C$ triangles that are not reached through an exterior doorway, but there must be *at least* as many as there are $A - B$ sides.

This can be made slightly more precise by considering all possible triangles that one enters through a doorway, as illustrated in Figure 7.33. Starting from a clockwise-oriented $A - B$ door, there are three possibilities: (a) one exits through $A - B$ door α, (b) one exits through $A - B$ door β, or (c) one ends in an $A - B - C$ triangle. But it can be seen that exiting through door α or β into another triangle will enter the next triangle with the same clockwise $A - B$ orientation, and the same three possibilities will arise in this triangle. The doors α and β cannot be exterior walls, however, because they would have a *counterclockwise* $A - B$ orientation, which

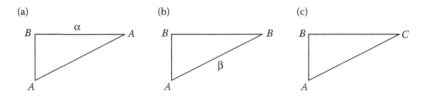

FIGURE 7.33 All possible outcomes on entering a doorway in a triangulated polygon.

we explicitly excluded as a possibility. Therefore, every entrance path must inevitably end in an $A - B - C$ triangle, and the lemma is proven.

This leads us readily to a more general lemma.

Lemma 7.2   With the boundary vertices of an arbitrary triangulated polygon labeled A, B, or C, any interior labeling of the polygon will produce at least d subtriangles $A - B - C$, where $d = |n_{AB} - n_{BA}|$ is the absolute difference in the number of $A - B$ and $B - A$ boundary edges.

This lemma follows readily from the previous one. Now, it is possible for a path through an $A - B$ door, however oriented, to exit through an $A - B$ door of opposite sense. In the worst-case scenario, every $A - B$ path leads to an exit, and these do not count toward $A - B - C$ triangles. The excess of $A - B$ sides of a particular sense, however, must end in $A - B - C$ triangles as before, and the theorem is proven.

So far, it seems that we have come nowhere near the hairy ball theorem. But let us now consider imagine a tangent vector field **v** on a spherical surface, and assume that this tangent vector field is continuous within a sufficiently small open disk D near the North Pole of the sphere. (It does not have to be the North Pole, of course, but this is convenient for the explanation.) The continuity of the vector field implies that we can make the disk sufficiently small that all of the lines of **v** are parallel within D, or as close to parallel as we like.

We now add a dipole field **d** to this sphere, where the singular point of the dipole lies within the disk D as well. Why do we choose a dipole? Because the field **d** is continuous and has a well-defined direction everywhere *outside* the disk D, we have therefore assumed that **v** is continuous inside D and work with a dipole field **d** that is continuous outside D. This dipole field is shown in Figure 7.34.

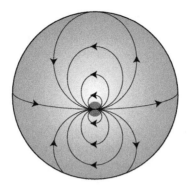

FIGURE 7.34 A dipole field on the surface of a sphere. The disk D is shown as a shaded circle.

At any point p on the exterior of the disk D, we can define the angle $\theta(p)$ between the vector field **v** and the dipole field **d**. We label each of these points A, B, or C as follows; if

$$0 \leq \theta(p) < 2\pi/3, \text{ label } A, \qquad (7.215)$$

$$2\pi/3 \leq \theta(p) < 4\pi/3, \text{ label } B, \qquad (7.216)$$

$$4\pi/3 \leq \theta(p) < 2\pi, \text{ label } C. \qquad (7.217)$$

If we consider the labeling of the points on the boundary of D, we get something similar to what is shown in Figure 7.35. What we see is that there are *two* continuous regions containing points of type A, B, and C. Let us now imagine taking the exterior of the disk D, which as a sphere with a hole may

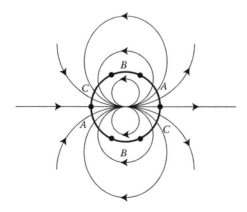

FIGURE 7.35 The points on the boundary of the disk D, labeled A, B, or C corresponding to the angle between **v** and **d**.

also be considered a disk that we will call D', and triangulating the area of this exterior disk. When a sufficient number of points are used to triangulate, we will find that there are exactly two sides of the polygon that are of counterclockwise $A - B$ type. This in turn means that the disk D' must possess at least two $A - B - C$ triangles.

In the limit that we make the number of triangles in D' infinite, these two $A - B - C$ triangles become infinitely small and the points A, B, and C of the triangle must come together to be the same point. However, each of these points represents a distinct angle of orientation of the vector field, which in turn implies that the limit points of the $A - B - C$ triangles are singularities of the vector field. We have therefore found that there are necessarily at least two singularities of a vector field on a sphere.

What is the nature of these singularities? A little geometric reasoning indicates that $A - B - C$ or $C - B - A$ triangles must have topological index $n = +1$; this is left as an exercise. We therefore find that a vector field on a sphere must have a total topological index of at least $n = +2$.

If we consider a tensor field instead of a vector field, we may apply a similar construction. The orientation of the major axis of a tensor field is only defined over a range from $-\pi/2 \leq \theta(p) < \pi/2$; so, we label each point on the sphere A, B, or C if

$$-\pi/2 \leq \theta(p) < -\pi/6, \text{ label } A, \tag{7.218}$$

$$-\pi/6 \leq \theta(p) < \pi/6, \text{ label } B, \tag{7.219}$$

$$\pi/6 \leq \theta(p) < \pi/2, \text{ label } C. \tag{7.220}$$

Then, the dipole construction given earlier indicates that there will in general be at least *four* singular points on the sphere, and each of these will have a topological index $n = +1/2$.

Our proof is not perfect as it leaves open the possibility of a topological index greater than $n = +2$ on the sphere, where it can be shown that it must be exactly 2. Nevertheless, it illustrates quite elegantly the necessity of singularities of a tangent vector field on a sphere.

7.14 EXERCISES

1. In this chapter, we have seen how the Stokes parameters could be calculated from a set of six measurements. Show that one can also determine the Stokes parameters from the following *four* measurements:

1. $\langle I(0,0)\rangle$,

2. $\langle I(\pi/4, 0)\rangle$,

3. $\langle I(-\pi/4, 0)\rangle$,

4. $\langle I(\pi/4, \pi/2)\rangle$.

Can you see a way to get the Stokes parameters from only three measurements?

2. Introducing a linearly polarized plane wave normally incident on a perfect mirror and the reflected plane wave, derive the interference pattern and therefore Equations 7.93 and 7.94 of Wiener's experiment. Assuming a wavelength of $\lambda = 600$ nm, estimate the angle δ needed to separate the peaks of the interference pattern on the plate by 0.5 mm. Also estimate how thin the photographic layer must be in order to be able to resolve the interference patterns of the plane waves. (This latter condition was a real challenge in Wiener's time.)

3. Derive the expressions for the major and minor semi-axes of the polarization ellipse, Equations 7.29 and 7.30, from the expressions for their inverses, Equations 7.27 and 7.28. To do this, it will be helpful to apply the trigonometric formulas

$$\cos^2 \psi = \frac{1}{2}[1 + \cos(2\psi)],$$

$$\sin^2 \psi = \frac{1}{2}[1 - \cos(2\psi)],$$

and write the inverse expressions in terms of the Stokes parameters. Also, the use of the orientation angle equation, Equation 7.31, will be crucial.

4. Prove that the quantity $\delta \equiv ad - bc$ is invariant under rotations of the coordinate system by an angle θ.

5. Prove Equation 7.164, which relates the Stokes parameters to the determinant of the polarization matrix **J**.

6. Using the following form of the electric field:

$$\mathbf{E} = \mathbf{e}_1 \cos(\theta/2)e^{-i\phi/2} + \mathbf{e}_2 \sin(\theta/2)e^{i\phi/2},$$

reproduce the result of Pancharatnam's theorem from Section 7.11 and show that it is consistent. What has changed in the evolution of the geometric phase from the previous derivation?

7. Argue that the $A - B - C$ and $C - B - A$ triangles of Section 7.13 must have topological index of $n = +1$, and describe what types of singularities (left-hand center, source, sink, etc.) each type of triangle encompasses.

8. Let us work in a circular polarization basis instead of a Cartesian basis, for which

$$E_+ = \frac{1}{\sqrt{2}}(E_x - iE_y), \quad E_- = \frac{1}{\sqrt{2}}(E_x + iE_y).$$

Show that the Stokes parameters can be written in this basis in the form

$$S_0 = |E_+|^2 + |E_-|^2,$$

$$S_1 = 2\text{Re}\{E_+^* E_-\},$$

$$S_2 = -2\text{Im}\{E_+^* E_-\},$$

$$S_3 = |E_-|^2 - |E_+|^2.$$

CHAPTER 8

Singularities of the Poynting Vector

I N CHAPTER 5, WE DISCUSSED in some detail the relationship between phase singularities and OAM. Though this relationship is not in general a simple one, we gained much physical insight by investigating it. In this chapter, we consider what may be learned from the power flow of electromagnetic fields, as characterized by the Poynting vector, and singularities associated with it.

In fact, the earliest known electromagnetic vortex was discovered theoretically by Wolter [Wol50] in an investigation of the Goos–Hänchen shift on reflection. The vortex of the power flow in this case provided a new understanding of the physics of the Goos–Hänchen shift, as we will see.

This idea—that vortices of power flow can provide insight into physical problems—will inspire much of our discussion in this chapter. We will see that such vortices, though difficult to observe directly, are often connected with the macroscopic properties of electromagnetic wave propagation. We begin by introducing the Poynting vector, and consider some conceptual difficulties associated with it. We then look at Wolter's original vortex discovery, along with some other presingular optics Poynting singularities, and then talk about the role of Poynting singularities in nanoscale systems.

8.1 POWER FLOW IN ELECTROMAGNETIC WAVES

A discussion of energy conservation in electromagnetic waves begins in the same place that the discussion of momentum conservation did, with the Lorentz force law, Equation 5.1,

$$\mathbf{F}(\mathbf{r}, t) = q\left[\mathbf{E}(\mathbf{r}, t) + \mathbf{v} \times \mathbf{B}(\mathbf{r}, t)\right]. \tag{8.1}$$

The magnetic force, which is always perpendicular to the path of motion of a charged particle, plays no role in energy transfer between fields and charges. The infinitesimal work dW done by an electric field on a charge q as it moves a distance $d\mathbf{x}$ is $dW = q\mathbf{E} \cdot d\mathbf{x}$; if we divide by an infinitesimal time dt and use $\mathbf{v} = d\mathbf{x}/dt$, we find that the mechanical power dE_{mech}/dt imparted to the charge is given by

$$\frac{dE_{mech}}{dt} = q\mathbf{v} \cdot \mathbf{E}. \tag{8.2}$$

This expression can be converted into a macroscopic form by using $q\mathbf{v} = \mathbf{J}d^3r$, where d^3r is an infinitesimal volume element and $\mathbf{J}$ is again the current density. The rate of energy transfer to a system of currents confined to a volume V can then be written as

$$\frac{dE_{mech}}{dt} = \int_V \mathbf{J}(\mathbf{r}, t) \cdot \mathbf{E}(\mathbf{r}, t)d^3r. \tag{8.3}$$

Next, we eliminate the explicit mention of the current by using the Ampere–Maxwell law

$$\nabla \times \mathbf{H}(\mathbf{r}, t) = \mathbf{J}(\mathbf{r}, t) + \frac{\partial \mathbf{D}(\mathbf{r}, t)}{\partial t}. \tag{8.4}$$

Solving this for $\mathbf{J}$, we substitute from this into Equation 8.3, with the result

$$\frac{dE_{mech}}{dt} = \int_V \left[\mathbf{E}(\mathbf{r}, t) \cdot [\nabla \times \mathbf{H}(\mathbf{r}, t)] - \mathbf{E}(\mathbf{r}, t) \cdot \frac{\partial \mathbf{D}(\mathbf{r}, t)}{\partial t}\right] d^3r. \tag{8.5}$$

We now have an expression that is entirely in terms of fields. It can be written in an even more suggestive form by using the vector calculus identity

$$\nabla \cdot (\mathbf{E} \times \mathbf{H}) = \mathbf{H} \cdot (\nabla \times \mathbf{E}) - \mathbf{E} \cdot (\nabla \times \mathbf{H}). \tag{8.6}$$

With this, and an application of Faraday's law,

$$\nabla \times \mathbf{E}(\mathbf{r}, t) = -\frac{\partial \mathbf{B}(\mathbf{r}, t)}{\partial t}, \tag{8.7}$$

we may write the equation for the change in mechanical energy as

$$\frac{dE_{mech}}{dt} = -\int_V \nabla \cdot [\mathbf{E}(\mathbf{r}, t) \times \mathbf{H}(\mathbf{r}, t)] d^3 r$$

$$-\int_V \left[\mathbf{H}(\mathbf{r}, t) \cdot \frac{\partial \mathbf{B}(\mathbf{r}, t)}{\partial t} + \mathbf{E}(\mathbf{r}, t) \cdot \frac{\partial \mathbf{D}(\mathbf{r}, t)}{\partial t} \right] d^3 r. \tag{8.8}$$

Up to this point, our calculation has been exact. To proceed further, we must make some seemingly reasonable physical interpretations of terms in Equation 8.8; it should be emphasized, however, that the correctness of these interpretations is questionable. We will consider them further in the next section.

The first integral in Equation 8.8 can be written, with the help of Gauss' theorem, as a surface integral, namely,

$$\int_V \nabla \cdot [\mathbf{E}(\mathbf{r}, t) \times \mathbf{H}(\mathbf{r}, t)] d^3 r = \oint_S [\mathbf{E}(\mathbf{r}, t) \times \mathbf{H}(\mathbf{r}, t)] \cdot d\mathbf{a}, \tag{8.9}$$

where S is the surface bounding the volume V and $d\mathbf{a}$ is the infinitesimal vector area element normal to the surface. The vector that appears in the surface integral is the *Poynting vector* **S**, defined as

$$\mathbf{S}(\mathbf{r}, t) \equiv \mathbf{E}(\mathbf{r}, t) \times \mathbf{H}(\mathbf{r}, t). \tag{8.10}$$

To evaluate the second integral in Equation 8.8, we make the assumption that the constitutive relations between the microscopic and macroscopic fields take on the simple form

$$\mathbf{D}(\mathbf{r}, t) = \epsilon \mathbf{E}(\mathbf{r}, t), \tag{8.11}$$

$$\mathbf{B}(\mathbf{r}, t) = \mu \mathbf{H}(\mathbf{r}, t), \tag{8.12}$$

that is, that there is a linear relationship and that the material parameters ϵ and μ are time independent. This is a very strong assumption and one that will not hold in general, but it is sufficient for our present purposes. We may then write the second integral in Equation 8.8 as

$$
\int_V \left[\mathbf{H}(\mathbf{r}, t) \cdot \frac{\partial \mathbf{B}(\mathbf{r}, t)}{\partial t} + \mathbf{E}(\mathbf{r}, t) \cdot \frac{\partial \mathbf{D}(\mathbf{r}, t)}{\partial t} \right] d^3 r
$$

$$
= \frac{1}{2} \frac{\partial}{\partial t} \int_V \left[\epsilon |\mathbf{E}(\mathbf{r}, t)|^2 + \mu |\mathbf{H}(\mathbf{r}, t)|^2 \right] d^3 r. \tag{8.13}
$$

In this form, the integral has units of energy, and suggests that we may associate an *electromagnetic energy density* $U(\mathbf{r}, t)$ with the expression

$$
U(\mathbf{r}, t) \equiv \frac{1}{2} \left[\mathbf{H}(\mathbf{r}, t) \cdot \mathbf{B}(\mathbf{r}, t) + \mathbf{E}(\mathbf{r}, t) \cdot \mathbf{D}(\mathbf{r}, t) \right]. \tag{8.14}
$$

The total rate of change of electromagnetic energy in our volume may then be written as

$$
\frac{dE_{field}}{dt} = \frac{\partial}{\partial t} \int_V U(\mathbf{r}, t) d^3 r. \tag{8.15}
$$

Expressing the total energy in our volume as $E_{tot} = E_{mech} + E_{field}$, we may finally write Equation 8.8 in the suggestive form

$$
\frac{dE_{tot}}{dt} = - \oint_S \mathbf{S}(\mathbf{r}, t) \cdot d\mathbf{a}. \tag{8.16}
$$

Equation 8.16 is evidently a law of energy conservation in electromagnetic systems. If we interpret the Poynting vector as a flux of energy per unit area, our law suggests that the only way the total energy can change in our volume is if electromagnetic energy flows across the surface. A net outward flow ($\mathbf{S}$ primarily parallel to $d\mathbf{a}$) results in a decrease in total energy within the system.

For future reference, it is worth noting that this conservation law can also be written in a differential form. If we consider a region of space free of charge, then $E_{mech} = 0$ and $E_{tot} = E_{field}$. We may use the divergence theorem to write the right-hand side of Equation 8.16 in terms of a volume integral; then, equating integrands, we find that

$$
\frac{\partial U}{\partial t} = -\nabla \cdot \mathbf{S}. \tag{8.17}
$$

We have evaluated the law of energy conservation in the time domain because of the clear interpretation of all terms involved; however, this has resulted in us having to make strong and generally unrealistic assumptions about the properties of macroscopic materials. If we work in the frequency domain, we can find an expression of broader applicability. Also, it should be clear at this point that our preference is to work in the frequency domain for most optical problems.

We assume a time dependence $\exp[-i\omega t]$, and consider the cycle-averaged power flow into a system of charges, which, following Equation 2.19, can be written as

$$\left\langle \frac{dE_{mech}}{dt} \right\rangle = \frac{1}{2} \int_V \mathbf{J}^*(\mathbf{r}, \omega) \cdot \mathbf{E}(\mathbf{r}, \omega) d^3 r, \tag{8.18}$$

where, as always, it is implied that the physical result is found by taking the real part of the equation. We may use Maxwell's curl equations in the frequency domain, namely,

$$\nabla \times \mathbf{E}(\mathbf{r}, \omega) = i\omega \mathbf{B}(\mathbf{r}, \omega), \tag{8.19}$$

$$\nabla \times \mathbf{H}(\mathbf{r}, \omega) = \mathbf{J}(\mathbf{r}, \omega) - i\omega \mathbf{D}(\mathbf{r}, \omega), \tag{8.20}$$

to write this equation entirely in terms of fields. Following the steps similar to those taken in the time domain, we arrive at the result

$$\left\langle \frac{dE_{mech}}{dt} \right\rangle = -\frac{1}{2} \oint_S \mathbf{E}(\mathbf{r}, \omega) \times \mathbf{H}^*(\mathbf{r}, \omega) \cdot d\mathbf{a}$$

$$- \frac{i\omega}{2} \int_V \left[\mathbf{D}^*(\mathbf{r}, \omega) \cdot \mathbf{E}(\mathbf{r}, \omega) - \mathbf{B}(\mathbf{r}, \omega) \cdot \mathbf{H}^*(\mathbf{r}, \omega) \right] d^3 r. \tag{8.21}$$

We now adopt as definitions for the cycle-averaged Poynting vector, electric energy density and magnetic energy density in the following expressions:

$$\mathbf{S}(\mathbf{r}, \omega) \equiv \frac{1}{2} \mathbf{E}(\mathbf{r}, \omega) \times \mathbf{H}^*(\mathbf{r}, \omega), \tag{8.22}$$

$$U_e(\mathbf{r}, \omega) \equiv \frac{1}{4} \mathbf{D}^*(\mathbf{r}, \omega) \cdot \mathbf{E}(\mathbf{r}, \omega), \tag{8.23}$$

$$U_m(\mathbf{r}, \omega) \equiv \frac{1}{4} \mathbf{H}^*(\mathbf{r}, \omega) \cdot \mathbf{B}(\mathbf{r}, \omega). \tag{8.24}$$

With these definitions, our energy conservation law for monochromatic fields may be written as

$$\left\langle \frac{dE_{mech}}{dt} \right\rangle = -2i\omega \int_V [U_e(\mathbf{r}, \omega) - U_m(\mathbf{r}, \omega)] d^3r - \oint_S \mathbf{S}(\mathbf{r}, \omega) \cdot d\mathbf{a}. \quad (8.25)$$

Let us assume for simplicity that the materials are linear, that is, $\mathbf{D}(\mathbf{r}, \omega) = \epsilon(\mathbf{r}, \omega)\mathbf{E}(\mathbf{r}, \omega)$, and $\mathbf{B}(\mathbf{r}, \omega) = \mu(\mathbf{r}, \omega)\mathbf{H}(\mathbf{r}, \omega)$. In general, we expect that $\epsilon(\mathbf{r}, \omega)$ and $\mu(\mathbf{r}, \omega)$ may be complex numbers, for example, $\epsilon = \epsilon_R + i\epsilon_I$, with a positive imaginary part representing absorption. If we make this substitution, we find that

$$\left\langle \frac{dE_{mech}}{dt} \right\rangle = -\frac{i}{2}\omega \int_V \left[(\epsilon_R - i\epsilon_I)|\mathbf{E}|^2 - (\mu_R + i\mu_I)|\mathbf{H}|^2 \right] d^3r$$

$$- \oint_S \mathbf{S}(\mathbf{r}, \omega) \cdot d\mathbf{a}. \quad (8.26)$$

As only the real part of this expression physically contributes to the conservation law, it may be simplified to the form

$$\left\langle \frac{dE_{mech}}{dt} \right\rangle + \frac{\omega}{2} \int_V \left[\epsilon_I|\mathbf{E}|^2 + \mu_I|\mathbf{H}|^2 \right] d^3r = -\oint_S \mathbf{S}(\mathbf{r}, \omega) \cdot d\mathbf{a}. \quad (8.27)$$

Again, this is readily interpretable. The second term on the left of this equation is the rate of change of the electromagnetic energy in the volume, which is lost from absorption; the pair of terms on the left therefore represents the total rate of change of energy in the volume. Any net change of energy within the volume must be accounted for by a flux through the surface.

If the materials are nonabsorptive, then, ϵ and μ are real valued, that is, $\epsilon_I = 0$ and $\mu_I = 0$. Consequently, the second term on the left of Equation 8.27 vanishes. In the absence of absorption, then, the energy density plays no role in monochromatic energy conservation. On further consideration, this is not terribly surprising: because the system is assumed to be monochromatic, the cycle average of all quantities is independent of time. The energy density does not change in our system, and there is no change in the total electromagnetic energy in the volume.

A particularly important special case of Equation 8.27 occurs in the absence of any free charges. Using Gauss' theorem, we may then write this equation in differential form as

$$\nabla \cdot \mathbf{S}(\mathbf{r}, \omega) = -\frac{1}{2} \left[\epsilon_I |\mathbf{E}|^2 + \mu_I |\mathbf{H}|^2 \right]. \tag{8.28}$$

As a negative divergence of a vector field represents a location of a "sink" of the field, this expression agrees with the interpretation of an imaginary part of ϵ and μ representing absorption of energy. We will find Equation 8.28 useful in the discussion of singularities of power flow later in this chapter.

Even in the absence of absorption, Equation 8.27 suggests that there can generally be a nonzero net flow of energy into or out of the volume. You might wonder where this energy goes, given that we are working with a time-harmonic system that, by definition, is cyclic in all of its properties. In this case, the energy essentially "disappears," implicitly by some other mechanism not accounted for in our formalism. It might be said that it is a mild contradiction to talk about energy flow in a closed time-harmonic system.

It is important to note that the Poynting vector therefore only appears in a physical problem through its divergence or its integral over a closed surface; this will be of some significance in the next section.

When calculated for simple electromagnetic waves, the Poynting vector has reasonable, physically intuitive behavior. If we consider an electromagnetic plane wave propagating through free space in the $\hat{\mathbf{z}}$-direction with polarization along $\hat{\mathbf{x}}$, the fields may be written as

$$\mathbf{E}(\mathbf{r}, \omega) = \hat{\mathbf{x}} E_0 e^{ikz}, \tag{8.29}$$

$$\mathbf{H}(\mathbf{r}, \omega) = \hat{\mathbf{y}} \frac{E_0}{\mu_0 c} e^{ikz}. \tag{8.30}$$

On substituting these from these expressions into Equation 8.22, we find that

$$\mathbf{S}(\mathbf{r}, \omega) = \frac{|E_0|^2}{2\mu_0 c} \hat{\mathbf{z}}. \tag{8.31}$$

As one would expect, the Poynting vector points along the $\hat{\mathbf{z}}$-direction, the direction of wavefront propagation. If we consider a Hertzian, point-like electric dipole [Pap88, Section 3.2] of length d and current I pointing along the $\hat{\mathbf{z}}$-direction, the Poynting vector can be shown to have the far-zone form

$$\mathbf{S}(\mathbf{r}, \omega) = \hat{\mathbf{r}} \frac{1}{2} \sqrt{\frac{\mu_0}{\epsilon_0}} \left| \frac{kId}{4\pi r} \right|^2 \sin^2 \theta, \tag{8.32}$$

where θ is the angle from the z-axis. Here, the radiation propagates radially outward, and there is a null of radiation along the z-axis; this is intuitively reasonable because we cannot produce a transverse electromagnetic wave that propagates along the line of the current.

Considering the amount of time spent in Section 5.6 on the proper form of the electromagnetic momentum in matter, it is natural to ask why we have not similarly agonized over the form of the Poynting vector. We settled, in fact, rather quickly on the form $\mathbf{S} = (\mathbf{E} \times \mathbf{H}^*)/2$. One good motivation for this choice comes from looking at the boundary conditions for electromagnetic waves between two media, which have the form [BW99, Section 1.1.3]

$$\hat{\mathbf{n}} \cdot (\mathbf{B}_2 - \mathbf{B}_1) = 0, \tag{8.33}$$

$$\hat{\mathbf{n}} \cdot (\mathbf{D}_2 - \mathbf{D}_1) = \sigma_f, \tag{8.34}$$

$$\hat{\mathbf{n}} \times (\mathbf{E}_2 - \mathbf{E}_1) = 0, \tag{8.35}$$

$$\hat{\mathbf{n}} \times (\mathbf{H}_2 - \mathbf{H}_1) = \mathbf{K}_f, \tag{8.36}$$

where $\hat{\mathbf{n}}$ is the surface normal pointing from medium 1 to medium 2, σ_f is the free surface charge density on the boundary, and $\mathbf{K}_f$ is the free surface current density on the boundary. The energy flow across the surface will be determined by the normal component of the Poynting vector, which will in turn be determined by the transverse components of the field. Neglecting the surface current density at first, we can see that the transverse components of $\mathbf{E}$ and $\mathbf{H}$ will be continuous across the surface, which means that the normal component of the Poynting vector will itself be continuous across the surface; we can see that the same will *not* be true of the fields $\mathbf{B}$ and $\mathbf{D}$. The presence of a surface current will generally result in a discontinuous $\mathbf{S}$, implying that some energy is taken up/emitted by the oscillating surface current.

This continuity in electromagnetic energy flow is not essential—one can imagine an alternative formulation where the jump in flow is accounted for by energy uptake in matter—but it seems to be the most elegant description.

8.2 CONCEPTUAL DIFFICULTIES WITH THE POYNTING VECTOR

With all that has been said in the previous section, however, it is important to note that the Poynting vector can lead to seemingly paradoxical situations if one is not careful. The simplest of these, popularized by Feynman,

Leighton, and Sands [FLS64, Section 27-5], is to consider a bar magnet sitting stationary next to a static electric charge. There will be an electrostatic field **E** and a magnetostatic field **H** and therefore the Poynting vector will be in general nonzero, implying that there is an electromagnetic power flow around these static objects! This is a very odd result, as there are no electromagnetic waves produced by the system. One way around this difficulty is to note that the divergence of the Poynting vector is identically zero outside the source region, as can be shown by elementary vector calculus identities. This implies that the net energy flow into or out of any infinitesimal region of space is zero: this static Poynting energy, whatever it is, never builds up anywhere and is therefore not observable.

The fact that only the divergence of the Poynting vector is of physical significance helps us somewhat with the static case, but it also raises additional problems. Let us define a well-behaved vector field $X(r, \omega)$, and define a new vector field $Y(r, \omega) \equiv \nabla \times X(r, \omega)$. We can add $Y(r, \omega)$ to our formal definition of the Poynting vector, Equation 8.22, without changing the energy conservation law at all, as $\nabla \cdot Y(r, \omega) = 0$. Our naive definition of the Poynting vector is perhaps the simplest one, but is completely nonunique: an infinite number of other vector fields [characterized by $Y(r, \omega)$] are equally valid as an expression of "power flow."

Even simple problems involving the Poynting vector can yield surprising results. In a paper with the (translated) title "Is the Poynting vector observable?" Fritz Bopp [Bop63] introduced the following example. We consider a pair of monochromatic plane waves propagating in the xz-plane, one with TE polarization and one with TM polarization, as illustrated in Figure 8.1.

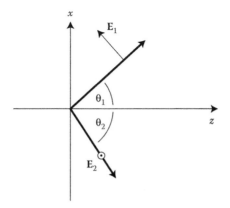

FIGURE 8.1 The propagation direction and field polarizations for Bopp's Poynting vector illustration.

The electric fields may be written in the form

$$E_1(r) = E_0 \left[\hat{x} \cos \theta_1 - \hat{z} \sin \theta_1 \right] e^{ik(\sin \theta_1 x + \cos \theta_1 z)}, \tag{8.37}$$

$$E_2(r) = E_0 \hat{y} e^{ik(\sin \theta_2 x + \cos \theta_2 z)}. \tag{8.38}$$

The corresponding magnetic fields are found to be

$$H_1(r) = \frac{E_0}{\mu_0 c} \hat{y} e^{ik(\sin \theta_1 x + \cos \theta_1 z)}, \tag{8.39}$$

$$H_2(r) = \frac{E_0}{\mu_0 c} \left[-\hat{x} \cos \theta_2 + \hat{z} \sin \theta_2 \right] e^{ik(\sin \theta_2 x + \cos \theta_2 z)}. \tag{8.40}$$

Individually, of course, the two plane waves have Poynting vectors that propagate in the xz-plane with angles θ_1 and θ_2. If we calculate the Poynting vector of the *total* field S_{tot}, however, we find that there is now a component out of the plane, given by

$$S_{tot}|_y = -\frac{|E_0|^2}{\mu_0 c} \left[\cos \theta_1 \sin \theta_2 - \sin \theta_1 \cos \theta_2 \right]$$

$$\times \cos[k(\sin \theta_1 + \sin \theta_2)x + k(\cos \theta_1 + \cos \theta_2)z]. \tag{8.41}$$

This component is of standing-wave form, suggesting that the combined electromagnetic wave "wiggles" up and down around the xz-plane as it propagates, something we would not have naively expected—how can waves traveling along x and z, which also produce no interference pattern, suddenly oscillate in the y-direction?

The mystery is somewhat resolved when we realize that we have been interpreting the Poynting vector as a geometric flow of electromagnetic power, even though light is inherently a wave phenomenon. When we analyze the combination of multiple waves, interference and polarization effects can make strange, nongeometric behaviors appear. This will be even more evident in the next few sections.

8.3 THE FIRST OPTICAL VORTEX

Though the birth of singular optics as a subfield in its own right began in the 1970s with the foundational paper of Nye and Berry [NB74], unusual singularities of the Poynting vector had already been discovered decades earlier. The earliest example, we have said, is found in Hans Wolter's 1950 paper [Wol50]. As this is illustrative of many effects related to Poynting singularities, we take some time to consider it here.

Wolter's work was inspired by the 1947 experimental paper of Fritz Goos and Hilda Hänchen [GH47], in which they observed what is now known as the Goos–Hänchen shift. In short, when a beam of light with finite cross section is reflected via total internal reflection at a surface, it experiences a slight transverse shift in its position.

The mathematical origin of the Goos–Hänchen shift is straightforward: in total internal reflection, the Fresnel reflection coefficient is a complex number of unit modulus. Any plane wave reflected at an angle exceeding the critical angle is therefore perfectly reflected, but with a phase change. A beam of light (which may be mathematically decomposed into a series of plane waves) will experience a direction-dependent phase change, and those changes will combine to produce a lateral shift of the beam.

What is not so straightforward, however, is coming up with a physical interpretation: what happens to the beam that makes it shift? The original explanation given is that the beam travels some distance in the evanescent region before reflection, but the smallness of the shift compared to the largeness of the beam make this hypothesis difficult to test.

In a strategy that curiously anticipated some of the applications of singular optics, Wolter attempted to study the Goos–Hänchen shift by following a line of zero intensity as it reflects internally at an interface. He considered the reflection of a pair of plane waves, the simplest field that has a transverse spatial structure.

The geometry is illustrated in Figure 8.2. Wolter's original paper[*] is somewhat hard to follow, as it relies a lot on his previous work; we derive his results somewhat differently here.

Two plane waves are incident from the positive z-direction in the denser medium with refractive index n_1 and, if beyond the critical angle, are totally internally reflected from the medium with index n_0. In the dense medium, the electric fields may be written as

$$\mathbf{E}_1(\mathbf{r}) = E_0 \hat{\mathbf{y}} e^{ikn_1[-\cos(\overline{\varphi}-\Delta/2)z+\sin(\overline{\varphi}-\Delta/2)x]}, \tag{8.42}$$

$$\mathbf{E}_2(\mathbf{r}) = E_0 \hat{\mathbf{y}} e^{ikn_1[-\cos(\overline{\varphi}+\Delta/2)z+\sin(\overline{\varphi}+\Delta/2)x]}, \tag{8.43}$$

where $\overline{\varphi}$ is the average angle of incidence and Δ is the angular difference in propagation between the two waves.

---

[*] Wolter's paper has been translated by U.T. Schwarz and M.R. Dennis [Wol09].

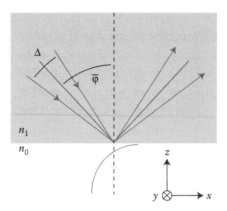

FIGURE 8.2 Illustrating the geometry used by Wolter to study the Goos–Hänchen shift.

This combined incident field will have zeros whenever the arguments of the two exponents differ by an odd multiple of π, that is,

$$
\begin{aligned}
&-\cos(\overline{\varphi} - \Delta/2)z + \sin(\overline{\varphi} - \Delta/2)x \\
&= -\cos(\overline{\varphi} + \Delta/2)z + \sin(\overline{\varphi} + \Delta/2)x + \frac{(2l+1)\pi}{kn_1},
\end{aligned}
\tag{8.44}
$$

where l is an integer. This defines the equation of a line, which can be written explicitly as

$$
z = \frac{\sin\varphi_+ - \sin\varphi_-}{\cos\varphi_+ - \cos\varphi_-}x + \frac{(2l+1)\pi}{kn_1[\cos\varphi_+ - \cos\varphi_-]}.
\tag{8.45}
$$

We have introduced $\varphi_\pm \equiv \overline{\varphi} \pm \Delta/2$ for brevity. It can be readily shown that these zero lines make an angle $\overline{\varphi}$ with respect to the z-axis and intercept the interface at $z = 0$ at positions

$$
x_l = -\frac{(2l+1)\pi}{2kn_1 \cos\overline{\varphi} \sin(\Delta/2)}.
\tag{8.46}
$$

The reflected waves can similarly be written in the form

$$
\mathbf{E}_{1r}(\mathbf{r}) = E_0\hat{\mathbf{y}}r_- e^{ikn_1[+\cos(\overline{\varphi}-\Delta/2)z+\sin(\overline{\varphi}-\Delta/2)x]},
\tag{8.47}
$$

$$
\mathbf{E}_{2r}(\mathbf{r}) = E_0\hat{\mathbf{y}}r_+ e^{ikn_1[+\cos(\overline{\varphi}+\Delta/2)z+\sin(\overline{\varphi}+\Delta/2)x]},
\tag{8.48}
$$

where

$$r_\pm = \frac{n_1 \cos \varphi_\pm - n_0 \cos \varphi_{t\pm}}{n_1 \cos \varphi_\pm + n_0 \cos \varphi_{t\pm}} \qquad (8.49)$$

is the Fresnel reflection formula [Gri13, Section 9.3.3], with $\varphi_{t\pm}$ as the (complex) angle of transmission into the medium of index n_0. The cosine of this angle may be written as

$$\cos \varphi_{t\pm} = i\sqrt{\frac{n_1^2}{n_0^2} \sin^2 \varphi_\pm - 1}. \qquad (8.50)$$

It can be seen by taking the absolute value of Equation 8.49 that the reflection coefficients are complex numbers of unit modulus. We may therefore write $r_\pm = e^{i\phi_\pm}$, and seek the zero lines of the combined reflected field. They again appear when the phase of the waves differs by π, and the equation for the line is

$$z = x \cot \overline{\varphi} + \frac{(2l+1)\pi + (\phi_+ - \phi_-)}{2kn_1 \sin \overline{\varphi} \sin(\Delta/2)}. \qquad (8.51)$$

The reflected zero line propagates at the angle $\overline{\varphi}$, and the zero intercept is

$$x_{rl} = -\frac{(2l+1)\pi}{2kn_1 \cos \overline{\varphi} \sin(\Delta/2)} + \frac{(\phi_- - \phi_+)}{2kn_1 \cos \overline{\varphi} \sin(\Delta/2)}. \qquad (8.52)$$

If we compare the positions of the zeros of the incident and reflected waves for $l = 1$, we see a clear manifestation of the Goos–Hänchen shift. The lateral shift is given by

$$\Delta x \equiv x_{rl} - x_l = \frac{(\phi_+ - \phi_-)}{2kn_1 \cos \overline{\varphi} \sin(\Delta/2)}. \qquad (8.53)$$

It is important to note that the zeros of the incident and reflected waves are not present in the total field, which is a superposition of all four plane waves.

With some manipulation, the total electric field may be written in the form

$$\mathbf{E}_{tot}(\mathbf{r}) = 2E_0\hat{\mathbf{y}}\left[e^{ikn_1 \sin \varphi_- x}e^{i\phi_-/2} \cos(kn_1 \cos \varphi_- z + \phi_-/2)\right.$$

$$\left. + e^{ikn_1 \sin \varphi_+ x}e^{i\phi_+/2} \cos(kn_1 \cos \varphi_+ z + \phi_+/2)\right]. \qquad (8.54)$$

In the forbidden region $z < 0$, the field consists of only two evanescent waves

$$\mathbf{E}_{1t}(\mathbf{r}) = E_0\hat{\mathbf{y}}t_-e^{ikn_0[-\cos\varphi_{t-}z+\sin\varphi_{t-}x]}, \tag{8.55}$$

$$\mathbf{E}_{2t}(\mathbf{r}) = E_0\hat{\mathbf{y}}t_+e^{ikn_0[-\cos\varphi_{t+}z+\sin\varphi_{t+}x]}, \tag{8.56}$$

where, by Snell's law, $n_1\sin\varphi_+ = n_0\sin\varphi_{t+}$, and $t_\pm$ is the Fresnel formula for the transmitted waves

$$t_\pm = \frac{2n_1\cos\varphi_+}{n_1\cos\varphi_+ + n_0\cos\varphi_{t+}}. \tag{8.57}$$

Looking at the electric field in this region, Wolter did not find a continuation of the zero line that connects the incident line with the reflected line, defeating his original purpose. When he looked at the Poynting vector, however, he found something quite surprising: in the denser medium, the Poynting vector has a complicated circulating flow near the interface. In the center of this circulation, the magnitude of the Poynting vector is zero: it is a singularity of the direction of the Poynting vector.

This flow is illustrated in Figure 8.3, using Wolter's original choice of parameters. Two different visualizations are presented; in (a), connected

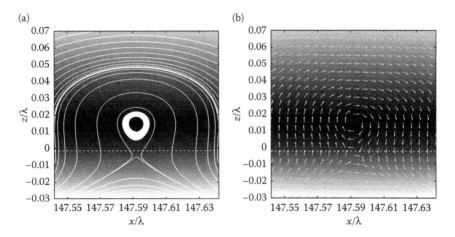

FIGURE 8.3 Illustrating the Poynting vector that appears in the Goos–Hänchen shift, both as (a) streamlines and (b) a vector field. Here, $n_1 = 1.52$, $n_0 = 1$, $\Delta = 3 \times 10^{-3}$, and $\overline{\varphi} = \varphi_c + 6.5 \times 10^{-3}$.

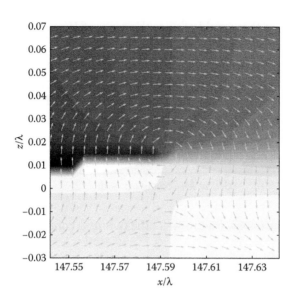

FIGURE 8.4 Illustrating the phase of E_y with the Poynting vector.

streamlines[*] of the Poynting vector are shown, while in (b), the local direction of the Poynting vector is illustrated by arrows. In both images, the Poynting vector flow is superimposed on the intensity of the electric field, that is, $|E_y|^2$. It is to be noted that the incident wave is coming from the upper left and reflected to the upper right. There are in fact two singularities in the range of the figure: the vortex of power flow in the dense medium and a saddle point of flow in the rare medium, whose structure will be elaborated upon in the next few sections.

What is the origin of these singularities? It can be seen from the figure that the vortex seems centered on a minimum of intensity of the electric field. This can be confirmed by plotting the phase of the electric field along with the Poynting vector, as done in Figure 8.4.

We are now on familiar ground, as in this case both components of the Poynting vector are proportional to the y-component of the electric field, that is,

$$S_x = \frac{1}{2}\text{Re}\left\{E_y H_z^*\right\}, \quad S_z = -\frac{1}{2}\text{Re}\left\{E_y H_x^*\right\}. \tag{8.58}$$

A zero of the electric field is also a zero of the Poynting vector. Also, as we have discussed in Chapter 7, we do not expect to see any zeros of the

---

[*] We emphasize again that streamlines do not represent geometric propagation of energy. It can be seen, for example, that the intensity is not constant along such lines.

magnetic field, as that would require the zeros of H_x and H_z to coincide, a nongeneric occurrence.

We can therefore find the location of the Poynting singularities simply by searching for zeros of the electric field. A zero of the electric field will simultaneously satisfy the equations $\text{Re}\{E_y(\mathbf{r})\} = 0$ and $\text{Im}\{E_y(\mathbf{r})\} = 0$; let us introduce two new variables for the region $z > 0$

$$\theta_x^\pm \equiv kn_1 \sin \varphi_\pm x + \varphi_\pm/2, \qquad (8.59)$$

$$\theta_z^\pm \equiv kn_1 \cos \varphi_\pm z + \varphi_\pm/2. \qquad (8.60)$$

With these coordinates, we can write the two zero conditions as

$$\cos[\theta_x^-] \cos[\theta_z^-] = -\cos[\theta_x^+] \cos[\theta_z^+], \qquad (8.61)$$

$$\sin[\theta_x^-] \cos[\theta_z^-] = -\sin[\theta_x^+] \cos[\theta_z^+]. \qquad (8.62)$$

To determine the actual locations of zeros, we first take the ratio of these two equations; the result is of the form

$$\tan[\theta_x^-] = \tan[\theta_x^+]. \qquad (8.63)$$

Since the tangent function is π-periodic, this implies that the zeros are located at $\theta_x^\pm$-values such that

$$\theta_x^- = \theta_x^+ - l\pi, \qquad (8.64)$$

with l as an integer, or at x-values such that

$$x = \frac{\varphi_- - \varphi_+}{4kn_1 \cos \overline{\varphi} \sin(\Delta/2)} + \frac{l\pi}{2kn_1 \cos \overline{\varphi} \sin(\Delta/2)}. \qquad (8.65)$$

On comparison with Equations 8.46 and 8.52, we find that the zero for $l = 0$ lies midway between the incident and reflected zero lines.

For the location of the zero along the z-direction, we apply Equation 8.64 to our original zero conditions, Equations 8.61 and 8.62. We readily find that they reduce to the same condition, namely,

$$(-1)^l \cos \theta_z^- = -\cos \theta_z^+. \qquad (8.66)$$

It is noteworthy that there are, in fact, two different sets of zeros here, depending on the value of l and, correspondingly, to the specific x position. We will focus on the zeros for $l = 0$, so that we need to find solutions of the equation

$$\cos \theta_z^- + \cos \theta_z^+ = 0. \tag{8.67}$$

With the use of trigonometric identities, we can rewrite this expression in the form

$$\cos \left[\frac{\theta_z^+ + \theta_z^-}{2} \right] \cos \left[\frac{\theta_z^+ - \theta_z^-}{2} \right] = 0. \tag{8.68}$$

We find that there are two subsets of zeros along the z-direction, each with a different spatial period. These subsets are given by z values such that

$$z_m^{(1)} = -\frac{\phi_+ + \phi_-}{4kn_1 \cos \overline{\phi} \cos(\Delta/2)} + \frac{(2m+1)\pi}{2kn_1 \cos \overline{\phi} \cos(\Delta/2)}, \tag{8.69}$$

$$z_m^{(2)} = \frac{\phi_+ - \phi_-}{4kn_1 \sin \overline{\phi} \sin(\Delta/2)} - \frac{(2m+1)\pi}{2kn_1 \sin \overline{\phi} \sin(\Delta/2)}, \tag{8.70}$$

where m is an integer. Looking at the denominators of the m-dependent term, we can see that $z^{(1)}$ will have a short spatial period (due to the small value of Δ) and $z^{(2)}$ will have a long spatial period. Wolter himself observed the first set of short spatial period singularities, simply by looking at a larger region of space, as illustrated in Figure 8.5.

The phase singularities can be clearly seen in this figure as those positions on the central axis where all phase contours converge. It can be seen that the singularities alternate in sign as one proceeds along the axis.

We can also consider an incident wave with **H** transverse to the plane of reflection. In this case, the zeros of the Poynting vector will now coincide with the generic zeros of the magnetic field. This alternate case illustrates that there is not necessarily a simple relationship between the fields and the singularities of the Poynting vector; this will be made clear in the upcoming sections.

8.4 OTHER EARLY OBSERVATIONS OF POYNTING SINGULARITIES

Before the formal introduction of singular optics, several other groups of researchers also observed circulation of the Poynting vector in different electromagnetic systems. Here, we briefly discuss each of these systems, highlighting the similarities and differences among them.

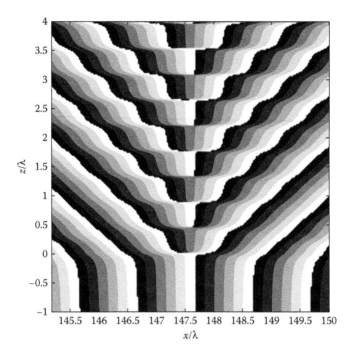

FIGURE 8.5 Illustrating the phase of E_y with the Poynting vector over an extended region.

In 1896, Arnold Sommerfeld [Som96] published the first rigorous solution of a diffraction problem, examining the two-dimensional case of a plane wave incident on an infinitely thin, perfectly conducting half-plane. It distinguished itself from previous formulations of diffraction that relied on mathematically imprecise notions of perfectly "black" screens. Remarkably, the solution can be written in a closed mathematical form with the help of special functions, though visualizing these solutions typically requires computational work. A detailed discussion of Sommerfeld's calculation can be found in Born and Wolf [BW99, Chapter 11].

One significant advantage of Sommerfeld's result is the ability to calculate the electromagnetic fields in the near zone of the screen. In 1952, Braunbek and Laukien [BL52] studied the amplitude, phase, and power flow of a normally incident plane wave and observed singularities of phase coinciding with those of the Poynting vector.

Their results are reproduced in Figure 8.6. A plane wave is normally incident from above in the figures, and has a magnetic field polarized out of the page. Figure 8.6a shows the phase contours of H_z, while Figure 8.6b shows the flow lines of the Poynting vector superimposed over the intensity distribution. By now, we shouldn't have to convince the reader that these

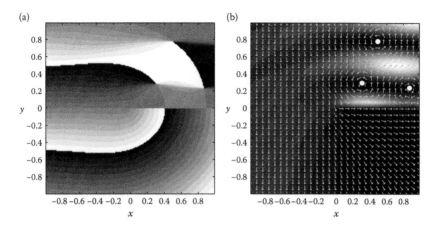

FIGURE 8.6 Diffraction of light by a half-plane. (a) The phase of H_y. (b) The intensity of the field and the power flow. White dots have been placed at the vortices of power flow.

singularities correspond to a zero of amplitude of H_y. As one would expect, the singularities appear on the illuminated side of the screen, where multiple waves interfere. In this case, the total field may be formally broken into three parts: the incident field, the field directly reflected from the screen, and a field diffracted from the edge of the screen. Because we have more than two waves interfering, the singularities that appear are of a generic form. It is to be noted that the singularities shown in the figure have neighbors of alternate handedness. We can also see, between the uppermost and leftmost singularities, that there is a simultaneous saddle point of phase and power flow.

A quite different example of vortices in power flow was provided by Boivin, Dow, and Wolf in 1967 [BDW67]. In this case, the Poynting vector was analyzed in the neighborhood of geometric focus of an electromagnetic wave focused through a hard aperture, building on earlier results by Richards and Wolf [RW59] and Boivin and Wolf [BW65].

The system geometry is illustrated in Figure 8.7. In this system, the focal length is f and the origin is placed at the geometrical focus. The angle 2α represents the angular aperture on the image side of the system, and positions within the focal region are specified by the normalized coordinates

$$u = kz \sin^2 \alpha, \tag{8.71}$$
$$v = k\rho \sin \alpha, \tag{8.72}$$

with $\rho = \sqrt{x^2 + y^2}$.

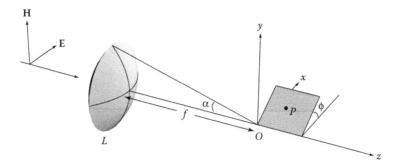

FIGURE 8.7 Notation relating to the focusing of electromagnetic waves. (Adapted from D.W. Diehl and T.D. Visser. *J. Opt. Soc. Am. A*, 21:2103–2108, 2004.)

The results of their calculation are shown in Figure 8.8, with $\alpha = 45°$. In (a), the flow lines are shown within a large portion of the focal region, while in (b), the plot is centered on the first transverse singularity at $u = 0, v = 3.67$.

It is perhaps not surprising at this point to note that the singularities closely correspond to the familiar Airy rings, rings of zero intensity, in the

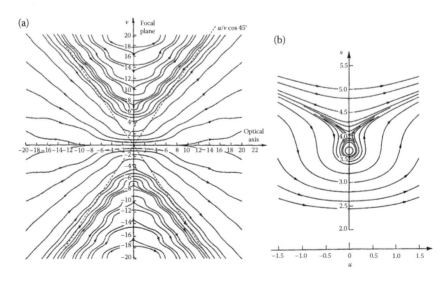

FIGURE 8.8 Streamlines of the Poynting vector in the neighborhood of geometric focus, with $\alpha = 45°$, (a) within a large portion of the focal region and (b) centered on the first transverse singularity at $u = 0, v = 3.67$. (Taken from A. Boivin, J. Dow, and E. Wolf. *J. Opt. Soc. Am.*, 57:1171–1175, 1967.)

focal plane. It is also to be noted that the pattern of Figure 8.8b looks strikingly similar to the flow lines of Wolter's system, as illustrated in Figure 8.3a. This is readily understood in that both systems consist of a superposition of waves converging on a single point in space; the role of the reflected waves in Wolter's system is played by the waves coming from below in the focusing system.

These results further demonstrate that there is a regular structure to the singularities of the Poynting vector, and that this structure manifests in a variety of optical systems.

8.5 GENERIC PROPERTIES OF POYNTING VECTOR SINGULARITIES

With some illustrative examples behind us, we now turn to a discussion of the generic features of Poynting vector singularities. Intuitively, we might guess that the Poynting vector will circulate around a zero line in three-dimensional space, in analogy with the behavior of optical vortices. However, we immediately run into a problem: the generic zeros of a three-dimensional vector field in three-dimensional space are points, and not lines. The Poynting vector will generally have three components, and in three-dimensional space, it is clear that zeros will typically only occur at isolated points where all three components are simultaneously zero.

Looking back at the previous section, this might at first seem to be a paradox, but we quickly resolve the discrepancy: all the systems previously considered were effectively two dimensional. Wolter's total internal reflection problem and Braunbek and Laukien's half-plane problem both involve systems that are invariant along the out-of-page axis, so to speak; Boivin, Dow, and Wolf's focusing problem has strong cylindrical symmetry, essentially reducing it to a problem described in terms of radial and axial coordinates. In two dimensions, the Poynting vector has a rich set of generic singularities, while in three dimensions, it does not.

For a general beam-like field propagating along the z-axis, however, we can extract a two-dimensional problem by considering only the components of the Poynting vector that lie in the transverse xy-plane. This is somewhat analogous to our consideration, in the previous chapter, of only the transverse electric field components in studying polarization singularities, and as in that case we will find some subtlety in the restriction. It should not be viewed as an artificial choice, though, as we may argue that the transverse components of the energy flow represent the most "interesting" part of the electromagnetic wave, departing from the purely z-propagating flow of a simple plane wave.

Following the same reasoning as above, we expect the generic singularities of a two-dimensional vector field in two-dimensional space to manifest as points. As we are now considering singularities of a vector field, which has a definite direction (arrow) associated with it at every point in space, we expect the singularities to have unit topological index, that is, $n = \pm 1$, in the generic case, as the vector must rotate a full 360 degrees when tracing out a path around a singular point. From here on, our work is in fact much easier than it might appear, as the mathematical tools of linear dynamical systems, or equivalently phase space problems in differential equations [Gbu11, Section 14.3], may be directly applied. This connection was noted and explored in detail by Bekshaev and Soskin [BS07], and then later by Novitsky and Barkovsky [NB09]. A brief summary of the methods is given here.

We proceed as follows. Near a generic singular point, assumed to be located at $x = y = 0$, we expect the components of the Poynting vector to be linear in x and y, namely,

$$S_x = ax + by, \tag{8.73}$$

$$S_y = cx + dy, \tag{8.74}$$

where a, b, c, and d are real-valued constants. If we interpret the Poynting vector as a direction of "motion" of energy, we may write these equations in the dynamic form

$$\frac{dx}{dt} = ax + by, \tag{8.75}$$

$$\frac{dy}{dt} = cx + dy, \tag{8.76}$$

where t is a parameter on which the streamlines of the Poynting vector are defined. We now, however, have a system of linear differential equations, that is,

$$\frac{d\,|\mathbf{x}\rangle}{dt} = \mathbf{A}\,|\mathbf{x}\rangle, \tag{8.77}$$

where

$$|\mathbf{x}\rangle = \begin{bmatrix} x \\ y \end{bmatrix}, \tag{8.78}$$

and

$$\mathbf{A} = \begin{bmatrix} a & b \\ c & d \end{bmatrix}. \tag{8.79}$$

We look for solutions to this system of equations of the form

$$|\mathbf{x}\rangle = e^{\lambda t} |\mathbf{x}_0\rangle, \tag{8.80}$$

where $|\mathbf{x}_0\rangle$ represents the solution at $t = 0$. On substitution of this trial solution, we are led to an ordinary eigenvalue problem

$$\mathbf{A} |\mathbf{x}\rangle = \lambda |\mathbf{x}\rangle, \tag{8.81}$$

and the acceptable values of λ are given by the determinant

$$\begin{vmatrix} a - \lambda & b \\ c & d - \lambda \end{vmatrix} = \lambda^2 - \lambda(a + d) + (ad - bc) = 0. \tag{8.82}$$

This equation for λ depends entirely on the trace and determinant of the matrix $\mathbf{A}$, which we express as

$$\mathrm{Tr}(\mathbf{A}) = A_T = a + d, \tag{8.83}$$

$$\mathrm{Det}(\mathbf{A}) = A_D = ad - bc. \tag{8.84}$$

The two solutions $\lambda_\pm$ are readily found

$$\lambda_\pm = \frac{1}{2} \left[A_T \pm \sqrt{A_T^2 - 4A_D} \right]. \tag{8.85}$$

As these eigenvalues appear in an exponential, the real parts relate to exponential decay toward or growth away from the singularity, while the imaginary parts relate to periodic motion around the singularity. To explore the possibilities further, we begin with the special case in which there is no absorption in the system; then, Equation 8.28 reduces to the simple form

$$\nabla \cdot \mathbf{S}(\mathbf{r}, \omega) = 0. \tag{8.86}$$

Using our linear approximation for $\mathbf{S}$, this zero-divergence condition results in the requirement that $d = -a$. This, in turn, implies that $A_T = 0$ and that we may write

$$\lambda_\pm = \pm \frac{1}{2} \sqrt{-4A_D}. \tag{8.87}$$

If $A_D > 0$, we get a pair of pure imaginary eigenvalues. The streamlines of the Poynting vector therefore form closed loops around the singularity, and we have what is known as a *center* in dynamic systems and a *vortex* in singular optics. If $A_D < 0$, we get a pair of real eigenvalues of opposite sign. Along one axis, the streamlines approach the singular point, while along the other, they diverge away; we have a *saddle*. In the absence of absorption, these are the only two possibilities.

In the case where absorption is present, we return to the general form of Equation 8.28

$$\nabla \cdot \mathbf{S}(\mathbf{r}, \omega) = -\frac{1}{2} \left[\epsilon_I |\mathbf{E}|^2 + \mu_I |\mathbf{H}|^2 \right] \equiv -\alpha, \tag{8.88}$$

where it is assumed that $\alpha \geq 0$. This condition is equivalent to $d = -a - \alpha$, or $d \leq -a$. We therefore expect the $A_T \leq 0$. As the $A_T = 0$ case is covered above, we concentrate strictly on $A_T < 0$.

Looking back at Equation 8.85, we may first divide the behaviors into two cases: $A_T^2 \leq 4A_D$ and $A_T^2 > 4A_D$. Beginning with the former, we find that the eigenvalues reduce to the form

$$\lambda_{\pm} = \frac{1}{2} \left[A_T \pm i\sqrt{4A_D - A_T^2} \right]. \tag{8.89}$$

The eigenvalues are complex, with a negative real part; the streamlines of the Poynting vector therefore spiral toward the singularity, and the singularity is known as a *spiral point*. In the special case that $4A_D = A_T^2$, all the streamlines proceed radially to the singularity as a *star* in dynamical systems, or a *sink* in singular optics.

For $A_T^2 > 4A_D$, the eigenvalues are entirely real. We may further divide our results into two cases. If $\sqrt{A_T^2 - 4A_D} < A_T/2$, both eigenvalues are negative. We have what is known as a *node* in dynamical systems, which is also called a sink in singular optics. If $\sqrt{A_T^2 - 4A_D} > A_T/2$, one eigenvalue is positive, and one is negative; we again have a saddle.

All these possibilities are illustrated in Figure 8.9. It is to be noted that it is possible to have a center even in a system with absorption. If $\mu_I = 0$ but $\epsilon_I \neq 0$, for example, and we consider a point where $\mathbf{E} = 0$, then, $\alpha = 0$ and the point must be a center or a saddle.

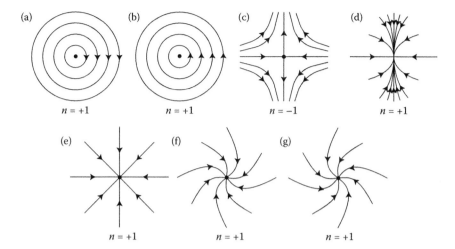

FIGURE 8.9 Illustration of the generic singularities of the transverse Poynting vector, with topological index listed: (a) right-hand center, (b) left-hand center, (c) saddle, (d) inward node with y eigenvalue more negative than x eigenvalue, (e) sink, (f) right-hand spiral, and (g) left-hand spiral.

If the medium possesses gain instead of loss, for example, $\epsilon_I < 0$ instead of $\epsilon_I > 0$, then, the possibility of sinks and inward spiral points are replaced with sources and outward spiral points.

Creation and annihilation events must conserve topological index, which would at first glance seem to only require a single saddle annihilating with one of any other type of singularity. However, we will see in the examples that the typical reaction involves two vortices of opposite handedness combined with two saddles, as also happens in the creation and annihilation of phase vortices. This is not surprising, as singularities of the Poynting vector are often associated with points where either the electric or magnetic field is zero, and there is consequently a singularity in the phase of that field component.

It should be noted, however, that there are other possible circumstances in which the transverse Poynting vector may vanish. We may identify four distinct possibilities:

- The electric field $\mathbf{E} = 0$.

- The magnetic field $\mathbf{H} = 0$.

- The cross-product of $\mathbf{E}$ and $\mathbf{H}^*$ vanishes, that is, $\mathbf{E}$ is parallel to $\mathbf{H}^*$.

- The product $\mathbf{E} \times \mathbf{H}^*$ is purely imaginary.

To investigate these, it is helpful to have the explicit expression of the complex Poynting vector on hand

$$\mathbf{S} = \frac{1}{2}\left[\hat{\mathbf{x}}(E_y H_z^* - E_z H_y^*) - \hat{\mathbf{y}}(E_x H_z^* - E_z H_x^*) + \hat{\mathbf{z}}(E_x H_y^* - E_y H_x^*)\right].$$
(8.90)

The first possibility appears in Wolter's simulations of total internal reflection, as the singularities appear when $E_y = 0$. The second possibility appears in Sommerfeld's half-plane problem, as the singularities appear when $H_y = 0$. However, it is readily apparent that both of these are special cases, where the complex fields in question only have a single component throughout space. In general, we expect the electric and magnetic fields to each have three complex components, and therefore three conditions must be satisfied in three-dimensional space to get a zero of one field or the other. Such points of zero Poynting vector are evidently generic, but do not result in a circulation of power flow.

The third possibility requires the complex electric field to be proportional to the complex magnetic field, which involves the two *complex* equations

$$E_y H_z^* - E_z H_y^* = 0, \quad E_x H_z^* - E_z H_x^* = 0.$$
(8.91)

Because the fields are complex, the above equation constitutes *four* conditions that must be satisfied in three-dimensional space, and we expect that this is a nongeneric situation.

The fourth possibility is a generalization of the third, and involves the two real conditions

$$\text{Re}\{E_y H_z^* - E_z H_y^*\} = 0, \quad \text{Re}\{E_x H_z^* - E_z H_x^*\} = 0.$$
(8.92)

This possibility involves two conditions that must be satisfied in three-dimensional space, and it is expected that the zeros are lines in this space. It encompasses all the previous possibilities as well, as satisfying those former possibilities automatically satisfies Equation 8.92 as well. We expect, then, to find singularities of the transverse Poynting vector for a general three-dimensional electromagnetic wave on lines for which Equation 8.92 is satisfied.

We still have the problem that the topological index alone does not force the annihilation of two vortices of power flow. Curiously, this problem

TABLE 8.1   Angles of Rotation for Planes Waves in the Poynting Vector Example

Plane Wave	ϕ	θ	η
1	0	$\pi/6$	0
2	$\pi/2$	$\pi/6$	$2\pi/3$
3	$\pi/2$	$\pi/6$	$4\pi/3$

does not seem to have been explicitly addressed as yet. It seems clear that a vortex of the Poynting vector must have a discrete topological charge associated with it, related to its handedness, but a proper definition of this charge has not been found. A major problem, as noted by Berry [Ber09], is that it is not possible to find a scalar wave that can be used to reproduce the Poynting vector flow. As we will see in Section 10.8, it is quite easy to define a topological charge for any complex scalar field; however, electromagnetic waves cannot be represented in this form. One attempted solution was suggested in the form of the so-called Riemann–Silberstein vector, which we will discuss in Section 12.2; the results, however, are not directly applicable to the Poynting vector.

We can nevertheless show by example that Poynting singularities annihilate in a way strictly analogous to phase singularities. We do so by looking at a superposition of electromagnetic plane waves. We take three plane waves of unit amplitude propagating in the $\hat{\mathbf{z}}$-direction, with polarization along the $\hat{\mathbf{x}}$-direction, and rotate each of them by an angle ϕ about the z-axis, an angle θ about the x-axis, and then an angle η about the z-axis again. The choice of these parameters is given in Table 8.1. Three plane waves together give, as we would expect at this point, generic singularities of the transverse Poynting vector. To generate annihilation events, we add a fourth normally incident plane wave with $\hat{\mathbf{x}}$-polarization, and increase its amplitude E_4 from zero.

The results of this calculation are shown in Figure 8.10. With the fourth plane wave not included, we can see sets of four Poynting/phase singularities. One particular set is labeled, with p and q vortices and r and s saddles. As the amplitude of the fourth wave increases, the four singularities come together to annihilate.

Evidently the Poynting singularities must annihilate in pairs of vortices with opposite handedness along with two saddles, just like phase singularities. A rigorous accounting of such creation and annihilation events in the transverse Poynting vector has yet to be done, however.

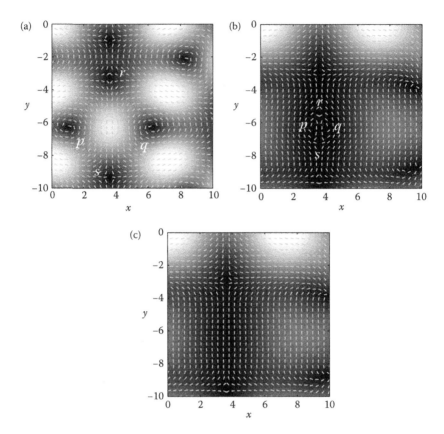

FIGURE 8.10 Transverse Poynting vector direction and amplitude for the combination of four plane waves, with the amplitude of the fourth being (a) $E_4 = 0.0$, (b) $E_4 = 2.4$, and (c) $E_4 = 3.0$.

8.6 SINGULARITIES, TRANSMISSION, AND RADIATION

As we have already seen in Sections 8.3 and 8.4, Poynting vector singularities can provide insight into a variety of optical systems and phenomena. A recent and illuminating example of this is the study of light transmission through narrow slits in flat dielectric or metal plates, which we consider in some detail in this section.

The system geometry is illustrated in Figure 8.11. A monochromatic electromagnetic wave, polarized along the y-direction, is incident upon a slit in a plate of thickness d, width w, and generally complex index of refraction n. The entire system is invariant along the y-direction, making it similar to the Poynting vector examples of the earlier sections. Unlike those previous sections, however, the electromagnetic field cannot be solved for

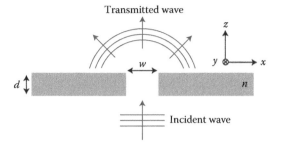

FIGURE 8.11   Illustrating the notation and geometry related to light transmission through an infinitely long slit.

analytically, due to the finite thickness of the plate and its complex material parameters. Instead, the fields are found numerically through the solution of the electromagnetic integral equation

$$\mathbf{E}(\mathbf{r}) = \mathbf{E}^{(inc)}(\mathbf{r}) - i\omega\Delta\epsilon \int \mathbf{G}(\mathbf{r}, \mathbf{r}') \cdot \mathbf{E}(\mathbf{r}')d^2r', \qquad (8.93)$$

where $\mathbf{E}(\mathbf{r})$ is the total electric field, $\mathbf{E}^{(inc)}(\mathbf{r})$ is the incident electric field (field present in the absence of the slit in the plate, but still in the presence of the plate), and $\mathbf{G}(\mathbf{r}, \mathbf{r}')$ is the electric Green's tensor for the plate without the slit. This Green's tensor can be calculated analytically to within a single Fourier transform; the technique of solution is often referred to as the *domain integral equation method*, and is described in some detail in the work of Visser, Blok, and Lenstra [VBL99].

Research on the transmission properties of subwavelength slits was motivated by the observation in 1998 of "extraordinary optical transmission" through an array of subwavelength holes in a silver plate; this enhanced transmission was identified as being due to the effects of surface plasmons [ELG+98]. The term "extraordinary" deserves some clarification in this context. When considering light transmission through a large hole in an opaque screen, it is expected that light will roughly satisfy a geometrical optics approximation: the amount of light transmitted will be equal to the amount of light directly impinging on the hole. For subwavelength size holes, however, it was long thought that the amount of light transmitted will always be much smaller than the geometric prediction, approaching zero rapidly as the hole size decreases.[*] The extraordinary transmission

---

[*] See, for instance, Section 8 of Bouwkamp [Bou54].

observed, however, was not only larger than the theoretical subwavelength prediction, but also much larger than the geometrical prediction.

In 2003, Schouten et al. [SVLB03] investigated whether extraordinary transmission is possible even in the absence of plasmons (which are not present when the electric field is parallel to the slit). Because of the use of an infinite plane wave as the illuminating field, and the possibility of light directly tunneling through the plate itself, they assessed the transmission using a normalized transmission coefficient T defined as

$$T \equiv \frac{\int_{slit} S_z(x, d) dx + \int_{plate} [S_z(x, d) - S_z^{(inc)}(x, d)] dx}{\int_{slit} S_z^{(0)}(x, 0) dx}, \qquad (8.94)$$

where S_z is the Poynting vector of the total field, $S_z^{(inc)}$ is the Poynting vector of the incident field in the presence of the plate, and $S_z^{(0)}$ is the Poynting vector of the plane wave illuminating the plate. The first integral in the numerator is the total amount of power directly flowing through the slit; the second integral in the numerator is the total power flowing through the solid plate *minus that power that would otherwise flow through in the absence of the slit*. The denominator normalizes the transmission so that $T = 1$ is the amount of transmission predicted by geometrical optics.

Figure 8.12 shows the result of the calculation for aluminum and silver at particular wavelengths. It is to be noted that the transmission through the slit for both materials has an oscillatory character, and can be significantly above or below the geometric prediction. This oscillation can be attributed to the birth of new waveguide modes in the slit as its width is increased. For a perfect conductor, it is predicted that these modes arise when $w \approx 0.5\lambda$, $1.5\lambda, \ldots$, though for silver it can be shown that the first mode appears at a slightly smaller width of $w = 0.4\lambda$.

The birth of a waveguide mode does not, however, give an intuitive picture of why the transmission would be greater than the geometric prediction, $T = 1.33$ in the case of silver at $w = 0.5\lambda$. This intuition is provided by an examination of the power flow in the neighborhood of a slit of width 0.4λ, as in Figure 8.13. The transmission here is $T = 1.11$, above the geometric prediction, and one can see that there are vortices and saddles present in the flow. In fact, it may be roughly said that the vortices c and d "funnel" light that would otherwise strike the plate and guide it into the hole.

Even further, it can be shown that there are families of Poynting singularities on the illuminated side of the plate, and that a maximum of

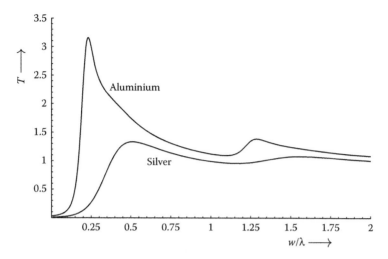

FIGURE 8.12 The transmission coefficient of a narrow slit as a function of the normalized slit width w/λ for two different materials. For aluminum, $d = 100$ nm, $\lambda = 91.8$ nm, and $n = 0.041 + i0.517$, while for silver $d = 100$ nm, $\lambda = 500$ nm, and $n = 0.05 + i2.87$. (Taken from H.F. Schouten et al. *Phys. Rev. E*, 67:036608, 2003, Figure 1.)

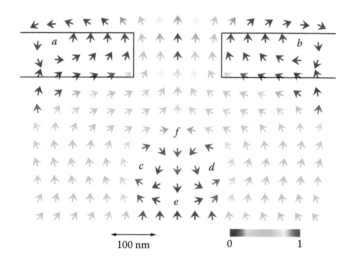

FIGURE 8.13 Behavior of the Poynting vector near a 200-nm-wide slit in a 100-nm-thick silver plate with $\lambda = 500$ nm. The points a, b, c, and d indicate vortices of power flow, while e and f indicate saddles. (Taken from H.F. Schouten et al. *Phys. Rev. E*, 67:036608, 2003, Figure 3.)

(a)

| LV | | | | RV |
| S | S LV | RV S | S |

			S	
LV	RV	LV	RV LV	RV	LV	RV
S	S		S	S	S	S
			S			
RV	LV		RV LV	RV	LV
S	S		S	S	S
RV	LV		RV LV	RV	LV
S			S	S	
			S		
LV 250 nm		RV LV	RV	

(b)

| LV | | | | RV |
| S | S LV | RV S | S |

LV	RV	LV		RV	LV	RV
S	S			S	S
RV		LV		RV	LV
S		S		S	S
RV		LV		RV	LV
S				S	
LV 250 nm				RV

FIGURE 8.14 Positions of Poynting singularities for light transmission through the silver plate when (a) $w = 0.4\lambda$ and (b) $w = 0.5\lambda$. (Taken from H.F. Schouten et al. *Phys. Rev. E*, 67:036608, 2003, Figures 4 and 5.)

transmission coincides with an annihilation of those singularities directly in front of the hole. This is illustrated in Figure 8.14. When the slit width reaches $w = 0.5\lambda$, the Poynting singularities in line with the slit annihilate in groups of four consisting of two opposite-handed vortices and a pair of saddles. The result is a smooth flow of power directly into the slit.

This initial observation was followed up by several others; see, for instance, Schouten et al. and Schouten, Visser, and Lenstra [SGV$^+$03, SVL04,SVG$^+$04a]. The latter paper is of particular interest because it was demonstrated that certain system configurations can result in *negative* transmission, that is, less light being transmitted with the slit present than in the absence of it.

The power flow for such a case is illustrated in Figure 8.15. For light transmitted through a 100-nm-wide slit in a 100-nm-thick silicon plate at $\lambda = 500$ nm, it is calculated that $T = -0.05$. The handedness of vortices b and c are the same as those in Figure 8.13; however, because they are on the dark side of the plate, they tend to funnel light back toward the aperture, rather than increasing throughput. Furthermore, one can see a saddle and sink within the material on either side of the slit.

It is important to note that these singularities should not be considered the *cause* of enhanced or suppressed transmission; rather, they may be said to be a symptom of the complicated light–matter interactions that occur in the neighborhood of the slit. It is also worth noting that singularities of the Poynting vector in the near field of antennas were studied quite some time above by Landstorfer, Meinke, and Niedermair [LMN72].

Most of the Poynting singularities considered so far in this section have been on the illuminated side of the plate. When the size of the slit is increased above a wavelength, however, additional singularities appear on

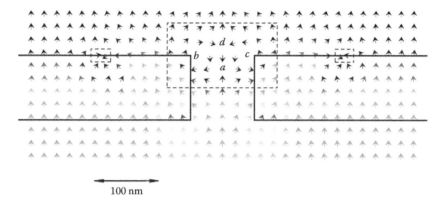

100 nm

FIGURE 8.15 Behavior of the Poynting vector near a 100-nm-wide slit in a 100-nm-thick silicon plate with $\lambda = 500$ nm. The points a and b indicate vortices of power flow, while c and d indicate saddles. The other dashed regions indicate the presence of a saddle and a sink. (Taken from H.F. Schouten et al. *J. Opt. A*, 6:S277–S280, 2004, Figure 4.)

the transmission side of the plate, and these singularities can be directly related to the radiation pattern of the aperture. This was studied using the domain integral equation by Schouten et al. [SVG⁺04b], and two examples of their results are shown in Figure 8.16. A polar diagram of the radiation pattern is illustrated along with the positions of two near-zone vortices. It can be seen that the radiation pattern has minima in directions that are "obstructed" by the singularities; furthermore, the further the singularities

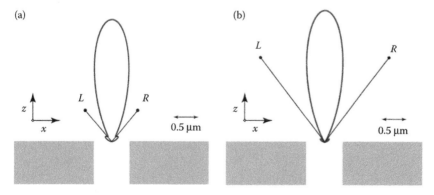

FIGURE 8.16 Behavior of the radiation pattern far from a subwavelength slit in a silver plate, correlated with the presence of zeros in the near zone, for (a) $w = 750$ nm, $d = 860$ nm, and $\lambda = 510$ nm and (b) $w = 750$ nm, $d = 860$ nm, and $\lambda = 500$ nm. (Taken from H.F. Schouten et al. *Phys. Rev. Lett.*, 93:173901, 2004, Figures 3 and 4.)

are from the slit, the deeper the minima are. It is argued that an exact zero of the radiation pattern appears when the Poynting singularity reaches the point at infinity.

8.7 EXERCISES

1. Starting from Equation 8.18, derive Equation 8.21, the energy conservation law for monochromatic fields.

2. Derive Equations 8.46 and 8.52 for the intercepts of the zero lines of Wolter's model from the expressions for the incident and reflected electric fields.

3. Returning to Equation 8.66, determine the location of the two sets of zeros for the $l = 1$ case, and sketch the relative positions of the $l = 0$ and $l = 1$ zeros.

CHAPTER 9

Coherence Singularities

U P TO THIS POINT, we have primarily considered singularities in monochromatic wavefields. However, even the most stable laser has random fluctuations in phase as it evolves in time, and most natural sources of light have large fluctuations in both amplitude and phase. This raises a significant question: what happens to phase singularities when the phase is scrambled? We have seen that optical vortices are stable under perturbations of the wavefield; such perturbations presumably include passing a vortex beam through a random phase screen, turning it into a partially coherent field.

The study of fields with random fluctuations in space and time, known as *partially coherent fields*, is known as *optical coherence theory*, and it has now been demonstrated both theoretically and experimentally that the coherence functions that describe wavefields can possess phase singularities of their own. These *coherence singularities* may be considered generalizations of the phase singularities we have considered previously, and they exhibit a number of new and potentially useful properties.

To explain coherence singularities, we must first introduce some concepts and mathematics of coherence theory. It is a mature and rather complicated topic, and by necessity, we will attempt to explain only the bare minimum needed to elucidate the related singular optics. Full explanations can be found in Mandel and Wolf [MW95], and the introductory text by Wolf [Wol07].

9.1 OPTICAL COHERENCE

Let us step gingerly away from monochromatic fields and imagine a field that is quasi-monochromatic. A precise definition of this term will be given momentarily, but for the moment, we take it to mean a field with center frequency ω, which has a phase $\phi(\mathbf{r}, t)$ that varies randomly in time on a scale that is slow compared to $1/\omega$. We may therefore write its time dependence as

$$U(\mathbf{r}, t) = U(\mathbf{r}) \exp[i\omega t + i\phi(\mathbf{r}, t)]. \tag{9.1}$$

Though the phase fluctuates randomly in time, we make the important assumption that the underlying statistics of the process does not. In other words, the average behavior of the field is independent of the time of the measurement. Such a field is referred to as *statistically stationary*, and includes those produced by many sources of interest, including stars, incandescent bulbs, and continuous-wave (CW) lasers.

In Equation 9.1, we have already written the field in complex form, something that we have so far only done for purely monochromatic fields. The physical field will be the real part of this quantity, that is,

$$V(\mathbf{r}, t) = \mathrm{Re}\{U(\mathbf{r}, t)\}. \tag{9.2}$$

The instantaneous intensity of the field will be given by the square of the real field

$$I(\mathbf{r}, t) = [V(\mathbf{r}, t)]^2 = \frac{1}{4}\left[U(\mathbf{r}, t) + U^*(\mathbf{r}, t)\right]^2. \tag{9.3}$$

As was discussed in Section 2.1, the oscillations of optical fields are much too fast for detectors to resolve, and these detectors instead measure the average properties of the field. However, in our present case, it is insufficient to average over a single cycle: the slowly varying phase $\phi(\mathbf{r}, t)$ wrecks the simple cosine and sine oscillation of the field. Assuming that there are many oscillations of the field during the detection process, we instead formally average over an infinite time interval, defining

$$\langle I(\mathbf{r}, t)\rangle \equiv \lim_{T\to\infty} \frac{1}{2T} \int_{-T}^{T} [V(\mathbf{r}, t)]^2 dt. \tag{9.4}$$

If we write this expression in terms of the complex field, we have

$$\langle I(\mathbf{r}, t) \rangle \lim_{T \to \infty} \frac{1}{8T} \int\limits_{-T}^{T} \left\{ [U(\mathbf{r}, t)]^2 + 2U(\mathbf{r}, t)U^*(\mathbf{r}, t) + [U^*(\mathbf{r}, t)]^2 \right\} dt.$$

(9.5)

Referring to Equation 9.1, we see that the first and last terms of this average must be extremely small due to the rapid oscillation of the terms at frequencies 2ω and -2ω. Even though $\phi(\mathbf{r}, t)$ is also varying, we have assumed that it is varying so slowly that it remains constant over many cycles at the center frequency. We are left with

$$\langle I(\mathbf{r}, t) \rangle = \text{Re}\{\langle U(\mathbf{r}, t)U^*(\mathbf{r}, t) \rangle\}/2.$$

(9.6)

We have found that, similar to the monochromatic case, the average intensity can be expressed in terms of the product of the *complex field* $U(\mathbf{r}, t)$ with its complex conjugate. For brevity, we will drop the factor of $1/2$ from the definition of intensity; it can be absorbed into the definition of $U(\mathbf{r}, t)$.

The point of this discussion is to demonstrate that, even for nonmonochromatic fields, we can use complex wavefields and that there are advantages of simplicity in doing so. There is a rigorous process for doing this for any field, in fact—not just quasi-monochromatic fields—and the complex field that results is known as the *complex analytic signal*. In almost all discussions of optical coherence, the complex analytic signal is used instead of the real fields.

With this background, we now introduce a new function, the *mutual coherence function*, which depends on the products of fields at two different times and at two different points in space; its significance will become clear momentarily. We define it as

$$\Gamma(\mathbf{r}_1, \mathbf{r}_2, \tau) \equiv \langle U^*(\mathbf{r}_1, t)U(\mathbf{r}_2, t + \tau) \rangle.$$

(9.7)

Because of the assumption of statistical stationarity, the mutual coherence function only depends on the difference τ in times, and not on the origin of time t.

Up to this point, we have calculated average quantities such as the mutual coherence function by averaging over time. This is the physical operation that a detector would perform, but it is often more convenient mathematically to perform an average over a large number of hypothetically identical experimental configurations, known as an *ensemble*. We label each experiment by an integer k. Because of the randomness of the field, each experimental realization $^{(k)}U(\mathbf{r}, t)$ is different, and the average properties are calculated in a straightforward manner

$$\langle U^*(\mathbf{r}_1, t)U(\mathbf{r}_2, t+\tau)\rangle = \lim_{N\to\infty} \sum_{k=1}^{N} \frac{1}{N}\,^{(k)}U^*(\mathbf{r}_1, t)\,^{(k)}U(\mathbf{r}_2, t+\tau)\,.$$

(9.8)

It is by no means obvious that this ensemble average will produce the same result as the time average previously given. A field for which such an identity can be made is known as *ergodic*, and all experiments to date have confirmed that ergodicity is upheld.

With an ensemble average, we can easily prove a fundamental result of coherence theory, now often referred to as the *Wolf equations*. We note that a scalar wavefield $U(\mathbf{r}_i, t_i)$ propagating in vacuum satisfies the wave equation

$$\nabla_i^2 U(\mathbf{r}_i, t_i) - \frac{1}{c^2}\frac{\partial^2}{\partial t_i^2}U(\mathbf{r}_i, t_i) = 0,$$

(9.9)

where ∇_i is the gradient with respect to $\mathbf{r}_i$, $i = 1, 2$. If we take this equation with $i = 2$, and premultiply it by $U^*(\mathbf{r}_1, t_1)$, we get

$$\nabla_2^2 U^*(\mathbf{r}_1, t_1)U(\mathbf{r}_2, t_2) - \frac{1}{c^2}\frac{\partial^2}{\partial t_2^2}U^*(\mathbf{r}_1, t_1)U(\mathbf{r}_2, t_2) = 0.$$

(9.10)

If we take the ensemble average of this equation, we readily get the expression

$$\nabla_2^2 \Gamma(\mathbf{r}_1, \mathbf{r}_2, t_2 - t_1) - \frac{1}{c^2}\frac{\partial^2}{\partial t_2^2}\Gamma(\mathbf{r}_1, \mathbf{r}_2, t_2 - t_1) = 0,$$

(9.11)

and a similar calculation starting with Equation 9.9 with $i = 1$ gives

$$\nabla_1^2 \Gamma(\mathbf{r}_1, \mathbf{r}_2, t_2 - t_1) - \frac{1}{c^2}\frac{\partial^2}{\partial t_1^2}\Gamma(\mathbf{r}_1, \mathbf{r}_2, t_2 - t_1) = 0.$$

(9.12)

Finally, we introduce the new variable $\tau = t_2 - t_1$; the pair of equations can be written entirely in terms of τ in the form

$$\nabla_1^2 \Gamma(\mathbf{r}_1, \mathbf{r}_2, \tau) - \frac{1}{c^2} \frac{\partial^2}{\partial \tau^2} \Gamma(\mathbf{r}_1, \mathbf{r}_2, \tau) = 0, \tag{9.13}$$

$$\nabla_2^2 \Gamma(\mathbf{r}_1, \mathbf{r}_2, \tau) - \frac{1}{c^2} \frac{\partial^2}{\partial \tau^2} \Gamma(\mathbf{r}_1, \mathbf{r}_2, \tau) = 0. \tag{9.14}$$

These equations demonstrate that the mutual coherence function satisfies a pair of wave equations; it may be loosely stated that the statistical properties of a partially coherent field propagate as a wave. Though the calculation leading to these equations was straightforward, they were not considered obvious when first derived.[*] The wave properties of the mutual coherence function will be of great importance in understanding the singular optics related to it.

Defining the mutual coherence function is one thing; calculating it is quite another. Even for our simple model, given by Equation 9.1, it is necessary to calculate the average of $\exp[i(\phi(\mathbf{r}_1, t) - \phi(\mathbf{r}_2, t + \tau))]$, something that is often not possible even if the statistical model is known. Even worse, however, the statistics of the source are typically not derivable from first principles, and models of the mutual coherence function must be constructed.

As an example of the use and derivation of the mutual coherence function, we consider the classic Michelson interferometer, a simplified illustration of which is shown in Figure 9.1. A beam of light $U_0(t)$ incident from the left (from a star, for instance, or a thermal light source) is split into two at a 50/50 beamsplitter B and each half is sent to a reflect off of a mirror. After reflection, the two beams combine again at the beamsplitter and are sent to an observation screen S. The distance to mirror 1 is taken to be d_1 and the distance to mirror 2 is taken to be d_2. The ith beam is time delayed during the round trip by a time $\tau_i = 2d_i/c$, where c is again the speed of light. Neglecting time delays that are common between the two beams, the intensity at the observation screen may be written as

$$\langle I(t) \rangle = \langle [U_0(t - \tau_1) + U_0(t - \tau_2)]^* [U_0(t - \tau_1) + U_0(t - \tau_2)] \rangle. \tag{9.15}$$

---

[*] When Emil Wolf first explained the result to his employer Max Born, Born replied, "Wolf, you have always been such a sensible fellow, but now you have become completely crazy!" [Wol83].

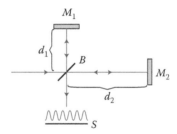

FIGURE 9.1 Simplified schematic of a Michelson interferometer. The differences between d_1 and d_2 have been greatly exaggerated.

Multiplying out all terms, we find that

$$\langle I(t) \rangle = \langle |U_0(t - \tau_1)|^2 + |U_0(t - \tau_2)|^2 + 2\mathrm{Re}\{U_0^*(t - \tau_1)U_0(t - \tau_2)\} \rangle. \tag{9.16}$$

The first two terms are each equal to the average intensity of the incident field, which may be written as the mutual coherence function at zero time delay, $\Gamma(0)$. The intensity at the observation screen is therefore

$$\langle I(t) \rangle = 2\Gamma(0) + 2\mathrm{Re}\left\{ \Gamma(\tau_1 - \tau_2) \right\}. \tag{9.17}$$

This expression may be further simplified by introducing a new quantity, the *complex degree of coherence*, defined generally as

$$\gamma(\mathbf{r}_1, \mathbf{r}_2, \tau) \equiv \frac{\Gamma(\mathbf{r}_1, \mathbf{r}_2, \tau)}{\sqrt{\Gamma(\mathbf{r}_1, \mathbf{r}_1, 0)\Gamma(\mathbf{r}_2, \mathbf{r}_2, 0)}}. \tag{9.18}$$

We then can write

$$\langle I(t) \rangle = 2\Gamma(0) \left[1 + \mathrm{Re}\left\{ \gamma(\tau_1 - \tau_2) \right\} \right]. \tag{9.19}$$

Clearly, the behavior of the intensity depends crucially on the statistical properties of the illuminating light, characterized by $\gamma(\tau)$. To understand the implications of the equation, we need a reasonable model for the mutual coherence function. We can construct such a model using the *Wiener–Khintchine theorem* [MW95, Section 2.4], which suggests that the power spectrum of a beam of light is the temporal Fourier transform of the mutual coherence function, that is,

$$S(\omega) = \frac{1}{2\pi} \int_{-\infty}^{\infty} \Gamma(\tau) e^{i\omega\tau} d\tau. \tag{9.20}$$

Conversely, the mutual coherence function may be determined from the power spectrum by the inverse transform

$$\Gamma(\tau) = \int\limits_{-\infty}^{\infty} S(\omega)e^{-i\omega\tau}d\omega. \tag{9.21}$$

We may therefore derive an expression for the mutual coherence function by working backward from a physically reasonable model for the power spectrum. Let us assume that the power spectrum $S_0(\omega)$ of light entering the Michelson interferometer is a single Gaussian spectral line centered on frequency ω_0 and of width Δ, such that

$$S_0(\omega) = S_0 \exp[-(\omega - \omega_0)^2/2\Delta^2]. \tag{9.22}$$

On substitution into Equation 9.21, it is found that

$$\Gamma(\tau) = \sqrt{2\pi}S_0 e^{-i\omega_0\tau}e^{-\tau^2\Delta^2/2}. \tag{9.23}$$

Furthermore, the degree of coherence is given by

$$\gamma(\tau) = e^{-i\omega_0\tau}e^{-\tau^2\Delta^2/2}. \tag{9.24}$$

It can be seen for this case that the absolute value of $\gamma(\tau)$ has a maximum value of unity and a minimum value of zero. Though we will not prove it here, it can be shown that these limits are true for $\gamma(\tau)$ in general. At a time delay such that $|\gamma(\tau)| = 0$, the field is said to be *incoherent*; at a time delay such that $|\gamma(\tau)| = 1$, the field is said to be *coherent*.

The variation of the intensity as the path length of the Michelson interferometer is changed is shown in Figure 9.2a. As the path length difference is made larger, the variation of the intensity at the detector plane drops off; we say that the *temporal coherence* of the field drops off with increasing path length. We may roughly characterize the intensity oscillations as disappearing once the time delay approaches a value of $\tau_c = 1/\Delta$; this time is known as the *coherence time*, and it is inversely related to the bandwidth of light. It should be noted that we have implicitly treated the field in our interferometer as being perfectly planar; in a realistic system, the incident field is beam like, and different directional components of the beam travel slightly different path lengths on propagation, resulting in a ring-like interference

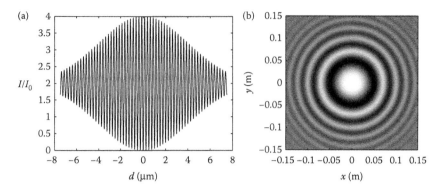

FIGURE 9.2 (a) Intensity of field on-axis at the observation plane in Michelson interferometer, as a function of path length difference d. (b) Simulation of the realistic interference pattern in the observation plane, for a fixed path length difference $d = 25\lambda$. Here, we have $d_1 \approx d_2 = 0.5$ m, $\lambda_0 = 500$ nm, and $\Delta = 0.01\omega_0$.

pattern, as illustrated in Figure 9.2b. The visibility of the fringes—the contrast between light and dark in the pattern—is then directly related to the complex degree of coherence. It can be seen in Figure 9.2b that the visibility drops away from the center of the pattern, resulting from lower coherence at longer path lengths.

Just as it is easier to work with monochromatic fields when solving ordinary wave equations, it can be advantageous to work in the frequency domain when investigating problems involving coherence. To that end, we now introduce the *cross-spectral density* $W(\mathbf{r}_1, \mathbf{r}_2, \omega)$, defined as

$$W(\mathbf{r}_1, \mathbf{r}_2, \omega) \equiv \frac{1}{2\pi} \int_{-\infty}^{\infty} \Gamma(\mathbf{r}_1, \mathbf{r}_2, \tau) e^{i\omega\tau} d\tau. \tag{9.25}$$

This expression may be considered to be a generalization of the Wiener–Khintchine theorem, Equation 9.20, to coherence functions of two spatial positions. If $\mathbf{r}_1 = \mathbf{r}_2 \equiv \mathbf{r}$, we recover the power spectrum of the field at point $\mathbf{r}$, that is,

$$S(\mathbf{r}, \omega) = W(\mathbf{r}, \mathbf{r}, \omega). \tag{9.26}$$

By taking the temporal Fourier transform of the Wolf equations, Equations 9.13 and 9.14, we find that the cross-spectral density satisfies its own propagation laws in the form of a pair of Helmholtz equations

$$\nabla_1^2 W(\mathbf{r}_1, \mathbf{r}_2, \omega) + k^2 W(\mathbf{r}_1, \mathbf{r}_2, \omega) = 0, \tag{9.27}$$

$$\nabla_2^2 W(\mathbf{r}_1, \mathbf{r}_2, \omega) + k^2 W(\mathbf{r}_1, \mathbf{r}_2, \omega) = 0, \tag{9.28}$$

where $k = \omega/c$. Again, these equations will prove to be extraordinarily useful in understanding the singular optics of partially coherent fields.

Because of the Fourier transform operation used to derive it, it is not obvious what observable property of the field the cross-spectral density represents. In the 1980s, however, Wolf [Wol82] demonstrated that the cross-spectral density at a given frequency ω may always be represented as the ensemble average of a new, fictitious, frequency ensemble of monochromatic fields, as

$$W(\mathbf{r}_1, \mathbf{r}_2, \omega) = \langle U^*(\mathbf{r}_1, \omega) U(\mathbf{r}_2, \omega) \rangle_\omega. \tag{9.29}$$

Here, the brackets $\langle \cdots \rangle_\omega$ represent an average over the new ensemble. The importance of this result is twofold: (1) it gives us a way to easily derive results of coherence theory in the space–frequency domain and (2) it provides a method to construct models of partially coherent fields.

We will see the second of these advantages in the next section. As for the first advantage, we return to Young's two-pinhole interferometer of Section 3.1, but now consider the illuminating field to be fluctuating. For simplicity, we imagine that the light emerging from the pinholes is filtered to an extremely narrow spectral range around ω, so that the cross-spectral density represents to a good approximation the complete behavior of the field. Alternatively, we could imagine that the filtering is done at the measurement plane.

The system is illustrated in Figure 9.3. Because we already know the monochromatic propagation laws for Young's interferometer, we can use them in Equation 9.29 to determine the spectrum of light that reaches the observation screen. To begin, we have

$$S(\mathbf{r}, \omega) = \left\langle \left[U(\mathbf{r}_1, \omega) \frac{e^{ikR_1}}{R_1} + U(\mathbf{r}_2, \omega) \frac{e^{ikR_2}}{R_2} \right]^* \right.$$

$$\left. \times \left[U(\mathbf{r}_1, \omega) \frac{e^{ikR_1}}{R_1} + U(\mathbf{r}_2, \omega) \frac{e^{ikR_2}}{R_2} \right] \right\rangle_\omega. \tag{9.30}$$

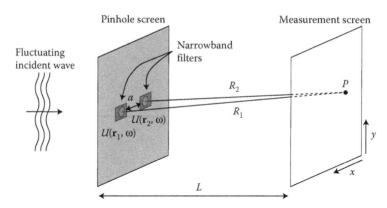

FIGURE 9.3 Young's experiment with filtered light.

Using the ensemble definition of the cross-spectral density, we may write

$$S(\mathbf{r}, \omega) = \frac{S(\mathbf{r}_1, \omega)}{R_1^2} + \frac{S(\mathbf{r}_2, \omega)}{R_2^2} + \frac{2}{R_1 R_2} \operatorname{Re}\left\{ W(\mathbf{r}_1, \mathbf{r}_2, \omega) e^{ik(R_2 - R_1)} \right\}.$$

$$(9.31)$$

In direct analogy with coherence theory in the time domain, we now introduce a new normalized function, the *spectral degree of coherence*

$$\mu(\mathbf{r}_1, \mathbf{r}_2, \omega) = \frac{W(\mathbf{r}_1, \mathbf{r}_2, \omega)}{\sqrt{W(\mathbf{r}_1, \mathbf{r}_1, \omega) W(\mathbf{r}_2, \mathbf{r}_2, \omega)}}.$$

$$(9.32)$$

As with the complex degree of coherence, it can be shown that $0 \le |\mu(\mathbf{r}_1, \mathbf{r}_2, \omega)| \le 1$. Applying the definition of μ to our two-pinhole spectrum, we have

$$S(\mathbf{r}, \omega) = \frac{S(\mathbf{r}_1, \omega)}{R_1^2} + \frac{S(\mathbf{r}_2, \omega)}{R_2^2} + \frac{2\sqrt{S(\mathbf{r}_1, \omega)S(\mathbf{r}_2, \omega)}}{R_1 R_2}$$

$$\times \operatorname{Re}\left\{ \mu(\mathbf{r}_1, \mathbf{r}_2, \omega) e^{ik(R_2 - R_1)} \right\}.$$

$$(9.33)$$

To better understand this expression, we make a few more simplifying assumptions. First, assuming that the observation screen is sufficiently far from the pinholes and that we are interested in the interference pattern only near the axis, we may take $R_1 \approx R_2 \equiv R$ in the denominators of Equation 9.33. We further assume that the spectrum of light is the same at both pinholes, that is, $S(\mathbf{r}_1, \omega) = S(\mathbf{r}_2, \omega) = S_0(\omega)$. Finally, we write

$\mu(\mathbf{r}_1, \mathbf{r}_2, \omega)$ in terms of its amplitude and phase

$$\mu(\mathbf{r}_1, \mathbf{r}_2, \omega) = |\mu(\mathbf{r}_1, \mathbf{r}_2, \omega)| \exp[i\phi_\mu(\mathbf{r}_1, \mathbf{r}_2, \omega)]. \quad (9.34)$$

With these conditions, the expression for the spectral density on the observation plane becomes

$$S(\mathbf{r}, \omega) = \frac{2S_0(\omega)}{R^2} \left\{ 1 + |\mu(\mathbf{r}_1, \mathbf{r}_2, \omega)| \cos\left[\phi_\mu(\mathbf{r}_1, \mathbf{r}_2, \omega) + k(R_2 - R_1) \right] \right\}. \quad (9.35)$$

From this expression, it can be seen that the power spectrum only possesses significant interference fringes when $|\mu(\mathbf{r}_1, \mathbf{r}_2, \omega)| \approx 1$, and possesses no fringes at all when $|\mu(\mathbf{r}_1, \mathbf{r}_2, \omega)| = 0$. The spectral degree of coherence is a measure of the *spatial coherence* of the field at points $\mathbf{r}_1$ and $\mathbf{r}_2$, and represents the strength of field correlations between these points as well as the ability of those fields to produce interference patterns.

This spectral density is plotted for several values of $|\mu|$ in Figure 9.4. As in the time-domain case, it can be seen that the contrast of the interference fringes is directly related to the magnitude of μ. In fact, complete destructive interference only occurs when $|\mu| = 1$, that is, when the field emerging from the pinholes is spatially coherent. However, it should be remembered that a two-pinhole interferometer produces nongeneric interference fringes; we will see in Section 9.4 that things change for more complicated systems.

One more result, also from Wolf [Wol82], will be useful in our discussion. The cross-spectral density can be shown to be Hermitian, that is,

$$W(\mathbf{r}_1, \mathbf{r}_2, \omega) = W^*(\mathbf{r}_2, \mathbf{r}_1, \omega), \quad (9.36)$$

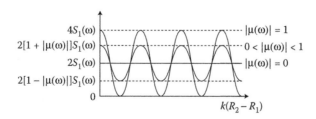

FIGURE 9.4   The observed spectral density for several values of $|\mu|$. Here, $S_1(\omega) = S_0(\omega)/R^2$, the spectrum observed from a single pinhole on the observation screen.

and nonnegative definite, such that

$$\int\limits_{-\infty}^{\infty} \int\limits_{-\infty}^{\infty} W(\mathbf{r}_1, \mathbf{r}_2, \omega) f^*(\mathbf{r}_1) f(\mathbf{r}_2) d^2 r_1 d^2 r_2 \geq 0, \qquad (9.37)$$

for any well-behaved choice of function $f(\mathbf{r})$ in the domain of integration (here taken to be a plane transverse to the direction of propagation). From these two properties, and in direct analogy to a theory of linear algebra, it can be shown that the cross-spectral density can be written in a series form

$$W(\mathbf{r}_1, \mathbf{r}_2, \omega) = \sum_{n}^{N} \lambda_n(\omega) \phi_n^*(\mathbf{r}_1, \omega) \phi_n(\mathbf{r}_2, \omega), \qquad (9.38)$$

where $\lambda_n(\omega) > 0$ are the eigenvalues of the cross-spectral density, and the functions $\phi_n(\mathbf{r}, \omega)$ are its orthonormal eigenfunctions, that is,

$$\int\limits_{-\infty}^{\infty} \int\limits_{-\infty}^{\infty} W(\mathbf{r}_1, \mathbf{r}_2, \omega) \phi_n(\mathbf{r}_1, \omega) d^2 r_1 = \lambda_n(\omega) \phi_n(\mathbf{r}_2, \omega), \qquad (9.39)$$

$$\int\limits_{-\infty}^{\infty} \phi_n^*(\mathbf{r}, \omega) \phi_m(\mathbf{r}, \omega) d^2 r = \delta_{mn}. \qquad (9.40)$$

This sum may be over a finite number of orthonormal modes N or over one or more indices of an infinite sum of modes. Expression 9.38 is known as the *coherent mode representation* of a partially coherent wavefield. As a general rule, the more modes with nonzero eigenvalues $\lambda_n(\omega)$ that a field possesses, the less spatially coherent it is. Conversely, if a partially coherent field possesses only a single mode ($N = 1$), it is fully coherent; this latter case includes ordinary monochromatic wavefields.

The coherent mode representation may be readily understood as the functional equivalent to finding the diagonal representation of a matrix. The eigenfunctions $\phi_n(\mathbf{r}, \omega)$ are analogous to the eigenvectors of a matrix, while the eigenvalues $\lambda_n(\omega)$ are directly analogous to their finite matrix counterparts. Just as a Hermitian matrix may always be written in diagonal form, so a Hermitian cross-spectral density may be expressed in a coherent mode representation.

This whirlwind tour of coherence theory now equips us to consider what type of wave singularities might exist in correlation functions.

9.2 SINGULARITIES OF CORRELATION FUNCTIONS

The discussion of Young's two-pinhole experiment for partially coherent fields suggested that only fully coherent fields can produce zeros of intensity, and consequently singularities of phase, in any interference experiment. We will later see that this is not entirely true, but we can also readily show that intensity zeros are nongeneric features of a wavefield.

To do so, we look for such zeros in the context of the coherent mode expansion, Equation 9.38. From this equation, we can see that the power spectrum will only be zero if

$$S(\mathbf{r}, \omega) = \sum_{n}^{N} \lambda_n(\omega) |\phi_n(\mathbf{r}, \omega)|^2 = 0. \tag{9.41}$$

However, each mode has a real and imaginary part that must simultaneously go to zero. This implies that, to get a zero of the power spectrum, we must have $2N$ functions that are simultaneously zero at the same position in three-dimensional space. If $N = 1$, that is, if the field is fully coherent, we must satisfy two equations with three spatial degrees of freedom, and the typical result is a set of lines. For $N > 1$, however, we must satisfy $2N$ equations with three spatial degrees of freedom; except in very special cases, *there will be no zeros of the power spectrum in a field which has partial spatial coherence*. Zeros of the power spectrum are nongeneric.

This observation would seem to bode ill for the singular optics of partially coherent fields, but we only need to broaden our sights a bit to find singularities of a significantly different character. We recall from Equations 9.27 and 9.28 that the cross-spectral density satisfies a pair of independent Helmholtz equations in $\mathbf{r}_1$ and $\mathbf{r}_2$. If we fix one of the spatial points, say $\mathbf{r}_1$, the cross-spectral density propagates exactly like a monochromatic wave with respect to the variable $\mathbf{r}_2$. Because we already know that optical vortices are typical features of a monochromatic wave, we can see that the cross-spectral density itself possesses generic vortices when one observation point is fixed; we call them *correlation vortices*.

But what do these correlation vortices represent? From the definition of the spectral degree of coherence, we may always write

$$W(\mathbf{r}_1, \mathbf{r}_2, \omega) = \sqrt{S(\mathbf{r}_1, \omega) S(\mathbf{r}_2, \omega)} \, \mu(\mathbf{r}_1, \mathbf{r}_2, \omega). \tag{9.42}$$

We have just seen that the spectral density is, in general, nonzero in partially coherent fields. This implies that a correlation vortex is a location for which $\mu(\mathbf{r}_1, \mathbf{r}_2, \omega) = 0$; in other words, the fields at points $\mathbf{r}_1$ and $\mathbf{r}_2$ are uncorrelated at a correlation vortex.

A correlation vortex is a two-point singularity; that is, its location depends on the values of both $\mathbf{r}_1$ and $\mathbf{r}_2$, and if we change the fixed observation point $\mathbf{r}_1$, the location of the vortex in $\mathbf{r}_2$ changes accordingly. This statement, however, raises the question of what exactly is meant by an "observation point" and how one measures singularities of the cross-spectral density. The most straightforward technique is to use Young's two-pinhole experiment, as in Figure 9.3. Leaving one pinhole fixed at $\mathbf{r}_1$ to provide a reference field, we can deduce the magnitude and phase of $W(\mathbf{r}_1, \mathbf{r}_2, \omega)$ from the interference pattern produced by the light coming from a (movable) pinhole at $\mathbf{r}_2$.

The Hermitian nature of $W(\mathbf{r}_1, \mathbf{r}_2, \omega)$, as given by Equation 9.36, suggests at least one interesting symmetry of the singularities. Let us suppose, with observation point $\mathbf{r}_1 = \mathbf{r}_A$, that there exists a right-handed correlation singularity at $\mathbf{r}_2 = \mathbf{r}_B$. Then, due to Hermiticity, there will be a left-handed singularity at point $\mathbf{r}_A$ with observation point at $\mathbf{r}_B$, as $W(\mathbf{r}_A, \mathbf{r}_B, \omega) = W^*(\mathbf{r}_B, \mathbf{r}_A, \omega)$.

To further explore the properties of correlation vortices, it will help to have a model to work with, ideally one that characterizes the generic features of the vortices. The natural option is to take one of the Laguerre–Gauss beams that we anatomically dissected in Chapter 2 and apply some sort of randomization to it. We consider a left-handed beam of order $n = 0, m = 1$, and restrict our study to its behavior in the $z = 0$ plane. The field then has the form

$$U(x,y) = \sqrt{\frac{4}{\pi}} \frac{x + iy}{w_0^2} \exp[-(x^2 + y^2)/w_0^2], \tag{9.43}$$

where w_0 is the width of the beam. To randomize it, we assume that the central axis of the beam, situated at $\rho_0 = (x_0, y_0)$, is allowed to wander in the transverse plane, and that the probability of being located at position ρ_0 is given by

$$f(\rho_0) = \frac{1}{\pi\delta^2} \exp[-\rho_0^2/\delta^2]. \tag{9.44}$$

The fields for different values of ρ_0 represent different members of a monochromatic ensemble; we may then average over this ensemble, as formally shown in Equation 9.29, to get the cross-spectral density. In terms of this specific example, this average may be written as

$$W(\mathbf{r}_1, \mathbf{r}_2, \omega) = \iint f(\boldsymbol{\rho}_0) U^*(\boldsymbol{\rho}_1 - \boldsymbol{\rho}_0) U(\boldsymbol{\rho}_2 - \boldsymbol{\rho}_0) d^2 \boldsymbol{\rho}_0. \tag{9.45}$$

This model has been dubbed the *beam wander model* of a partially coherent vortex field; it was first introduced in Gbur, Visser, and Wolf [GVW04b]. When $\delta = 0$, the beam does not wander at all and the field is fully coherent. As δ increases, then, the spatial coherence of the field decreases.

The integral can be evaluated analytically, and the result takes the form

$$W(\boldsymbol{\rho}_1, \boldsymbol{\rho}_2, \omega) = Q \left\{ \left[\frac{1}{\alpha^2}(x_1 - iy_1) - \frac{1}{w_0^2}(x_2 - iy_2) \right] \right.$$
$$\left. \times \left[\frac{1}{\alpha^2}(x_2 + iy_2) - \frac{1}{w_0^2}(x_1 + iy_1) \right] + \frac{1}{\beta^2} \right\}, \tag{9.46}$$

where

$$\frac{1}{\alpha^2} \equiv \frac{1}{\delta^2} + \frac{1}{w_0^2}, \tag{9.47}$$

$$\frac{1}{\beta^2} \equiv \frac{1}{\delta^2} + \frac{2}{w_0^2}, \tag{9.48}$$

and Q is a real-valued factor given by

$$Q = \frac{4\beta^6}{\pi\delta^2} \exp\left[-\frac{\beta^2}{w_0^2 \delta^2}(\rho_1^2 + \rho_2^2) \right] \exp\left[-\frac{\beta^2}{w_0^2}(\boldsymbol{\rho}_1 - \boldsymbol{\rho}_2)^2 \right], \tag{9.49}$$

which plays no role in the location of zeros.

We can use Equation 9.46 to study the behavior of correlation vortices for various states of coherence. We first look at the behavior of $W(\boldsymbol{\rho}_1, \boldsymbol{\rho}_2, \omega)$ as the spatial coherence is decreased.

Figure 9.5 gives an example. Several notable things occur as the spatial coherence is decreased. First, it can be seen that the phase singularity of a coherent field automatically becomes a correlation singularity of a partially coherent field; the two objects, seemingly distinct, are in fact different manifestations of the same wavefield circulation. Second, it can be seen that the original vortex moves away from the origin along the line containing the origin and the observation point $\boldsymbol{\rho}_1$. Third, a vortex of opposite handedness

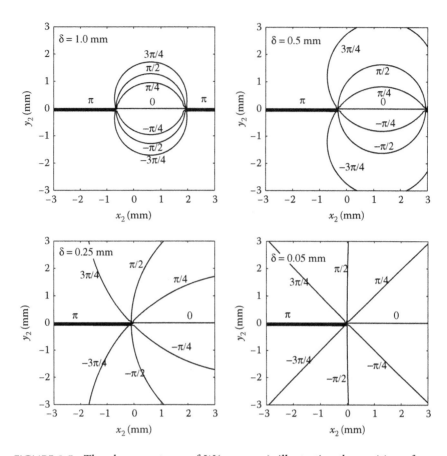

FIGURE 9.5 The phase contours of $W(\rho_1, \rho_2, \omega)$, illustrating the position of correlation vortices as δ increases. Here, we have taken $w_0 = 1.0$ mm, $x_1 = 0.5$ mm, and $y_1 = 0$.

comes from infinity along the same line. The phase lines of the two vortices become increasingly compressed along a direction transverse to their axis. As $\delta \to \infty$, it can be shown that both singularities asymptotically lie on a circle centered on the observation point.

Figure 9.6 shows how the position of the correlation vortices changes as the observation point is moved. It can be shown that the singularities move hardly at all for a highly coherent field, while they move significantly for low coherence fields.

These observations answer the question we asked at the beginning of this chapter: what happens to phase singularities when the phase is scrambled? We now see that those optical vortices become correlation vortices. In

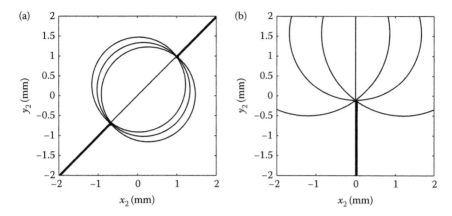

FIGURE 9.6 The position of correlation vortices as the observation point is changed, with (a) $x_1 = 0.1$ mm, $y_1 = 0.1$ mm and (b) $x_1 = 0$, $y_1 = 1.0$ mm. In both cases, $w_0 = 1.0$ mm and $\delta = 0.75$ mm.

fact, as no field is perfectly coherent, we may say that we are almost always observing correlation vortices in experiments, as all fields necessarily possess some slight degree of randomness. If the field is sufficiently coherent, however, those correlation vortices will stay localized near zeros of intensity and be indistinguishable from phase singularities.

Correlation singularities were first theoretically predicted [SGVW03] in the (nongeneric) Young's two-pinhole experiment with partially coherent light. Not long afterward, a number of different research groups independently investigated such singularities, in the context of partially coherent beams [GV03b], optical Lissajous fields [Fre03] (to be discussed in detail in Section 12.6), separable phase vortex beams [BFP+03], and the cross-correlation function [PMM+04].

This latter case can be illustrated using the beam wander model. In our case, the cross-correlation function $C(\mathbf{r})$ is simply the cross-spectral density evaluated at anti-diagonal positions

$$C(\mathbf{r}) = W(\mathbf{r}, -\mathbf{r}, \omega). \tag{9.50}$$

Using Equation 9.46, the cross-correlation function can be shown to take the form

$$C(\mathbf{r}) = -Q\left\{\frac{1}{\beta^4}(x^2 + y^2) - \frac{1}{\beta^2}\right\}. \tag{9.51}$$

This expression is completely real-valued, and only goes to zero when

$$x^2 + y^2 = \frac{\delta^2}{1 + \dfrac{2\delta^2}{w_0^2}}. \tag{9.52}$$

This is the equation of a circle; the cross-correlation function $C(\mathbf{r})$ exhibits a ring dislocation centered on the origin, and its phase jumps by π across the dislocation. It can be seen from Equation 9.52 that the radius of this dislocation increases monotonically from zero as δ increases from zero, and that it approaches a limit of $w_0^2/2$ as $\delta \to \infty$. This is roughly the behavior found by Palacios et al. [PMM$^+$04], both theoretically and experimentally; more detail can be found in Maleev et al. [MPMS04]. We will discuss the experimental side of this research in Section 9.6.

Spatial correlation singularities are not the only type of singularities possible in a partially coherent field. Assuming a polychromatic vortex beam, researchers have also demonstrated [SS04] that so-called *temporal correlation vortices* that exist at zeros of the complex degree of coherence $\gamma(\mathbf{r}_1, \mathbf{r}_2, \tau)$, where τ is fixed.

Spatially coherent but polychromatic beams can exhibit optical vortices at each frequency, and this leads to unusual changes of spectra in the neighborhood of such vortices. This observation was first made in Gbur, Visser, and Wolf [GVW02a], looking at dramatic spectral changes in the neighborhood of focus of a polychromatic field, and has spurred much investigation between phase singularities and spectral changes.

The effect can be summarized, in short, as follows. Let us assume that an optical system produces a phase singularity at a spatial location $\mathbf{r}_0$ at the center frequency ω_0 of the illuminating light. The spectrum at $\mathbf{r}_0$ will not contain the frequency ω_0 at all, and the overall transmitted spectrum will have a minimum at that frequency. Other illuminating frequencies, however, will propagate differently through the system (due, for instance, to diffractive and dispersion effects) and will produce singularities at locations different from $\mathbf{r}_0$, leading to different spectral modifications in the neighborhood of this point.

These changes can be dramatic, as we can illustrate with a simple example taken from Gbur, Visser, and Wolf [GVW04b]. Let us consider the coherent superposition of three polychromatic plane waves with propagation directions $\mathbf{s}_1 = \hat{\mathbf{z}}$, $\mathbf{s}_2 = \hat{\mathbf{z}} \cos \theta_0 + \hat{\mathbf{x}} \sin \theta_0$, and $\mathbf{s}_3 = \hat{\mathbf{z}} \cos \theta_0 - \hat{\mathbf{x}} \sin \theta_0$ and with identical spectra $S_0(\omega)$. The field at a single frequency can be written in the form

$$U(\mathbf{r}, \omega) = \sqrt{S_0(\omega)} \left[e^{i k \mathbf{s}_1 \cdot \mathbf{r}} + e^{i k \mathbf{s}_2 \cdot \mathbf{r}} + e^{i k \mathbf{s}_3 \cdot \mathbf{r}} \right], \qquad (9.53)$$

where $k = \omega/c$. Using this as the sole member of the space–frequency ensemble defined by Equation 9.29, we may write

$$W(\mathbf{r}_1, \mathbf{r}_2, \omega) = U^*(\mathbf{r}_1, \omega) U(\mathbf{r}_2, \omega). \qquad (9.54)$$

The spectral density may then be written as

$$S(\mathbf{r}, \omega) = S_0(\omega) \{3 + 2 \cos [kz(1 - \cos \theta_0) - kx \sin \theta_0]$$
$$+ 2 \cos [kz(1 - \cos \theta_0 + kx \sin \theta_0] + 2 \cos [2kx \sin \theta_0] \}. \qquad (9.55)$$

An example of the result is shown in Figure 9.7. A Gaussian spectral density of width σ_0 is assumed, and part (a) of the figure shows the effective red-shift as a function of position. It can be seen that there are dramatic swings of the spectrum from red to blue shift in isolated regions; it can be demonstrated that the center of these regions are edge dislocations at the center frequency ω_0. The origin of these changes is illustrated in Figure 9.7b: the spectral location of the singularity varies with position, leading to the red, blue, or central part of the spectrum being suppressed. The splitting of the spectrum also leads to an effective spike in the variance of the spectrum, as was illustrated in Gbur, Visser, and Wolf [GVW02b].

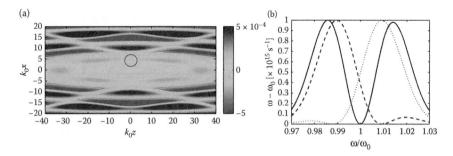

FIGURE 9.7 (a) Effective redshifts and blueshifts of the spectrum as a function of position, showing the shift in average frequency ω. (b) Detail of the normalized spectrum at selected points in the neighborhood of a singularity. The solid line is $k_0 x = 4.189$, the short dashes are $k_0 x = 4.240$, and the long dashes are $k_0 x = 4.150$. For this figure, $\omega_0 = 10^{15}$ Hz, $\sigma_0/\omega_0 = 0.01$, and $\theta_0 = \pi/6$. (Based on results of G. Gbur, T.D. Visser, and E.Wolf. *J. Opt. A*, 6:S239–S242, 2004.)

FIGURE 9.8 Universal color pattern near a polychromatic optical vortex. (After M.V. Berry. *New J. Phys.*, 4:74.1–74.14, 2002.)

This behavior of the spectrum near a phase singularity is generic. Not long after these results were released, Berry [Ber02a,Ber02b] approached the problem from an interesting perspective: what would the colors of such a singularity look like to the human eye? By mapping the spectrum near the singularity to the human trichromatic sensitivity and adjusting the luminosity, he found a simple and elegant pattern of red, orange, and blue. An example of this is shown in Figure 9.8.

9.3 GENERIC STRUCTURE OF A CORRELATION SINGULARITY

One significant limitation of our discussion of spatial correlation singularities so far is its focus on a simple model. This calculation gave no indication of whether the singular structure found was generic or a special case. Furthermore, our discussion of this structure was limited to simple projections of the cross-spectral density—either one point fixed or the cross-correlation function—and gave no indication of the complete structure in the transverse plane. In this section, we clarify these issues, first making an argument for the genericity of the beam wander vortex structure and then describing that structure in the most comprehensive way possible.

To investigate genericity, we note that many of our methods of producing optical vortices, as discussed in Chapter 4, involve the transformation of a beam via a linear optical system. A linear optical system is one whose output field $U(\mathbf{r}, \omega)$ is directly proportional to the input, $U_0(\mathbf{r}, \omega)$, through the relation

$$U(\mathbf{r}, \omega) = \int U_0(\mathbf{r}', \omega) f(\mathbf{r}, \mathbf{r}', \omega) d^2 r', \qquad (9.56)$$

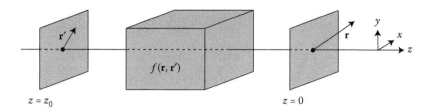

$z = z_0$ $z = 0$

FIGURE 9.9 Illustrating the notation for a linear optical system.

where $f(\mathbf{r}, \mathbf{r}', \omega)$ is the *kernel* (or point-spread function) of the optical system and represents the output due to a spatially localized input. The notation for such a system is given in Figure 9.9. The plane $z = z_0$ is taken as the input plane of the system, and $z = 0$ is the output plane. The vector $\mathbf{r}'$ represents a position in the plane $z = z_0$, while $\mathbf{r}$ represents a three-dimensional position beyond the system.

Using, once again, the space–frequency ensemble, we can immediately write the cross-spectral density of the output in terms of the input as

$$W(\mathbf{r}_1, \mathbf{r}_2, \omega) = \iint W_0(\mathbf{r}_1', \mathbf{r}_2', \omega) f^*(\mathbf{r}_1, \mathbf{r}_1', \omega) f(\mathbf{r}_2, \mathbf{r}_2', \omega) d^2 r_1' d^2 r_2'. \quad (9.57)$$

We now assume that the input field is *Schell-model* [MW95, Section 5.2], which is defined as a field whose spectral degree of coherence only depends on the difference between the measurement points, and therefore satisfies the relation

$$\mu_0(\mathbf{r}_1', \mathbf{r}_2', \omega) = \mu_0(\mathbf{r}_2' - \mathbf{r}_1', \omega). \quad (9.58)$$

The beam is referred to as "Schell-model" after Schell [Sch61], who first studied such translation-invariant correlations.

Because it possesses this simple form, μ_0 may be written simply in terms of an inverse Fourier transform as

$$\mu_0(\mathbf{R}, \omega) = \int \tilde{\mu}_0(\mathbf{K}, \omega) e^{i\mathbf{K}\cdot\mathbf{R}} d^2 K. \quad (9.59)$$

If we substitute this expression into Equation 9.57, and define a modified system kernel as

$$F(\mathbf{r}, \mathbf{r}', \omega) \equiv f(\mathbf{r}, \mathbf{r}', \omega) \sqrt{S_0(\mathbf{r}', \omega)}, \quad (9.60)$$

we may write the output cross-spectral density of the system in the form

$$W(\mathbf{r}_1, \mathbf{r}_2, \omega) = (2\pi)^4 \int \tilde{\mu}_0(\mathbf{K}, \omega) \tilde{F}^*(\mathbf{r}_1, \mathbf{K}, \omega) \tilde{F}(\mathbf{r}_2, \mathbf{K}, \omega) d^2 K, \quad (9.61)$$

where we have defined

$$\tilde{F}(\mathbf{r}, \mathbf{K}, \omega) \equiv \frac{1}{(2\pi)^2} \int F(\mathbf{r}, \mathbf{r}', \omega) e^{-i\mathbf{K}\cdot\mathbf{r}'} d^2 r'. \quad (9.62)$$

Let us next assume that the function $\mu_0(\mathbf{R}, \omega)$ is real-valued, and can be expressed in the form

$$\mu_0(\mathbf{R}, \omega) \equiv \xi_0 \left[|\mathbf{R}|/\Delta \right], \quad (9.63)$$

where ξ_0 is a function for which $\xi_0(0) = 1$ and which decays rapidly for large values of $|\mathbf{R}|$. These assumptions give us the simplified case that the input field has a constant phase in the plane $z = z_0$ and that the effective correlation length in that plane is Δ. When $\Delta = \infty$, we then have $\mu_0(\mathbf{r}, \omega) = 1$, and $\tilde{\mu}_0(\mathbf{K}, \omega) = \delta^{(2)}(\mathbf{K})$, where $\delta^{(2)}$ is the two-dimensional Dirac delta function. We find that, in the coherent limit, the output of our linear optical system is

$$W^{(coh)}(\mathbf{r}_1, \mathbf{r}_2, \omega) = (2\pi)^4 \tilde{F}^*(\mathbf{r}_1, 0, \omega) \tilde{F}(\mathbf{r}_2, 0, \omega), \quad (9.64)$$

with the coherent field given by $U^{(coh)}(\mathbf{r}, \omega) = (2\pi)^2 \tilde{F}(\mathbf{r}, 0, \omega)$.

We are now in a position to evaluate how the output of a linear optical system changes when the spatial coherence of the input field is decreased. Assuming that Δ is extremely large, we expect that $\tilde{\mu}_0(\mathbf{K}, \omega)$ will be extremely localized in $\mathbf{K}$-space. We are then justified in approximating $\tilde{F}(\mathbf{K}, \omega)$ by the lowest two terms of its Taylor series expansion in $\mathbf{K}$, as

$$\tilde{F}(\mathbf{r}, \mathbf{K}, \omega) \approx \tilde{F}(\mathbf{r}, 0, \omega) + \gamma(\mathbf{r}, \omega) \cdot \mathbf{K}, \quad (9.65)$$

where

$$\gamma(\mathbf{r}, \omega) \equiv \nabla_K \tilde{F}(\mathbf{r}, \mathbf{K})\big|_{\mathbf{K}=0}, \quad (9.66)$$

and ∇_K is the gradient with respect to the vector $\mathbf{K}$. The general expression for $\tilde{F}(\mathbf{r}, \mathbf{K}, \omega)$ may then be written as

$$\tilde{F}(\mathbf{r}, \mathbf{K}, \omega) \approx \frac{U^{(coh)}(\mathbf{r}, \omega)}{(2\pi)^2} + \gamma(\mathbf{r}, \omega) \cdot \mathbf{K}. \quad (9.67)$$

We may now substitute from Equation 9.67 into Equation 9.61. The Fourier integrals involving $\mathbf{K}$ can be evaluated explicitly, and the result is of the form

$$W(\mathbf{r}_1, \mathbf{r}_2, \omega) = U^{(coh)*}(\mathbf{r}_1, \omega)U^{(coh)}(\mathbf{r}_2, \omega)$$
$$+ (2\pi)^4 \gamma^*(\mathbf{r}_1, \omega) \cdot \gamma(\mathbf{r}_2, \omega)D_0/\Delta^2, \tag{9.68}$$

where

$$D_0 \equiv -\frac{\Delta^2}{2}\nabla_\perp^2 \mu_0(\mathbf{R})\bigg|_{R=0}, \tag{9.69}$$

and $\nabla_\perp^2$ is the transverse Laplacian. It can be immediately seen that the cross-spectral density reduces to the coherent limit as $\Delta \to \infty$.

Now, let us assume that, when the input field is spatially coherent, the output field produces a phase singularity at point $\mathbf{r}_0$. If points $\mathbf{r}_1$ and $\mathbf{r}_2$ are sufficiently close to the singularity, we can approximate $\tilde{F}(\mathbf{r}, 0, \omega)$ and $\gamma(\mathbf{r}_2, \omega)$ by the lowest nonzero term of their Taylor expansions

$$\tilde{F}(\mathbf{r}, 0, \omega) \approx \boldsymbol{\beta} \cdot (\mathbf{r}_1 - \mathbf{r}_0), \tag{9.70}$$

$$\gamma(\mathbf{r}, \omega) \approx \gamma(\mathbf{r}_0, \omega), \tag{9.71}$$

where

$$\boldsymbol{\beta} \equiv \nabla\tilde{F}(\mathbf{r}, 0, \omega)\big|_{\mathbf{r}=\mathbf{r}_0}. \tag{9.72}$$

The cross-spectral density can be approximated by

$$W(\mathbf{r}_1, \mathbf{r}_2, \omega) = (2\pi)^4 \left\{ [\boldsymbol{\beta} \cdot (\mathbf{r}_1 - \mathbf{r}_0)]^* [\boldsymbol{\beta} \cdot (\mathbf{r}_2 - \mathbf{r}_0)] \right.$$
$$\left. + \gamma^*(\mathbf{r}_0, \omega) \cdot \gamma(\mathbf{r}_0, \omega)D_0/\Delta^2 \right\}. \tag{9.73}$$

We now specialize this result to the case of a screw dislocation located at the origin, so we may write

$$U^{(coh)}(\mathbf{r}, \omega) = x + iy. \tag{9.74}$$

We come to the final result

$$W(\mathbf{r}_1, \mathbf{r}_2, \omega) = (2\pi)^4 \left\{ [x_1 + iy_1]^* [x_2 + iy_2] \right.$$
$$\left. + \gamma^*(0, \omega) \cdot \gamma(0, \omega)D_0/\Delta^2 \right\}. \tag{9.75}$$

If we compare this result to Equation 9.46 for the beam wander model, we find that there is excellent agreement for highly coherent fields, that is, $\delta \ll 1$. This suggests that the beam wander model is an archetypical representation of what happens to a screw dislocation when the spatial coherence of the beam carrying it is decreased.

With this in mind, we can now elaborate somewhat on the generic features of this partially coherent screw dislocation. As we have seen, the singularity of this system appears as a pair of correlation vortices when measured with one observation point fixed, and appears as a ring dislocation when the cross-correlation function is measured. Our goal is now to unify these two seemingly different results and determine the general structure of the field with respect to all 4 degrees of freedom, namely, x_1, x_2, y_1, and y_2.

Returning to Equation 9.46, we recall that the cross-spectral density will only be zero if its real and imaginary parts are simultaneously zero. The condition $\text{Im}\{W\} = 0$ readily reduces to the form

$$x_1 y_2 = y_1 x_2. \tag{9.76}$$

If we rewrite this in polar coordinates, that is, $x_1 + i y_1 = \rho_1 e^{i\phi_1}$, we find that

$$\tan(\phi_1) = \tan(\phi_2). \tag{9.77}$$

We come to the important observation that singularities of the cross-spectral density only appear for pairs of points that are collinear with the origin. This is in agreement with the behavior of correlation vortices seen in Figure 9.6. Applying the condition that $\phi_1 = \phi_2$ to the condition $\text{Re}\{W\} = 0$, we get the simple constraint

$$\eta \rho_1^2 + \eta \rho_2^2 - (1 + \eta^2)\rho_1 \rho_2 = \frac{\alpha^4}{\beta^2}, \tag{9.78}$$

where

$$\eta \equiv \frac{\alpha^2}{w_0^2}. \tag{9.79}$$

Equation 9.78 is the equation of a hyperbola, examples of which are shown for several values of the wander parameter δ in Figure 9.10a.

This structure was first derived in Gbur and Swartzlander [GGS08], and it reconciles the different projections of the correlation singularity

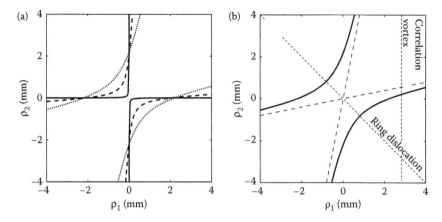

FIGURE 9.10 (a) The zero surfaces for a correlation singularity for different values of δ. The solid line is $\delta = 0.1$ mm, long dashed line is $\delta = 0.5$ mm, and short dashed line is $\delta = 1.0$ mm. For all curves, $w_0 = 1$ mm. (b) Illustrating the relationship between a correlation singularity and its projections, with $\delta = 1.0$ mm and $w_0 = 2$ mm.

described earlier. For both correlation vortices and the ring dislocations, the singularities lie along the line $\phi_1 = \phi_2$. Figure 9.10b shows how these projections manifest in terms of the radial variables ρ_1 and ρ_2.

The model presented so far only considers the singularity in what is effectively the waist plane of a partially coherent beam. On propagation, the basic hyperbolic feature of the correlation singularity is maintained, though other parts of the singularity undergo nontrivial evolution, as was investigated by Stahl and Gbur [SG14].

The generic nature of correlation singularities has been confirmed for a specific case by Fischer and Visser [FV04], who demonstrated such singularities in the focusing of partially coherent light.

9.4 PHASE SINGULARITIES IN PARTIALLY COHERENT FIELDS

We noted at the beginning of Section 9.2 that zeros of the spectral density in partially coherent fields are nongeneric. We therefore do not typically expect to see such zeros, as they normally are replaced by zeros of the cross-spectral density. However, it is possible to design a partially coherent wavefield that has zeros of the spectrum, and in this section we see the implications of this result.

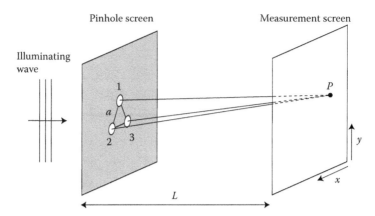

FIGURE 9.11   Illustration of "Young's three-pinhole experiment."

We have seen that Young's two-pinhole interferometer will only produce intensity zeros when the illuminating field is fully coherent; the next natural step is to consider Young's *three*-pinhole interferometer and see if the same result holds. The geometry is illustrated again in Figure 9.11.

The spectral density at the position $P(\mathbf{r})$ on the screen is given by the expression

$$S(\mathbf{r}, \omega) = \langle |U_1(\mathbf{r}, \omega) + U_2(\mathbf{r}, \omega) + U_3(\mathbf{r}, \omega)|^2 \rangle_\omega, \qquad (9.80)$$

where

$$U_i(\mathbf{r}, \omega) = U_0(\mathbf{r}_i, \omega) \frac{e^{ikR_i}}{R_i}, \qquad (9.81)$$

and $i = 1, 2, 3$. If we make the identification

$$\langle U_0^*(\mathbf{r}_i, \omega) U_0(\mathbf{r}_j, \omega) \rangle_\omega = \sqrt{S_i}\sqrt{S_j}\mu_{ij}, \qquad (9.82)$$

with S_i the spectral density at the ith pinhole and μ_{ij} the spectral degree of coherence between the ith and jth pinholes, we may write

$$S(\mathbf{r}, \omega) = \sum_{i=1}^{3} \frac{S_i}{R_i^2} + \sum_{i<j\leq 3} \frac{\sqrt{S_iS_j}}{R_iR_j}\left[\mu_{ij}e^{ik(R_j-R_i)} + \mu_{ij}^*e^{-ik(R_j-R_i)}\right]. \qquad (9.83)$$

We again make the now-familiar assumption that $R_1 \approx R_2 \approx R_3 \approx R$ in the denominators of this equation. Once we have done so, the spectral density can be written in an exceedingly simple matrix form

$$R^2 S(\mathbf{r}, \omega) = \mathbf{x}^\dagger \mathbf{M} \mathbf{x}, \tag{9.84}$$

where

$$\mathbf{M} \equiv \begin{bmatrix} 1 & \mu_{12} & \mu_{13} \\ \mu_{12}^* & 1 & \mu_{23} \\ \mu_{13}^* & \mu_{23}^* & 1 \end{bmatrix}, \tag{9.85}$$

and

$$\mathbf{x} \equiv \begin{pmatrix} \sqrt{S_1} e^{ikR_1} \\ \sqrt{S_2} e^{ikR_2} \\ \sqrt{S_3} e^{ikR_3} \end{pmatrix}. \tag{9.86}$$

As the position $\mathbf{r}$ is varied over the observation screen, the phase of the vector $\mathbf{x}$ will oscillate rapidly. From basic linear algebra, we know that a necessary condition for Equation 9.84 to vanish is for the matrix $\mathbf{M}$ to possess a zero eigenvalue or, equivalently,

$$\det [\mathbf{M}] = 0. \tag{9.87}$$

Explicitly, this determinant may be written as

$$1 - |\mu_{12}|^2 - |\mu_{23}|^2 - |\mu_{13}|^2 + \mu_{12}\mu_{23}\mu_{13}^* + \mu_{12}^*\mu_{23}^*\mu_{13} = 0. \tag{9.88}$$

The values of μ_{ij} are only constrained to lie within the range $0 \leq |\mu_{ij}| \leq 1$, which leaves a lot of freedom of choice. We limit ourselves to the special case where $\mu_{12} = \mu_{23} = \mu_{13} = \mu_0$, and μ_0 is real-valued. Then Equation 9.88 has the simple form

$$1 - 3\mu_0^2 + 2\mu_0^3 = 0. \tag{9.89}$$

Though cubic equations are in general not easy to solve, we already know that at least one solution to the equation must be $\mu_0 = 1$, that is, full coherence. In fact, there are three roots to this equation, namely,

$$\mu_0 = +1, +1, -1/2. \tag{9.90}$$

Obviously, the nontrivial result is the case $\mu_0 = -1/2$, and an example of the spectral density and phase of the cross-spectral density on the observation screen for this case is shown in Figure 9.12. It can be seen that the

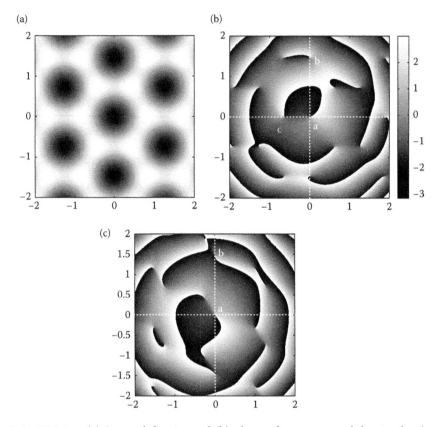

FIGURE 9.12 (a) Spectral density and (b) phase of cross-spectral density for the case when $\mu_0 = -1/2$, with $L = 2$ m, $a = 1$ mm, $k = 9921$ mm^{-1}, $x_1 = 1.0$ mm, and $y_1 = 0$. (c) Phase of cross-spectral density with observation point changed to $x_1 = 0.5$ mm and $y_1 = 0.5$ mm. It can be seen that the singularities at a and b do not move, while the correlation singularity at c does.

points of zero spectral density possess optical vortices, while locations of nonzero spectral density can also contain singularities, namely, correlation vortices. The distinction can be seen in Figure 9.12c; when the observation point is moved, the correlation vortices move but the optical vortices do not.

This result was first demonstrated in Gbur, Visser, and Wolf [GVW04a]. One important result from this paper was the observation that the condition for complete destructive interference is simply the extreme zero value of the nonnegative definiteness condition of Equation 9.37 earlier.

More recently, the result was generalized [GG07] to an N-pinhole interferometer, and it was demonstrated that complete destructive interference can be achieved when $\mu_0 = -1/(N-1)$. Even more recently [RGG12], complete destructive interference was examined for more general values of μ_{ij}, real and complex.

At first glance, this result seems quite paradoxical: we have grown accustomed to believing that complete destructive interference is not possible unless all the fields are mutually coherent. However, as shown in Gbur, Visser, and Wolf [GVW04a], the condition $\mu_0 = -1/2$ is equivalent to the situation where the *sum* of fields from pinholes 1 and 2 is fully coherent with the field from pinhole 3. It is therefore possible for all fields to be mutually partially coherent, yet still produce an interference pattern.

Systems such as these allow the study of the interaction between correlation singularities and phase singularities; this work is still ongoing.

9.5 ELECTROMAGNETIC CORRELATION SINGULARITIES

We have, up to this point, worked only with scalar wavefields. As noted way back in Section 2.1, this is equivalent to assuming that we are only working with fields that are uniformly polarized and beam like. However, as we have discussed in Chapter 7, there are many interesting effects associated with full electromagnetic beams, and it is therefore useful to see how our discussion of correlation singularities extends to the electromagnetic case.

It is clear that the problem is significantly more complicated than the scalar one, even for paraxial fields. Assuming that the field has both $\hat{\mathbf{x}}$-polarized and $\hat{\mathbf{y}}$-polarized components of the electric field $\mathbf{E}$, there are now four cross-spectral density functions, representing the correlations between each combination of polarizations, which can be written in matrix form as

$$\mathbf{W}(\mathbf{r}_1, \mathbf{r}_2, \omega) \equiv \begin{bmatrix} W_{xx}(\mathbf{r}_1, \mathbf{r}_2, \omega) & W_{xy}(\mathbf{r}_1, \mathbf{r}_2, \omega) \\ W_{yx}(\mathbf{r}_1, \mathbf{r}_2, \omega) & W_{yy}(\mathbf{r}_1, \mathbf{r}_2, \omega) \end{bmatrix}, \tag{9.91}$$

with

$$W_{ij}(\mathbf{r}_1, \mathbf{r}_2, \omega) = \langle E_i^*(\mathbf{r}_1, \omega) E_j(\mathbf{r}_2, \omega) \rangle_\omega. \tag{9.92}$$

It is not immediately clear, however, how the components of this electromagnetic cross-spectral density matrix play a role in interference and singular optics.

To understand this, we return once again to Young's two-pinhole experiment as illustrated in Figure 9.3, but now consider a general partially coherent electromagnetic wave illuminating the screen. Using the space–frequency representation, we can write the intensity of an electromagnetic field at a point $\mathbf{r}$ beyond the screen as

$$I(\mathbf{r}, \omega) = \langle \mathbf{E}^*(\mathbf{r}, \omega) \cdot \mathbf{E}(\mathbf{r}, \omega) \rangle_\omega = \text{Tr}\{\mathbf{W}(\mathbf{r}, \mathbf{r}, \omega)\}, \qquad (9.93)$$

where Tr indicates the trace of the matrix.

Assuming only paraxial propagation to the observation screen, the electric field that reaches the screen at point $\mathbf{r}$ from the pinhole at point $\mathbf{r}_1$ will be of the form

$$\mathbf{E}(\mathbf{r}, \omega) = \mathbf{E}_0(\mathbf{r}_1, \omega)\frac{e^{ikR_1}}{R_1}, \qquad (9.94)$$

where $\mathbf{E}_0(\mathbf{r}_1, \omega)$ is the illuminating electric field at the first pinhole. Calculating the intensity at position $\mathbf{r}$, we find that

$$I(\mathbf{r}, \omega) = \frac{I_0(\mathbf{r}_1, \omega)}{R_1^2} + \frac{I_0(\mathbf{r}_2, \omega)}{R_2^2}$$
$$+ \frac{2}{R_1 R_2} \text{Re}\langle \mathbf{E}_0^*(\mathbf{r}_1, \omega) \cdot \mathbf{E}_0(\mathbf{r}_2, \omega)e^{ik(R_2-R_1)} \rangle_\omega. \qquad (9.95)$$

In analogy with previous work on Young's experiment, we pull out $\sqrt{I_0(\mathbf{r}_1, \omega)I_0(\mathbf{r}_2, \omega)}$ from the complex term on the right, which leaves us the expression

$$I(\mathbf{r}, \omega) = \frac{I_0(\mathbf{r}_1, \omega)}{R_1^2} + \frac{I_0(\mathbf{r}_2, \omega)}{R_2^2} + \frac{2\sqrt{I_0(\mathbf{r}_1, \omega)I_0(\mathbf{r}_2, \omega)}}{R_1 R_2}$$
$$\times \text{Re}\{\eta(\mathbf{r}_1, \mathbf{r}_2, \omega)e^{ik(R_2-R_1)}\}, \qquad (9.96)$$

where we have introduced a new *electromagnetic degree of coherence* of the form

$$\eta(\mathbf{r}_1, \mathbf{r}_2, \omega) \equiv \frac{\text{Tr}[\mathbf{W}(\mathbf{r}_1, \mathbf{r}_2, \omega)]}{\sqrt{\text{Tr}[\mathbf{W}(\mathbf{r}_1, \mathbf{r}_1, \omega)]\text{Tr}[\mathbf{W}(\mathbf{r}_2, \mathbf{r}_2, \omega)]}}. \qquad (9.97)$$

We again assume that the intensity of the field at the two pinholes is equal, that is, $I_0(\mathbf{r}_1, \omega) = I_0(\mathbf{r}_2, \omega) \equiv I_0$, and that the distances to the observation screen are approximately equal near the center axis, $R_1 \approx R_2 \equiv R$. We may then simplify our formula for the intensity on the observation screen to

$$I(\mathbf{r}, \omega) = 2I_0 + 2I_0|\eta(\mathbf{r}_1, \mathbf{r}_2, \omega)| \cos\left[\phi_\eta(\mathbf{r}_1, \mathbf{r}_2, \omega) + k(R_2 - R_1)\right], \qquad (9.98)$$

where ϕ_η is the phase of η.

From Equation 9.98, it is clear that the visibility of interference fringes in Young's interference experiment, and indeed any interference experiment, depends on the magnitude of $|\eta|$, which again can be shown to be restricted to the range $0 \leq |\eta| \leq 1$. This electromagnetic degree of coherence was first introduced by Wolf [Wol03] in a unified theory of coherence and polarization; more details can be found in Wolf [Wol07, Chapter 9].

Zeros of intensity for a partially coherent electromagnetic beam are, in a sense, even less generic than those of scalar fields. In the electromagnetic case, the real and imaginary parts of both W_{xx} and W_{yy} must vanish for a zero to exist; if the functions have N_x and N_y coherent modes, respectively, this requires $2(N_x + N_y)$ constraints to be simultaneously satisfied in three-dimensional space, something that is not even in general possible for fully coherent fields ($N_x = 1$, $N_y = 1$). However, correlation singularities of $\eta(\mathbf{r}_1, \mathbf{r}_2, \omega)$, called "eta singularities," are generic. The quantity $\text{Tr}[\mathbf{W}(\mathbf{r}_1, \mathbf{r}_2, \omega)]$ is mathematically equivalent to the sum of two scalar cross-spectral densities, which in turn is equivalent to a single scalar cross-spectral density. We therefore expect to see correlation vortices of $\eta(\mathbf{r}_1, \mathbf{r}_2, \omega)$ in a partially coherent electromagnetic field that are similar to those of $\mu(\mathbf{r}_1, \mathbf{r}_2, \omega)$.

When the polarization of a partially coherent field is changed from a uniform to a nonuniform state, we expect the correlation singularities present to evolve into electromagnetic correlation singularities. This is illustrated in Figure 9.13 for a three-pinhole interferometer in which the polarization of one of the pinholes is rotated. Referring back to Figure 9.11, we take the fields emerging from pinholes 1 and 3 to be uncorrelated, while the fields emerging from pinholes 1 and 2 and 2 and 3 are taken to be correlated with a spectral degree of coherence $\mu = 0.8$. Pinholes 2 and 3 are assumed to have linear polarization, while pinhole 1 has a polarization oriented at an angle θ to the horizontal. With $\theta = 0$, the field is effectively scalar, and we can see multiple eta singularities in the field, which are in this case equivalent to correlation singularities. When the polarization of the first pinhole is rotated, the eta singularities move but are not immediately annihilated.

There is one significant difference between scalar and electromagnetic correlation singularities: the electromagnetic versions can exist even for fully coherent fields. This is possible because $\text{Tr}[\mathbf{W}(\mathbf{r}_1, \mathbf{r}_2, \omega)]$ is the sum of two complex terms, and these two terms—W_{xx} and W_{yy}—can cancel even when they are fully coherent.

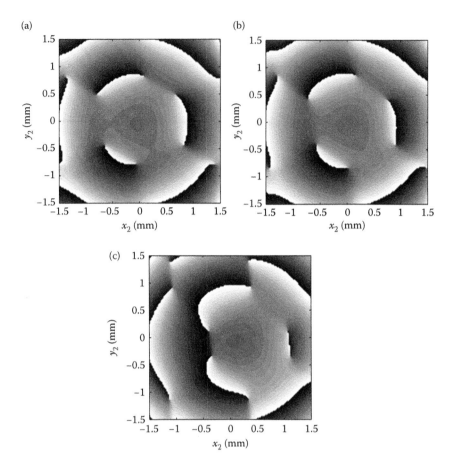

FIGURE 9.13 Evolution of a correlation singularity into an electromagnetic correlation singularity, plotting the phase of $\eta(\mathbf{r}_1, \mathbf{r}_2, \omega)$ for (a) $\theta = 0$, (b) $\theta = 0.2\pi$, and (c) $\theta = 0.4\pi$. Here, $x_1 = -0.5\,\text{mm}$, $x_2 = 0.5\,\text{mm}$, $a = 1\,\text{mm}$, $L = 2\,\text{m}$, and $k = 9921\,\text{mm}^{-1}$.

Relatively little work has been done in the study of η-singularities to date. A paper by Raghunathan, Schouten, and Visser [RSV12] demonstrated that such singularities are generic in partially coherent beams. A later paper by the same authors [RSV13] studied topological reactions of such singularities. These singularities have also been observed in the process of Mie scattering, as discussed in a series of papers [MPP10,MPPA12].

9.6 EXPERIMENTS AND APPLICATIONS

Many of the results discussed in this chapter have been confirmed by experiment. The earliest experimental work was done by Bogatyryova et al.

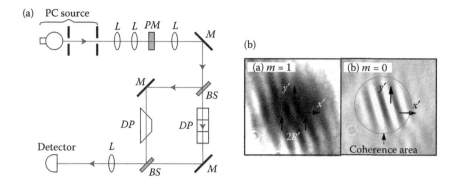

FIGURE 9.14 (a) Experimental setup and (b) example of correlation ring dislocation. (Experimental result taken from D.M. Palacios et al. *Phys. Rev. Lett.*, 92:143905, 2004.)

[BFP⁺03], who introduced a partially coherent vortex beam with correlation singularities and noted circular edge dislocations of the spectral degree of coherence.

The first experimental work dedicated entirely to correlation singularities was done by Palacios et al. [PMM⁺04]; their experimental arrangement is illustrated in Figure 9.14a. A halogen bulb and a pair of apertures serve as a partially coherent source, which is then given a vortex "twist" with a spiral phase mask (PM). A wavefront-folding interferometer, consisting of two Dove prisms (DP) in a Mach–Zender interferometer, performs the cross-correlation of the field, resulting in an interference pattern at the detector. A typical example of the measured result is shown in Figure 9.14b. With the phase plate present ($m = 1$), a dark ring is observed whose radius increases monotonically with decrease in coherence; it can be seen from the fringes that the phase jumps across this dark ring. With no phase plate present ($m = 0$), no phase jump or ring is observed.

Correlation vortices have also been demonstrated experimentally by Wang et al. [WDH⁺06]. In their arrangement, a coherent vortex beam was imaged onto a rotating ground glass plate to produce an incoherent vortex source. This source was interfered with itself via a Michelson interferometer and, with an appropriate path length difference, singularities of the correlation function could be seen in the interference pattern.

Phase singularities of partially coherent fields, as discussed in Section 9.4, have also been demonstrated experimentally. In 2005, an acoustic experiment was performed [BO05] using three loudspeakers, providing excellent agreement with the theory described in Gbur, Visser, and Wolf

[GVW04a]. It is to be noted that the experiment had to be performed over extremely short time periods, to avoid echoes from the surrounding walls. An optical experiment was performed at about the same time [AGP05] using a pair of mutually incoherent beams illuminating a trio of pinholes; again, good agreement was found with the theoretical results.

Correlation singularities have not yet found their way into practical applications, though a number of possibilities have been suggested. Returning to Palacios et al. [PMM⁺04], the experimenters created a partially coherent field, then passed it through a PM, and found that the resulting correlation singularity is dependent on the initial state of coherence. This suggests that such an experimental arrangement could be used as a simple measure of the degree of coherence of any partially coherent light wave.

Correlation singularities also have potential in probing random media, as was recently suggested by Gu and Gbur [GG12]. Because a light beam will be randomized as it passes through fluctuating atmospheric turbulence, it is possible to pass a vortex beam through a region of such turbulence and, from the radius of the correlation ring that results, determine the strength of turbulence.

9.7 TWISTED GAUSSIAN SCHELL-MODEL BEAMS

Up to this point, we have introduced partially coherent vortex beams only as randomized versions of coherent vortex beams, as done in Section 9.2. However, there is at least one broad class of partially coherent beams with a phase "twist" that cannot be directly derived in this manner. These *twisted Gaussian Schell-model* (tGSM) beams were first introduced by Simon and Mukunda [SM93], and have been elaborated upon in much detail since then [SSM93a,SSM93b].

Such beams may be considered a generalization of standard *Gaussian Schell-model* (GSM) beams that have found much use in coherence theory due to their simplicity and analytic tractability. In GSM beams, both the spectral density $S_0(\mathbf{r}, \omega)$ and the spectral degree of coherence $\mu_0(\mathbf{r}_1, \mathbf{r}_2, \omega)$ are taken to be of Gaussian form in the source plane $z = 0$, that is, we express the cross-spectral density as

$$W_0(\mathbf{r}_1, \mathbf{r}_2, \omega) = \sqrt{S_0(\mathbf{r}_1, \omega) S_0(\mathbf{r}_2, \omega)} \mu_0(\mathbf{r}_1, \mathbf{r}_2, \omega), \qquad (9.99)$$

with

$$S_0(\mathbf{r}, \omega) = S_0 \exp[-r^2/2\sigma_S^2], \qquad (9.100)$$

$$\mu_0(\mathbf{r}_1, \mathbf{r}_2, \omega) = \exp[-(\mathbf{r}_2 - \mathbf{r}_1)^2/2\sigma_\mu^2]. \qquad (9.101)$$

In these expressions, σ_S is the width of the spectrum and σ_μ may be interpreted as a transverse correlation length. As noted in Section 9.3, "Schell-model" refers to a degree of coherence that only depends on the difference in transverse coordinates, that is, $\mu(\mathbf{r}_1, \mathbf{r}_2, \omega) = \mu(\mathbf{r}_2 - \mathbf{r}_1, \omega)$. All values of σ_S and σ_μ are permitted, as the cross-spectral density will be Hermitian and nonnegative definite for any choice of these parameters. The limit $\sigma_\mu \to \infty$ represents a fully coherent source, while the limit $\sigma_\mu \to 0$ represents an incoherent source.

It can be shown with some effort that GSM beams remain Gaussian and Schell-model on propagation. By directly propagating the field using Fresnel diffraction formulas,[*] it can be shown that the cross-spectral density in a plane of constant z is given by

$$W(\mathbf{r}_1, \mathbf{r}_2, z, \omega) = \frac{S_0}{[\Delta(z)]^2} \exp\left[-\frac{(\mathbf{r}_1 + \mathbf{r}_2)^2}{8\sigma_S^2[\Delta(z)]^2}\right] \exp\left[-\frac{(\mathbf{r}_2 - \mathbf{r}_1)^2}{2\gamma^2[\Delta(z)]^2}\right]$$

$$\times \exp\left[\frac{ik(r_2^2 - r_1^2)}{2R(z)}\right], \tag{9.102}$$

where

$$R(z) = z\left[1 + \left(\frac{k\sigma_S\gamma}{z}\right)^2\right], \tag{9.103}$$

$$\Delta(z) = [1 + (z/k\sigma_S\gamma)^2]^{1/2}, \tag{9.104}$$

$$\frac{1}{\gamma^2} = \frac{1}{(2\sigma_S)^2} + \frac{1}{\sigma_\mu^2}. \tag{9.105}$$

A tGSM beam may then be introduced by asking the following question: is it possible to generalize the GSM beam in such a way that it retains its rotational symmetry about the z-axis? A standard GSM beam depends on the three quantities $|\mathbf{r}_1|^2$, $|\mathbf{r}_2|^2$, and $\mathbf{r}_1 \cdot \mathbf{r}_2$, all of which are invariant under such a rotation. A little thought suggests that a fourth vector combination is rotationally invariant, namely,

$$\mathbf{r}_1 \times \mathbf{r}_2 = (x_1 y_2 - y_1 x_2)\hat{\mathbf{z}}, \tag{9.106}$$

---

[*] See, for instance, Friberg and Sudol [FS82], or the discussion in Mandel and Wolf [MW95], Section 5.6.

or the cross-product of $\mathbf{r}_1$ and $\mathbf{r}_2$. As the cross-product is a "handed" vector product, satisfying the right-hand rule, the inclusion of such a term in our beam gives it a definite handedness or "twist."

We may incorporate this term (more accurately, the dot product of it with $\hat{\mathbf{z}}$) into an exponent and include it in our cross-spectral density via multiplication; in order to remain Hermitian, however, we must also make the exponent imaginary. The final expression for a tGSM beam in the source plane is therefore

$$W(\mathbf{r}_1, \mathbf{r}_2, \omega) = S_0 \exp[-r_1^2/2\sigma_S^2] \exp[-r_2^2/2\sigma_S^2] \exp[-(\mathbf{r}_2 - \mathbf{r}_1)^2/2\sigma_\mu^2]$$
$$\times \exp[-iku(x_1 y_2 - y_1 x_2)]. \tag{9.107}$$

We have introduced a parameter u, labeled the twist parameter, that characterizes the strength of the beam twist. In order for the entire cross-spectral density to be nonnegative definite, it has been shown [SM93] that the twist parameter must satisfy the constraint

$$k|u| \le \frac{1}{\sigma_\mu^2}. \tag{9.108}$$

A fully coherent beam, for which $\sigma_\mu \to \infty$, cannot possess a twist at all. This indicates that the helicity in this case is inherently associated with the partial coherence of light, and not a hidden vortex structure. It is to be noted that the beam possesses no zero of spectral density.

Insight into the physics of tGSM beams can be found by constructing them as an incoherent superposition of coherent Gaussian beams, as was done in Ambrosini et al. [ABGS94]. Similar[*] to the beam wander construction of Section 9.2, we consider the cross-spectral density as the incoherent sum of modes $U(\mathbf{r}, \mathbf{r}_0)$ of the form

$$W(\mathbf{r}_1, \mathbf{r}_2, \omega) = \iint U^*(\mathbf{r}_1, \mathbf{r}_0) U(\mathbf{r}_2, \mathbf{r}_0) P(\mathbf{r}_0) d^2 r_0. \tag{9.109}$$

For this case, however, the modes are taken to have a field distribution of the type

$$U(\mathbf{r}, \mathbf{r}_0) = \exp\left[-\frac{(x - x_0)^2 + (y - y_0)^2}{X^2} + 2\pi i\alpha(x_0 y - y_0 x)\right], \tag{9.110}$$

---

[*] Though this tGSM model was constructed much earlier.

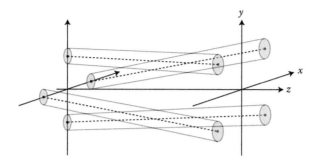

FIGURE 9.15 Illustration of four of the tilted Gaussian modes used to construct a tGSM, for positive α.

and X is a positive constant and α is real-valued. The quadratic part of the exponent represents a Gaussian beam with width X and centered on the point (x_0, y_0). The linear part of the exponent is a phase tilt of the beam and results in its direction of propagation being characterized by the vector $\mathbf{k}$ with components

$$k_x = -2\pi\alpha y_0, \quad k_y = 2\pi\alpha x_0. \tag{9.111}$$

This construction is illustrated in Figure 9.15. For positive α, each of the modes is tilted in a left-handed sense around the central axis, making the field overall left handed. We take our weight function to be of Gaussian form as well, that is,

$$P(\mathbf{r}_0) = A \exp\left[-\frac{x_0^2 + y_0^2}{2\sigma^2}\right], \tag{9.112}$$

with A and σ positive constants.

The integrals of Equation 9.109 can be evaluated by completing the squares in the exponent, and the result after much work is of the form

$$W(\mathbf{r}_1, \mathbf{r}_2, \omega) = A\frac{2\pi\sigma^2 X^2}{4\sigma^2 + X^2} \exp\left[-\frac{r_1^2 + r_2^2}{4\sigma^2 + X^2}\right]$$

$$\times \exp\left[-\frac{2\sigma^2(1 + \pi^2 X^4 \alpha^2)}{X^2(4\sigma^2 + X^2)}(\mathbf{r}_2 - \mathbf{r}_1)^2\right]$$

$$\times \exp\left[-4\pi i\alpha\frac{2\sigma^2}{4\sigma^2 + X^2}(x_1 y_2 - x_2 y_1)\right]. \tag{9.113}$$

On comparison of this expression with Equation 9.107, we find that they are the same if the parameters σ, X, A, and α are taken to be

$$S_0 = \frac{2A\pi\sigma^2 X^2}{4\sigma^2 + X^2}, \quad \frac{1}{4\sigma_S^2} = \frac{1}{4\sigma^2 + X^2}, \quad \frac{1}{2\sigma_\mu^2} = \frac{2\sigma^2(1 + \pi^2 X^4 \alpha^2)}{X^2(4\sigma^2 + X^2)},$$

$$ku = \frac{8\pi\alpha\sigma^2}{4\sigma^2 + X^2}, \tag{9.114}$$

which can evidently always be done. Our beam superposition model can therefore be used to construct any tGSM beam.

The constraint on the twist phase, given in Equation 9.108, can be readily derived from the above results. From Equation 9.114, we may write

$$\sigma_\mu^2 k|u| = \frac{2\pi\alpha X^2}{1 + \pi^2 X^4 \alpha^2}. \tag{9.115}$$

Let us define the parameter $q \equiv \pi X^2 \alpha$. Then we may simply write

$$\sigma_\mu^2 k|u| = \frac{2q}{1 + q^2}. \tag{9.116}$$

The extreme value of the left-hand product is located at the stationary point with respect to q, that is,

$$\frac{\partial}{\partial q}\left(\frac{2q}{1 + q^2}\right) = 0, \tag{9.117}$$

from which we find that the only allowable solution is $q = 1$. The second derivative determines that this is the maximum value; on substitution, we find

$$\sigma_\mu^2 k|u| \leq 1. \tag{9.118}$$

The incoherent superposition model highlights the difference between tGSM beams and the "beam wander" vortex beams considered earlier. In the beam wander model, we considered the incoherent superposition of a collection of coherent vortex beams; that is, the vortex twist was inherent to the modes themselves. In a tGSM beam, the twist is imposed in the orientation of the modes, each of which is given a different inclination. This physical distinction will be even more evident in the following section.

It is to be noted, however, that it is also possible to construct a tGSM by an incoherent superposition of vortex modes. Gori and Santarsiero [GS15] have recently shown that an appropriate superposition of modified Bessel–Gauss vortex beams will also result in a tGSM beam.

9.8 OAM AND RANKINE VORTICES

Now that we have two different models of partially coherent beams with vorticity, it is worthwhile to consider the angular momentum of such beams. We return to Equation 5.64 of Section 5.3, which is of the form

$$M_{orbit}(x, y, \omega) = -\frac{\epsilon_0}{2k}\text{Im}\left\{yE_y^*\partial_x E_y - xE_x^*\partial_y E_x - xE_y^*\partial_y E_y + yE_x^*\partial_x E_x\right\}. \tag{9.119}$$

We may average this over an ensemble of monochromatic fields as in Equation 9.29 to immediately write the OAM in terms of the cross-spectral density as

$$M_{orbit}(x, y, \omega) = -\frac{\epsilon_0}{2k}\text{Im}\left\{y\partial_x' W_{yy}(\mathbf{r}, \mathbf{r}', \omega) - x\partial_y' W_{xx}(\mathbf{r}, \mathbf{r}', \omega)\right.$$
$$\left. - x\partial_y' W_{yy}(\mathbf{r}, \mathbf{r}', \omega) + y\partial_x' W_{xx}(\mathbf{r}, \mathbf{r}', \omega)\right\}\Big|_{\mathbf{r}=\mathbf{r}'}, \tag{9.120}$$

where ∂_x' represents the partial derivative with respect to x', and so forth. We simplify the calculation by assuming that we have an unpolarized field for which $W_{xx} = W_{yy} = W$, so that

$$M_{orbit}(x, y, \omega) = -\frac{\epsilon_0}{k}\text{Im}\left\{y\partial_x' W(\mathbf{r}, \mathbf{r}', \omega) - x\partial_y' W(\mathbf{r}, \mathbf{r}', \omega)\right\}, \tag{9.121}$$

and $W(\mathbf{r}, \mathbf{r}', \omega)$ is either a tGSM beam or a vortex beam in the beam wander model. From Equation 5.56 and the paraxial assumption that $\partial_z E_j \approx -ikE_j$, we may write the z-component of the Poynting vector as

$$S_z(\mathbf{r}, \omega) = \frac{k}{\mu_0\omega}W(\mathbf{r}, \mathbf{r}, \omega). \tag{9.122}$$

As in Section 5.4, we may introduce a normalized OAM $m_{orbit}(\mathbf{r})$—roughly the angular momentum density per photon—as a function of transverse position, by the expression

$$m_{orbit}(\mathbf{r}, \omega) = \frac{\hbar\omega M_{orbit}(\mathbf{r})}{S_z(\mathbf{r}, \omega)}. \tag{9.123}$$

For a tGSM beam, this is readily found to be of the form

$$m_{twist}(\mathbf{r}, \omega) = \hbar k u [x^2 + y^2].$$ (9.124)

The OAM density increases quadratically from the origin as a function of r. This is precisely the behavior one would expect in classical mechanics from a rotating rigid body. Such a body would rotate with constant angular frequency ω_r, with a velocity related to its radial position by $v = \omega_r r$. The angular momentum density, normalized by mass, would then be $rv = \omega_r r^2$. Therefore, tGSM beams behave like rigid rotators, at least with respect to their OAM flux density per photon. This is quite different from the behavior of a coherent Laguerre–Gauss vortex mode, which was seen in Section 5.4 to have a constant OAM flux density per photon, analogous to a rotating ideal fluid, for which the velocity is inversely proportional to the axial distance, that is, $v(r) = A/r$.

We now consider the flux density for the partially coherent vortex beam. With some effort, we find that

$$m_{vortex}(\mathbf{r}, \omega) = \hbar \frac{(x^2 + y^2)}{\frac{\beta^2}{\delta^2}(x^2 + y^2) + \delta^2}.$$ (9.125)

It can be seen from the equation that, for small values of $r = \sqrt{x^2 + y^2}$, the flux density will increase quadratically, like a rigid rotator. For large values of r, however, it will be constant, like an ideal fluid. Several examples of this phenomenon are plotted in Figure 9.16. In the coherent limit, that is, $\delta \to 0$, the flux density will be constant, as for a coherent vortex beam. In the incoherent limit, that is, $\delta \to \infty$, the flux density will be purely quadratic, as for a twisted Gaussian beam. A partially coherent vortex beam therefore represents an intermediate physical state, at least as far as OAM is concerned.

This intermediate state—in which the system behaves like a rigid rotator in the interior, and a fluid rotator in the exterior—appears elsewhere in physics and is known as a *Rankine vortex*. The most familiar example of such a vortex are tropical cyclones such as that illustrated in Figure 9.17, in which the wind speed in the core increases linearly with radius and decreases inversely with radius outside the core.

The Rankine vortex structure of a partially coherent vortex beam was first noted by Swartzlander and Hernandez-Aranda [JHA07], and they assessed this structure in terms of the average topological charge as a

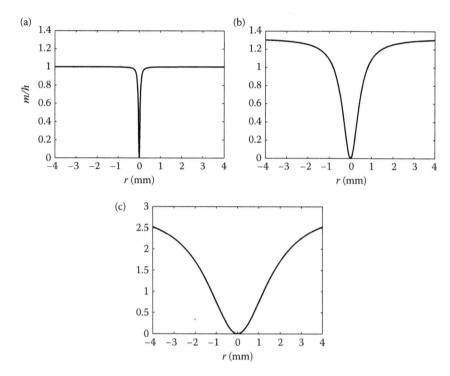

FIGURE 9.16 The normalized angular momentum flux density for a partially coherent vortex beam of width $w_0 = 1$ mm, for (a) $\delta = 0.04$ mm, (b) $\delta = 0.4$ mm, and (c) $\delta = 1.0$ mm.

FIGURE 9.17 Image of Hurricane Isabel as photographed by astronaut Ed Lu from the ISS in 2003. In the core (eye) of the hurricane, wind speeds roughly increase linearly with radius, while outside it decreases inversely with radius.

function of radial distance. Some time later, Kim and Gbur [KG12] demonstrated that the OAM has a Rankine behavior.[*]

9.9 EXERCISES

1. Using Equation 9.46, show that, as $\delta \to \infty$, any correlation singularities must lie on a circle centered on the (fixed) observation point (x_1, y_1). What is the radius of this circle?

2. By direct integration, derive the approximate cross-spectral density of a linear optical system, Equation 9.68, from Equations 9.61 and 9.67.

3. Derive expression (9.125) for the normalized angular momentum flux density of a randomized vortex beam.

---

[*] In this paper, however, the coherent and incoherent limits are mistakenly reversed.

CHAPTER 10

Singularities and Vortices in Quantum Optics

U P TO THIS POINT, we have almost entirely focused on singularities of classical wavefields, with the exception of a brief foray into photons in the discussion of angular momentum in Section 5.3. The wave–particle duality of light, however, also introduces additional effects that have implications for singular optics, and in this chapter we touch upon some of them.

It should be mentioned that vortices play a large role in nonoptics-related quantum physics, notably in Bose–Einstein condensates [MZW97, DCLZ98,MAH$^+$99] and in solutions to Schrödinger's equation [HCP74, HGB74,BBBBŠ00]. However, the study of vortices in the quantum theory is even older than its classical optical counterpart, stretching back to Dirac in 1931 [Dir31], and a detailed discussion of the full history would take us too far afield. With the exception of a brief discussion of vortices in Schrödinger's equation, and Dirac's fascinating work, we restrict ourselves to singular optics phenomena associated with the quantization of the electromagnetic field.

In the discussion that follows, we assume a significant familiarity with basic quantum physics. As the ideas introduced here are not used in the remainder of the text (which is, as we noted, primarily classical optics), this chapter may be skipped on a first reading.

We begin with a brief review of this *second quantization*, and use it as a launching point to discuss a number of effects related to singular optics in quantized fields.

10.1 QUANTIZATION OF THE ELECTROMAGNETIC FIELD

In traditional quantum mechanics, physical observables such as position, momentum, and energy are treated as operators on the wavefunction of a quantum particle; this is *first quantization*. With the recognition that light has particle-like as well as wave-like properties, it is natural to also treat electromagnetic fields themselves as operators, which is *second quantization*. Here, we introduce the fundamentals of this field quantization, which forms a basis for the rest of this chapter. We loosely follow the discussion in Loudon [Lou83]; see Grynberg, Aspect, and Fabre [GAF10] and Gerry and Knight [GK04] for alternative takes and more detail.

We begin again with Maxwell's equations in free space without sources, given by Equations 2.7 through 2.10, which we write again below for convenience:

$$\nabla \cdot \mathbf{E}(\mathbf{r}, t) = 0, \tag{10.1}$$

$$\nabla \cdot \mathbf{B}(\mathbf{r}, t) = 0, \tag{10.2}$$

$$\nabla \times \mathbf{E}(\mathbf{r}, t) = -\frac{\partial \mathbf{B}(\mathbf{r}, t)}{\partial t}, \tag{10.3}$$

$$\nabla \times \mathbf{B}(\mathbf{r}, t) = \frac{1}{c^2} \frac{\partial \mathbf{E}(\mathbf{r}, t)}{\partial t}. \tag{10.4}$$

Following standard electromagnetic theory, we may define the electric and magnetic fields in terms of a scalar potential $\Phi(\mathbf{r}, t)$ and a vector potential $\mathbf{A}(\mathbf{r}, t)$, as

$$\mathbf{E}(\mathbf{r}, t) = -\nabla \Phi(\mathbf{r}, t) - \frac{\partial \mathbf{A}(\mathbf{r}, t)}{\partial t}, \tag{10.5}$$

$$\mathbf{B}(\mathbf{r}, t) = \nabla \times \mathbf{A}(\mathbf{r}, t). \tag{10.6}$$

Because the fields are derived from vector derivatives of the potentials, we have significant freedom in choosing them, known as *gauge freedom* (see, for instance, Gbur [Gbu11, Chapter 2]). In particular, we work in the Coulomb gauge, for which $\nabla \cdot \mathbf{A} = 0$; in the absence of sources, we are also free to take $\Phi = 0$, which leaves us with a simplified relation between fields and a single potential,

$$\mathbf{B} = \nabla \times \mathbf{A}, \quad \mathbf{E} = -\frac{\partial \mathbf{A}}{\partial t}. \tag{10.7}$$

Substituting these definitions into Maxwell's equations, we may readily show (left as an exercise) that the vector potential satisfies the wave equation

$$\nabla^2 \mathbf{A} - \frac{1}{c^2} \frac{\partial \mathbf{A}}{\partial t} = 0. \tag{10.8}$$

We now seek a general solution to this equation in terms of some set of fundamental modes of the electromagnetic field. It is easiest to think of the potential as expanded in a spatial Fourier representation of the form

$$\mathbf{A}(\mathbf{r}, t) = \sum_{\mathbf{k}} \left[\mathbf{A_k}(t)e^{i\mathbf{k}\cdot\mathbf{r}} + \mathbf{A_k^*}(t)e^{-i\mathbf{k}\cdot\mathbf{r}} \right]. \tag{10.9}$$

The potential is assumed, for mathematical convenience, to be confined to a cube of side length L with periodic boundary conditions; this makes the complete possible set of modes countably infinite, and any integrals associated with them finite. One expects, in the end, that any physical results must be independent of L.

On substitution from Equation 10.9 into Equation 10.8, we immediately find that

$$\frac{\partial \mathbf{A_k^2}}{\partial t^2} + k^2 c^2 \mathbf{A_k} = 0. \tag{10.10}$$

The time-dependent components of the vector potential therefore satisfy a harmonic oscillator equation with frequency $\omega_k \equiv kc$. From our knowledge of plane waves, this should not be surprising. However, this result gives us a path to quantizing the field. We note that the vector potential can now be written as

$$\mathbf{A}(\mathbf{r}, t) = \sum_{\mathbf{k}} \left[\mathbf{A_k}e^{i(\mathbf{k}\cdot\mathbf{r}-\omega_k t)} + \mathbf{A_k^*}e^{-i(\mathbf{k}\cdot\mathbf{r}-\omega_k t)} \right], \tag{10.11}$$

where $\mathbf{A_k}$ is a set of complex constants.

To see how to quantize the field, we first use Equation 10.7 to write the electric and magnetic fields in terms of the vector potential as

$$\mathbf{E}(\mathbf{r}, t) = \sum_{\mathbf{k}} i\omega_{\mathbf{k}} \left[-\mathbf{A_k} e^{i(\mathbf{k}\cdot\mathbf{r}-\omega_{\mathbf{k}}t)} + \mathbf{A_k^*} e^{-i(\mathbf{k}\cdot\mathbf{r}-\omega_{\mathbf{k}}t)} \right], \quad (10.12)$$

$$\mathbf{B}(\mathbf{r}, t) = \sum_{\mathbf{k}} i\mathbf{k} \times \left[\mathbf{A_k} e^{i(\mathbf{k}\cdot\mathbf{r}-\omega_{\mathbf{k}}t)} - \mathbf{A_k^*} e^{-i(\mathbf{k}\cdot\mathbf{r}-\omega_{\mathbf{k}}t)} \right]. \quad (10.13)$$

From Equation 8.14 from Chapter 8, we may introduce the cycle-averaged energy $E_{\mathbf{k}}$ of each mode within the volume L^3 as

$$\langle E_{\mathbf{k}} \rangle = \frac{1}{2} \int \left[\epsilon_0 \langle \mathbf{E_k} \cdot \mathbf{E_k} \rangle + \mu_0^{-1} \langle \mathbf{B_k} \cdot \mathbf{B_k} \rangle \right] d^3 r, \quad (10.14)$$

where the angle brackets again indicate cycle averaging, as discussed in Chapter 2. On substitution, we readily find that

$$E_{\mathbf{k}} = 2\epsilon_0 V \omega_k^2 \mathbf{A_k} \cdot \mathbf{A_k^*}, \quad (10.15)$$

where we have dropped the angle brackets $\langle \cdots \rangle$ for convenience.

We are now close to being able to quantize the field. We first note that, because $\nabla \cdot \mathbf{A} = 0$, the modes must be transverse. We associate with each mode $\mathbf{k}$ a unit vector $\boldsymbol{\epsilon_k}$ such that $\mathbf{k} \cdot \boldsymbol{\epsilon_k} = 0$; it follows that for every $\mathbf{k}$, there are two orthogonal modes and we assume that the sum in Equation 10.9 includes them. We then write

$$\mathbf{A_k}(t) = \boldsymbol{\epsilon_k} A_{\mathbf{k}}(t). \quad (10.16)$$

We then decompose the complex time-dependent vector potential $\mathbf{A_k}(t)$ into a real and imaginary part in the form

$$\mathbf{A_k}(t) = (4\epsilon_0 V \omega_k^2)^{-1/2} \left[\omega_{\mathbf{k}} Q_{\mathbf{k}}(t) + i P_{\mathbf{k}}(t) \right] \boldsymbol{\epsilon_k}, \quad (10.17)$$

where $Q_{\mathbf{k}}(t)$ and $P_{\mathbf{k}}(t)$ are real-valued functions. We can readily show (left as an exercise) that

$$E_{\mathbf{k}} = \frac{1}{2} \left[P_{\mathbf{k}}^2(t) + \omega_k^2 Q_{\mathbf{k}}^2(t) \right]. \quad (10.18)$$

This expression is equal to the Hamiltonian of a classical harmonic oscillator, where $Q_{\mathbf{k}}(t)$ is the generalized position variable and $P_{\mathbf{k}}(t)$ is the generalized momentum. To quantize the electromagnetic field, then, we may evidently follow exactly the same steps that are taken to quantize the oscillator, as shown for example in quantum texts such as Shankar [Sha80, Chapter 7].

It may seem odd at first glance to simply apply the same rules that were used for a quantum particle in a harmonic potential to an electromagnetic field. We may treat it, however, as a reasonable guess that was inspired by the experimental observation of quantized light particles (photons). The agreement of this guess was verified by later experimental tests.

We rewrite the energy of the **k**th mode as a Hamiltonian operator, that is,

$$\hat{H} = \frac{1}{2}[\hat{p}^2 + \omega^2\hat{q}^2], \tag{10.19}$$

where we have dropped the **k** subscript for brevity. We will work, for a while, with a single mode of the field. This Hamiltonian does not represent an observable quantity by itself; only the expectation value of the Hamiltonian with respect to a given quantum state $|\psi\rangle$ gives us an observable energy.

The operators $\hat{p}$ and $\hat{q}$ satisfy the familiar commutation relation

$$[\hat{q}, \hat{p}] \equiv \hat{q}\hat{p} - \hat{p}\hat{q} = i\hbar. \tag{10.20}$$

This commutation relation reminds us that the order of operators is important! Whenever we deal with products of operators, we must take care in any rearrangement of the order of terms.

We now introduce a new pair of operators $\hat{a}$ and $\hat{a}^\dagger$, as

$$\hat{a} \equiv (2\hbar\omega)^{-1/2}(\omega\hat{q} + i\hat{p}), \tag{10.21}$$

$$\hat{a}^\dagger \equiv (2\hbar\omega)^{-1/2}(\omega\hat{q} - i\hat{p}). \tag{10.22}$$

For reasons that will be clear momentarily, we refer to these as the *destruction* and *creation* operators, respectively. By straightforward multiplication, we can show that

$$\hat{a}\hat{a}^\dagger = (\hbar\omega)^{-1}\left[\hat{H} - \frac{1}{2}\hbar\omega\right]. \tag{10.23}$$

Furthermore, using the relation

$$[\hat{a}, \hat{a}^\dagger] = \hat{a}\hat{a}^\dagger - \hat{a}^\dagger\hat{a} = 1, \tag{10.24}$$

derivable from Equation 10.20 above, we may write the Hamiltonian as

$$\hat{H} = \hbar w \left[\hat{a}^\dagger \hat{a} + \frac{1}{2} \right]. \tag{10.25}$$

The combination $\hat{n} \equiv \hat{a}^\dagger \hat{a}$ has particular significance; it is known as the *number operator*, and we will see that it is associated with the average number of quanta in a particular mode of the field. First, however, we consider the operator $\hat{a}$ and its adjoint in more detail. Let us assume that $|n\rangle$ is an energy eigenstate of the Hamiltonian, that is,

$$\hat{H} |n\rangle = E_n |n\rangle. \tag{10.26}$$

If we apply the creation operator to the left of both sides of this expression, we have

$$\hbar w \hat{a}^\dagger \left[\hat{a}^\dagger \hat{a} + \frac{1}{2} \right] |n\rangle = E_n \hat{a}^\dagger |n\rangle. \tag{10.27}$$

We may use Equation 10.24 to bring $\hat{a}^\dagger$ through to the right side of the Hamiltonian, with the result

$$\hbar w \left[\hat{a}^\dagger \hat{a} + \frac{1}{2} \right] \hat{a}^\dagger |n\rangle = (E_n + \hbar w) \hat{a}^\dagger |n\rangle. \tag{10.28}$$

On comparison with Equation 10.26, this suggests that state $\hat{a}^\dagger |n\rangle$ has energy $E_n + \hbar w$, or that the next highest quantum state may be written as

$$|n + 1\rangle \propto \hat{a}^\dagger |n\rangle. \tag{10.29}$$

Similarly, if we multiply Equation 10.26 by $\hat{a}$, we may then bring $\hat{a}$ through to the right of the Hamiltonian and conclude that

$$|n - 1\rangle \propto \hat{a} |n\rangle. \tag{10.30}$$

Our results suggest that the energy of a harmonic oscillator mode comes in discrete steps of $\hbar w$. If we assume that there exists a lowest state $|0\rangle$, called the ground state, such that

$$\hat{H} |0\rangle = E_0 |0\rangle, \tag{10.31}$$

we may apply the destruction operator $\hat{a}$ to both sides of this, to find

$$\hat{H}\hat{a}\,|0\rangle = (E_0 - \hbar w)\hat{a}\,|0\rangle. \qquad (10.32)$$

But, if $|0\rangle$ is the ground state, there cannot exist any state below it, as Equation 10.32 implies; therefore, $\hat{a}\,|0\rangle = 0$. Directly applying this condition to Equation 10.31 leads to $E_0 = \hbar w/2$. The energy of the nth state may then be built up by repeated applications of the creation operator, to find

$$E_n = \left(n + \frac{1}{2}\right)\hbar w. \qquad (10.33)$$

With this in mind, the energy eigenstates must also be eigenstates of the number operator, that is,

$$\hat{n}\,|n\rangle = \hat{a}^\dagger \hat{a}\,|n\rangle = n\,|n\rangle. \qquad (10.34)$$

We may now show (again, left as an exercise) that the proper normalization for these operations is given by

$$\hat{a}\,|n\rangle = \sqrt{n}\,|n-1\rangle, \qquad (10.35)$$

$$\hat{a}^\dagger\,|n\rangle = \sqrt{n+1}\,|n+1\rangle. \qquad (10.36)$$

Finally, we may introduce an operator form of the vector potential in terms of creation and destruction operators, namely

$$\hat{A}(\mathbf{r}, t) = \sum_{\mathbf{k}} \left(\frac{\hbar}{2\epsilon_0 V w_{\mathbf{k}}}\right)^{1/2} \epsilon_{\mathbf{k}} \left[\hat{a}_{\mathbf{k}} e^{i(\mathbf{k}\cdot\mathbf{r} - w_{\mathbf{k}} t)} + \hat{a}_{\mathbf{k}}^\dagger e^{-i(\mathbf{k}\cdot\mathbf{r} - w_{\mathbf{k}} t)}\right].$$

$$(10.37)$$

A comparison of our creation and destruction operators, defined by Equation 10.22, and our vector potential, defined by Equation 10.17, leads directly to this result. The electric field and magnetic field operators follow from Equations 10.7, to find

$$\hat{E}(\mathbf{r}, t) = i\left(\frac{\hbar w_{\mathbf{k}}}{2\epsilon_0 V}\right)^{1/2} \epsilon_{\mathbf{k}} \left[\hat{a}_{\mathbf{k}} e^{i(\mathbf{k}\cdot\mathbf{r} - w_{\mathbf{k}} t)} - \hat{a}_{\mathbf{k}}^\dagger e^{-i(\mathbf{k}\cdot\mathbf{r} - w_{\mathbf{k}} t)}\right], \qquad (10.38)$$

$$\hat{B}(\mathbf{r}, t) = i\left(\frac{\hbar}{2\epsilon_0 V w_{\mathbf{k}}}\right)^{1/2} \mathbf{k} \times \epsilon_{\mathbf{k}} \left[\hat{a}_{\mathbf{k}} e^{i(\mathbf{k}\cdot\mathbf{r} - w_{\mathbf{k}} t)} - \hat{a}_{\mathbf{k}}^\dagger e^{-i(\mathbf{k}\cdot\mathbf{r} - w_{\mathbf{k}} t)}\right].$$

$$(10.39)$$

Two particular sets of single-mode states are commonly used in quantum optics. The first of these is the set of number states $|n\rangle$, which we have seen are eigenstates of $\hat{n}$ and have a definite number of photons in them. One may readily show that $\langle n| \hat{n} |n\rangle = n$, $\langle n| \hat{n}^2 |n\rangle = n^2$, so that the root-mean-square (RMS) deviation in the photon number is

$$\Delta n = \sqrt{\langle n| \hat{n}^2 |n\rangle - [\langle n| \hat{n} |n\rangle]^2} = 0. \tag{10.40}$$

Loosely speaking, the number states are maximally quantum, in that their particle behavior is perfectly well defined. In the other extreme, we have the coherent states $|\alpha\rangle$, defined as

$$|\alpha\rangle = \exp\left[-\frac{1}{2}|\alpha|^2\right] \sum_{n=0}^{\infty} \frac{\alpha^n}{(n!)^{1/2}} |n\rangle. \tag{10.41}$$

These so-called *coherent states* are eigenstates of the destruction operator, and it can be shown that

$$\hat{a} |\alpha\rangle = \alpha |\alpha\rangle. \tag{10.42}$$

In this case, there is an RMS variation in the number of photons given by $\Delta n = |\alpha|$. These states are the ones that are most classical, in the sense that they most represent a classical electromagnetic plane wave. The average value of the electric field can be shown to be sinusoidal in space and time; however, the variance of the electric field is nonzero, due to fluctuations in photon number.

Of particular significance to us in the following sections are multimode states. The simplest form of multimode states are those that are a direct product of single-mode states; for instance, if a mode labeled 1 is in a number state with A photons and a mode labeled 2 is in a number state with B photons, the total ket $|\psi\rangle$ may be written as

$$|\psi\rangle = |A\rangle_1 |B\rangle_2. \tag{10.43}$$

We conclude this section by noting that the number operator $\hat{n}$ has a special significance in photodetection. Because an ideal photodetector has an output proportional to the number of photons incident upon it, the operator describing it must be proportional to the number operator. This suggests that the field intensity $\hat{I}$ of a mode, as measured by the detector, arises from that part of the product $\hat{\mathbf{E}}(\mathbf{r}, t) \cdot \hat{\mathbf{E}}(\mathbf{r}, t)$ that has the ordering $\hat{a}^\dagger \hat{a}$, or

$$\hat{I} \equiv \frac{\hbar\omega}{2\epsilon_0} \hat{a}^\dagger \hat{a}. \tag{10.44}$$

A photodetector does not therefore directly measure the square of the electric field, $|\hat{\mathbf{E}}(\mathbf{r}, t)|^2$, but rather only its *normally ordered* component. This point was stressed by Glauber [Gla63].

10.2 QUANTUM CORES OF OPTICAL VORTICES

A particularly striking new prediction of the quantum field theory is found in the average electric field fluctuations in the kth empty mode, that is,

$$\langle 0| \hat{\mathbf{E}} \cdot \hat{\mathbf{E}}^\dagger |0\rangle = \left(\frac{\hbar\omega_\mathbf{k}}{2\epsilon_0}\right) \langle 0| \left[\hat{a}_\mathbf{k} u_\mathbf{k}(\mathbf{r}, t) - \hat{a}_\mathbf{k}^\dagger u_\mathbf{k}^*(\mathbf{r}, t)\right]$$
$$\times \left[\hat{a}_\mathbf{k}^\dagger u_\mathbf{k}^*(\mathbf{r}, t) - \hat{a}_\mathbf{k} u_\mathbf{k}(\mathbf{r}, t)\right] |0\rangle, \tag{10.45}$$

where we have used $u_\mathbf{k}(\mathbf{r}, t)$ to represent the normalized mode amplitude for convenience, satisfying

$$\int_V |u_\mathbf{k}(\mathbf{r}, t)|^2 d^3r = 1. \tag{10.46}$$

Expanding out the bracketed terms, we find

$$\langle 0| \hat{\mathbf{E}} \cdot \hat{\mathbf{E}}^\dagger |0\rangle = \left(\frac{\hbar\omega_\mathbf{k}}{2\epsilon_0}\right) \langle 0| \left[\hat{a}_\mathbf{k} \hat{a}_\mathbf{k}^\dagger |u_\mathbf{k}(\mathbf{r}, t)|^2 - \hat{a}_\mathbf{k}^\dagger \hat{a}_\mathbf{k}^\dagger [u_\mathbf{k}(\mathbf{r}, t)]^2 \right.$$
$$\left. - \hat{a}_\mathbf{k} \hat{a}_\mathbf{k} [u_\mathbf{k}(\mathbf{r}, t)]^2 + \hat{a}_\mathbf{k}^\dagger \hat{a}_\mathbf{k} |u_\mathbf{k}(\mathbf{r}, t)|^2\right] |0\rangle. \tag{10.47}$$

The latter three terms within the bracket will be zero, due to the presence of an appropriate lowering operator acting on either the zero bracket of the ground state. The first term, however, will be nonzero, and in fact will result in the expression

$$\langle 0| \hat{\mathbf{E}} \cdot \hat{\mathbf{E}}^\dagger |0\rangle = \left(\frac{\hbar\omega_\mathbf{k}}{2\epsilon_0}\right) |u_\mathbf{k}(\mathbf{r}, t)|^2. \tag{10.48}$$

We therefore find that, in the quantum theory of light, even empty modes have nonzero fluctuations of the electric field. This is not simply a theoretical construct that can be ignored: these fluctuations are responsible for the spontaneous emission of light from atoms. Formally, the strength of fluctuations tends to infinity when summed over all modes; however, detectors are only sensitive to a finite range of frequencies, so summations are always over a finite interval in practice.

There is one important implication of these fluctuations, as noted by Berry and Dennis [BD04], which may be summarized as follows. Classically, an optical vortex is a point of zero intensity in the cross section of a wavefield. Quantum mechanically, however, there are always field fluctuations at any point. A vortex, therefore, will have a region around it where the quantum fluctuations are greater than the classical intensity of the field. This is called the *quantum core* of the vortex by Berry and Dennis, and we describe their estimate of the size of this core here.

We have so far considered the expansion of an electromagnetic field within a volume in terms of a plane-wave basis; however, it should be clear that we may use any complete set of fields as a basis. In this case, we consider a set that consists of a paraxial $m = 1$ vortex mode with scalar amplitude $u_v(\mathbf{r}, t)$ and a set of approximate plane waves $u_\mathbf{k}(\mathbf{r}, t)$ characterized by their wave vector $\mathbf{k}$. The total electric field may be written as

$$\hat{\mathbf{E}}(\mathbf{r}, t) = \sum_{n=v,\mathbf{k}} \left(\frac{\hbar\omega_n}{2\epsilon_0}\right)^{1/2} [\hat{a}_n u_n(\mathbf{r}, t) + \hat{a}_n^\dagger u_n^*(\mathbf{r}, t)]\boldsymbol{\epsilon}_n, \qquad (10.49)$$

where $\boldsymbol{\epsilon}_n$ is the polarization of each mode. The field is assumed to be in a coherent state with respect to the vortex mode, with average photon number $v \gg 1$, and in the vacuum state for every other mode. The electric energy density of the field, when averaged, is of the form

$$\epsilon_0\langle\hat{\mathbf{E}} \cdot \hat{\mathbf{E}}^\dagger\rangle = I_v(\mathbf{r}, t) + \sum_{n=\mathbf{k}} \frac{\hbar\omega_\mathbf{k}}{2}|u_n(\mathbf{r}, t)|^2, \qquad (10.50)$$

where $I_v(\mathbf{r}, t)$ is approximately the *classical* average intensity of the field for a large number of photons in that mode, we may neglect the spontaneous emission contributions.

As already noted, the field fluctuations are infinite when taken over all modes. We therefore imagine that we measure the intensity with a single-atom detector in an excited state that may be triggered by stimulated or spontaneous emission. This atom is only sensitive over a range $\Delta\omega$ with a center frequency ω; we may then estimate the contribution of the vacuum states as follows.

Our fields are confined to a cube of side length L, subject to periodic boundary conditions, as noted in Section 10.1. The allowed wavenumbers of plane wave modes in this volume are then given by

$$k_x = \frac{2\pi l}{L}, \quad k_y = \frac{2\pi m}{L}, \quad k_z = \frac{2\pi n}{L}, \tag{10.51}$$

with l, m, n integers. This indicates that, in three-dimensional $\mathbf{k}$-space, there is one mode in every $\mathbf{k}$-volume $(2\pi)^3/L^3 = (2\pi)^3/V$, and the average mode density in $\mathbf{k}$-space is

$$\rho(\mathbf{k}) = \frac{V}{(2\pi)^3}. \tag{10.52}$$

For a sufficiently large confining volume, these modes will be packed so densely as to be nearly continuous; the number of modes that will interact with our atom are those that lie within a thin spherical shell in $\mathbf{k}$-space of $\mathbf{k}$-volume $4\pi k^2 \Delta k$, where $k = \omega/c$ and $\Delta k = \Delta\omega/c$. The situation in $\mathbf{k}$-space is illustrated in Figure 10.1. The total number N of modes to be summed over in Equation 10.50 is therefore

$$N \approx \rho(\mathbf{k})4\pi k^2 \Delta k = \frac{V}{2\pi^2}\frac{\omega^2 \Delta\omega}{c^3}. \tag{10.53}$$

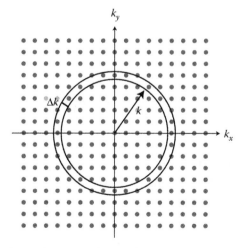

FIGURE 10.1 Modes in the k_x, k_y-plane in $\mathbf{k}$-space, with the thin spherical shell of modes that interact with the atom.

Noting that each mode is normalized, and that we may thus approximately write $|u_n(\mathbf{r}, t)|^2 \sim 1/V$, our expression for the energy density is of the form

$$\epsilon_0 \langle \hat{\mathbf{E}} \cdot \hat{\mathbf{E}}^\dagger \rangle = I_v(\mathbf{r}, t) + \frac{\hbar \omega^3 \Delta \omega}{4\pi^2 c^3} = I_v(\mathbf{r}, t) + I_Q. \tag{10.54}$$

This result already suggests that the average energy density of an electromagnetic field is never strictly zero, but has a finite quantum contribution I_Q. We now turn to the vortex mode, and note that near the core of a screw dislocation, the intensity is approximately

$$I_v(\mathbf{r}, t) = \alpha r^2. \tag{10.55}$$

Sufficiently close to the vortex center, we will find $I_v < I_Q$, and the quantum effects will dominate. The threshold radius R for this transition is

$$R = \sqrt{\frac{\hbar \omega^3 \Delta \omega}{4\pi^2 c^3 \alpha}}. \tag{10.56}$$

To estimate this radius, we need a reasonable value for α. If we assume that the vortex is the core of an LG_{01} mode, as given by Equation 2.37, its energy density in the source plane near the core may be given by

$$I_{LG}(r) \approx \frac{p_0}{c} \frac{4}{\pi w_0^4} r^2, \tag{10.57}$$

where p_0 is the total power flux across the plane $z = 0$. We can readily determine α from this expression, and then determine that

$$R = w_0^2 \sqrt{\frac{\hbar \omega^3 \Delta \omega}{16 p_0 \pi c^2}} = w_0^2 \pi \sqrt{\frac{\hbar c \Delta \omega}{2 p_0 \lambda^3}}. \tag{10.58}$$

As an estimate of the size of this core, we consider a He–Ne laser with central wavelength $\lambda = 633$ nm, linewidth $\Delta \omega = 1500$ MHz, $w_0 = 100$ μm, and $p_0 = 5$ mW. Solving for R/w_0, we find

$$R \sim 0.04 w_0. \tag{10.59}$$

Owing to the $\lambda^{-3/2}$ factor in the estimate for R, we generally expect that shorter-wavelength fields will have a correspondingly larger quantum, that is, particle-like, core. This is in general agreement with the observation that short-wavelength electromagnetic waves such as x-rays have a strong particle-like nature, manifested for instance in Compton scattering of photons from matter.

This calculation may seem somewhat artificial, in that it requires a very specific detection scheme in the form of an excited atom that is subject to both spontaneous and stimulated emission. As later shown by Barnett [Bar08], this scheme is only necessary because the calculation implicitly treats the atom as stationary and point like, whereas the Heisenberg uncertainty relation between position and momentum requires both quantities to be generally uncertain. This means that an atom centered on the vortex core will experience quantum fluctuations due to its exposure to the surrounding vortex field. Barnett shows that the vortex core determined in this manner also has a radius proportional to $\sqrt{\hbar}$.

Throughout this book, we have introduced various sorts of generic singularities, only to reveal that they are not as typical as they first appear. After introducing scalar phase singularities, we noted that polarization singularities are more generic, and beyond this, all coherent singularities are special cases of correlation singularities. In this section, we see perhaps the ultimate realization of this trend: all singularities in fact possess a quantum core that obscures any true singularity of the field. In practice, though, this quantum core will generally not be evident.

10.3 INTRODUCTION TO ENTANGLEMENT

Though it may not seem like it, so far we have considered only a small subset of all optical quantum states that are realizable. For a single mode of the field, we have primarily considered eigenstates of a particular field operator, such as the eigenstates $|n\rangle$ of the number operator $\hat{n}$. More generally, we may consider what are known as *pure states* of the field, which may be represented as a superposition of eigenstates in some particular basis. For instance, a state $|\psi\rangle$ that is equally likely to possess one or two photons, with no other possibilities, may be written as

$$|\psi\rangle = \frac{1}{\sqrt{2}}[|1\rangle + e^{i\phi}|2\rangle], \tag{10.60}$$

where ϕ is the relative phase of the superposition. The probability that a single photon is measured is therefore $|\langle 1|\psi\rangle|^2 = 1/2$, with a similar relation for two photons. As another example, the coherent states of Equation 10.41 are pure states, written as a superposition of number states.

It is possible to further generalize to systems in a *mixed state* that must be characterized by a *density operator* $\hat{\rho}$ rather than a single wavefunction. We consider a density operator as written with respect to a set of orthonormal pure states $|\psi_i\rangle$, which is expressed as

$$\hat{\rho} = \sum_i p_i |\psi_i\rangle \langle \psi_i|, \tag{10.61}$$

where p_i is the probability of finding the system in the state $|\psi_i\rangle$. More generally, the states $|\psi_i\rangle$ do not need to be orthogonal; however, since the matrix of $\hat{\rho}$ is Hermitian with respect to any complete orthonormal basis, it is always possible to find a diagonal representation in which the pure states are orthogonal. For a pure state $|\psi_k\rangle$, the density operator is just a factorized outer product, that is, $\hat{\rho} = |\psi_k\rangle \langle \psi_k|$.

The classical average value of an operator $\hat{O}$ can be written in terms of probabilities as

$$\overline{O} = \sum_i p_i \langle \psi_i| \hat{O} |\psi_i\rangle. \tag{10.62}$$

With some simple manipulations, we may write this average in terms of the density operator. Using the identity operator $\hat{I} = \sum_j |\eta_j\rangle\langle\eta_j|$ with respect to an orthonormal basis $|\eta_j\rangle$, we have

$$\overline{O} = \sum_{i,j} p_i \langle \psi_i| \hat{O} |\eta_j\rangle\langle\eta_j| \psi_i\rangle = \sum_{i,j} p_i \langle\eta_j| \psi_i\rangle \langle\psi_i| \hat{O} |\eta_j\rangle$$

$$= \sum_j \langle\eta_j| \hat{\rho}\hat{O} |\eta_j\rangle. \tag{10.63}$$

This latter expression is simply the trace (Tr) of the operator $\hat{\rho}$, so we may write

$$\overline{O} = \mathrm{Tr}\{\hat{\rho}\hat{O}\}. \tag{10.64}$$

A density operator is needed when a quantum system can only be represented by a classical ensemble of states. This analogy becomes clear by comparing the density operator with the coherent mode representation of optical coherence theory given by Equation 9.38. In both cases, the average properties of the system come from an incoherent superposition of modes. For a system with only two states, the polarization matrix of partially polarized light, Equation 7.135, is also mathematically similar. It will be important in what follows to distinguish between *classical correlations*, such as those studied in coherence theory, and *quantum correlations*, which we discuss next.

For multimode fields, we have so far also restricted ourselves to what are known as *separable* states, such as the state $|\psi\rangle = |A\rangle_1 |B\rangle_2$ given in Equation 10.43. Separable states are ones that may be factorized into a direct product of independent modal states; in contrast, however, we may also form *entangled* states, such as the following:

$$|\psi\rangle = \frac{1}{\sqrt{2}}[|A\rangle_1 |B\rangle_2 + |B\rangle_1 |A\rangle_2], \qquad (10.65)$$

where states A and B are assumed to be orthonormal eigenstates of a Hermitian operator $\hat{A}$. This quantum state cannot be factorized into a direct product of a mode-1 state with a mode-2 state.

For such an entangled state, the outcome of any individual measurement of mode 1 or mode 2 is unpredictable, but the relation between them is completely deterministic. For example, we cannot predict in advance whether the measurement of mode 1 will result in A or B, each of which has a probability $1/2$ of occurring. But if we measure A for mode 1, then, mode 2 must automatically have the value B, and vice versa.

Such a perfect correlation between modes, however, can also be introduced in a mixed state, by considering a density operator of the form

$$\hat{\rho} = \frac{1}{2}[|A\rangle_1 |B\rangle_2 \langle B|_2 \langle A|_1 + |A\rangle_2 |B\rangle_1 \langle B|_1 \langle A|_2]. \qquad (10.66)$$

The difference between the entangled state of Equation 10.65 and the mixed state of Equation 10.66 can be seen by writing the density operator of the former, that is,

$$\hat{\rho}_\psi = \frac{1}{2} \, [|A\rangle_1 |B\rangle_2 \langle B|_2 \langle A|_1 + |A\rangle_2 |B\rangle_1 \langle B|_1 \langle A|_2$$

$$+ |A\rangle_1 |B\rangle_2 \langle B|_1 \langle A|_2 + |A\rangle_2 |B\rangle_1 \langle B|_2 \langle A|_1]. \qquad (10.67)$$

The density operator for the entangled state possesses two additional terms, which we may describe as interference terms associated with the entangled state. A mixed state is an incoherent superposition of correlated states, while an entangled state is a coherent superposition.

Physically, we may describe the difference as follows. A mixed state, which we treat as a large ensemble, is a mixture of modes with *definite* states and relations to one another. For any particular experiment, either mode 1 is produced in state A and mode 2 in state B, or vice versa. In the entangled state, however, according to the classical Copenhagen interpretation of quantum mechanics, neither mode 1 nor mode 2 has a definite value until it is measured.

This entangled state behavior was considered as an argument against the Copenhagen interpretation of quantum mechanics in the early days of the theory. In a famous 1935 paper, Einstein, Podolsky, and Rosen [EPR35] argued that such states necessarily imply that quantum mechanics is an incomplete theory of nature. Their argument, condensed to its simplest form, may be considered as follows. Let us suppose that we have a charge-less, spinless elementary particle, such as a pion π^0 that decays into an electron and positron, each of spin 1/2, in the reaction $\pi^0 \rightarrow e^+ + e^-$. Because the pion is spinless, and total angular momentum is conserved, the z-component of the spins of the decay products must be opposite to each other; however, since the overall system is rotationally symmetric, there is no preferred choice for the spin of either particle. This implies a combined state for the electron–positron pair as

$$|\psi\rangle = \frac{1}{\sqrt{2}}[|\uparrow\rangle_+ |\downarrow\rangle_- - |\downarrow\rangle_+ |\uparrow\rangle_-], \qquad (10.68)$$

where $\uparrow$ and $\downarrow$ represent spin up and spin down, respectively, and subscripts $+$ and $-$ represent the positron and electron, respectively. This is known as the angular momentum singlet state, of net angular momentum zero.

We now imagine that the decay products are allowed to propagate arbitrarily far away from each other. Provided they undergo no interactions that can influence their spins along the way, their entangled state will be conserved. If the spin of the electron is measured at some distant station, the outcome of the measurement is not predictable, and is equally likely to be up or down. Let us suppose that it is measured to be up. According to the Copenhagen interpretation of quantum mechanics, the entire wavefunction immediately collapses into the "up" eigenstate of the electron. However, because of angular momentum conservation, the positron must have instantly collapsed into a "down" eigenstate. This collapse must, evidently, happen instantaneously, that is, faster than the vacuum speed of light; otherwise, a measurement of the positron in the meantime could also return an "up" value, which would violate angular momentum conservation. Einstein referred to this seeming relativity-violating change as "spooky action at a distance," and he and his coauthors argued that it implied that the particles must in fact carry additional information about their spin and other quantum properties with them, and that these "hidden variables" are simply inaccessible to known experimental techniques. (This view would change dramatically, though, as we will discuss in Section 10.7.)

Another peculiarity of entangled states was introduced by Schrödinger [Sch35], also in 1935, and is the famous "Schrödinger's cat paradox." He noted that it is possible, in principle, to force the behavior of a macroscopic system, such as a living cat, to depend on a quantum system, such as a radioactive atom. As he tells it (translation by Trimmer [Sch80]),

> One can even set up quite ridiculous cases. A cat is penned up in a steel chamber, along with the following diabolical device (which must be secured against direct interference by the cat): in a Geiger counter there is a tiny bit of radioactive substance, *so* small, that *perhaps* in the course of one hour one of the atoms decays, but also, with equal probability, perhaps none; if it happens, the counter tube discharges and through a relay releases a hammer which shatters a small flask of hydrocyanic acid. If one has left this entire system to itself for an hour, one would say that the cat still lives *if* meanwhile no atom has decayed. The first atomic decay would have poisoned it. The ψ-function of the entire system would express this by having in it the living and the dead cat (pardon the expression) mixed or smeared out in equal parts.

A proper understanding of such issues, namely, the relationship between the quantum and the classical worlds, is still being actively investigated. However, disregarding some significant and subtle questions of interpretation, we may say that the existence of entangled states has been verified. Furthermore, such entanglement is a key ingredient in a number of mind-boggling and potentially world-changing technologies, such as quantum cryptography and quantum computation.[*]

Entanglement schemes involving photons often entangle the polarization states of photon pairs. Just as we have seen that OAM may be used to increase the information transfer of communication systems in Section 6.2, OAM may also be used to broaden the number of possible entangled states. We will spend a number of sections in this chapter exploring the implications of OAM in entangled systems.

---

[*] See, for instance, the book by Bouwmeester, Ekert, and Zeilinger [BEZ00].

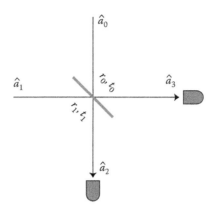

FIGURE 10.2 Quantum description of a beamsplitter, where $\hat{a}_0$ and $\hat{a}_1$ represent the modes entering the splitter, and $\hat{a}_2$ and $\hat{a}_3$ represent the modes exiting. r_0, t_0 represent the transmission and reflection of the 0th mode, with similar definition for r_1, t_1.

How does one generally create an entangled state? The pion example above gives us a general guide. Loosely speaking, it can be said that entanglement arises when physical conservation laws force a relationship between quantum particles, but do not constrain the individual behavior of either particle. As a simple warm-up example of entanglement, we consider the quantum properties of an ordinary beamsplitter.

An illustration of such a beamsplitter* is illustrated in Figure 10.2. Quantized light may illuminate the beamsplitter from above in a mode with destruction operator $\hat{a}_0$ or from the left in a mode with destruction operator $\hat{a}_1$; the reflection and transmission amplitudes for each path are given by r_j and t_j, with $j = 0, 1$. Light from these two input modes couples into output modes $\hat{a}_2$ and $\hat{a}_3$.

In a classical system, with light only coming through branch 1 of the beamsplitter, we would completely ignore branch 0. However, as discussed in Section 10.2, in the quantum picture, there are always vacuum fluctuations of the field in this branch, and the operators $\hat{a}_2$ and $\hat{a}_3$ describing the output fields must include this effect. We may write

$$\hat{a}_2 = r_1\hat{a}_1 + t_0\hat{a}_0, \tag{10.69}$$

$$\hat{a}_3 = t_1\hat{a}_1 + r_0\hat{a}_0. \tag{10.70}$$

---

* A more detailed description can be found in Gerry and Knight [GK04, Chapter 6], which our discussion loosely follows.

The values of r_i, t_i are not arbitrary. From Equation 10.24, we expect that the operators $\hat{a}_2$ and $\hat{a}_3$ must satisfy the commutation relations

$$[\hat{a}_i, \hat{a}_j^\dagger] = \delta_{ij}, \quad i, j = 2, 3. \tag{10.71}$$

From these, we can readily derive familiar relationships about beamsplitters, namely,

$$|t_0|^2 + |r_0|^2 = 1, \tag{10.72}$$

$$|t_1|^2 + |r_1|^2 = 1, \tag{10.73}$$

$$r_0 t_0^* + t_1 r_1^* = 0, \tag{10.74}$$

$$|t_0|^2 = |t_1|^2, \tag{10.75}$$

$$|r_0|^2 = |r_1|^2. \tag{10.76}$$

An important case that satisfies these conditions is a 50/50 beamsplitter, with $t_j = 1/\sqrt{2}$, $r_j = i/\sqrt{2}$, and $j = 0, 1$. We then have the simplified relations

$$\hat{a}_2 = \frac{1}{\sqrt{2}}[i\hat{a}_1 + \hat{a}_0], \tag{10.77}$$

$$\hat{a}_3 = \frac{1}{\sqrt{2}}[\hat{a}_1 + i\hat{a}_0]. \tag{10.78}$$

We now consider the action of the beamsplitter on a single-photon state incident from branch 1. The state in question may be written in terms of number states as

$$|\psi\rangle = |0\rangle_0 |1\rangle_1 = \hat{a}_1^\dagger |0\rangle_0 |0\rangle_1, \tag{10.79}$$

or a single creation operator $\hat{a}_1^\dagger$ acting on the total vacuum state. We solve Equations 10.77 and 10.78 in terms of $\hat{a}_0$, $\hat{a}_1$, finding

$$\hat{a}_1 = \frac{1}{\sqrt{2}}[-i\hat{a}_2 + \hat{a}_3], \tag{10.80}$$

$$\hat{a}_0 = \frac{1}{\sqrt{2}}[\hat{a}_2 - i\hat{a}_3]. \tag{10.81}$$

We may therefore express the output of the beamsplitter as

$$\hat{a}_1^\dagger |0\rangle_2 |0\rangle_3 = \frac{1}{\sqrt{2}}[i\hat{a}_2^\dagger + \hat{a}_3^\dagger] |0\rangle_2 |0\rangle_3 = \frac{1}{\sqrt{2}}[i |1\rangle_2 |0\rangle_3 + |0\rangle_2 |1\rangle_3].$$

(10.82)

We have an entangled state. If a photon is measured in branch 2, it will not be measured in branch 3, and vice versa. Here, we may loosely say that the entanglement comes from particle number being conserved, but the choice of output branch being undetermined.

The single-photon case does not produce any results that are distinct from quasi-classical expectations. A distinctly quantum effect arises, however, when branch 0 and branch 1 are simultaneously filled with single photons. The input state is then

$$|\psi\rangle = |1\rangle_0 |1\rangle_1 = \hat{a}_0^\dagger \hat{a}_1^\dagger |0\rangle_0 |0\rangle_1.$$

(10.83)

Again, substituting from Equations 10.80 and 10.81 into this expression, we find that

$$|\psi\rangle = \frac{1}{2} \left[i\hat{a}_2^\dagger \hat{a}_2^\dagger + i\hat{a}_3^\dagger \hat{a}_3^\dagger + \hat{a}_2^\dagger \hat{a}_3^\dagger - \hat{a}_3^\dagger \hat{a}_2^\dagger \right] |0\rangle_2 |0\rangle_3.$$

(10.84)

But since $\hat{a}_2^\dagger$ and $\hat{a}_3^\dagger$ commute, the last two terms of the product cancel, and we may write[*] the output state as

$$|\psi\rangle = \frac{i}{\sqrt{2}}[|2\rangle_2 |0\rangle_3 + |0\rangle_2 |2\rangle_3].$$

(10.85)

We again have an entangled state but, surprisingly, the two photons will always arrive at the same detector; a coincidence measurement of branch 2 and branch 3 will detect no events. This curious prediction was measured experimentally by Hong, Ou, and Mandel [HOM87]. Interference effects have caused the two possible ways in which one photon arrives at each detector to cancel each other out. In order for this to occur, however, the wavefunctions of the photons must significantly overlap, so that they are *indistinguishable*; in the Hong–Ou–Mandel experiment, it was found that the temporal duration of a photon pulse was roughly 100 femtoseconds.

These simple examples of entanglement are instructive, but not terribly useful, as we may loosely say that they involve the entanglement of

---

[*] Why square root of 2 in the denominator instead of 2? Recall that $\hat{a}^\dagger |n\rangle = \sqrt{n+1} |n+1\rangle$.

"something" with "nothing." Much more significant are states where pairs of photons have their energy, momentum, polarization, or OAM states entangled. The latter will, of course, be of most interest to us in the next few sections.

For more descriptions of entanglement and its proposed applications, see the article by Terhal, Wolf, and Doherty [TWD03]. One topic we have avoided in this section is how one *quantifies* the degree to which a state is entangled. See the review by Bruß for a discussion on measures of entanglement [Bru02].

10.4 NONLINEAR OPTICS AND ANGULAR MOMENTUM

The production of entangled photons is most commonly done through the use of nonlinear optical processes, in particular spontaneous parametric down conversion (SPDC), in which a higher-energy photon is split into two lower-energy photons through an interaction with a nonlinear medium. Before discussing this phenomenon, however, it will be helpful to review some basics of nonlinear optics, including a discussion of angular momentum conservation in the simplest nonlinear effects. For a full, quantitative discussion of nonlinear optics, we refer the reader to Boyd's classic text [Boy08].

When an oscillating electric field interacts with a dielectric medium, it induces oscillating dipoles in that medium, creating a net polarization density. This oscillating polarization then emits its own electromagnetic field, and this simple physical picture may be used to explain refraction, reflection, and scattering. In traditional linear optics, we may write the following relationship:

$$P(t) = \chi^{(1)} E(t), \qquad (10.86)$$

where $P(t)$ is a component of the polarization density, $E(t)$ is a component of the electric field, and $\chi^{(1)}$ is the generally anisotropic electric susceptibility. The superscript "(1)" indicates that this is the linear susceptibility. If the material is illuminated with light of a frequency far from the material's resonance frequencies, we may for purposes of illustration treat the susceptibility as a real-valued constant.

This linear relationship between the electric field and the polarization is only approximate, however; when a medium is illuminated with light with a sufficiently high density of photons, nonlinear susceptibilities become important. We may then write

$$P(t) = \chi^{(1)}E(t) + \chi^{(2)}E^2(t) + \chi^{(3)}E^3(t) + \cdots$$
$$= P^{(1)}(t) + P^{(2)}(t) + P^{(3)}(t) + \cdots, \qquad (10.87)$$

where $\chi^{(m)}$ is the mth-order susceptibility and $P^{(n)}$ is the nth-order polarization density.

We may now ask what sort of new effects arise from the nonlinear contributions. Let us first imagine a monochromatic field illuminating our nonlinear medium, of the form

$$E(t) = Ee^{-i\omega t} + E^*e^{i\omega t}, \qquad (10.88)$$

where E is a generally complex constant. It is clear from Equation 10.86 that, in linear optics, the excited polarization density is of the same frequency as the illuminating field, and therefore, the field scattered or refracted from the medium will also have the same frequency.

If second-order polarization effects are included, however, the situation changes dramatically. On substitution, one can readily show that

$$P^{(2)}(t) = \chi^{(2)}\left[2EE^* + E^2e^{-2i\omega t} + (E^*)^2e^{2i\omega t}\right]. \qquad (10.89)$$

The first term on the right-hand side is constant with respect to time, and results in a static electric field within the material in a process known as *optical rectification*. The latter two terms oscillate at frequency 2ω, and can result in the generation of electromagnetic radiation at this frequency. From a quantum perspective, this phenomenon may be described as two photons of frequency ω interacting with the nonlinear medium to produce a single photon of frequency 2ω, and it is known as *second-harmonic generation* (SHG). A simple schematic of the process is shown in Figure 10.3a.

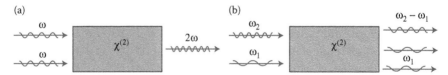

FIGURE 10.3 Two types of nonlinear processes that can occur within a material with a second-order susceptibility $\chi^{(2)}$. (a) SHG. (b) Difference-frequency generation.

If light composed of two frequencies is input to the nonlinear medium, the possible interactions become much more complicated. We consider a field of the form

$$E(t) = E_1 e^{-i\omega_1 t} + E_2 e^{-i\omega_2 t} + E_1^* e^{i\omega_1 t} + E_2^* e^{i\omega_2 t}. \tag{10.90}$$

From this, we get the lengthy expression

$$P^{(2)}(t) = \chi^{(2)} \Big[E_1^2 e^{-2i\omega_1 t} + (E_1^*)^2 e^{2i\omega_1 t} + E_2^2 e^{-2i\omega_2 t} + (E_2^*)^2 e^{2i\omega_2 t}$$
$$+ 2(E_1 E_1^* + E_2 E_2^*) + 2E_1 E_2 e^{-i(\omega_1 + \omega_2)t} + 2E_1^* E_2^* e^{i(\omega_1 + \omega_2)t}$$
$$+ 2E_1^* E_2 e^{-i(\omega_2 - \omega_1)t} + 2E_1 E_2^* e^{i(\omega_2 - \omega_1)t} \Big]. \tag{10.91}$$

The first four terms are associated with SHG for frequencies ω_1 and ω_2; the fifth term is associated with optical rectification again. The sixth and seventh terms represent what is known as *sum-frequency generation*, and it is similar to SHG in that two photons are combined into one.

The final two terms are somewhat different, in that the output wave has a frequency equal to the difference $\omega_2 - \omega_1$ of the two input waves; this process is known as *difference-frequency generation*. In a mechanism that is a form of stimulated emission, a photon of frequency ω_2 (known as a *pump photon*), through interaction with a photon of frequency ω_1 (known as a *signal photon*), breaks into photons of energy $\omega_2 - \omega_1$ (the *idler photon*) and ω_1 (another signal photon). The stimulating photon of frequency ω_1 is still present, as well, meaning that difference-frequency generation serves as a way to amplify a pump signal; the process is illustrated in Figure 10.3b. In short, a pump beam + a signal beam = an amplified signal beam + an idler beam.

Even more complicated combinations of frequencies can be created using the third-order susceptibility $\chi^{(3)}$, including third-harmonic generation; we restrict ourselves to the second-order case for simplicity.

All the cases described here are known as *parametric* processes, those in which the material system begins and ends in the same quantum state. This implies that the total energy and momentum are conserved in the beginning and ending states of the photons. An example of a nonparametric process is Raman scattering, in which energy is transferred to or from a vibrational mode of a molecule in the light–matter interaction. Processes are parametric in spectral regions where there is little absorption, usually at frequencies much lower than the first resonance, and where the susceptibilities only weakly depend on frequency.

Let us first focus on second-harmonic and sum-frequency generation, and refer to the frequencies of the two input photons as ω_1 and ω_2 (though, in the case of SHG, $\omega_1 = \omega_2$). Because of energy conservation, we have

$$\omega_1 + \omega_2 = \omega_{out}, \tag{10.92}$$

where ω_{out} is the frequency of the output photon. Because the input photons are essentially parallel, the output photon must be colinear with them; we may then write the momentum conservation formula as a scalar relation as well

$$k_1 + k_2 = k_{out}. \tag{10.93}$$

In terms of the linear refractive index, we have

$$\omega_1 n(\omega_1) + \omega_2 n(\omega_2) = \omega_{out} n(\omega_{out}). \tag{10.94}$$

Now we have a problem, however; below resonance, the refractive index $n(\omega)$ tends to increase with frequency, meaning that Equations 10.92 and 10.94 cannot be satisfied simultaneously for a single-index medium. However, in a uniaxial crystal, in which there is an ordinary index $n_o(\omega)$ for one polarization and a distinct extraordinary index $n_e(\omega)$ for an orthogonal polarization, it is possible to satisfy both the energy and momentum conservation equations.

There are two general types[*] of SHG:

- Type-I SHG: Two photons with parallel polarization propagating in index n_o combine to produce an output photon with orthogonal polarization in index n_e.

- Type-II SHG: Two photons with orthogonal polarization, one in index n_o, one in index n_e, combine to produce an output photon with either ordinary or extraordinary polarization, depending on whether the crystal is positive uniaxial or negative uniaxial, respectively.

---

[*] There is also a "type 0" SHG, in which the nonlinear medium is given a periodic structure to achieve momentum balance.

For Type-I SHG, for instance, Equation 10.94 leads to the following momentum balance equation:

$$n_o(2\omega) = \frac{1}{2}n_e(\omega) + \frac{1}{2}n_e(\omega). \qquad (10.95)$$

Provided a material is chosen such that $n_o(2\omega) = n_e(\omega)$, momentum and energy conservation will be simultaneously achieved. For Type-II SHG that outputs an ordinary photon, we have

$$n_o(2\omega) = \frac{1}{2}n_o(\omega) + \frac{1}{2}n_e(\omega). \qquad (10.96)$$

Since n_o increases with frequency, we must necessarily have $n_e(\omega) > n_o(\omega)$, which is the definition of a positive uniaxial crystal.

SHG was first experimentally observed[*] by Franken et al. in 1961 [FHPW61]. Light from a ruby laser operating at 694.3 nm was passed through crystalline quartz, producing a second-harmonic signal at 347.2 nm. The laser produced roughly 3 joules of energy in a pulse that was 1-ms long. A quartz prism spectrometer was used to separate out the different frequencies, and red-insensitive photographic plates were used to record the second harmonic while simultaneously reducing the primary signal. An illustration of a similar experiment, performed by Savage and Miller [SM62] in 1962 to study SHG in a variety of crystals, is shown in Figure 10.4.

Our discussion up to this point may have given the impression that harmonic generation is a relatively weak process, that is, that only a small fraction of the input beam gets frequency doubled. In general, however, the opposite is true: for an appropriate material and system geometry, the conversion of photons can be nearly 100%. In fact, SHG is now routinely used to construct lasers that operate at new frequencies. Many green lasers, for instance, use the process of converting a Nd:YAG light source with wavelength 1064 nm into a 532-nm output through SHG in a potassium dihydrogen phosphate (KDP) crystal. Even inexpensive green laser pointers use this process, though they usually use KTP instead of KDP.

KDP forms a crystalline structure with tetragonal symmetry; however, the unit cells of the crystal are noncentrosymmetric: that is, they do not

---

[*] This paper is famous not only for its results but also due to a mistake by the copy editor, who removed the image of the second-harmonic signal from the paper's photograph, thinking it was a speck of dirt! The first article demonstrating SHG therefore contains no actual evidence of SHG.

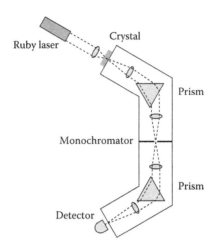

FIGURE 10.4 Rough illustration of the experiment by Savage and Miller [SM62]. The second-harmonic signal is separated from the pump signal by the use of a prism and a monochromator (i.e., an appropriately placed aperture). The spreading of the filtered beam due to the first prism is corrected by a second prism.

possess inversion symmetry about their origin. In fact, SHG can *only* occur in noncentrosymmetric crystals, and the lowest nonlinear process that will appear in a centrosymmetric crystal is third-harmonic generation.

We may roughly see why this must be the case by considering the Lorentz model of the atom. In a linear approximation, the displacement $x(t)$ of an electron from its equilibrium position around the nucleus satisfies the equation

$$m\ddot{x}(t) = -\kappa x(t) - g\dot{x}(t) + qE(t), \qquad (10.97)$$

where m is the mass of the electron, q is the charge, κ is the effective "spring constant" of the atom, and g is a damping constant. $E(t)$ represents the applied electric field. If we include a quadratic restoring force, this expression becomes

$$m\ddot{x}(t) = -\kappa_1 x(t) - \kappa_2 [x(t)]^2 - g\dot{x}(t) + qE(t), \qquad (10.98)$$

where κ_1 and κ_2 represent the linear and nonlinear parts of the restoring force, respectively. However, the linear term in $x(t)$ is associated with a quadratic potential, proportional to $[x(t)]^2$, which is symmetric around the origin. The quadratic term in $x(t)$ will be associated with a *cubic* potential, which is antisymmetric about the origin; the total potential, then, will have no symmetry. An atom possessing a quadratic force term, which we would naturally associate with the second-order polarization, must therefore be noncentrosymmetric. If we instead add a force term depending on the cube of the displacement, that is, $[x(t)]^3$, it will have a quartic potential and the combined linear and cubic forces will remain centrosymmetric. Because of this, third-harmonic generation can be achieved in most materials, regardless of their symmetry.

With this elementary review of nonlinear optics behind us, we may now turn to a discussion of OAM in nonlinear processes. For the parametric processes discussed here, we expect that OAM should be conserved between the initial and final photon states, as is energy and momentum. However, the conservation law for OAM is much simpler than those for energy and momentum, as OAM is a discrete quantity that is independent of both frequency and the index of refraction. In short, if we consider the process of sum-frequency generation for two input photons with angular momenta l_1 and l_2, the angular momentum l_{out} of the output photon should be

$$l_{out} = l_1 + l_2. \tag{10.99}$$

Such an outcome was first experimentally confirmed by Basistiy et al. [BBSV93] in 1993. They passed an LG_{01} mode from a Nd:YAG pulsed laser, wavelength 1064 nm, through a frequency-doubling crystal[*]; the second-harmonic output beam at wavelength 532 nm possessed an $m = 2$ vortex. A more general study was performed several years later by Dholakia et al. [DSPA96]. They used a cylindrical lens mode converter, as described in Section 4.1.2, to transform a variety of Hermite–Gauss beams into Laguerre–Gauss vortex beams. Their source was also a Nd:YAG laser, and they used both a KTP crystal (with Type II phase matching) and a lithium triborate (LBO) crystal (with Type I phase matching) to produce second harmonics. The resulting field structure was analyzed by the use of another mode converter, which returned the Laguerre–Gauss modes to Hermite–Gauss modes. Looking at modes with azimuthal indices from $m = 0$ to $m = 7$, they found that this index was doubled in the second-harmonic field in each case.

These authors also noticed a curious symmetry in the frequency-doubling process. If we consider an LG_{0m} beam in the waist plane, of

---

[*] The specific crystal used is not mentioned in the paper, but is presumably KDP or KTP.

the form

$$E(\rho, \phi) = E_0 \left(\frac{\sqrt{2}\rho}{w_0} \right)^m e^{im\phi} e^{-\rho^2/w_0^2}, \tag{10.100}$$

the second-order polarization, proportional to $[E(\rho, \phi)]^2$, will have the functional form

$$[E(\rho, \phi)]^2 = \frac{E_0^2}{2^{m/2}} \left(\frac{2\rho}{w_0} \right)^{2m} e^{2im\phi} e^{-2\rho^2/w_0^2}. \tag{10.101}$$

This expression is essentially a pure $LG_{0,2m}$ state with a reduced beam width $w_0/\sqrt{2}$. With the polarization taking on the transverse cross section of a pure state, it is reasonable to expect that the second-harmonic beam will also be a pure state. We may then say that an LG_{0m} beam gets frequency doubled into an $LG_{0,2m}$ state. This is not true for modes with a nonzero azimuthal index; however, an LG_{nm} mode will get frequency doubled into a superposition of various radial modes, all with the same azimuthal order $2m$.

More generally, Equation 10.99 is expected to hold for any parametric sum-frequency generation process. This was tested and confirmed by Beržanskis et al. [BMP+97], who suggested that this effect could be used as a simple "optical processor" that performs integer addition with input vortices of any order.

Parametric difference-frequency generation will also generally conserve OAM, though the means by which this occurs is slightly more subtle. As noted earlier, the process of difference-frequency generation involves a pump photon and a signal photon interacting in a nonlinear medium; the result is that the pump photon breaks into a signal photon and an idler photon. Because this is a stimulated process, we would expect that the generated signal photon would have the same angular momentum as the stimulating signal photon. Angular momentum conservation therefore takes the form

$$l_{pump} + l_{signal} = l_{idler} + 2l_{signal}, \tag{10.102}$$

or, canceling the signal photon from each side and rearranging,

$$l_{idler} = l_{pump} - l_{signal}. \tag{10.103}$$

Difference-frequency generation therefore allows one to perform integer subtraction with input vortices of any order. If the pump beam is taken to

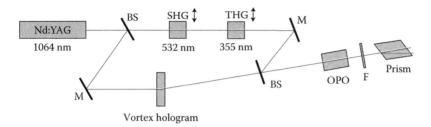

Vortex hologram

FIGURE 10.5 Schematic of the difference-frequency experiment performed by Beržanskis et al. [BMP$^+$97]. Here, SHG = second-harmonic generator, THG = third-harmonic generator, M = mirror, F = filter, BS = beamsplitter, and OPO = optical parametric oscillator.

have $l_{pump} = 0$, then, the idler beam will have $l_{idler} = -l_{signal}$, providing a means to reverse the vortex charge of the input signal.

This was experimentally tested by Beržanskis et al. [BMP$^+$97] as well, using an experimental arrangement shown in Figure 10.5. A beam from a Nd:YAG laser was split into two parts, one of which was sent through a vortex hologram and the other of which was sent through either a second-harmonic generating or a third-harmonic generating material, resulting in a beam with wavelength 532 nm or 355 nm, respectively. These vortex-free harmonics were then combined with the vortex beam, serving as the signal, and sent into an OPO, a nonlinear crystal combined with a resonant cavity tuned to one or more of the nonlinear outputs.*

With the third harmonic serving as the pump and the vortex beam as the signal, Type II matching was achieved, outputting an idler wave of $\lambda = 532$ nm; with the second harmonic serving as the pump, Type I matching was achieved, outputting an idler wave of $\lambda = 1064$ nm. In both cases, the output beams satisfied the arithmetic of Equation 10.103, in particular having an idler with opposite OAM to the signal. The possibilities for difference-frequency generation were later summarized by Soskin and Vasnetsov, in a review of nonlinear singular optics [SV98].

10.5 ENTANGLEMENT OF ANGULAR MOMENTUM STATES

In the previous section, we referred to difference-frequency generation as a *stimulated* process, in which a photon of frequency ω_1 induces a photon of frequency ω_2 to break into two photons with frequencies $\omega_2 - \omega_1$ and ω_1.

---

* For a different use of an OPO, see the end of Section 4.1.6.

The existence of a stimulated process implies that there must be a corresponding spontaneous process, in which a photon ω_2 spontaneously breaks into two photons with frequencies $\omega_2 - \omega_1$ and ω_1. This phenomenon does in fact exist, and is known as *parametric fluorescence* or, more commonly, spontaneous parametric down conversion (SPDC). The physics of SPDC is significantly different from difference-frequency generation, and this has important implications for quantum optics as a whole, as well as for experiments with OAM, as we will see.

Some of the earliest experiments on parametric fluorescence used collinear phase matching, in which the signal and idler photons propagate in the same, or nearly the same, direction as the pump photons. The earliest example by Harris, Oshman, and Byer [HOB67] used a crystal of LiNbO$_3$ and an argon laser with pump wavelength $\lambda = 488$ nm. The crystal was placed in an oven and the output wavelength ranged from 540 to 660 nm, depending on the oven temperature. The temperature change modified the momentum conservation condition and allowed different down-conversion processes to occur. At the same time, Magde and Mahr [MM67] observed a broad range of output frequencies in an ammonium dihydrogen phosphate (ADP) crystal by rotating the extraordinary axis of the crystal with respect to the incoming beam.

Much more commonly employed these days is a noncolinear phase-matching geometry, in which the signal and idler photons do not have to propagate in the same direction as the pump photon. The energy and momentum conservation laws are therefore of the form

$$\omega_p = \omega_s + \omega_i, \tag{10.104}$$

$$\mathbf{k}_p = \mathbf{k}_s + \mathbf{k}_i. \tag{10.105}$$

Because refractive index tends to increase with frequency below resonance, we must use anisotropic crystals to take advantage of the difference between the ordinary and extraordinary indices. We again have two general types of processes, which are in essence the reverse of those of SHG:

- Type I SPDC: A photon with extraordinary polarization breaks into two photons with ordinary polarization propagating in index n_o.

- Type II SPDC: A photon with extraordinary polarization breaks into two photons with orthogonal polarization, one in index n_o, and one in index n_e.

If the signal and idler photons are noncolinear, this implies that the sum of the magnitudes of their momenta must be greater than that of the the the pump photon; for Type I SPDC with signal and idler photons of the same frequency, we have

$$n_e(2\omega) < \frac{1}{2}n_o(\omega) + \frac{1}{2}n_o(\omega). \tag{10.106}$$

This implies that the crystal is necessarily negative uniaxial; a similar result applies to Type II SPDC.

To explore the behavior of the photons in SPDC in more detail, we need to introduce some concepts from the optics of anisotropic media. In an anisotropic medium, the relation between the displacement field **D** and the electric field **E** is of a tensor nature

$$D_i = \epsilon_{ij}E_j, \tag{10.107}$$

where summation over the repeated index $j = 1, 2, 3$ is implied. In parametric systems, the tensor ϵ_{ij} is real-valued and symmetric, and therefore it is clear that we can always find a coordinate system x, y, z in which it is diagonal, that is,

$$\epsilon = \begin{bmatrix} \epsilon_x & 0 & 0 \\ 0 & \epsilon_y & 0 \\ 0 & 0 & \epsilon_z \end{bmatrix}, \tag{10.108}$$

where ϵ_x, ϵ_y, and ϵ_z are known as the *principal dielectric constants*.

Let us assume that there exists a monochromatic plane wave solution to Maxwell's equations in this anisotropic medium of the form

$$\mathbf{D}(\mathbf{r}) = \mathbf{D}_0 e^{ikn\mathbf{s}\cdot\mathbf{r}}, \tag{10.109}$$

where **s** is the direction of wave propagation, k is the free-space wavenumber, and n is a refractive index *to be determined*. We assume that there exist plane wave forms of **H**, **E**, and **B** with the same functional dependence. If we take the curl of Faraday's law,

$$\nabla \times \mathbf{E} = i\omega\mathbf{B}, \tag{10.110}$$

and apply the current-free form of the Maxwell–Ampére law,

$$\nabla \times \mathbf{H} = -i\omega\mathbf{D}, \tag{10.111}$$

with $\mathbf{B} = \mu_0\mathbf{H}$, we can derive a relation between $\mathbf{D}$ and $\mathbf{E}$ of the form

$$\mathbf{D} = \frac{1}{\mu_0\omega^2}[\nabla(\nabla \cdot \mathbf{E}) - \nabla^2\mathbf{E}]. \tag{10.112}$$

For a plane wave, we may substitute $\nabla \leftrightarrow ikn\mathbf{s}$, and find

$$\mathbf{D} = \frac{n^2k^2}{\mu_0\omega^2}[\mathbf{E}_0 - \mathbf{s}(\mathbf{s} \cdot \mathbf{E}_0)]. \tag{10.113}$$

Now, if we use Equations 10.107 and 10.108, we may write

$$E_j = \frac{\dfrac{n^2k^2}{\mu_0\omega^2}s_j(s_lE_l)}{\dfrac{n^2k^2}{\mu_0\omega^2} - \epsilon_j}. \tag{10.114}$$

Ideally, we would like to find an expression for the refractive index as a function of the direction of propagation; to do so, we need to eliminate E_j from both sides of this expression. To do so, we take the dot product of both sides with $\mathbf{s}$, which gives us an expression s_jE_j on both sides of the equation that may be divided out. Finally, we may write

$$\frac{1}{n^2} = \frac{s_x^2}{n^2 - \epsilon_x/\epsilon_0} + \frac{s_y^2}{n^2 - \epsilon_y/\epsilon_0} + \frac{s_z^2}{n^2 - \epsilon_z/\epsilon_0}. \tag{10.115}$$

This equation, quadratic in n, suggests that there are two possible refractive indices for any direction of propagation, presumably related to two orthogonal polarizations of the displacement field. We may write this in a more useful form by multiplying both sides by n^2, and substituting in $s_x^2 + s_y^2 + s_z^2 = 1$; with some rearrangement (left as an exercise), we have

$$\frac{s_x^2}{1/\epsilon_x - 1/\epsilon_0 n^2} + \frac{s_y^2}{1/\epsilon_y - 1/\epsilon_0 n^2} + \frac{s_z^2}{1/\epsilon_z - 1/\epsilon_0 n^2} = 0. \tag{10.116}$$

We then introduce the phase velocity of the plane wave $v_p = c/n$, and the *principal velocities* of the crystal $v_j = 1/\sqrt{\mu_0\epsilon_j}$; we finally arrive at the expression

$$\frac{s_x^2}{v_p^2 - v_x^2} + \frac{s_y^2}{v_p^2 - v_y^2} + \frac{s_z^2}{v_p^2 - v_z^2} = 0. \tag{10.117}$$

This is known as *Fresnel's equation of wave normals*. We may multiply out all terms in the denominator to get it in a more useful form

$$(v_p^2 - v_y^2)(v_p^2 - v_z^2)s_x^2 + (v_p^2 - v_x^2)(v_p^2 - v_z^2)s_y^2$$
$$+ (v_p^2 - v_x^2)(v_p^2 - v_y^2)s_z^2 = 0. \qquad (10.118)$$

Now, let us assume that we are working with a uniaxial crystal, for which two of the principal velocities are the same, that is, $v_x = v_y = v_o$, $v_z = v_e$. We then have

$$(v_p^2 - v_o^2)\left[(v_p^2 - v_e^2)(s_x^2 + s_y^2) + (v_p^2 - v_o^2)s_z^2\right] = 0. \qquad (10.119)$$

There are two possible wave velocities, and corresponding refractive indices, given by

$$v_p = v_o, \qquad (10.120)$$

$$v_p = \sqrt{v_e^2 \sin^2 \theta + v_o^2 \cos^2 \theta}, \qquad (10.121)$$

where θ is the angle between the wave direction s and the extraordinary axis.

We may at last, at least roughly, describe the output behavior of the signal and idler photons; for simplicity, we will consider the case where $\omega_p = 2\omega$, $\omega_s = \omega_i = \omega$. For Type I SPDC, we must match the longitudinal and transverse momenta of the input and output states, that is,

$$2n_e(2\omega) = n_o(\omega)s_s \cdot \hat{z} + n_o(\omega)s_i \cdot \hat{z}, \qquad (10.122)$$

$$n_o(\omega)\hat{z} \times (s_s \times \hat{z}) = -n_o(\omega)\hat{z} \times (s_i \times \hat{z}). \qquad (10.123)$$

Using θ_s and θ_i as the angle between the z-axis and the signal and idler photons, respectively, we readily see that $\theta_s = \theta_i = \theta$ and $\phi_s = \phi_i + \pi$ in the second equation; the first equation, then, reduces to

$$\cos\theta = \frac{n_e(2\omega)}{n_o(\omega)}. \qquad (10.124)$$

The absolute azimuthal angle of the output photons is unspecified by the conservation laws. We therefore find that, when many pairs of SPDC photons are considered, they must emerge on a cone of angle θ. This cone is

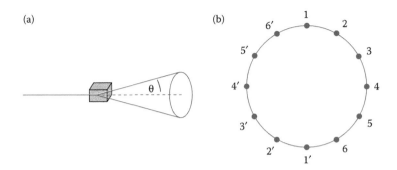

(a)   (b)

FIGURE 10.6 (a) Cone of photons emitted in Type I SPDC. (b) Positions of entangled photons on the cone in Type I SPDC.

illustrated in Figure 10.6a. Because $n_e(2\omega)$ depends on the angle between the pump beam and the extraordinary axis, the opening angle of the cone depends on the orientation of the crystal.

But what is the behavior of individual photon pairs created in the process? Here, we have the situation described loosely in Section 10.3: the physics of the problem defines a relationship between the directions of the photons, but does not uniquely specify the direction of any individual photon. The photons are produced in an entangled state with respect to their propagation direction. We may illustrate this with the help of Figure 10.6b, in which we assume for simplicity that the photons may only appear at 12 distinct locations. The output state in this basis may be written as

$$|\psi\rangle = \frac{1}{\sqrt{6}} \sum_{j=1}^{6} e^{i\psi_j} |j\rangle |j'\rangle, \qquad (10.125)$$

where ψ_j is a phase that depends on the output state. As the photons are otherwise indistinguishable, it is to be noted that there is no distinction between, say, a signal photon in state 3 and an idler photon in state 3' or vice versa.

This sort of entangled state can be useful as a single-photon source; it was shown experimentally by Burnham and Weinberg [BW70] that the SPDC photons produced in an ADP crystal pumped at 325 nm were generated within 4 ns of each other. The idler photon, therefore, can be used as a trigger for an experiment working with single signal photons. However, the entanglement otherwise is similar to the beamsplitter output discussed earlier, which is sort of an "all or nothing" result; of more interest are entangled states where the positions of the photons are more or less determined but an internal variable such as spin or OAM is entangled.

Type II SPDC can provide a polarization-entangled pair of photons. In this case, we have an extraordinary pump photon that breaks into a signal and idler that have ordinary and extraordinary polarization. The momentum conservation equations seemingly change little from the Type I case

$$2n_e(2\omega) = n_e(\omega)s_e \cdot \hat{\mathbf{z}} + n_o(\omega)s_o \cdot \hat{\mathbf{z}}, \qquad (10.126)$$

$$n_o(\omega)\hat{\mathbf{z}} \times (\mathbf{s}_o \times \hat{\mathbf{z}}) = -n_e(\omega)\hat{\mathbf{z}} \times (\mathbf{s}_e \times \hat{\mathbf{z}}), \qquad (10.127)$$

where we now label the output photons as "ordinary" and "extraordinary" for clarity. However, these equations are significantly more difficult to solve because $n_e(\omega)$ is a function of the propagation direction of the extraordinary photon itself. From Equation 10.121, we may write

$$\frac{1}{n_e^2} = \frac{\epsilon_0}{\epsilon_e}[\mathbf{s}_e \times \mathbf{e}_e]^2 + \frac{\epsilon_0}{\epsilon_o}[\mathbf{s}_e \cdot \mathbf{e}_e]^2, \qquad (10.128)$$

where $\mathbf{e}_e$ is the unit vector pointing along the extraordinary crystal axis. Solving these equations for the propagation directions of the photons is quite difficult, and we will only describe the results. The ordinary and extraordinary photons can be shown to propagate along two overlapping cones, as depicted in Figure 10.7a. The output photons now have orthogonal polarizations; again, treating a finite number of cases for simplicity, as in Figure 10.7b, the output state may be written as

$$|\psi\rangle = \frac{1}{\sqrt{12}} \sum_{j=1}^{12} e^{i\psi_j} |j\rangle_o |j'\rangle_e, \qquad (10.129)$$

where we have labeled the states as ordinary or extraordinary polarization.

Let us imagine that we position detectors to only collect photons from the overlap positions, namely, 1, 9′ and 1′, 9. Any photon pairs coming from these directions will be in the (renormalized) state

$$|\psi\rangle = \frac{1}{\sqrt{2}}[|1\rangle_o |1'\rangle_e + |9\rangle_o |9'\rangle_e] = \frac{1}{\sqrt{2}}[|1\rangle_o |1'\rangle_e + |1'\rangle_o |1\rangle_e]. \quad (10.130)$$

We have a pair of photons in an entangled polarization state!

This configuration was first proposed by Kwiat et al. [KESC94], and soon after experimentally demonstrated by Kwiat et al. [KMW+95]. They used

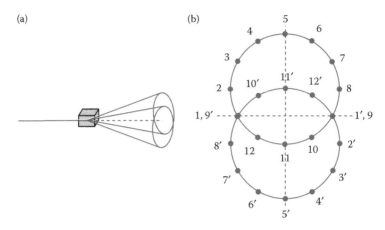

FIGURE 10.7 (a) Cones of photons emitted in Type II SPDC. (b) Positions of entangled photons on the cone in Type II SPDC.

a beta-barium-borate (BBO) crystal with a 351-nm pump beam from an argon ion laser; the output photons were used to test Bell inequalities, which we will discuss in Section 10.7. Type II SPDC remains a standard technique to produce entangled photons in quantum optics for applications such as cryptography, teleportation, or investigation of the fundamentals of quantum theory. More information on applications can be found in the book by Bouwmeester, Ekert, and Zeilinger [BEZ00].

With some understanding of SPDC and its connection to entanglement, we now return to the question of OAM. In the process of difference-frequency generation, the OAM is conserved and, furthermore, the OAM of the output photons is uniquely determined by the constraint that the stimulated photon at frequency ω_1 must have the same OAM as the input photon of the same frequency. In SPDC, however, this constraint is gone; so what happens?

An early experiment by Arlt et al. [ADAP99] concluded that OAM is in fact *not* conserved in SPDC. They reasoned that a photon with even OAM, say $2\hbar l$, with l an integer, would break into two photons each with OAM $\hbar l$. On a macroscopic scale, a Laguerre–Gauss beam of azimuthal order $2l$ would presumably produce a beam of azimuthal order l. They measured the transverse intensity profile as a function of distance from the crystal; though the characteristic donut of a vortex beam appeared near the exit of the crystal, it became washed out on propagation over a short distance. From this, they initially argued that OAM is not conserved in the process of SPDC.

Let us assume, however, that OAM is conserved. Without a constraint on the idler photon OAM, however, the conservation law takes the form

$$l_{pump} = l_{signal} + l_{idler}. \tag{10.131}$$

Given only l_{pump}, this equation has two unknowns: l_{signal} and l_{idler}. We again have a situation where the physics of the problem forces a definite relation between two photons, but does not force a definite value on either of them. We therefore expect the photons to be in an entangled OAM state. For a pump beam with zero angular momentum, we would predict the output two-photon state to be of the form

$$|\psi\rangle = c_{0,0}|A,0\rangle|B,0\rangle + c_{1,-1}|A,1\rangle|B,-1\rangle + c_{-1,1}|A,-1\rangle|B,1\rangle$$
$$+ c_{2,-2}|A,2\rangle|B,-2\rangle + c_{-2,2}|A,-2\rangle|B,2\rangle + \cdots, \tag{10.132}$$

where the constants $c_{p,q}$ are the probability amplitudes for measuring a photon along a path A with OAM $\hbar p$ and a photon along a second path B with OAM $\hbar q$. It is important to note that this result applies for both Type I and Type II SPDC, unlike polarization entanglement.

This result was experimentally tested by Mair et al. [MVWZ01]; a schematic of their experiment is shown in Figure 10.8. A pump beam from an argon-ion laser of wavelength 351 nm undergoes Type I SPDC in a BBO crystal, and the two photons are passed through computer-generated vortex holograms that are designed to diffract a particular vortex mode toward the detectors. Lenses concentrate the light and couple it into optical fibers, which lead to detectors and a coincidence circuit.

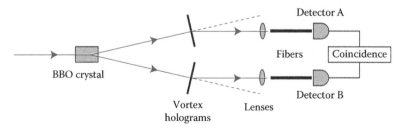

FIGURE 10.8 Illustration of the experiment by Mair et al. [MVWZ01] to detect the OAM entanglement of photons.

Using combinations of various vortex holograms, the researchers were able to determine how often coincidence of photons with different combinations of l_{signal} and l_{idler} occur. They found that Equation 10.131 is satisfied.

However, as noted in Section 10.3, the existence of a correlation in angular momentum does not necessarily imply entanglement. In every SPDC event from a vortex-free beam, for instance, a pair of photons could be produced in a pure state $|l\rangle\,|-l\rangle$, with the value of l random for each event. Such a physical system would be described by a density operator such as given by Equation 10.66, which is missing the interference terms of an entangled state. To prove that the state was truly entangled, Mair et al. displaced a vortex hologram of order $l = 1$ leading to detector A slightly, resulting in a mixture of $l = 0$ and $l = 1$ modes arriving at the detector; the hologram was removed in the path to detector B, allowing the intensity distribution to be mapped as a function of position. For an entangled state, we expect that this will result in a coherent superposition of the complementary modes 0 and -1 at the second detector.

To see how this works, let us consider the entangled state of Equation 10.132. We first put a hologram in path A that deflects an OAM state $|A, 1\rangle$ to the state $|A\rangle$ which reaches that detector; the action of this hologram may be described by a projection operator

$$\hat{P} \equiv |A\rangle\,\langle A, 1|. \tag{10.133}$$

The action of this on the entangled state is readily found to be

$$\hat{P}\,|\psi\rangle = c_{1,-1}\,|A\rangle\,|B, -1\rangle, \tag{10.134}$$

and the density matrix $\hat{\rho}_e$ of this system is given by the outer product

$$\hat{\rho}_e = |c_{1,-1}|^2\,|A\rangle\,|B, -1\rangle\,\langle B, -1|\,\langle A|. \tag{10.135}$$

For comparison, let us consider a statistical mixture of states with conserved OAM. The density operator in this case is given by

$$\hat{\rho}_s = \sum_q |c_{q,-q}|^2\,|A, q\rangle\,|B, -q\rangle\,\langle B, -q|\,\langle A, q|. \tag{10.136}$$

The action of the hologram on this mixture can be found by a similarity transformation, namely,

$$\hat{P}\hat{\rho}_s\hat{P}^\dagger = |A\rangle \langle A, 1| \left[\sum_q |c_{q,-q}|^2 |A, q\rangle |B, -q\rangle \langle B, -q| \langle A, q| \right] |A, 1\rangle \langle A|$$

$$= |c_{1,-1}|^2 |A\rangle |B, -1\rangle \langle B, -1| \langle A|. \tag{10.137}$$

The result is identical to that of the entangled state.

Now, let us imagine shifting the hologram slightly off the axis of beam propagation. This change causes the hologram to partly diffract photons from orders $l = 0$ and $l = 1$; the projection operator now takes the form

$$\hat{P} = \alpha |A\rangle \langle A, 0| + \beta |A\rangle \langle A, 1|, \tag{10.138}$$

where α and β are complex constants. We can readily show that the entangled state after this projection is reduced to

$$\hat{P}|\psi\rangle = |A\rangle \left[c_{0,0}\alpha |B, 0\rangle + c_{1,-1}\beta |B, -1\rangle \right], \tag{10.139}$$

with a corresponding density matrix

$$\hat{P}\hat{\rho}_e\hat{P}^\dagger = |A\rangle \langle A| \{ |c_{0,0}|^2 |\alpha|^2 |B, 0\rangle \langle B, 0| + |c_{1,-1}|^2 |\beta|^2 |B, -1\rangle \langle B, -1|$$

$$+ c_{0,0}c_{1,-1}^* |B, 0\rangle \langle B, -1| + c_{1,-1}^* c_{0,0} |B, -1\rangle \langle B, 0| \}. \tag{10.140}$$

With a little work, the corresponding density matrix for the mixed state is found to be

$$\hat{P}\hat{\rho}_s\hat{P}^\dagger = |A\rangle \langle A| \{ |c_{0,0}|^2 |\alpha|^2 |B, 0\rangle \langle B, 0| + |c_{0,0}|^2 |\beta|^2 |B, -1\rangle \langle B, -1| \}. \tag{10.141}$$

On comparison of the density matrices for the entangled state and the mixed state, we can see that the entangled state possesses interference terms that the mixed state does not. To interpret these results a little more clearly, we look at the probability amplitude of measuring an event at detector A and a photon at a point $\mathbf{r}$ in the plane of the detector B; this can be done by taking the inner product of Equation 10.139 with $\langle A| \langle B, \mathbf{r}|$, which gives

$$\langle A| \langle B, \mathbf{r}| \hat{P}|\psi\rangle = c_{0,0}\alpha U_{00}(\mathbf{r}) + c_{1,-1}\beta U_{0,-1}(\mathbf{r}), \tag{10.142}$$

with $U_{nm}(\mathbf{r})$ the Laguerre–Gauss mode of order n, m and we have used $\langle B, \mathbf{r} | B, l \rangle = U_{0l}$. The measurement produces a coherent superposition of the $l = 0$ and $l = -1$ vortex modes. For the mixed state, we operate on both sides of the density matrix of Equation 10.141 with $\langle A | \langle B, \mathbf{r} |$, getting

$$\langle A | \langle B, \mathbf{r} | \hat{P} \hat{\rho}_m \hat{P}^\dagger | A \rangle | B, \mathbf{r} \rangle = |c_{0,0}|^2 |\alpha|^2 |U_{00}(\mathbf{r})|^2 + |c_{1,-1}|^2 |\beta|^2 |U_{01}(\mathbf{r})|^2.$$
(10.143)

The intensity detected would be an incoherent superposition of the $l = 0$ and $l = -1$ modes. In the experiment by Mair et al., a coherent superposition was found, verifying entanglement.

We have not yet said anything about the relative magnitudes of the coefficients $c_{q,-q}$ of the entangled wavefunction. Intuitively, one expects them to be largest for those states in which the OAM of the pump is divided equally among the photons, and progressively smaller for situations where the signal and idler photons have large OAM different from the pump. For a pump of OAM $l = 0$, for instance, we would expect that $c_{0,0}$ would be the largest amplitude in the entangled state.

However, this suggests that the state is only partially entangled. Imagine, for example, a case where $c_{0,0} \approx 1$ and all other $c_{q,-q} \approx 0$. Such a state is very close to a pure state and only weakly entangled, and far from ideal for proposed quantum applications. Vaziri et al. [VPJ⁺03] have introduced a scheme to adjust the relative weights of the coefficients $c_{q,-q}$ and produce a maximally entangled state in which all weights are approximately equal. The scheme, which we will not discuss in detail here, is conceptually similar to the adjustment of the hologram position in the projection operator of Equation 10.138. By adjusting the position of the hologram, we were able to control the amount of each mode that passes through to the detector, with controllable relative amplitudes α and β. In the experiment by Vaziri et al., lenses were used to modify the relative amplitudes of the different OAM modes. In particular, they were able to modify an initial state

$$|\psi\rangle_i = 0.80 \, |A, 0\rangle \, |B, 0\rangle + 0.44 \, |A, 1\rangle \, |B, -1\rangle + 0.41 \, |A, 2\rangle \, |B, -2\rangle,$$
(10.144)

and concentrate it to a state

$$|\psi\rangle_f = 0.60 \, |A, 0\rangle \, |B, 0\rangle + 0.56 \, |A, 1\rangle \, |B, -1\rangle + 0.57 \, |A, 2\rangle \, |B, -2\rangle.$$
(10.145)

These states, which include three distinct OAM pairs, were referred to as "qutrit states."

With an appropriate system, it is possible to create entanglement between photons with very high angular momenta. Fickler et al. [FLP$^+$12] were able to demonstrate entanglement in which the photon's OAM differed by 600, that is,

$$|\psi\rangle = \frac{1}{\sqrt{2}}[|A, 300\rangle\, |B, -300\rangle + |A, -300\rangle\, |B, 300\rangle]. \qquad (10.146)$$

One final study of OAM and entanglement is worth noting here. Very recently, Hiesmayr, de Dood, and Löffler demonstrated [HdDL16] OAM entanglement between *four* photons.

10.6 A NONLOCAL OPTICAL VORTEX

Up to this point, our discussion of entanglement has relied primarily on very abstract two-photon quantum states. We may also, however, consider the two-photon wavefunction $\psi(\mathbf{r}_1, \mathbf{r}_2)$ as represented in position space. From this representation, we will see that it is possible to make a new type of nonlocal optical vortex, "nonlocal" in this case indicating that the vortex is manifested only in the combined wavefunction, and not in the wavefunction of either of the individual photons.

We must first consider deriving an expression for $\psi(\mathbf{r}_1, \mathbf{r}_2)$ from the physics of SPDC. This wavefunction represents the probability amplitude of finding one photon at position $\mathbf{r}_1$ and another at position $\mathbf{r}_2$; the probability of such an occurrence is, of course, given by $|\psi(\mathbf{r}_1, \mathbf{r}_2)|^2$. Such a derivation was first done rigorously by Hong and Mandel [HM85]; we consider a more heuristic approach as given by Franke-Arnold et al. [FABPA02]. For slightly different takes on the derivation, see Monken, Souto Ribeiro, and Pádua [MRP98] and Walborn et al. [WdOTM04].

Our approach is somewhat similar to that used to describe a beamsplitter in Section 10.3. In that case, we related the creation operators $\hat{a}_0$, $\hat{a}_1$ for an input state directly to the creation operators $\hat{a}_2$, $\hat{a}_3$ for a photon in the output state. Here, *assuming that SPDC occurs*, we relate the creation operators for a pump state to the creation operators for the signal and idler states.

In a fixed z-plane of the crystal, we decompose the pump field $U_p(\mathbf{r})$ into a set of plane waves, of the form

$$U_p(\mathbf{r}) = \int \tilde{U}_p(\mathbf{k}_p)e^{i\mathbf{k}_p \cdot \mathbf{r}}d^2k_p. \qquad (10.147)$$

The quantity $\tilde{U}_p(\mathbf{k}_p)$ represents the amplitude of the plane wave with transverse wavevector $\mathbf{k}_p$ of the pump beam. For each plane wave of the pump beam, it will spawn signal and idler plane waves satisfying the operator relation

$$\hat{a}^\dagger(\mathbf{k}_p) = \iint F(\mathbf{k}_s, \mathbf{k}_i)\hat{a}^\dagger(\mathbf{k}_s)\hat{a}^\dagger(\mathbf{k}_i)d^2k_s d^2k_i, \tag{10.148}$$

where $\mathbf{k}_s$ and $\mathbf{k}_i$ are the transverse wavevectors of the signal and idler photons, and $F(\mathbf{k}_s, \mathbf{k}_i)$ is the probability amplitude of finding the signal and idler photons with those particular wavevectors.

But what is the form of $F(\mathbf{k}_s, \mathbf{k}_i)$? To satisfy momentum conservation, we expect that we must have, to a good approximation, $\mathbf{k}_p = \mathbf{k}_s + \mathbf{k}_i$; this can be represented by $\delta^{(2)}(\mathbf{k}_p - \mathbf{k}_s - \mathbf{k}_i)$. To satisfy energy conservation, we expect that $\mathbf{k}_s$ and $\mathbf{k}_i$ must be quite similar in value; any large difference between them would generally require one of the signal or idler frequencies to be very large. We model this behavior by the function $\tilde{\Delta}(\mathbf{k}_s - \mathbf{k}_i)$, where

$$\int |\tilde{\Delta}(\mathbf{K})|^2 d^2K = 1, \tag{10.149}$$

and $\tilde{\Delta}(\mathbf{K})$ is a narrow function centered on the origin. This suggests that we may write

$$\hat{a}^\dagger(\mathbf{k}_p) = \iint \delta^{(2)}(\mathbf{k}_p - \mathbf{k}_s - \mathbf{k}_i)\tilde{\Delta}(\mathbf{k}_s - \mathbf{k}_i)\hat{a}^\dagger(\mathbf{k}_s)\hat{a}^\dagger(\mathbf{k}_i)d^2k_s d^2k_i. \tag{10.150}$$

The total output wavefunction $|\psi\rangle$, however, must be summed over all possible values of $\mathbf{k}_p$; we may write

$$|\psi\rangle = \int \tilde{U}_p(\mathbf{k}_p)\hat{a}^\dagger(\mathbf{k}_p)|0\rangle \, d^2k_p. \tag{10.151}$$

On substitution from Equation 10.150 into this expression, and integrating over $\mathbf{k}_p$, we have

$$|\psi\rangle = \iint \tilde{U}_p(\mathbf{k}_s + \mathbf{k}_i)\tilde{\Delta}(\mathbf{k}_s - \mathbf{k}_i)\hat{a}^\dagger(\mathbf{k}_s)\hat{a}^\dagger(\mathbf{k}_i)|0\rangle \, d^2k_s d^2k_i. \tag{10.152}$$

This represents a momentum space decomposition of a pump photon into signal and idler photons. We would now like to write this instead in position space; to do so, we apply the Fourier representations of the creation operators

$$\hat{a}^\dagger(\mathbf{K}) = \frac{1}{2\pi} \int \hat{a}^\dagger(\mathbf{r}) e^{i\mathbf{K}\cdot\mathbf{r}} d^2 r. \tag{10.153}$$

On substitution, the wavefunction may be written as

$$|\psi\rangle = \frac{1}{(2\pi)^2} \iint \tilde{U}_p(\mathbf{k}_s + \mathbf{k}_i) \tilde{\Delta}(\mathbf{k}_s - \mathbf{k}_i) \hat{a}^\dagger(\mathbf{r}_s) \hat{a}^\dagger(\mathbf{r}_i) |0\rangle$$
$$\times e^{i(\mathbf{k}_s\cdot\mathbf{r}_s + \mathbf{k}_i\cdot\mathbf{r}_i)} d^2 k_s d^2 k_i d^2 r_s d^2 r_i. \tag{10.154}$$

We may now evaluate the integrals over the wavevectors. It is convenient to do so with the variables

$$\mathbf{K} \equiv \mathbf{k}_s + \mathbf{k}_i, \quad \mathbf{k} \equiv \mathbf{k}_s - \mathbf{k}_i, \tag{10.155}$$

which leads to the expression

$$|\psi\rangle = \frac{1}{(2\pi)^2} \iint \tilde{U}_p(\mathbf{K}) \tilde{\Delta}(\mathbf{k}) \hat{a}^\dagger(\mathbf{r}_s) \hat{a}^\dagger(\mathbf{r}_i) |0\rangle \, e^{i\mathbf{K}\cdot(\mathbf{r}_s+\mathbf{r}_i)/2}$$
$$\times e^{i\mathbf{k}\cdot(\mathbf{r}_s-\mathbf{r}_i)/2} d^2 k_s d^2 k_i d^2 r_s d^2 r_i. \tag{10.156}$$

From the definition of inverse Fourier transforms, we may then write

$$|\psi\rangle = \frac{1}{\pi^2} \iint U_p\left(\frac{\mathbf{r}_s + \mathbf{r}_i}{2}\right) \Delta\left(\frac{\mathbf{r}_s - \mathbf{r}_i}{2}\right) \hat{a}^\dagger(\mathbf{r}_s) \hat{a}^\dagger(\mathbf{r}_i) |0\rangle \, d^2 r_s d^2 r_i. \tag{10.157}$$

The position wavefunction is the projection of this state onto the position basis, or in other words simply the amplitude of the state at position $\mathbf{r}_s$, $\mathbf{r}_i$,

$$\psi(\mathbf{r}_s, \mathbf{r}_i) = \frac{1}{\pi^2} U_p\left(\frac{\mathbf{r}_s + \mathbf{r}_i}{2}\right) \Delta\left(\frac{\mathbf{r}_s - \mathbf{r}_i}{2}\right). \tag{10.158}$$

Those familiar with classical coherence theory may be surprised at this point, because the functional form of $\psi(\mathbf{r}_1, \mathbf{r}_2)$ is strikingly similar to the cross-spectral density of a so-called *quasi-homogeneous source* [Wol07, Section 5.3.2]. It has, in fact, been noted [SAST00] that there is a strong mathematical similarity between classical correlation functions and two-photon wavefunctions. This similarity has been referred to as a "duality" because the correlation function of a highly *coherent* field looks like the wavefunction of an *unentangled* photon pair. We point out this duality here, however, to emphasize that correlation functions and two-photon wavefunctions represent physically very different things.

With Equation 10.158, we may now explore the curious phenomenon referred to as a *nonlocal optical vortex* by Gomes et al. [GST⁺09]. We first introduce, for convenience, the one-variable Hermite–Gauss modes $u_0(x/\delta)$ and $u_1(x/\delta)$, of the form

$$u_0(x/\delta) = c_0 e^{-x^2/\delta^2}, \tag{10.159}$$

$$u_1(x/\delta) = c_1 [x/\delta] e^{-x^2/\delta^2}, \tag{10.160}$$

where c_0 and c_1 are normalization constants, and δ is the width of the modes. Let us assume that the function $\Delta(\mathbf{R})$ is a simple Gaussian, that is,

$$\Delta\left(\frac{\mathbf{r}_s - \mathbf{r}_i}{2\delta}\right) = u_0\left(\frac{x_s - x_i}{2\delta}\right) u_0\left(\frac{y_s - y_i}{2\delta}\right), \tag{10.161}$$

and that we take as our pump beam a Hermite–Gauss beam of order $m = 1$, $n = 0$, so that

$$U_p\left(\frac{\mathbf{r}_s + \mathbf{r}_i}{2}\right) = u_1\left(\frac{x_s + x_i}{2\delta}\right) u_0\left(\frac{y_s + y_i}{2\delta}\right). \tag{10.162}$$

The widths of U_p and Δ are taken to be the same; assuming that the width of Δ is fixed by the physics of down conversion, the width of the pump beam can be adjusted to match.

The two-photon wavefunction will therefore be the product

$$\psi(\mathbf{r}_s, \mathbf{r}_i) = \frac{1}{\pi^2} u_0\left(\frac{x_s - x_i}{2\delta}\right) u_1\left(\frac{x_s + x_i}{2\delta}\right) u_0\left(\frac{y_s - y_i}{2\delta}\right) u_0\left(\frac{y_s + y_i}{2\delta}\right). \tag{10.163}$$

It is not difficult to rearrange these products into separable functions of $\mathbf{r}_i$ and $\mathbf{r}_s$ alone; for example,

$$u_0\left(\frac{y_s - y_i}{2\delta}\right) u_0\left(\frac{y_s + y_i}{2\delta}\right) = u_0\left(y_s/\sqrt{2}\delta\right) u_0\left(y_i/\sqrt{2}\delta\right). \tag{10.164}$$

A similar separation may be done for the $u_1 u_0$ product; the final result is that we may write

$$\psi(\mathbf{r}_s, \mathbf{r}_i) = \frac{1}{\pi^2} \frac{1}{\sqrt{2}} \left[u_1 \left(x_s/\sqrt{2}\delta \right) u_0 \left(x_i/\sqrt{2}\delta \right) + u_1 \left(x_i/\sqrt{2}\delta \right) \right.$$
$$\left. \times u_0 \left(x_s/\sqrt{2}\delta \right) \right] u_0 \left(y_s/\sqrt{2}\delta \right) u_0 \left(y_i/\sqrt{2}\delta \right). \quad (10.165)$$

It is to be noted that the output state is entangled in the coordinates x_i, x_s.

We now imagine taking the Fourier transform of the wavefunction of only one of the photons, say the idler. This can be done by using a lens in a $2f$-focusing configuration in the idler beam. We may evaluate these transforms using the expressions

$$\frac{1}{2\pi} \int u_0 \left(x/\sqrt{2}\delta \right) e^{-ik_x x} dx = \frac{\delta}{\sqrt{2\pi}} u_0 \left(k_x \delta/\sqrt{2} \right), \quad (10.166)$$

$$\frac{1}{2\pi} \int u_1 \left(x/\sqrt{2}\delta \right) e^{-ik_x x} dx = -i\frac{\delta}{\sqrt{2\pi}} u_1 \left(k_x \delta/\sqrt{2} \right), \quad (10.167)$$

and similar expressions for the y-transforms, as can be demonstrated by direct calculation (left as an exercise). The final wavefunction, in terms of $\mathbf{r}_s$ and $\mathbf{k}_i$, is

$$\psi(\mathbf{r}_s, \mathbf{k}_i) = \frac{1}{\pi^2} \frac{1}{\sqrt{2}} \frac{\delta^2}{2\pi} \left[u_1 \left(x_s/\sqrt{2}\delta \right) u_0 \left(k_x \delta/\sqrt{2} \right) - iu_1 \left(k_x \delta/\sqrt{2} \right) \right.$$
$$\left. \times u_0 \left(x_s/\sqrt{2}\delta \right) \right] u_0 \left(y_s/\sqrt{2}\delta \right) u_0 \left(k_y \delta/\sqrt{2} \right). \quad (10.168)$$

If we remove the constant terms and the common Gaussian factors, the bracketed term may be simplified to the form

$$\psi(x_s, k_x) \sim [x_s/\delta - ik_x\delta]. \quad (10.169)$$

In other words, the wavefunction has a vortex structure with respect to the signal variable x_s/δ and the idler variable $k_x\delta$. This vortex is not present in either the signal or idler wavefunctions alone.

Gomes et al. [GST⁺09] measured such a nonlocal vortex experimentally using the configuration shown in Figure 10.9. Light from a 441.6-nm HeCd laser is bisected by a glass slide to produce the π-phase shift of a

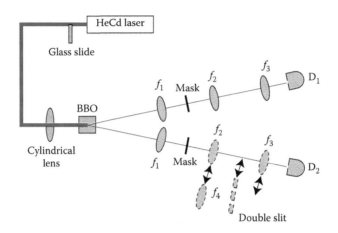

FIGURE 10.9   Illustration of the experiment to measure a nonlocal optical vortex. (Adapted from R.M. Gomes et al. *Phys. Rev. Lett.*, 103:033602, 2009.)

Hermite–Gauss beam. A cylindrical lens focuses the beam into a BBO crystal, producing an output two-photon wavefunction for which the widths of U_p and Δ are approximately the same. Lenses f_1 perform an optical Fourier transform onto Gaussian transmission masks, which force the function Δ to have a Gaussian shape. The fields output from these masks are then imaged using lenses f_2 and f_3 onto the detectors.

The detectors were scanned horizontally first to verify, as in Equation 10.163, that the two-photon wavefunction has a functional dependence of $u_1 [(x_s + x_i)/2]$. Then, the lenses f_2 and f_3 were removed and replaced with a lens f_4 that results in the Fourier transform of the idler photon, giving a two-photon wavefunction as in Equation 10.168. In scanning the detectors over the horizontal again, they found, as expected, an intensity pattern given by a Laguerre–Gauss beam. To demonstrate that this is a true phase vortex, a double slit was inserted into the path of detector 2 and the interference pattern was recorded.

This rather surprising result is an illustration of how the nonlocal quantum properties of light can introduce new types of singularities into a carefully prepared optical system.

10.7 BELL'S INEQUALITIES FOR ANGULAR MOMENTUM STATES

The history of quantum physics reads very much like a mystery novel, filled with many surprising and unexpected twists. One of those twists, as we have seen in Section 10.3, is the "spooky action at a distance" implication of entanglement. In the Copenhagen interpretation of quantum mechanics, a pair of entangled particles are in an indefinite state only defined by the probabilities of outcomes, and only collapse to a definite state when measured. This collapse, however, must be instantaneous, and a measurement on one particle will result in an instantaneous change in the other. Einstein, Podolsky, and Rosen (EPR) argued that this seeming violation of special relativity implied that quantum theory must be *incomplete*, not probabilistic, and that the properties of entangled particles must be perfectly well defined by hidden variables that we do not know how to measure.

This question—hidden variables or probability?—seemed for many years to be a truly metaphysical one, with no experimental difference between the interpretations. In 1964, however, Bell [Bel64] demonstrated, with a simple argument, that any local hidden variable theory (known collectively as local realism) is fundamentally inconsistent with the predictions of quantum mechanics. Bell's observation, and the theoretical and experimental work that followed, led to the first real tests of the philosophical implications of quantum theory, and this work is ongoing today.

The original research into what are now known as *Bell inequalities* was concerned only with 2-state particle systems, for example spin-1/2 particles. It has been shown more recently, however, that stronger and clearer violations of local realism can be achieved with higher-dimensional N-state systems. Naturally, researchers have turned to OAM states to provide the additional dimensionality. In this section, we discuss some of this work.

We begin by deriving a Bell inequality for 2-state systems. The most practical and commonly cited inequality is one first derived by Clauser, Horne, Shimony, and Holt (CHSH) [CHSH69], and we loosely follow their work.

Let us imagine, as depicted in Figure 10.10, that a pair of correlated 2-state particles are sent in different directions to analyzers with preferred measurement axes $\hat{a}$ and $\hat{b}$. The values A, B of the measurements are binary;

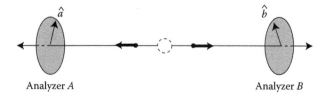

FIGURE 10.10 Illustration of a hypothetical experiment on correlated 2-state particles.

that is, regardless of the orientation of the analyzers, we have $A(\hat{a}) = \pm 1$, $B(\hat{b}) = \pm 1$.

We further assume, as EPR did, that the measurement depends on one or more hidden variables, represented by a single symbol λ; that is, we have $A(\hat{a}, \lambda)$, $B(\hat{b}, \lambda)$. In an ensemble of experiments, these hidden variables may be represented by a probability density $\rho(\lambda)$, such that

$$\int \rho(\lambda) d\lambda = 1. \tag{10.170}$$

We note, however—and it is the key to this entire argument—that $A(\hat{a}, \lambda)$ does not depend on the orientation $\hat{b}$, and vice versa. This simple statement is the imposition of *locality* to our model: the measurement of A does not affect the outcome of the measurement of B. This is at odds with the quantum picture, which suggests that the measurement of A collapses the wavefunction, and hence dramatically influences, the measurement of B. Our goal will be to ask whether *any* hidden variable theory can reproduce the same results as quantum physics.

Let us introduce a correlation function $C(\hat{a}, \hat{b})$ that characterizes the average correlations between measurements. We may express this as

$$C(\hat{a}, \hat{b}) = \int \rho(\lambda) A(\hat{a}, \lambda) B(\hat{b}, \lambda) d\lambda. \tag{10.171}$$

We now imagine a second pair of analyzer orientations, $\hat{a}'$ and $\hat{b}'$, and then consider the difference in correlation functions given by

$$C(\hat{a}, \hat{b}') - C(\hat{a}', \hat{b}') = \int \rho(\lambda) A(\hat{a}, \lambda) B(\hat{b}', \lambda) d\lambda - \int \rho(\lambda) A(\hat{a}', \lambda) B(\hat{b}', \lambda) d\lambda. \tag{10.172}$$

To the right side of this expression, we add and subtract the quantity

$$\pm \int \rho(\lambda) A(\hat{a}, \lambda) B(\hat{b}, \lambda) A(\hat{a}', \lambda) B(\hat{b}', \lambda) d\lambda.$$

After some rearranging, Equation 10.172 becomes

$$C(\hat{a}, \hat{b}') - C(\hat{a}', \hat{b}') = \int \rho(\lambda) A(\hat{a}, \lambda) B(\hat{b}', \lambda) [1 \pm A(\hat{a}', \lambda) B(\hat{b}', \lambda)] d\lambda$$

$$- \int \rho(\lambda) A(\hat{a}', \lambda) B(\hat{b}', \lambda) [1 \pm A(\hat{a}, \lambda) B(\hat{b}, \lambda)] d\lambda. \tag{10.173}$$

Taking the absolute value of this expression, we then apply the triangle inequality, $|x + y| \leq |x| + |y|$, along with the observation that $|A| = 1$, $|B| = 1$, to get

$$|C(\hat{a}, \hat{b}') - C(\hat{a}', \hat{b}')| \leq \int \rho(\lambda)[1 \pm A(\hat{a}', \lambda)B(\hat{b}', \lambda)]d\lambda$$

$$+ \int \rho(\lambda)[1 \pm A(\hat{a}, \lambda)B(\hat{b}, \lambda)]d\lambda. \qquad (10.174)$$

We may apply Equation 10.170 and the definition of the correlation function to write

$$|C(\hat{a}, \hat{b}') - C(\hat{a}', \hat{b}')| \mp [C(\hat{a}', \hat{b}) + C(\hat{a}, \hat{b})] \leq 2. \qquad (10.175)$$

The choice of $\mp$ is dictated by whichever option gives us an overall positive value for the bracketed terms. Then, the triangle equality may be applied in reverse, to get

$$|Q| \equiv |C(\hat{a}, \hat{b}') - C(\hat{a}', \hat{b}') + C(\hat{a}', \hat{b}) + C(\hat{a}, \hat{b})| \leq 2. \qquad (10.176)$$

This is the *CHSH inequality*. It sets an upper bound of 2 for the given combination of four correlation functions of a pair of 2-state variables, under the assumption of locality.

But what is the largest possible value of the CHSH combination Q? It was shown by Cirel'son [Cir80] that the maximum possible value of the CHSH combination for 2-state variables, without any assumptions of locality, is $|Q| = 2\sqrt{2}$. This can be demonstrated in a straightforward manner, though the calculation is long and tedious, so we do not provide it here. It would appear that an entangled state, with its nonlocal correlations, would be able to exceed the CHSH bound and approach the maximum.

A proof that states exist that cause a maximal violation was given by Gisin and Peres [GP92] and further quantified by Braunstein, Mann, and Revsen [BMR92]. Those proofs focused on true spin-1/2 particles; here, we consider the optical analog, in which we have entangled photons from SPDC. We consider states of the form

$$|\psi_\pm\rangle = \frac{1}{\sqrt{2}}[|H_a\rangle |V_b\rangle \pm |V_a\rangle |H_b\rangle], \qquad (10.177)$$

where H_i and V_i represent horizontal and vertical polarization of the ith photon, respectively. We imagine a polarization-sensitive analyzer placed into each of the output paths of the photons; a simple polarizing beamsplitter works for this purpose. The kets $|H_i\rangle$ and $|V_i\rangle$ can be represented by Jones vectors; then, the operator representing an analyzer along the x-axis is given by

$$\hat{\sigma} = \begin{bmatrix} 1 & 0 \\ 0 & -1 \end{bmatrix}. \tag{10.178}$$

For an arbitrary orientation θ of the analyzer, we simply perform a similarity transformation of the matrix σ; this gives us

$$\hat{\sigma}(\theta) = \begin{bmatrix} \cos\theta & -\sin\theta \\ \sin\theta & \cos\theta \end{bmatrix} \begin{bmatrix} 1 & 0 \\ 0 & -1 \end{bmatrix} \begin{bmatrix} \cos\theta & \sin\theta \\ -\sin\theta & \cos\theta \end{bmatrix}$$

$$= \begin{bmatrix} \cos(2\theta) & \sin(2\theta) \\ \sin(2\theta) & -\cos(2\theta) \end{bmatrix}. \tag{10.179}$$

The correlation function for a pair of analyzer positions θ_a and θ_b has an operator of the form

$$\hat{C}(\theta_a, \theta_b) \equiv \hat{\sigma}(\theta_a)\hat{\sigma}(\theta_b), \tag{10.180}$$

where each operator only acts upon the component of the state $|\psi\rangle$ in its path. One can readily show that the correlation function then has the average value

$$C(\theta_a, \theta_b) = -\cos[2(\theta_1 \mp \theta_2)]. \tag{10.181}$$

Let us consider a particular set of orientations for the analyzers. We choose $\theta_a = 0, \theta_{a'} = 45°, \theta_b = 22.5°$, and $\theta_{b'} = -22.5°$. It is then straightforward to show that

$$C(\theta_a, \theta_{b'}) - C(\theta_{a'}, \theta_{b'}) + C(\theta_{a'}, \theta_b) + C(\theta_a, \theta_b) = \pm 2\sqrt{2}. \tag{10.182}$$

We see that the two states $|\psi_\pm\rangle$ give the two extreme limits of $|Q| = 2\sqrt{2}$ for an appropriate choice of orientations. These are two of what are known as *Bell states*; the other two, given by

$$|\phi_\pm\rangle = \frac{1}{\sqrt{2}}[|H_a\rangle\,|H_b\rangle \pm |V_b\rangle\,|V_a\rangle], \tag{10.183}$$

give a value of $|Q| = 0$ for the same orientations.

Since the description of the CHSH inequality, there have been many tests of them, with later tests taking advantage of better sources and more efficient detectors; almost all of these tests have shown clear violations of the CHSH inequalities. An early one by Freedman and Clauser [FC72] used correlated photons produced in an atomic cascade of calcium; an improved test with a similar source was done by Aspect, Grangier, and Roger [AGR81]. The introduction of entangled photons via SPDC provided a new and extremely robust source for performing Bell tests; the earlier mentioned work by Kwiat et al. [KMW+95] showed violations of over 100 standard deviations in less than 5 min of data collection.

One limitation of the earliest experiments, however, was the need for supplementary assumptions. It can be shown that, when the efficiency η of a detector is taken into account, the correlation function of Equation 10.181 gets modified to $-\eta^2 \cos[2(\theta_1 \mp \theta_2)]$. This indicates that the largest violation of the CHSH inequality will be $|Q| = \eta^2 2\sqrt{2}$; in order to see a direct violation, then, one must have a detector with efficiency $\eta > 0.84$, and conventional detectors fall far short of this requirement.

One can get around this limitation by taking additional experimental measurements and introducing additional assumptions into the analysis, as discussed in detail by Garuccio and Rapisarda [GR81]; these additional assumptions, however, introduce potential loopholes in which locality falsely appears to be violated. Much of the work over the past 50 years has focused on closing these loopholes, and in 2015, several experiments, such as the one by Hensen et al. [HBD+15], appear to have finally shown loophole-free violation. This important work seems to have met with only a mild fanfare, as the wide variety and number of earlier experiments had already convinced most theorists that locality must be violated.

With this lengthy background out of the way, we may now return to the discussion of vortex beams! The aforementioned research only considers the simplest possible quantum states, in which a pair of 2-state particles ("qubits") are entangled. It is natural to ask whether stronger quantum effects, and stronger violations of locality, occur when either the number of entangled particles is increased or the dimensionality of the states is increased. The former case was shown to be true by Greenberger, Horne, and Zeilinger [GHZ89], in which they demonstrated that the difference between local realism and quantum mechanics is much stronger for three particles. The latter case was first shown to be true by Kaszlikowski et al. [KGZ+00], who used computational methods to show that higher-dimensional states have quantum effects more resistant to noise that can

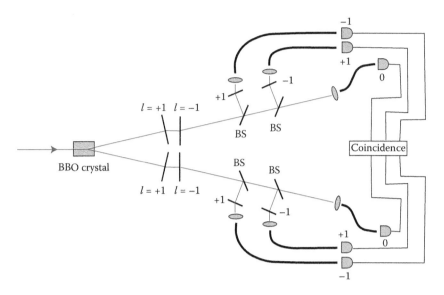

FIGURE 10.11 Illustration of the experiment of Vaziri, Weihs, and Zeilinger [VWZ02].

be observed with less-efficient detectors. Their work showed this to be true up to dimensionality $N = 9$, and not long after Durt, Kaszlikowski, and Zukowski [DKZ01] this was extended to $N = 16$.

The OAM of light, in principle, provides as large a dimensional space as desired for such Bell tests. One such test was first performed by Vaziri, Weihs, and Zeilinger [VWZ02], who constructed a three-dimensional state space with angular momentum values of $l = -1, 0, 1$; an illustration of their experiment is shown in Figure 10.11. Entangled photons from a BBO crystal are sent through a pair of displaced holograms with $l = +1$ and $l = -1$. As described in Section 10.5, the displaced holograms result in the output states being a mixture of the vortex modes and the pure Gaussian mode; the result of the combined holograms is an entangled state that is primarily mixed between the $l = -1, 0, 1$ modes. Coincidences between pairs of modes are then measured by the use of beamsplitters and additional holograms. A later hologram of order $+1$, for instance, will convert an $l = -1$ mode into a pure Gaussian mode, which can be focused into a fiber for detection and correlation. In this experiment, a violation of a CHSH inequality by more than 18 standard deviations was observed.

In testing higher-dimensional entanglement, it is not possible to use the standard CHSH inequality, which was specifically derived for a two-dimensional state space. Instead, one must use generalized inequalities that were first derived by Collins et al. [CGL$^+$02]; we briefly outline the idea behind such inequalities here.

We assume, as in the CHSH case, that we have two distant observers, A and B, each able to carry out two possible measurements, A_1 and A_2 and B_1 and B_2. Each measurement has d possible outcomes, that is, $A_1 = 0, 1, \ldots, d-1$, and so forth. For future calculations, we treat these variables as cyclic, that is, $j = d$ is the same as $j = 0$. In a local hidden variable theory, there is a definite probability for each of these four outcomes, which do not depend on each other; we label this probability $c_{jk,lm}$, where (jk) is the label for measurements A_1 and A_2, and (lm) is the label for measurements B_1 and B_2. The joint probabilities for any pair of measurements may be found by summing over the extra indices, for example,

$$P(A_1 = j, B_1 = l) = \sum_{km} c_{jk,lm}. \tag{10.184}$$

Let us consider a specific set of outcomes (jk, lm), which occurs with probability $c_{jk,lm}$. We introduce a dependent set of variables

$$r' \equiv B_1 - A_1 = l - j, \tag{10.185}$$

$$s' \equiv A_2 - B_1 = k - l, \tag{10.186}$$

$$t' \equiv B_2 - A_2 = m - k, \tag{10.187}$$

$$u' \equiv A_1 - B_2 = j - m. \tag{10.188}$$

We may readily see that

$$r' + s' + t' + u' = 0, \tag{10.189}$$

which implies that only three of the four dependent variables may be freely chosen; the fourth is constrained by the preceding expression. This must generally be true of any local hidden variable theory, and it forms the basis of the generalized CHSH inequalities.

For example, consider the expression

$$I \equiv P(A_1 = B_1) + P(B_1 = A_2 + 1) + P(A_2 = B_2) + P(B_2 = A_1), \tag{10.190}$$

where

$$P(A_\alpha = B_\beta + k) \equiv \sum_{j=0}^{d-1} P(A_\alpha = j, B_\beta = j + k), \qquad (10.191)$$

and we emphasize again that we treat the variables as cyclic. Because of the aforementioned constraint, the four probabilities within I cannot be satisfied simultaneously: we can satisfy three with $P = 1$, but the fourth must then have $P = 0$. We must conclude that

$$I \leq 3 \text{ for local realism.} \qquad (10.192)$$

For a nonlocal quantum theory, none of the variables has a definite value before measurement; it is possible, therefore, for the probabilities of all four to be nonzero. The upper limit on the quantum case, then, is $I \leq 4$.

The expression given by Equation 10.190 is somewhat different from the traditional CHSH inequality because it is a relationship between probabilities, rather than correlation functions as in Equation 10.176. For the two-dimensional case, however, we can see the structural similarity by noting that $P(B_1 = A_2 + 1) = 1 - P(B_1 = A_2)$; on substitution, Equation 10.190 becomes

$$P(A_1 = B_1) + P(A_2 = B_2) + P(B_2 = A_1) - P(B_1 = A_2) \leq 2, \quad (10.193)$$

which has the same limit as the CHSH inequality, with one negative and three positive terms contributing.

The advantage of the probability formalism of Bell inequalities is that they can be generalized to more complicated and robust expressions. The first generalization proposed by Collins et al. is of the form

$$\begin{aligned} I_3 \equiv &+[P(A_1 = B_1) + P(B_1 = A_2 + 1) + P(A_2 = B_2) + P(B_2 = A_1)] \\ &- [P(A_1 = B_1 - 1) + P(B_1 = A_2) + P(A_2 = B_2 - 1) \\ &+ P(B_2 = A_1 - 1)]. \end{aligned} \qquad (10.194)$$

As all the probabilities are positive, a nonlocal theory can have a maximum value of $I_3 = 4$ because it is possible to have all four positive terms equal to unity. For a local variable theory, we can have at most $I_3 = 2$, because the positive bracket can be at most 3, while the negative bracket will always contribute at least -1.

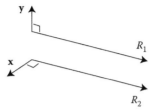

FIGURE 10.12 Illustration of classical entanglement between two orthogonally polarized beams.

We briefly mention a few other articles related to entanglement, nonlocality, and OAM. Aiello et al. [AOEW05] proposed a test of non-locality using fractional vortex states (to be discussed in Section 12.3); an experiment was conducted to demonstrate entanglement of such states [OMV$^+$05], but a violation of a CHSH inequality was not tested. More recently, Leach et al. [LJR$^+$09] demonstrated that one can violate a standard CHSH inequality by restricting the entanglement to two-dimensional OAM state spaces.

We have noted that quantum physics is much like a mystery novel, and in that spirit we present one final twist: it is possible to create entangled states of classical wavefields, and measurements of those entangled states can violate CHSH inequalities! This was first noted* some time ago by Spreeuw [Spr98], and an experimental demonstration of such an effect was done more recently by Qian et al. [QLHE15]. Examples of such entanglements are ridiculously easy to come by. For instance, let us suppose that we have a pair of spatially separated and orthogonally polarized beams, as shown in Figure 10.12. If we label the state of the upper beam as R_1 and the lower beam as R_2, we may write a classical state for the total field as

$$|\psi\rangle = |\hat{y}\rangle |R_1\rangle + |\hat{x}\rangle |R_2\rangle. \tag{10.195}$$

We cannot factor this state, so it is mathematically "entangled." As noted by Spreeuw, a polarizer large enough to block the entire beam pair will have positions where the upper beam can be totally transmitted, and the lower beam totally blocked, and vice versa. It is possible to design experiments to measure maximal CHSH violations of such classically entangled pairs.

---

* Spreeuw's paper is a joy to read; he introduces a classical version of "bra-ket" notation as "paren-thesis" notation.

One may immediately wonder if the entire history of Bell's inequalities needs to be rewritten. The key difference, however, is that such classical entanglement may be considered as entanglement between degrees of freedom of a *single* photon, rather than entanglement between two or more photons. That is, if we were to lower the intensity of the optical field of Figure 10.12 until only one photon at a time passes through the system, we would find the single-photon state to be given by Equation 10.195. Measurements of such entanglement are inherently local, as the particle can only be measured at one place at a time.

Such classical entangled states, however, demonstrate that one must take care in the interpretation of experiments designed to measure the boundary between the classical and quantum worlds.

10.8 VORTICES IN SCHRÖDINGER'S EQUATION

In Chapter 1, it was noted that vortices are ubiquitous in nature, appearing in a wide variety of physical systems, from light to the tides. In fact, we would expect them to appear in any system in which a physical quantity can be described by a complex analytic field: analyticity guarantees that the zeros of the real and imaginary parts of the field appear as surfaces in three-dimensional space, which therefore intersect in lines. Schrödinger's equation for the wavefunction of a quantum particle fits these criteria, and will also therefore manifest the same sort of structures; in this section, we briefly discuss vortices in Schrödinger's equation, noting the similarities and differences between them and their classical optical counterparts. We assume that the reader is familiar with basic quantum physics such as that described in Shanker [Sha80].

In operator form, Schrödinger's equation for the evolution of a quantum particle may be written as

$$i\hbar \frac{d}{dt} |\psi\rangle = \hat{H} |\psi\rangle, \tag{10.196}$$

where $|\psi\rangle$ is the wavefunction of the particle and $\hat{H}$ is the Hamiltonian operator, which we initially consider in the form

$$\hat{H} = \frac{\hat{P}^2}{2m} + \hat{V}, \tag{10.197}$$

where $\hat{P}$ is the momentum operator and $\hat{V}$ is the potential operator. If we consider the Schrödinger's equation in the position basis, that is, we write

$\langle \mathbf{r}| \psi \rangle = \psi(\mathbf{r}, t)$, it reduces to the form

$$i\hbar \frac{d\psi(\mathbf{r}, t)}{dt} = -\frac{\hbar^2}{2m} \nabla^2 \psi(\mathbf{r}, t) + V(\mathbf{r})\psi(\mathbf{r}, t), \tag{10.198}$$

where we have made the association

$$\hat{P} \leftrightarrow -i\hbar\nabla, \quad \hat{V} \leftrightarrow V(\mathbf{r}). \tag{10.199}$$

Let us consider the simplest case first: a particle in a definite energy state. Similar to the case for monochromatic fields, we may write

$$\psi(\mathbf{r}, t) = \psi(\mathbf{r})e^{-iEt/\hbar}, \tag{10.200}$$

where $\psi(\mathbf{r})$ is a complex function of position. However, we do not take the real part of this expression, as Schrödinger's equation is inherently complex. With this substitution, Equation 10.198 becomes

$$\nabla^2 \psi(\mathbf{r}) + \frac{2m}{\hbar^2}[E - V(\mathbf{r})]\psi(\mathbf{r}) = 0. \tag{10.201}$$

If we now make the definitions

$$k_0^2 \equiv \frac{2mE}{\hbar^2}, \tag{10.202}$$

$$n^2(\mathbf{r}) \equiv \left[1 - \frac{V(\mathbf{r})}{E}\right], \tag{10.203}$$

then, the time-independent Schrödinger's equation is *exactly* the inhomogeneous Helmholtz equation, that is,

$$\nabla^2 \psi(\mathbf{r}) + k_0^2 n^2(\mathbf{r})\psi(\mathbf{r}) = 0. \tag{10.204}$$

Since we know that vortices are generic features in solutions to the Helmholtz equation, this suggests that solutions to Schrödinger's equation also possess generic singularities in the form of phase vortices. If the particle is propagating in free space, then, Equation 10.204 reduces to the homogeneous Helmholtz equation

$$\nabla^2 \psi(\mathbf{r}) + k_0^2 \psi(\mathbf{r}) = 0. \tag{10.205}$$

Since the entire discussion of scalar optical vortices has been based on the Helmholtz equation, we expect that *all* of our previous results can be applied directly to quantum particles in energy eigenstates. Since such vortices are stable under the presence of perturbations, they will also be generic in solutions with an inhomogeneous potential, that is, $n^2(\mathbf{r}) \neq 0$.

By our choice of labels, we have implicitly assumed above that the particle is not in a bound state, that is, E is greater than any boundaries produced by the potential $V(\mathbf{r})$. There is nothing, however, preventing vortex structures from appearing in bound-state wavefunctions, as they are also complex analytic functions. For example, if we consider the electron wavefunction of the hydrogen atom subject to the potential

$$V(\mathbf{r}) = -\frac{e^2}{4\pi\epsilon_0}\frac{1}{r}, \tag{10.206}$$

it can be shown [Gri95, Chapter 4] that the stationary states are specified by the quantum numbers n, l, and m, and of the functional form

$$\psi_{nlm}(\mathbf{r}) = \sqrt{\left(\frac{2}{na_B}\right)^3 \frac{(n-l-1)!}{2n[(n+l)!]^3}} e^{-r/na_B} \left(\frac{2r}{na_B}\right)^l$$
$$\times L_{n-l-1}^{2l+1}(2r/na_B)\, Y_l^m(\theta, \phi), \tag{10.207}$$

where L_m^n are Laguerre functions as discussed in Section 2.4 and $Y_l^m(\theta, \phi)$ are the spherical harmonics,[*] given by

$$Y_l^m(\theta, \phi) = (-1)^m \sqrt{\frac{2l+1}{4\pi}\frac{(l-m)!}{(l+m)!}} P_l^m(\cos\theta)e^{im\phi}. \tag{10.208}$$

The $P_l^m(\cos\theta)$ are the Legendre functions, and a_B is the Bohr radius. In these solutions, n is the principal quantum number, which must be nonnegative, l is the quantum number for the total angular momentum, which also must be nonnegative, and m is the quantum number for the z-component of angular momentum. It can be shown that $0 \leq |m| \leq l$. For $m \neq 0$, we clearly have an azimuthal phase such as that which appears in a vortex beam propagating along the z-axis, and we expect that we would have a similar

---

[*] There are many slight variants on the spherical harmonics; here, we have used the so-called Condon–Shortley phase, as discussed in Gbur [Gbu11], Section 17.8.

vortex structure around the z-axis of our hydrogenic system. As an example,

$$\psi_{2,1,\pm 1}(\mathbf{r}) \sim e^{-r/2a_B} r \sin\theta e^{\pm i\phi} = e^{-r/2a_B}(x \pm iy), \qquad (10.209)$$

which has the vortex structure we are familiar with, in addition to an exponential envelope.

In a curious example of synchronicity in physics, the first detailed papers on quantum mechanical singularities by Hirschfelder and collaborators [HCP74,HGB74] appeared in 1974, the same year that Nye and Berry published the foundational paper on singular optics [NB74]. Hirschfelder et al. noted many properties of quantum singularities that are by now familiar to us from optics: their vortex structure, quantized circulation, and stability under perturbation.

In these early papers, however, the authors characterized the singularities by a velocity field rather than a phase field. This is perhaps not surprising, as a 1-eV electron has a wavelength about $\lambda = 1.23$ nm, significantly smaller than a 1-eV photon with wavelength $\lambda = 1240$ nm. Phase measurements of electrons via interference are consequently much more difficult to do than corresponding measurements for photons.

To derive the so-called hydrodynamic model of a quantum particle, we consider Schrödinger's equation and its complex conjugate, that is,

$$-i\hbar\frac{\partial\psi}{\partial t} = -\frac{\hbar^2}{2m}\nabla^2\psi + V\psi, \qquad (10.210)$$

$$i\hbar\frac{\partial\psi^*}{\partial t} = -\frac{\hbar^2}{2m}\nabla^2\psi^* + V\psi^*. \qquad (10.211)$$

By multiplying Equation 10.210 by ψ^* and Equation 10.211 by ψ and taking the difference, we quickly find the expression

$$i\hbar\left[\psi^*\frac{\partial\psi}{\partial t} - \psi\frac{\partial\psi^*}{\partial t}\right] = -\frac{\hbar^2}{2m}\left\{\psi^*\nabla^2\psi - \psi\nabla^2\psi^*\right\}. \qquad (10.212)$$

The left side of this expression can be written as a single time derivative of $|\psi|^2$; using vector product rules, the right side may be written as the divergence of a single quantity, so that the complete expression is

$$\frac{\partial|\psi|^2}{\partial t} = \frac{i\hbar}{2m}\nabla\cdot\left\{\psi^*\nabla\psi - \psi\nabla\psi^*\right\}. \qquad (10.213)$$

This expression has the form of a conservation law, which can be made explicit by defining a *probability density* $P(\mathbf{r}, t)$ as

$$P(\mathbf{r}, t) \equiv |\psi(\mathbf{r}, t)|^2, \tag{10.214}$$

and a number flux $\mathbf{J}(\mathbf{r}, t)$ as

$$\mathbf{J}(\mathbf{r}, t) \equiv -\frac{i\hbar}{2m}\left\{\psi^*\nabla\psi - \psi\nabla\psi^*\right\}. \tag{10.215}$$

This gives us a differential probability conservation law of the form

$$\frac{\partial P}{\partial t} = -\nabla \cdot \mathbf{J}. \tag{10.216}$$

This should be compared with the differential form of the electromagnetic energy conservation law, Equation 8.17, which we recall to be

$$\frac{\partial U}{\partial t} = -\nabla \cdot \mathbf{S}, \tag{10.217}$$

with $U(\mathbf{r}, t)$ the electromagnetic energy density and $\mathbf{S}(\mathbf{r}, t)$ the Poynting vector. The analogy between this and Equation 10.216 suggests that the latter may be considered a law for the conservation of probability, or particle number: any change in the net probability of a particle remaining in a volume must be due to a net number flux across the surface.

The number flux $\mathbf{J}$ is therefore expected to have vortices of a similar nature to those studied in Chapter 8 for the Poynting vector $\mathbf{S}$. In fact, the first paper by Hirschfelder, Christoph, and Palke [HCP74] looked at the total internal reflection of a quantum particle from a planar potential barrier, and found a singularity essentially identical to that found by Wolter in optics [Wol50] decades earlier; recall Figure 8.3. They did not consider the flow of the number flux $\mathbf{J}$ itself, however, but rather a mean local velocity $\mathbf{v}$ of the field defined as

$$\mathbf{v} \equiv \frac{\mathbf{J}}{P}. \tag{10.218}$$

If we write the wavefunction $\psi(\mathbf{r}, t)$ in terms of an amplitude $|\psi(\mathbf{r}, t)|$ and phase $S(\mathbf{r}, t)$ in the form

$$\psi(\mathbf{r}, t) = |\psi(\mathbf{r}, t)|e^{iS(\mathbf{r}, t)/\hbar}, \tag{10.219}$$

we can readily show that

$$\mathbf{v} = \frac{\nabla S}{m}. \tag{10.220}$$

The *circulation* C of a vortex, or more broadly a particular region of a plane bounded by a path C, is defined as

$$C \equiv \int_C \mathbf{v} \cdot d\mathbf{r} = m^{-1} \int_C \nabla S \cdot d\mathbf{r}. \tag{10.221}$$

This expression is formally similar to that for topological charge t as given way back in Equation 3.36, namely,

$$t \equiv \frac{1}{2\pi} \oint_C \nabla \psi(\mathbf{r}) \cdot d\mathbf{r}, \tag{10.222}$$

where in this case $\psi(\mathbf{r})$ is the phase of a scalar wavefield. On comparison, we see that the circulation must also be quantized, so that

$$\int_C \mathbf{v} \cdot d\mathbf{r} = 2\pi(\hbar/m)n, \quad n = 0, \pm 1, \pm 2, \ldots. \tag{10.223}$$

It is worth noting that the number flux $\mathbf{J}$ and the local velocity $\mathbf{v}$ have significantly different structure in the neighborhood of a singularity. The number flux $\mathbf{J}$ will have a first-order zero at a singularity, while the local velocity $\mathbf{v}$ will have a first-order pole. This does not make a significant difference for the topology of the streamlines, however.

So far, we have focused on the similarities between quantum and optical singularities, but there are also significant differences. The wavefunction in Schrödinger's equation is inherently complex, even in the time domain, which means that there are well-defined phase vortices at any instant of time t. Furthermore, optical fields satisfy a wave equation in the time domain, whereas quantum fields satisfy Schrödinger's diffusion-like equation, making their evolution in time distinct and possibly their vortex evolution as well. Finally, Schrödinger's equation can be used to describe the interaction of charged particles with an electromagnetic field, which can also introduce new dynamics.

We briefly consider a couple of examples, following the work of Bialynicki-Birula, Bialynicka-Birula, and Šliwa [BBBBŠ00], who studied singularities of the phase instead of the velocity. In a very clever construction, they began with a plane wave solution to Schrödinger's equation of the form

$$\psi_{\mathbf{k}}(\mathbf{r}, t) = \exp[i\mathbf{k} \cdot \mathbf{r}] \exp[-i\hbar\mathbf{k}^2 t/2m], \qquad (10.224)$$

which can be verified by direct calculation. However, since Schrödinger's equation is linear, we may take derivatives of this solution with respect to any components of $\mathbf{k}$ and acquire a new solution. For instance, if we wish to make a simple screw dislocation at time $t = 0$, we use

$$\psi_{screw}(\mathbf{r}, t) = -i\left[\frac{\partial}{\partial k_x} + i\frac{\partial}{\partial k_y}\right]\psi_{\mathbf{k}}(\mathbf{r}, t)$$

$$= \left[\left(x - \frac{\hbar k_x t}{m}\right) + i\left(y - \frac{\hbar k_y t}{m}\right)\right]\psi_{\mathbf{k}}(\mathbf{r}, t). \qquad (10.225)$$

This represents a left-handed screw dislocation parallel to the z-axis. If $k_x, k_y \neq 0$, this dislocation line glides in the k_x, k_y direction as time evolves with velocity $(\hbar k_x/m, \hbar k_y/m)$.

Wavefunctions created in this manner may generally be written in the form

$$[W_R(\mathbf{r}, t) + iW_I(\mathbf{r}, t)]\psi_{\mathbf{k}}(\mathbf{r}, t), \qquad (10.226)$$

where the zeros of W_R and W_I represent surfaces in three-dimensional space; their intersection results in dislocation lines. With this in mind, we can design arbitrary singularity geometries for time $t = 0$ and see how they evolve. For instance, a vortex ring of radius R in the plane $z = 0$ at time $t = 0$ may be designed by the choice

$$W_R(\mathbf{r}, 0) = x^2 + y^2 + z^2 - R^2, \quad W_I(\mathbf{r}, 0) = az. \qquad (10.227)$$

Mapping $x \to -id/dk_x$, $y \to -id/dk_y$, $z \to -id/dk_z$, we may with some effort find a solution

$$\psi_{ring}(\mathbf{r}, t) = \left[x^2(t) + y^2(t) + z^2(t) - R^2 + ia\left(z(t) + \frac{3\hbar t}{ma}\right)\right]\psi_{\mathbf{k}}(\mathbf{r}, t), \qquad (10.228)$$

where $x(t) = x - \hbar k_x t/m$, and so forth.

We may simplify things considerably by taking the limit $\mathbf{k} \to 0$. Although there is no wave-like propagation in this case, there is still vortex evolution, due to the explicit appearance of time in Equation 10.228.

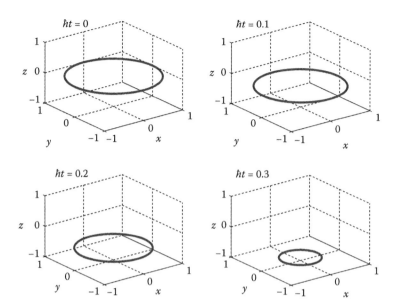

FIGURE 10.13   The evolution in time of a ring vortex in Schrödinger's equation. Here $R = 1$, $ma = 1$.

This is a striking departure from classical electromagnetics, in which the long-wavelength limit is a static field; apparently the time evolution in this case is a quantum effect. We find that the ring propagates in the $-z$-direction, disappearing completely when $t = maR/3\hbar$. This is illustrated in Figure 10.13.

We consider one more example from Bialynicki-Birula, Bialynicka-Birula, and Śliwa, the interaction of an electron wavefunction with a uniform magnetic field, showing behavior not typically seen in the optical case. Schrödinger's equation for this example, in the symmetric gauge, may be written in the form

$$i\frac{\partial \psi}{\partial t} = \left[-\frac{\hbar^2}{2m}\nabla^2 - \frac{i\hbar eB}{m}\left(x\frac{\partial}{\partial y} - y\frac{\partial}{\partial x} \right) + \frac{e^2 B^2}{8m}(x^2 + y^2) \right]\psi. \quad (10.229)$$

The magnetic field B is constant and along the z-direction. The solution at time $t = 0$, analogous to the plane wave in free space used before, is taken to be

$$\psi(\mathbf{r}, 0) = \exp[i\mathbf{k} \cdot \mathbf{r}] \exp[-eB(x^2 + y^2)/4\hbar]. \quad (10.230)$$

It is to be noted that the electron has a localized wavefunction because it classically will follow a circular path in the xy-plane. The solution to

Schrödinger's equation for all times has the complicated form

$$\psi_\mathbf{k}(\mathbf{r}, t) = \exp[-eB(x^2 + y^2)/4\hbar] \exp[-i\omega_c t/2] \exp[\hbar(e^{-i\omega_c t} - 1)$$
$$\times (k_x^2 + k_y^2)/2eB] \exp[i(e^{-i\omega_c t} + 1)(xk_x + yk_y)/2]$$
$$\times \exp[(e^{-i\omega_c t} - 1)(xk_y - yk_x)/2] \exp[izk_z - i\hbar k_z^2/2eB],$$

$$(10.231)$$

where $\omega_c = eB/m$ is the cyclotron frequency. As in the free-space case before, vortex fields can be built from this solution by appropriate derivatives with respect to k_x, k_y, and k_z. As an example, we consider a vortex that lies in the plane $y = a$ at time $t = 0$, making an angle ϕ with the field direction. It can then be found that the line position evolves in time according to the parametric equations

$$y(x) = \frac{2a + x\sin(\omega_c t)(1 + \sin\phi)}{1 - \sin\phi + (1 + \sin\phi)\cos(\omega_c t)},$$

$$(10.232)$$

$$z(x) = \frac{x\tan\phi + a\sin(\omega_c t)(\sec\phi + \tan\phi)}{1 - \sin\phi + (1 + \sin\phi)\cos(\omega_c t)}.$$

$$(10.233)$$

An example of the motion is shown in Figure 10.14. The vortex remains a straight line, while partially precessing around the field axis for part of

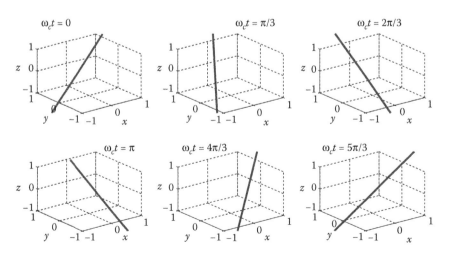

FIGURE 10.14 The evolution in time of a quantum vortex in a uniform magnetic field. Here $a = 0.4$, $\phi = \pi/3$.

the cyclotron period, before reversing its precession and returning to its original orientation.

10.9 UNCERTAINTY PRINCIPLE FOR ANGULAR MOMENTUM

One of the most profound results of early quantum theory is *Heisenberg's uncertainty relation,*[*] which suggests that there is a fundamental limitation on the precision with which certain complementary properties of a quantum particle can be known simultaneously. The most familiar example is the limit on the uncertainty Δx of the position and uncertainty Δp of the momentum of a particle, which takes on the form

$$(\Delta x)(\Delta p) \geq \hbar/2. \tag{10.234}$$

Considering the elegance of this theorem and its importance in physics, it is natural to consider whether a similar relation exists for the *angular* position ϕ and the *angular* momentum L_z. It is not quite so straightforward to formulate such a relation, however; where position and momentum have continuous eigenvalue spectra over an infinite range, the eigenvalues of the angular momentum operator are discrete. The angular position has a continuous spectra, but it must typically be bounded over a 2π range— and a naive definition of the angular uncertainty $\Delta\phi$ will depend on which 2π range is chosen.

There are a couple of reasons to pursue the topic, however, in spite of its conceptual challenges. One is that a good understanding of the angular uncertainty principle provides insight into the nature of angular momentum in optical and quantum fields. Another is that the same difficulties that arise in defining an angle operator also arise in defining a phase operator in quantum mechanics, and the corresponding *phase-number uncertainty relation* associated with it. Investigating one problem therefore helps with the other.

We begin this section by deriving the classic Heisenberg uncertainty relation. From this, we will be able to see clearly the difficulties that arise in the angular case, and discuss several strategies to resolve them. Experiments to measure such angular uncertainties are then discussed.

Let us suppose that we have two Hermitian operators $\hat{A}$ and $\hat{B}$ that satisfy the commutation relation

$$[\hat{A}, \hat{B}] = i\hat{C}, \tag{10.235}$$

---

[*] See, for instance, Shankar [Sha80], Chapter 9, or your favorite book on quantum mechanics.

where $\hat{C}$ must also be a Hermitian operator. We introduce a quantum wavefunction $|\psi\rangle$, and consider the uncertainties of the two operators $\hat{A}$ and $\hat{B}$, namely,

$$(\Delta A)^2 = \langle\psi|\,(\hat{A} - \overline{A})^2\,|\psi\rangle, \tag{10.236}$$

$$(\Delta B)^2 = \langle\psi|\,(\hat{B} - \overline{B})^2\,|\psi\rangle, \tag{10.237}$$

where $\overline{A}$ is the expectation value of $\hat{A}$,

$$\overline{A} = \langle\psi|\,\hat{A}\,|\psi\rangle, \tag{10.238}$$

and so forth. For simplicity, we introduce the zero-mean versions of the operators $\hat{A}$ and $\hat{B}$,

$$\hat{A}_0 \equiv (\hat{A} - \overline{A}), \quad \hat{B}_0 \equiv (\hat{B} - \overline{B}), \tag{10.239}$$

and we may write the product of uncertainties as

$$(\Delta A)^2(\Delta B)^2 = \langle\psi|\,\hat{A}_0^2\,|\psi\rangle \langle\psi|\,\hat{B}_0^2\,|\psi\rangle. \tag{10.240}$$

Because $\hat{A}_0$ and $\hat{B}_0$ are also Hermitian, we note that the right-hand side of the above equation is real-valued and nonnegative. If we define $\hat{A}_0\,|\psi\rangle = |V_A\rangle$, $\hat{B}_0\,|\psi\rangle = |V_B\rangle$, we may then write

$$(\Delta A)^2(\Delta B)^2 = |V_A|^2|V_B|^2. \tag{10.241}$$

We can now apply the classic *Schwartz inequality*, which states that

$$|V_A|^2|V_B|^2 \geq |\langle V_A|\,V_B\rangle\,|^2. \tag{10.242}$$

This inequality is a standard tool in both finite and infinite dimensional vector space calculations, so we briefly derive it here. We introduce the vector state $|V\rangle$ as follows:

$$|V\rangle \equiv |V_A\rangle - \langle V_B|\,V_A\rangle\,\frac{|V_B\rangle}{\langle V_B|\,V_B\rangle}. \tag{10.243}$$

This represents the state $|V_A\rangle$ with its projection along $|V_B\rangle$ removed. For any state $|V\rangle$, we always have

$$\langle V|\,V\rangle \geq 0. \tag{10.244}$$

If we substitute from Equation 10.243 into Equation 10.244, and apply the relation $\langle V_B | V_A \rangle^* = \langle V_A | V_B \rangle$, we may simplify and arrive immediately at inequality (10.242). Applying this to Equation 10.241, we arrive at the expression

$$(\Delta A)^2 (\Delta B)^2 \geq | \langle \psi | \hat{A}_0 \hat{B}_0 | \psi \rangle |^2. \tag{10.245}$$

We may then write the operator product $\hat{A}_0 \hat{B}_0$ in terms of the commutator and anti-commutator of $\hat{A}_0$ and $\hat{B}_0$, that is,

$$\hat{A}_0 \hat{B}_0 = \frac{1}{2} \{\hat{A}_0, \hat{B}_0\} + \frac{1}{2} [\hat{A}_0, \hat{B}_0], \tag{10.246}$$

with

$$\{\hat{A}_0, \hat{B}_0\} \equiv \hat{A}_0 \hat{B}_0 + \hat{B}_0 \hat{A}_0, \tag{10.247}$$

$$\left[\hat{A}_0, \hat{B}_0 \right] \equiv \hat{A}_0 \hat{B}_0 - \hat{B}_0 \hat{A}_0. \tag{10.248}$$

The expectation value of the anti-commutator, which is Hermitian, will be real-valued. The expectation value of the commutator will be purely imaginary. This implies that the absolute value squared of inequality (10.245) may be simplified to

$$(\Delta A)^2 (\Delta B)^2 \geq \frac{1}{4} | \langle \psi | \{\hat{A}_0, \hat{B}_0\} | \psi \rangle |^2 + \frac{1}{4} | \langle \psi | \hat{C} | \psi \rangle |^2, \tag{10.249}$$

where we have now used Equation 10.235. Because the first term on the right is necessarily nonnegative definite, we may finally write

$$(\Delta A)^2 (\Delta B)^2 \geq \frac{1}{4} | \langle \psi | \hat{C} | \psi \rangle |^2. \tag{10.250}$$

This general form of a quantum uncertainty relation is often referred to as the *Robertson uncertainty relation* [Rob29], after its discoverer. Equation 10.249 may be considered an even more precise inequality, and is known as the *Schrödinger uncertainty relation* [Sch30]. For the special case $\hat{A} = \hat{x}$, $\hat{B} = \hat{p}$, we have $\hat{C} = \hbar$ and we arrive at the classic form of the Heisenberg uncertainty relation, Equation 10.234.

More generally, we note that there are two special classes of states connected to inequality (10.250). Those states for which the equality is satisfied are said to be *intelligent states* with respect to the operators in question.

From the derivation, it is clear that intelligent states arise when the following conditions are met:

$$\hat{A}_0 \, |\psi\rangle = c\hat{B}_0 \, |\psi\rangle, \qquad (10.251)$$

with c as a constant, and

$$\langle\psi| \, \{\hat{A}_0, \hat{B}_0\} \, |\psi\rangle = 0. \qquad (10.252)$$

Because the right-hand side of inequality (10.250) also depends upon the particular state $|\psi\rangle$, an intelligent state is not necessarily also a *minimum uncertainty state* that provides a global minimum of the inequality. In the case of $\hat{x}$ and $\hat{p}$, and all operator pairs for which the commutator is proportional to the identity operator, intelligent states and minimum uncertainty states are the same.

Inequality (10.234) has been derived entirely by the use of operator algebra, and it may not be recognizable to those who are familiar with explicit wave calculations such as those used through the rest of this text. It is therefore instructive to convert the abstract expectation value equations into integral averages in position and momentum space. To do so, we employ the identity operators in each space, given by

$$\hat{I} = \int_{-\infty}^{\infty} |x'\rangle\langle x'| \, dx' \qquad (10.253)$$

in position space, and

$$\hat{I} = \int_{-\infty}^{\infty} |p'\rangle\langle p'| \, dp' \qquad (10.254)$$

in momentum space. Then, an operator expression may be written, for example, as

$$\langle\psi| \, \hat{x} \, |\psi\rangle = \int_{-\infty}^{\infty} \int_{-\infty}^{\infty} \langle\psi| \, x'\rangle\langle x'| \, \hat{x} \, |x\rangle \, \langle x| \, \psi\rangle \, dx \, dx'. \qquad (10.255)$$

In position space, we have $\langle x' | \hat{x} | x \rangle = x\delta(x - x')$, and in momentum space, we have $\langle p' | \hat{p} | p \rangle = p\delta(p - p')$. This indicates that we may write

$$(\Delta x)^2 = \int_{-\infty}^{\infty} \psi^*(x)(x - \bar{x})^2 \psi(x) dx, \tag{10.256}$$

$$(\Delta p)^2 = \int_{-\infty}^{\infty} \psi^*(p)(p - \bar{p})^2 \psi(p) dp, \tag{10.257}$$

where $\psi(x)$ is the wavefunction in position space, and $\psi(p)$ is the wavefunction in momentum space. To find the relation between these two representations, we note that, in position space,

$$\langle x | \hat{p} | x \rangle = -i\hbar \frac{\partial}{\partial x}, \tag{10.258}$$

and so the eigenfunctions $\psi_p(x)$ of the $\hat{p}$ operator in position space must satisfy

$$-i\hbar \frac{\partial \psi_p}{\partial x} = p\psi_p. \tag{10.259}$$

It is straightforward to show that

$$\psi_p(x) = \frac{1}{\sqrt{2\pi}} e^{ipx/\hbar}. \tag{10.260}$$

The factor of $1/\sqrt{2\pi}$ is chosen to normalize the states to a delta function, that is,

$$\int_{-\infty}^{\infty} \psi_{p'}^*(x)\psi_p(x) dx = \delta(p - p'). \tag{10.261}$$

The wavefunction in position space may be then represented in terms of momentum eigenstates as

$$\psi(x) = \frac{1}{\sqrt{2\pi}} \int_{-\infty}^{\infty} \psi(p) e^{ipx/\hbar} dp. \tag{10.262}$$

The representations in position and momentum space are therefore Fourier transforms of each other. With this in mind, it should now be emphasized that there is nothing inherently quantum mechanical about the uncertainty relation, which applies to any Fourier transform pair. The most familiar optical example of this is in diffraction theory, in which the angular spreading of light diffracted by an aperture is inversely related to the aperture width.

We now consider an uncertainty relation for the angular position operator $\hat{\phi}$ and angular momentum operator $\hat{L}_z$. In a state space of angle eigenfunctions $|\phi\rangle$, we can readily determine that $\langle\phi|\hat{\phi}|\phi\rangle = \phi$ and $\langle\phi|\hat{L}_z|\phi\rangle = -i\hbar\partial/\partial\phi$. It would seem at first glance that one would, by direct analogy with the $\hat{x}, \hat{p}$ case, arrive at a relation of the form $(\Delta L_z)(\Delta\phi) \geq \hbar/2$, by making the association $\phi \leftrightarrow x$. However, we immediately run into nontrivial differences. Though an eigenfunction $\psi_L(\phi)$ of the angular momentum operator satisfies

$$-i\hbar\frac{\partial\psi_L}{\partial\phi} = l_z\psi_L, \tag{10.263}$$

leading to solutions

$$\psi_L(\phi) = \frac{1}{\sqrt{2\pi}}e^{il_z\phi/\hbar}, \tag{10.264}$$

these solutions must be 2π periodic in ϕ, which results in the familiar constraint that $l_z = m$, where m is an integer. The eigenvalue spectrum of the $\hat{L}_z$ operator is discrete, not continuous like the $\hat{p}$ operator.

From Equation 10.264, we see that these states are uniformly distributed in ϕ, that is, $|\psi_L(\phi)|^2 = 1/2\pi$. If we naively define the nth moment of ϕ as

$$\overline{\phi^n} = \int_{-\pi}^{\pi} \psi^*(\phi)\phi^n\psi(\phi)d\phi, \tag{10.265}$$

then, we may introduce the variance of a wavefunction in the usual manner as

$$(\Delta\phi)^2 \equiv \overline{\phi^2} - [\overline{\phi}]^2. \tag{10.266}$$

For the uniform distribution, the result is

$$(\Delta\phi)^2 = \frac{1}{2\pi}\int_{-\pi}^{\pi} \phi^2 d\phi - \left[\frac{1}{2\pi}\int_{-\pi}^{+\pi} \phi\, d\phi\right]^2 = \frac{\pi^2}{3}. \tag{10.267}$$

But since $(\Delta L_z)^2 = 0$ for the uniform distribution, we find that $(\Delta L_z)^2(\Delta\phi)^2 = 0$ for angular momentum eigenstates. Apparently angular momentum and angular position do *not* satisfy an uncertainty relation directly analogous to momentum and position.

From Equation 10.265, we see yet another difficulty: we chose a particular integration range, $-\pi \le \phi < \pi$, but in general the nth moment of a wavefunction will depend explicitly on the choice of this integration range. The exception of this is the uniform distribution.

This occurs because the functions ϕ^n are not periodic themselves, and therefore fall outside the class of physical, single-valued functions that we are interested in. The result is a completely unsatisfying result that $(\Delta\phi)^2$ depends explicitly on a choice of coordinates, a property no physical quantity should have. In fact, we should go further and note that any point in the 2π interval works equally well as the origin of the function ϕ, meaning that the definition of the second moment has, in general, two degrees of freedom, that is,

$$V(\gamma, \alpha) = \int_{\gamma-\pi}^{\gamma+\pi} \psi^*(\phi)(\phi - \alpha)^2\psi(\phi)d\phi. \tag{10.268}$$

The function $V(\gamma, \alpha)$ therefore depends on the choice of integration range as well as the relative angle α between the wavefunction $\psi(\phi)$ and the quadratic term $(\phi - \alpha)^2$.

The earliest solution to this ambiguity was presented by Judge [Jud63], who proposed that $(\Delta\phi)^2$ for any wavefunction should be defined as the minimum value of $V(\gamma, \alpha)$. One can readily show that $\partial V/\partial\alpha = 0$ leads to the result $\alpha = \gamma$, so that

$$(\Delta\phi)^2 = \min\left[\int_{\gamma-\pi}^{\gamma+\pi} \psi^*(\phi)(\phi - \gamma)^2\psi(\phi)d\phi\right]. \tag{10.269}$$

A simple change of variables leads to the modified expression

$$(\Delta\phi)^2 = \min\left[\int_{-\pi}^{\pi} \psi^*(\phi + \gamma)\phi^2\psi(\phi + \gamma)d\phi\right]. \tag{10.270}$$

For a given γ, we may derive the uncertainty relation for $(\Delta\phi)^2$ and $(\Delta L_z)^2$ by the use of Equation 10.250. The commutator for $\hat{\phi}$ and $\hat{L}_z$ may be written as

$$\langle\psi| [\hat{\phi}, \hat{L}_z] |\psi\rangle = -i\hbar \int_{-\pi}^{\pi} \left[\psi^*(\phi+\gamma)\frac{\partial}{\partial\phi}\psi(\phi+\gamma) \right.$$

$$\left. + \psi(\phi+\gamma)\frac{\partial}{\partial\phi}\psi^*(\phi+\gamma) \right] d\phi. \qquad (10.271)$$

By the use of integration by parts, this may be reduced to the form

$$\langle\psi| [\hat{\phi}, \hat{L}_z] |\psi\rangle = -i\hbar[1 - 2\pi|\psi(\pi+\gamma)|^2], \qquad (10.272)$$

so that our final uncertainty relation becomes

$$(\Delta\phi)^2(\Delta L_z)^2 \geq \frac{\hbar^2}{4}\min\left[1 - 2\pi|\psi(\pi+\gamma)|^2\right]^2. \qquad (10.273)$$

It can be seen that, for a uniform distribution with $|\psi|^2 = 1/2\pi$, the minimum uncertainty is zero.

What is the meaning of the minimization in inequality (10.273)? If we explicitly require $\partial V(\gamma,\gamma)/\partial\gamma = 0$, we readily find

$$-2 \int_{-\pi}^{\pi} \psi^*(\phi+\gamma)\phi\psi(\phi+\gamma)d\phi = 0. \qquad (10.274)$$

This implies that γ must be chosen to make the "mean angular position" of the distribution equal to zero.

As the problems with angular uncertainty arise in large part from the nonperiodicity of the operator $\hat{\phi}$, a completely periodic angle operator was proposed by Judge and Lewis [JL63]. In the angular position basis, such an operator $\hat{\Phi}$ takes the form

$$\Phi \equiv \phi - 2\pi \sum_{n=-\infty}^{\infty} S(\phi - 2\pi n), \qquad (10.275)$$

where $S(x)$ is the Heaviside step function and $\hat{\Phi}$ is designed to be linear over the range $0 \leq \phi < 2\pi$. Careful inspection suggests that this operator is missing an infinite constant term that counteracts the contributions of an infinite number of steps; however, this constant does not affect the commutator. Calculating the commutator $[\hat{\Phi}, \hat{L}_z]$ in the position basis readily yields the result

$$\langle \psi | [\hat{\Phi}, \hat{L}_z] | \psi \rangle = -i\hbar[1 - 2\pi|\psi(0)|^2]. \qquad (10.276)$$

On comparison with Equation 10.272, we find that we have arrived at essentially the same commutator, and therefore the same uncertainty relation. By shifting the origin of the step functions to γ in Equation 10.275, we could match the earlier result exactly.

For future reference, we note that this commutator may also be written in the angular position basis as

$$\langle \phi | [\hat{\Phi}, \hat{L}_z] | \phi \rangle = -i\hbar \left[1 - 2\pi\delta(\phi)\right]. \qquad (10.277)$$

The delta function arises due to the discontinuous phase operator, and it feels somewhat uncomfortable to have such a singular distribution arise in what would appear to be a simple commutator relation.

Barnett and Pegg [BP90] took a more systematic approach to derive the quantum angle operator that clearly shows the limitations of the earlier derivations. Their method is based on generating functions[*] for position and momentum, and by analogy with these, they construct an appropriate and well-behaved angle operator. If $|x\rangle$ is an eigenstate of position, then it can be shown that

$$\exp[-i\hat{p}\eta/\hbar] |x\rangle = |x + \eta\rangle, \qquad (10.278)$$

and if $|p\rangle$ is an eigenstate of momentum, it can also be shown that

$$\exp[i\hat{x}\beta/\hbar] |p\rangle = |p + \beta\rangle. \qquad (10.279)$$

Barnett and Pegg start with a $(2l + 1)$-dimensional angular momentum space, with $m = -l, \ldots, 0, \ldots, l$. Then, by analogy with the above generators, they suggest that a pure angle state $|\phi\rangle$ must satisfy

$$\exp[-i\hat{L}_z\eta/\hbar] |\phi\rangle = |\phi + \eta\rangle. \qquad (10.280)$$

---

[*] We defer the derivation of these relations as an exercise, though they can also be found in many quantum books.

If we define $|\phi_0\rangle$ as the zero-angle state, then, a state of angle ϕ will be related to this state by

$$|\phi\rangle = \exp[-i\hat{L}_z\phi/\hbar]\,|\phi_0\rangle\,. \tag{10.281}$$

We now label the pure angular momentum states as $|m\rangle$, where m is an integer. As shifts in angular momentum must be integer values, that is, $n\hbar$, we expect that the generator for angular momentum must give the relation

$$\exp[in\hat{\phi}]\,|m\rangle = |m+n\rangle\,. \tag{10.282}$$

We now relate the eigenstates in angle and angular momentum. The zero-angle state must be expressible in terms of angular momentum as

$$|\phi_0\rangle = \sum_{m=-l}^{l} c_m\,|m\rangle\,. \tag{10.283}$$

If we operate on both sides of this expression with the generating function $\exp[in\hat{\phi}]$, we also immediately find that

$$|\phi_0\rangle = \sum_{m=-l}^{l} c_m\,|m+n\rangle\,. \tag{10.284}$$

It is assumed that the angular momentum states "wrap around" at the extreme values of m, that is, $|l+1\rangle = |-l\rangle$. Then, on comparison of Equations 10.283 and 10.284, we find that the constants c_m must be identical and independent of m. We choose the normalization $c_m = 1/\sqrt{2l+1}$; operating on the resulting zero-angle state with the generator of Equation 10.281, we find that

$$|\phi\rangle = \frac{1}{\sqrt{2l+1}} \sum_{m=-l}^{l} \exp[-im\phi]\,|m\rangle\,. \tag{10.285}$$

This expression would imply a continuum of angle states $|\phi\rangle$. However, the inner product of two of these states shows that all of them are not orthogonal, and therefore the set is overcomplete, that is,

$$\langle\phi'|\,\phi\rangle = \frac{1}{\sqrt{2l+1}} \frac{\sin[(2l+1)(\phi-\phi')/2]}{\sin[(\phi-\phi')/2]}\,. \tag{10.286}$$

Here, we have used the sum of the geometric series to simplify this expression. Clearly, this product is nonzero for most values of $\phi \neq \phi'$, making the continuum of angle states only approximately orthogonal. This is, in hindsight, not surprising: having used a finite set of angular momentum states, one would expect only a finite set of distinct angle states. The situation is analogous to the discrete Fourier transform [Gbu11, Chapter 13], in which the same number of states are used in the direct and Fourier space.

We therefore choose $2l + 1$ angle states to match the angular momentum states; if we choose angles ϕ_n such that

$$\phi_n \equiv \gamma + \frac{2\pi n}{2l + 1}, \quad n = 0, 1, \ldots, 2l, \tag{10.287}$$

the states $|\phi_n\rangle$ will be mutually orthonormal. Here, again γ represents the choice of origin of angular position.

With a complete set of angle states, we may define an angle operator $\hat{\phi}_\gamma$ with respect to a particular origin as

$$\hat{\phi}_\gamma \equiv \sum_{n=0}^{2l} \phi_n |\phi_n\rangle \langle \phi_n| = \gamma + \sum_{n=0}^{2l} \frac{2\pi n}{2l + 1} |\phi_n\rangle \langle \phi_n|. \tag{10.288}$$

In the latter relation, we have applied the result that

$$\sum_{n=0}^{2l} |\phi_n\rangle \langle \phi_n| = 1. \tag{10.289}$$

Using Equation 10.285, we can determine the matrix elements of this operator in the angular momentum basis

$$\langle m' | \hat{\phi}_\gamma | m \rangle = \frac{1}{2l + 1} \sum_{n=0}^{2l} \phi_n \exp[i(m - m')\phi_n], \tag{10.290}$$

and therefore determine the form of the operator in this basis

$$\hat{\phi}_\gamma = \gamma + \frac{2\pi l}{2l + 1} + \frac{2\pi}{2l + 1} \sum_{m \neq m'} \frac{\exp[i(m - m')\gamma] |m'\rangle \langle m|}{\exp[i(m - m')2\pi/(2l + 1)] - 1}, \tag{10.291}$$

where the sum is over both m and m', excluding those cases where $m = m'$. This operator looks very complex but can readily be seen to be Hermitian. The angular momentum operator in the same basis is simply given by

$$\hat{L}_z = \sum_{m=-l}^{l} m \, |m\rangle \langle m| . \tag{10.292}$$

In the angular momentum basis, these two results can be combined to find the commutator

$$[\hat{\phi}_\gamma, \hat{L}_z] = \frac{2\pi\hbar}{2l+1} \sum_{m \neq m'} \frac{(m - m') \exp[i(m - m')\gamma] \, |m'\rangle \langle m|}{\exp[i(m - m')2\pi/(2l+1)] - 1} . \tag{10.293}$$

It does not take much work to show that we may expand $|m\rangle$ in the angle states $|\phi_n\rangle$, in the form

$$|m\rangle = \frac{1}{\sqrt{2l+1}} \sum_{n=0}^{2l} \exp[im\phi_n] \, |\phi_n\rangle , \tag{10.294}$$

which allows us to also write the angular momentum operator in the angle basis

$$\hat{L}_z = -i\frac{\hbar}{2} \sum_{n \neq n'} \frac{(-1)^{n-n'} \, |\phi_{n'}\rangle \langle \phi_n|}{\sin[(n - n')\pi/(2l+1)]} . \tag{10.295}$$

Finally, we may write the commutator in the angle-state basis as

$$[\hat{\phi}_\gamma, \hat{L}_z] = i\frac{\hbar\pi}{2l+1} \sum_{n \neq n'} \frac{(n - n')(-1)^{n-n'} \, |\phi_{n'}\rangle \langle \phi_n|}{\sin[(n - n')\pi/(2l+1)]} . \tag{10.296}$$

Following Barnett and Pegg, we have derived most of the relevant results with respect to a finite set of $2l + 1$ angular momentum states. In principle, though, l may be taken as large as needed without fundamentally changing the behavior of the result. To compare with our previous results, however, we wish to take the limit $l \to \infty$, though this immediately comes with some caveats. Since an arbitrary state may now be written in the form

$$|\psi\rangle = \sum_{m=-\infty}^{\infty} c_m \, |m\rangle , \tag{10.297}$$

it is easy to construct states that are normalized, that is,

$$\langle \psi | \psi \rangle = \sum_{m=-\infty}^{\infty} |c_m|^2 < \infty, \tag{10.298}$$

but which have diverging moments of the angular momentum. Such moments would be of the form

$$\langle \psi | (\hat{L}_z)^N | \psi \rangle = \sum_{m=-\infty}^{\infty} m^N |c_m|^2. \tag{10.299}$$

For instance, if $c_m \sim 1/m$, the state will have a finite normalization but will have a diverging first moment. We therefore restrict our attention to states referred to by Barnett and Pegg as *physical states*, for which all moments of $\hat{L}_z$ are finite. It is important to immediately note that the angle states $|\phi\rangle$ of Equation 10.285 are *not* physical states; we will return to this point momentarily.

In the limit $l \to \infty$, also assuming that elements with $m - m' \gg 1$ are of less significance, we can approximate Equation 10.291 as a "physical" operator $\hat{\phi}_p$, written as

$$\hat{\phi}_p = \gamma + \pi - i \sum_{m \neq m'} \frac{\exp[i(m - m')\gamma]}{m - m'} |m'\rangle \langle m|. \tag{10.300}$$

The commutator of the physical operators can then be found as

$$[\hat{\phi}_\gamma, \hat{L}_z]_p = -i\hbar \sum_{m \neq m'} \exp[i(m - m')\gamma] |m'\rangle \langle m|. \tag{10.301}$$

In angle space, we may use Equation 10.285, with $\phi_0 = \gamma$, to show that the same physical commutator will be given by

$$[\hat{\phi}_\gamma, \hat{L}_z]_p = i\hbar[1 - (2l + 1) |\phi_0\rangle \langle \phi_0|]. \tag{10.302}$$

For a physical state $|P\rangle$, we may write the expectation value as

$$\langle P | [\hat{\phi}_\gamma, \hat{L}_z]_p | P \rangle = i\hbar[1 - (2l + 1)| \langle P | \phi_0 \rangle |^2]. \tag{10.303}$$

For a physical state, we may not choose a pure angle state, however, as already noted. We try a state that has a Gaussian spread around a central angle θ, such that

$$|P\rangle = \frac{1}{N} \sum_{n=0}^{2l} \exp[(\theta - \phi_n)^2/2\sigma^2] |\phi_n\rangle, \qquad (10.304)$$

where the normalization is

$$N^2 = \sum_{n=0}^{2l} \exp[(\theta - \phi_n)^2/\sigma^2]. \qquad (10.305)$$

We note that, from Equation 10.287, the discrete angular positions are spaced by an angular distance

$$\delta\phi \equiv \frac{2\pi}{2l+1}. \qquad (10.306)$$

The normalization may therefore be written as

$$N^2 = \frac{1}{\delta\phi} \sum_{n=0}^{2l} \exp[(\theta - \phi_n)^2/\sigma^2]\delta\phi. \qquad (10.307)$$

In the limit of large l, we may approximate this by a Gaussian integral, that is,

$$N^2 = \frac{1}{\delta\phi} \sum_{n=0}^{2l} \exp[(\theta - \phi_n)^2/\sigma^2]\delta\phi \approx \frac{1}{\delta\phi} \int_{-\infty}^{\infty} \exp[(\theta - \phi)^2/\sigma^2]d\phi$$

$$= \frac{2l+1}{2\pi} \sqrt{\pi}\sigma. \qquad (10.308)$$

With this, and the definition of $|P\rangle$, we may write

$$\langle P| [\hat{\phi}_\gamma, \hat{L}_z]_p |P\rangle = i\hbar[1 - 2\pi f(\theta)], \qquad (10.309)$$

where we introduce

$$f(\theta) = \frac{1}{\sqrt{\pi}\sigma} \exp[-(\theta - \gamma)^2/\sigma^2]. \qquad (10.310)$$

This latter function, however, is simply $\delta(\theta - \gamma)$ in the limit of an extremely small σ, as it is a delta sequence [Gbu11, Chapter 6]. In this limit, then, we get

$$\langle P| [\hat{\phi}_\gamma, \hat{L}_z]_p |P\rangle = i\hbar[1 - 2\pi\delta(\theta - \gamma)], \qquad (10.311)$$

which is equivalent to Equation 10.277 of Judge and Lewis.

This more careful derivation of the angular operator, the angular eigenstates, and the commutation relation demonstrate that there are subtle differences between a proper operator formulation of angular states and the somewhat ad hoc methods mentioned earlier. The angular eigenstates are distinct from states that are arbitrarily well localized in angle, and care must be taken in using the physical operators and their associated commutators. In fact, Barnett and Pegg argue that the limit $l \to \infty$ should generally only be taken *after* the expectation values of operators have been calculated, as inconsistencies arise in the interchange of the limit and the integrations associated with the averages.

The study of the relation between angle and angular momentum was originally inspired by research by Barnett and Pegg [BP89] into the relation between the number operator $\hat{n}$ and phase operator $\hat{\theta}$ of a single mode of an optical field, and the two systems are closely analogous. Classical investigations of the number and phase operators suggested that they should satisfy the commutation relation $[\hat{\theta}, \hat{n}] = -i$, but this cannot be true because it would imply an uncertainty relation of the form $(\Delta n)(\Delta\theta) \geq 1/2$, and we already know that a number state with $\Delta n = 0$ will have a bounded $\Delta\theta$.

They resolved these issues by a nearly identical construction to that used above for angle. An eigenstate of phase is written in an $(s + 1)$-dimensional space as

$$|\theta\rangle = \frac{1}{s+1} \sum_{n=0}^{s} \exp[in\theta] |n\rangle, \qquad (10.312)$$

where $|n\rangle$ are again the number states. In this finite basis, only $s + 1$ phase states can be orthogonal, of the form

$$\theta_m = \theta_0 + \frac{2\pi m}{s+1}, \quad m = 1, 0, \dots, s. \qquad (10.313)$$

The number states can then be written as

$$|n\rangle = \frac{1}{s+1} \sum_{m=0}^{s} \exp[-in\theta_m] |\theta_m\rangle. \qquad (10.314)$$

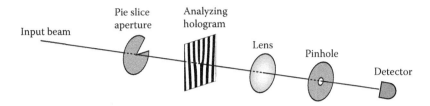

Input beam

Pie slice aperture

Analyzing hologram

Lens

Pinhole

Detector

FIGURE 10.15 Illustration of the experimental arrangement used by Franke-Arnold et al. [FABY+04].

These equations can be compared to Equations 10.285 and 10.294 to see the analogy.

Experiments to study the uncertainty principle for angular position and angular momentum were conducted by Franke-Arnold et al. [FABY+04]; an illustration of their experimental arrangement is shown in Figure 10.15. An input Gaussian beam is obstructed by a screen with a pie slice aperture cut out of it. This produces a finite angular spread and therefore a finite spread in angular momentum. An adjustable SLM is used to create a vortex hologram that selects a particular momentum state and sends it to the detector; all others are blocked by the pinhole screen. They verified that the angular uncertainty relation, such as that given by Equation 10.273, holds. They also determined the intelligent states of the angular uncertainty relation; in later work, a slightly different collaboration studied the minimum uncertainty states of this relation [PBZ+05]. We will not discuss these special states here, but note that the corresponding intelligent and minimum states for the number-phase relation have been studied in detail by Vaccaro and Pegg [VP90].

One other interesting measure of the uncertainty associated with periodic functions is worth mentioning here. We have seen that much of our difficulty comes from somewhat artificially imposing periodicity on a single-variable function, resulting in delta function jumps in commutators and wavefunction-dependent uncertainty. Forbes and Alonso [FA01b] instead treat the squared wavefunction $|\psi(\phi)|^2$ as a distribution around the rim of the unit disk, and calculate the variance of the vector $\mathbf{u} = (\cos\phi, \sin\phi)$ with respect to it; this is illustrated in Figure 10.16.

The variance with respect to ϕ is defined as

$$(\Delta\phi)^2 = \int_0^{2\pi} [\mathbf{u}(\phi) - \overline{\mathbf{u}}]^2 |\psi(\phi)|^2 d\phi = 1 - (\overline{\mathbf{u}})^2. \qquad (10.315)$$

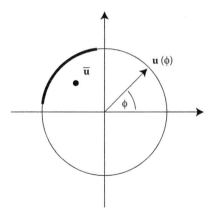

FIGURE 10.16 Illustration of the vector **u** and its average **ū** for a function distributed over the unit circle, as done by Forbes and Alonso [FA01b].

The last step readily follows because $\mathbf{u} \cdot \mathbf{u} = 1$. It is to be noted that such a representation is completely independent of the choice of the origin of ϕ. The Fourier series of $\psi(\phi)$ may be written as

$$\psi(\phi) = \frac{1}{\sqrt{2\pi}} \sum_{m=-\infty}^{\infty} \Psi_m e^{im\phi}, \tag{10.316}$$

with

$$\Psi_m = \frac{1}{\sqrt{2\pi}} \int_0^{2\pi} \psi(\phi) e^{-im\phi} d\phi. \tag{10.317}$$

The variance Δm in this conjugate space is taken with respect to the discrete variable m, and is defined as

$$(\Delta m)^2 = \sum_{m=-\infty}^{\infty} (m - \overline{m})^2 |\Psi_m|^2. \tag{10.318}$$

By an application of the Schwarz inequality again, and a significant amount of manipulation, it is found that

$$(\Delta\phi)^2 [(\Delta m)^2 + 1/4] \geq 1/4. \tag{10.319}$$

This expression differs from those found earlier in this section, but it is again a nontrivial departure from the familiar form of the Heisenberg uncertainty relation. Forbes and Alonso [FA01a] later generalized their results to make stronger inequalities both for Fourier series and for discrete Fourier transforms.

One may wonder how all these abstract angular calculations apply to singular optics. It should be noted that we can see the signature of the angular uncertainty relations in Laguerre–Gauss beams: all of them have a definite, discrete angular momentum and all of them have an intensity that is rotationally symmetric about the z-axis. Such beams are therefore minimum uncertainty states.

10.10 DIRAC'S MAGNETIC MONOPOLE

It would be almost criminal to conclude this chapter without a discussion of the earliest appearance of a vortex in quantum mechanics, an appearance that has spurred numerous experimental investigations and still captures the imagination of researchers. The vortex in question was introduced by Dirac in 1931 [Dir31]; he demonstrated that the hypothetical existence of even a single magnetic monopole in the universe, combined with quantum mechanics, leads to the quantization of electric charge. As the origins of charge quantization are still unknown, Dirac's hypothesis remains a possible explanation, one which is purely topological in origin. We will furthermore see that these arguments can also be unexpectedly connected to the quantization of angular momentum. Our argument loosely follows that of Jackson [Jac75, Section 6.13].

We begin by returning to the free-space Maxwell's equations with source terms, which may be written as

$$\nabla \cdot \mathbf{E}(\mathbf{r}, t) = \rho(\mathbf{r}, t)/\epsilon_0, \tag{10.320}$$

$$\nabla \cdot \mathbf{B}(\mathbf{r}, t) = 0, \tag{10.321}$$

$$\nabla \times \mathbf{E}(\mathbf{r}, t) = -\frac{\partial \mathbf{B}(\mathbf{r}, t)}{\partial t}, \tag{10.322}$$

$$\nabla \times \mathbf{B}(\mathbf{r}, t) = \mu_0 \mathbf{J}(\mathbf{r}, t) + \frac{1}{c^2} \frac{\partial \mathbf{E}(\mathbf{r}, t)}{\partial t}. \tag{10.323}$$

The beautiful symmetry of these equations is broken by the asymmetrical placement of sources and currents. Or, to put it another way, the symmetry is broken by the absence of magnetic sources and magnetic currents. Equation 10.321 above, which has no official name but may be called the "no magnetic monopoles" equation, quantifies the absence of magnetic sources. Nature seems to confirm this: all known magnetic sources are dipoles, and even if one breaks a bar magnet into half to separate its North and South poles, one simply gets a pair of dipole magnets, each with its own North and South poles.

But what would the existence of a magnetic monopole imply about the rest of physics? We may start to investigate simply by placing a source term in Equation 10.321, so that it has the form

$$\nabla \cdot \mathbf{B}(\mathbf{r}, t) = \mu_0 \rho_m(\mathbf{r}, t), \tag{10.324}$$

where $\rho_m(\mathbf{r}, t)$ is the magnetic charge density. By direct analogy with Gauss' law, we may write the magnetic field of a static monopole at the origin as

$$\mathbf{B}(\mathbf{r}) = \frac{\mu_0}{4\pi} \frac{g}{r^2} \hat{\mathbf{r}}, \tag{10.325}$$

where g is the magnetic charge.

Following Dirac, we now consider the behavior of a quantum particle such as an electron moving within the field of this monopole. But we immediately run into a problem; in Schrödinger's equation, the Hamiltonian of a particle exposed to a magnetic field has its momentum operator modified by the vector potential $\mathbf{A}(\mathbf{r}, t)$, with the transformation

$$\hat{P} \rightarrow \hat{P} - q\mathbf{A}(\mathbf{r}, t), \tag{10.326}$$

where q is the charge of the particle. To study quantum effects associated with monopoles, we must therefore determine the vector potential of a monopole.

It is not, however, immediately obvious how to determine the vector potential for a monopole; in electrostatics, fields with the functional form of Equation 10.325 are typically written by the use of a *scalar* potential $\phi(\mathbf{r})$. To solve this problem, Dirac considered an ingenious construction: a continuous line of magnetic dipoles running from the origin along the z-axis to the point $z = -\infty$. Because the poles of a macroscopic magnetic dipole manifest themselves at its ends, this moves one of the poles to infinity, leaving the other at the origin. One can imagine a finite-sized magnetic dipole made out of some sort of putty-like material; one end is fixed to the origin while the other end is pulled very far away, leaving an extremely thin, extremely long dipole.

To model this, we note that the vector potential of a point magnetic dipole with dipole moment **m** is well known to be of the form

$$\mathbf{A}(\mathbf{r}) = \frac{\mu_0}{4\pi} \mathbf{m} \times \nabla' \frac{1}{R}, \tag{10.327}$$

where $R = |\mathbf{r} - \mathbf{r}'|$, with $\mathbf{r}'$ the location of the point dipole. If we introduce a continuous line of dipoles, an infinitesimal dipole moment will be of the form $d\mathbf{m} = g\,d\mathbf{l}$, with $d\mathbf{l}$ the infinitesimal path element. A continuous distribution of such elements running from $z = -\infty$ to the origin has a vector potential that may be written as

$$\mathbf{A}(\mathbf{r}) = \frac{\mu_0 g}{4\pi} \int_{-\infty}^{0} \hat{\mathbf{z}} \times \nabla' \frac{1}{R} dz'. \tag{10.328}$$

The geometry is illustrated in Figure 10.17.

Using $\nabla(1/R) = \hat{\mathbf{R}}/R^3$, and noting that

$$\hat{\mathbf{z}} \times (\mathbf{r} - \mathbf{r}') = \hat{\mathbf{y}}(x - x') - \hat{\mathbf{x}}(y - y'), \tag{10.329}$$

we may write the vector potential in the form

$$\mathbf{A}(\mathbf{r}) = -\frac{\mu_0 g}{4\pi} \int_{-\infty}^{0} \frac{\hat{\mathbf{y}}x - \hat{\mathbf{x}}y}{[x^2 + y^2 + (z - z')^2]^{3/2}} dz', \tag{10.330}$$

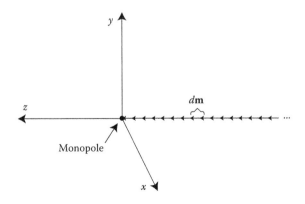

FIGURE 10.17 Illustration of Dirac's magnetic monopole construction.

where we have used $x' = y' = 0$. The numerator of this integrand is independent of z'; the integral is otherwise of a standard form

$$\int \frac{dx}{(a + cx^2)^{3/2}} = \frac{1}{a} \frac{x}{(a + cx^2)^{1/2}}. \tag{10.331}$$

With some effort, the vector potential may be written in cylindrical coordinates as

$$\mathbf{A}(\mathbf{r}) = -\frac{\mu_0 g}{4\pi} \frac{1}{\rho} \left[\frac{z}{r} - 1 \right] \hat{\phi}, \tag{10.332}$$

or in spherical coordinates as

$$\mathbf{A}(\mathbf{r}) = \frac{\mu_0 g}{4\pi} \frac{1}{r} \tan(\theta/2) \hat{\phi}. \tag{10.333}$$

The result of our calculation is a vector potential that circulates around the z-axis; we have, in essence, created a vortex of the vector potential. On the positive z-axis, where $\theta = 0$, we have $\mathbf{A} = 0$ and the result is a vector singularity analogous to the Poynting singularities of Chapter 8. On the negative z-axis, however, the magnitude of the vector potential diverges— we have an even more singular singularity! Nevertheless, one can readily show that $\mathbf{B} = \nabla \times \mathbf{A}$ reproduces the magnetic field of Equation 10.325 everywhere except on this exceptional singular line.

The presence of this singularity is troubling, and Dirac argued that the wavefield of a quantum particle must be zero on this line so that it doesn't "see" the singularity. The result is that solutions of Schrödinger's equation must have a field vortex along this line; in fact, such solutions were derived by Tamm [Tam31] and they have the characteristic $\exp[im\phi]$ phase of vortex lines. More troubling is the observation that the vector potential is not spherically symmetric like the magnetic field it represents; it is clear, however, that we have great freedom in choosing its location. For instance, the choice of placing it on the negative z-axis was arbitrary, and it may be oriented in any direction. Furthermore, the line need not be straight at all, and can be curved as long as one end is at the origin and the other is at infinity.

A change in the placement of the singular line, however, amounts to a *gauge transformation* of the potential. In classical electromagnetism, it is well known that the fields are unchanged[*] if the vector and scalar potentials

---

[*] See, for example, Gbur [Gbu11, Section 2.11].

are modified by the following transformation:

$$\mathbf{A}(\mathbf{r}, t) \rightarrow \mathbf{A}(\mathbf{r}, t) + \nabla\chi(\mathbf{r}, t), \quad \phi(\mathbf{r}, t) \rightarrow \phi(\mathbf{r}, t) - \frac{\partial\chi(\mathbf{r}, t)}{\partial t}, \quad (10.334)$$

where $\chi(\mathbf{r}, t)$ is a well-behaved scalar function. In order for the observables of Schrödinger's equation to remain unchanged under such a transformation, one must further replace [CTDL77, Complement H-III] the wavefunction $\psi(\mathbf{r}, t)$ by

$$\psi(\mathbf{r}, t) \rightarrow \psi(\mathbf{r}, t)e^{i\beta(\mathbf{r}, t)}. \quad (10.335)$$

To determine $\beta(\mathbf{r}, t)$, we require that Schrödinger's equation in the two gauges be consistent, that is, we must simultaneously satisfy

$$\frac{1}{2m}[\hat{P} - q\mathbf{A}]^2\psi + q\phi\psi = -i\hbar\frac{\partial\psi}{\partial t}, \quad (10.336)$$

$$\frac{1}{2m}[\hat{P} - q(\mathbf{A} + \nabla\chi)]^2\psi e^{i\beta} + q\left[\phi - \frac{\partial\chi}{\partial t}\right]\psi e^{i\beta} = -i\hbar\frac{\partial}{\partial t}\left(\psi e^{i\beta}\right). \quad (10.337)$$

With some effort, one can show that

$$\beta(\mathbf{r}, t) = \frac{q}{\hbar}\chi(\mathbf{r}, t). \quad (10.338)$$

We now ask what the gauge function $\chi(\mathbf{r}, t)$ looks like for a deflection of the path of the singularity. We imagine the situation shown in Figure 10.18,

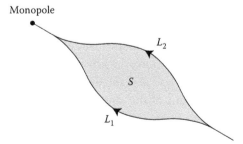

FIGURE 10.18 Displacement of the singularity path from L_1 to L_2, and the area S that the change bounds.

in which the initial singularity path L_1 is displaced to L_2. As the vector potential for the path L_j may be written in the form

$$A_j(r) = \frac{g\mu_0}{4\pi} \int_{L_j} dl \times \nabla' \frac{1}{R}, \tag{10.339}$$

the difference in vector potential is

$$\Delta A(r) = \frac{g\mu_0}{4\pi} \oint_{L_2-L_1} dl \times \nabla' \frac{1}{R}. \tag{10.340}$$

This integral can be evaluated to a particularly simple and useful form if we consider it component by component. The $\hat{x}$-component of ΔA is given by

$$\Delta A_x = \frac{g\mu_0}{4\pi} \oint \hat{x} \cdot \left[dl \times \nabla' \frac{1}{R} \right]. \tag{10.341}$$

Using the cyclic property of the scalar triple product, this may also be written as

$$\Delta A_x = \frac{g\mu_0}{4\pi} \oint dl \cdot \left(\nabla' \frac{1}{R} \times \hat{x} \right). \tag{10.342}$$

Now we apply Stokes' theorem, to write

$$\Delta A_x = \frac{g\mu_0}{4\pi} \int_S da \cdot \left[\nabla' \times \left(\nabla' \frac{1}{R} \times \hat{x} \right) \right]$$

$$= \frac{g\mu_0}{4\pi} \int_S da \cdot \left[(\hat{x} \cdot \nabla') \nabla' \frac{1}{R} - \hat{x} \nabla'^2 \frac{1}{R} \right]. \tag{10.343}$$

Provided our surface does not intersect the point r, the latter term of the integrand vanishes. We then have

$$\Delta A_x = \frac{g\mu_0}{4\pi} \int_S da \cdot \left(\frac{\partial}{\partial x'} \right) \nabla' \frac{1}{R}. \tag{10.344}$$

We note that, due to the symmetry of $R = |r - r'|$, we may write $\partial/\partial x' = -\partial/\partial x$. We may pull the x-derivative out of the integral, leaving

$$\Delta A_x = \frac{g\mu_0}{4\pi} \frac{\partial}{\partial x} \int_S da \cdot \nabla' \frac{1}{R}. \tag{10.345}$$

We are free to distort the surface S bounding the closed path $L_2 - L_1$ into any shape, provided it does not intersect the point $\mathbf{r}$. We project it onto a spherical surface of radius R centered on $\mathbf{r}$. Then, noting that the surface element $d\mathbf{a} = R^2 d\Omega \hat{\mathbf{R}}$, with $d\Omega$ the infinitesimal element of the solid angle, our integral reduces to

$$\Delta A_x = \frac{g\mu_0}{4\pi} \frac{\partial}{\partial x} \int_S d\Omega = \frac{\partial \Omega(\mathbf{r})}{\partial x}, \tag{10.346}$$

where $\Omega(\mathbf{r})$ is the solid angle of the surface subtended at the point $\mathbf{r}$.

A similar argument holds for the y and z components of $\Delta \mathbf{A}(\mathbf{r})$. We therefore have

$$\Delta \mathbf{A}(\mathbf{r}) = \frac{g\mu_0}{4\pi} \nabla \Omega(\mathbf{r}), \tag{10.347}$$

which in turn implies that

$$\chi(\mathbf{r}) = \frac{g\mu_0}{4\pi} \Omega(\mathbf{r}), \tag{10.348}$$

and that

$$\beta(\mathbf{r}) = \frac{qg\mu_0}{4\pi\hbar} \Omega(\mathbf{r}). \tag{10.349}$$

For those who have read Section 7.11, the change of phase related to a solid angle should look very much like an example of a *geometric phase*, and this is in fact the case. In those earlier examples, we considered the change in phase when the system undergoes a closed cycle in some sort of parameter space; we use this to guide our final discussion of the vector potential.

Let us now imagine that we make a continuous change of gauge by moving the singular line L_1 in a closed path around the point $\mathbf{r}$, returning it to its original position at the end of the motion. One can visualize this as the singular line swinging like a jump rope around the point $\mathbf{r}$. The total solid angle seen by the point $\mathbf{r}$ is therefore $\Omega = 4\pi$. This implies that, although we are now in exactly the same gauge as before, the quantum phase of the particle has changed by

$$\Delta\beta = \frac{qg\mu_0}{\hbar}. \tag{10.350}$$

This in turn suggests, however, that the phase of a wavefunction in the field of a magnetic monopole is ambiguous by a multiple of $\Delta\beta$. Because this phase can be measured in interference experiments, this ambiguity cannot be allowed. It can be removed if we assume that $\Delta\beta$ is always a multiple of 2π, that is,

$$\frac{qg\mu_0}{\hbar} = 2\pi n, \tag{10.351}$$

with n as an integer.

This result leads to the unavoidable conclusion that the product of q and g—the electric and magnetic charges—must be quantized. This further leads to the conclusion that they must be individually quantized. What is remarkable about this, and thematically in agreement with the rest of this book—is that the argument is a topological one. Simply the existence of a magnetic monopole, anywhere in the universe, constrains the topology of wavefields in such a way that charge must be quantized.

Dirac's argument introduces a circulating vortex line of the vector potential, suggesting that there is somehow a connection between electromagnetic angular momentum and charge quantization. This connection can be made explicit with an exceedingly simple example first put forth by Wilson [Wil48], and laid out explicitly by Jackson [Jac75, Section 6.13].

We consider a system that has an electric charge q at the origin and a magnetic monopole g at the point z_0 on the z-axis. The fields of these two static charges are given by

$$\mathbf{E}(\mathbf{r}) = \frac{q\hat{\mathbf{r}}}{4\pi\epsilon_0 r^2}, \tag{10.352}$$

$$\mathbf{B}(\mathbf{r}) = \frac{\mu_0 g\hat{\mathbf{r}}'}{4\pi r'^2}, \tag{10.353}$$

where $\mathbf{r}' = \mathbf{r} - z_0\hat{\mathbf{z}}$. Even though the fields are static, the presence of an electric and magnetic field indicates, as discussed in Section 8.2, that there is a nonzero Poynting vector and, consequently, a nonzero angular momentum density $\mathbf{l}$ throughout space. Recalling from Equation 5.22 that

$$\mathbf{l}(\mathbf{r}) = \epsilon_0 \mathbf{r} \times [\mathbf{E}(\mathbf{r}) \times \mathbf{B}(\mathbf{r})], \tag{10.354}$$

we may integrate this density throughout space to get the total angular momentum

$$\mathbf{L} = \epsilon_0 \int \mathbf{r} \times [\mathbf{E} \times \mathbf{B}]d^3r. \tag{10.355}$$

On substituting the expression for $\mathbf{E}$ into the integrand, we may write

$$\mathbf{L} = \frac{q}{4\pi} \int \frac{1}{r}\hat{\mathbf{r}} \times (\hat{\mathbf{r}} \times \mathbf{B}) d^3r = \frac{q}{4\pi} \int \frac{1}{r}[\hat{\mathbf{r}}(\hat{\mathbf{r}} \cdot \mathbf{B}) - \mathbf{B}] d^3r. \qquad (10.356)$$

At this point, we take advantage of a rather obscure vector identity

$$(\mathbf{a} \cdot \nabla)\hat{\mathbf{r}} = \frac{1}{r}[\mathbf{a} - \hat{\mathbf{r}}(\hat{\mathbf{r}} \cdot \mathbf{a})], \qquad (10.357)$$

and write

$$\mathbf{L} = -\frac{q}{4\pi} \int [\mathbf{B} \cdot \nabla]\hat{\mathbf{r}} d^3r. \qquad (10.358)$$

We now perform what amounts to an integration by parts over all space; the surface term vanishes by symmetry and we are left with

$$\mathbf{L} = \frac{q}{4\pi} \int (\nabla \cdot \mathbf{B})\hat{\mathbf{r}} d^3r. \qquad (10.359)$$

But, since $\mathbf{B}$ is the field of a point charge, we have

$$\nabla \cdot \mathbf{B} = \mu_0 g \delta^{(3)}(\mathbf{r} - z_0\hat{\mathbf{z}}). \qquad (10.360)$$

Finally, we arrive at the result

$$\mathbf{L} = \frac{qg\mu_0}{4\pi}\hat{\mathbf{z}}. \qquad (10.361)$$

We have argued in Chapter 5 that electromagnetic angular momentum is quantized in units of $\hbar$ or $\hbar/2$. If we therefore assume that $\mathbf{L} = \hat{\mathbf{z}}n\hbar/2$, we get

$$\frac{qg\mu_0}{2\pi\hbar} = n, \qquad (10.362)$$

which is precisely the same quantization condition of Equation 10.351, found by a quite different argument! It is worth noting that the mathematics of electric/magnetic monopole fields was first done by Thomson [Tho09, Section 284], and Wilson's arguments were further elaborated on by Goldhaber [Gol65].

The consistency and appeal of the arguments presented above have led to periodic searches for lone magnetic monopoles in high-energy physics experiments; a relatively recent review was made by Milton [Mil06]. The existence, or nonexistence, of such particles is still an open question.

There is one curious postscript to this discussion worth mentioning. Many may be uncomfortable with Dirac's rather artificial model for a monopole, in which a dipole is stretched to an arbitrary size, leaving an isolated pole at the origin. However, thanks to the development of cloaking optics and the theory of transformation optics [Gbu13], experimental work has been done that is strikingly close to Dirac's original idea. In 2007, Greenleaf et al. [GKLU07] suggested that it is in principle possible to make "electromagnetic wormholes," material structures that act, in effect, like a tunnel between two points in space. In practice, such a device would channel electromagnetic fields from one end of the hole to the other as if the intervening distance had not been traversed. They proposed that a wormhole devised for magnetostatic fields could work as a virtual monopole: by placing one end of a dipole magnet into the hole, the fields of that pole get "stretched" to the other end, artificially separating the poles. From this description, the similarity to Dirac's monopole model is quite clear. A magnetic-monopole-via-wormhole has in fact now been fabricated by Prat-Camps, Navau, and Sanchez [PCNS15]. The spherical structure falls far short of Dirac's ideal, infinitely thin and infinitely long dipole, but it shows that the construction of artificial monopoles is not beyond the realm of possibility.

10.11 EXERCISES

1. From Equation 10.7, the Coulomb gauge, and Maxwell's equations, show that the vector potential satisfies the wave equation

$$\nabla^2 \mathbf{A} - \frac{1}{c^2} \frac{\partial \mathbf{A}}{\partial t} = 0.$$

2. Derive Equations 10.18 from Equations 10.15 and 10.17. Show that, with $P = m\dot{x}$ and $Q = x$, this expression is simply the energy of a one-dimensional harmonic oscillator.

3. Derive Equations 10.35 and 10.36 by the following steps. Assume that the state $|n\rangle$ is normalized, that is, $\langle n| n \rangle = 1$, as well as the states $|n + 1\rangle$ and $|n - 1\rangle$. Assume, for the destruction operator, that $\hat{a} |n\rangle = C_n |n - 1\rangle$, where C_n is a constant, and a similar definition for the creation operator. Now, calculate the normalizations of $\langle n - 1| n - 1 \rangle$, $\langle n + 1| n + 1 \rangle$ to get the result.

4. Prove that the coherent states are eigenstates of the destruction operator, that is, prove Equation 10.42.

5. Prove Equation 10.220, that shows how the local velocity **v** is related to the wavefunction phase S.

6. Assuming a wavefunction with a local form $(x \pm iy)^n$, prove the statement that "the number flux **J** will have a first-order zero at a singularity, while the local velocity **v** will have a first-order pole."

7. Following the method of Section 10.8, design a wavefunction with a charge 2 screw dislocation in Schrödinger's equation at time $t = 0$. Describe the evolution of the singularity or singularities as time evolves.

8. Prove that the uniform distribution, $\psi(\phi) = 1/\sqrt{2\pi}$, has a variance $\overline{\phi^2} - [\overline{\phi}]^2$ that is independent of the integration range γ, using

$$\overline{\phi^n} = \int_{\gamma-\pi}^{\gamma+\pi} \psi^*(\phi)\phi^n\psi(\phi)d\phi. \qquad (10.363)$$

9. Prove the finite generator relation

$$\exp[-i\hat{p}\eta/\hbar]\,|x\rangle = |x+\eta\rangle,$$

by the following steps. Taylor expands the position space wavefunction $\psi(x - \epsilon)$, which represents a wavefunction shifted to the right by an amount ϵ, keeping only the first two terms (assuming ϵ is small). Next, rewrite the derivative in terms of the momentum operator, resulting in a two-term infinitesimal translation operator. Now, assume that a finite translation η is broken into a large number of steps N, each of size η/N. Use the relation

$$\lim_{N\to\infty} [1 + x/N]^N = e^x,$$

to arrive at the finite generator relation. Perform similar steps to derive the finite momentum generator.

10. Prove in Equations 10.166 and 10.167, that the Hermite modes $u_0(x)$ and $u_1(x)$ are eigenfunctions of the Fourier transform operation.

CHAPTER 11

Vortices in Random Wavefields

W E HAVE NOTED TIME and again that first-order optical vortices are generic features of wavefields, that is, that these vortices are the typical phase singularities that appear in an "unprepared" optical system. However, most of the examples and applications considered up to this point have been systems deliberately designed to produce a particular vortex state. An examination of the natural creation and evolution of vortices in wavefields is still lacking.

The ideal example of an unprepared system is the production of laser speckle through the scattering of coherent laser light from a rough surface. From a sufficiently far observation distance, this speckle appears as a dense packing of bright and dark spots; we will see that these dark spots are a dense collection of optical vortex lines.

The origin of optical speckle lies in the phase differences induced in the scattered light from variations in surface height. The scattered field at any point may be considered as the superposition of fields from a large number of small sources whose phases are essentially random. At some observation points, this superposition will produce strong constructive interference, resulting in a bright spot, while at others, it will produce complete destructive interference, resulting in an optical vortex.

In this chapter, we will investigate the statistics of vortices in such random speckle wavefields, considering both the density of vortices and their arrangement in space. We will find that there are surprising correlations between vortices in a speckle pattern, and will see that these correlations contain useful information about a wavefield.

11.1 SPECKLE STATISTICS AT A SINGLE POINT

We begin by considering the instantaneous field at a single point in a speckle pattern, and derive the probability density of the field at this point as well as its average properties.[*] This field may be envisioned to be the superposition of a large number of independent random complex fields u_n called *phasors*, where n labels the nth contribution to the total field. The total field u is written as

$$u \equiv u_1 + \cdots + u_n = u_R + iu_I, \tag{11.1}$$

where u_R and u_I are the real and imaginary parts of the total. For simplicity, the real and imaginary parts of the individual fields are assumed to have zero mean and equal variance σ_0^2; several examples of the construction of the total field from individual phasors are shown in Figure 11.1. It should be clear from the images that the process of combining phasors is equivalent to a two-dimensional random walk, of which much has been written (see, for instance, Rudnick and Gaspari [RG04]).

When the number N of combined phasors gets sufficiently large, we may apply the central limit theorem [PP02, Section 7.4], which indicates that the probability density for the real and imaginary parts of the total field is given by

$$p(u_R, u_I) = \frac{1}{2\pi\sigma^2} \exp\left[-\frac{u_R^2 + u_I^2}{2\sigma^2}\right], \tag{11.2}$$

where $\sigma^2 = N\sigma_0^2$. This probability density has the usual definition: the probability $P(u_R, u_I)$ that the real and imaginary parts of the field lie within

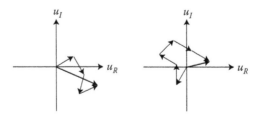

FIGURE 11.1   Illustration of a pair of phasor sums. The thick arrows indicate the total complex field.

---

[*] We give an abbreviated discussion of speckle in this chapter; a detailed description can be found in the book by Goodman [Goo07].

the infinitesimal ranges $(u_R, u_R + \delta u_R)$, $(u_I, u_I + \delta u_I)$ is

$$P(u_R, u_I) = p(u_R, u_I)\delta u_R \delta u_I, \tag{11.3}$$

and the integral of $p(u_R, u_I)$ over all possible values of u_R and u_I is unity, that is,

$$\int_{-\infty}^{\infty} \int_{-\infty}^{\infty} p(u_R, u_I) du_R \, du_I = 1. \tag{11.4}$$

Average properties of the total field may be directly found from the appropriate integrals of the probability density. The variance, for instance, may be found via

$$\overline{(u_R^2 + u_I^2)} \equiv \int_{-\infty}^{\infty} \int_{-\infty}^{\infty} (u_R^2 + u_I^2) p(u_R, u_I) du_R \, du_I = \sigma^2, \tag{11.5}$$

where the overline represents the average. When we calculated ensemble averages in the context of coherence theory in Chapter 9, we used angle brackets $\langle \cdots \rangle$ to denote the average. Here, the overline represents the same type of average, but we use a different notation to highlight that we are able to *explicitly* calculate the average from the probability density in this case.

We may always calculate averages by direct integration of the appropriate probability density function. However, average properties can be found in a more convenient manner by use of the *characteristic function* $M(\omega_R, \omega_I)$ of the probability distribution, defined as the Fourier transform of $p(u_R, u_I)$,

$$M(\omega_R, \omega_I) \equiv \int_{-\infty}^{\infty} \int_{-\infty}^{\infty} p(u_R, u_I) \exp\left[i(u_R \omega_R + u_I \omega_I)\right] du_R \, du_I. \tag{11.6}$$

This transform can be readily calculated for our case,

$$M(\omega_R, \omega_I) = \exp\left[-(\omega_R^2 + \omega_I^2)\sigma^2/2\right]. \tag{11.7}$$

We can readily show that the average properties of the field can be directly calculated from the characteristic function without integration, through the relation

$$\overline{u_R^n u_I^m} = \frac{1}{i^{n+m}} \frac{\partial^n}{\partial \omega_R^n} \frac{\partial^m}{\partial \omega_I^m} M(\omega_R, \omega_I)|_{\omega_R = \omega_I = 0}. \tag{11.8}$$

This can be demonstrated easily by direct calculation. Characteristic functions will play a larger role in the following section, and in particular, we will use the property that the characteristic function of a Gaussian process is itself Gaussian.

In light of the observation that an optical vortex arises when $u_R = u_I = 0$, the probability density of Equation 11.2 is particularly convenient. However, the real and imaginary parts of the field are usually not directly measured, but rather its amplitude A and phase ψ, where $u = A \exp[i\psi]$. These quantities may be related to u_R and u_I by the expressions

$$A = \sqrt{u_R^2 + u_I^2}, \tag{11.9}$$

$$\psi = \arctan[u_I/u_R]. \tag{11.10}$$

The probability densities $p(u_R, u_I)$ and $p(A, \psi)$ are related by a simple coordinate transformation

$$p(A, \theta) = p(u_R, u_I)\mathcal{J}, \tag{11.11}$$

where $\mathcal{J}$ is the Jacobian of the transformation, given by

$$\mathcal{J} = \begin{vmatrix} \frac{\partial u_R}{\partial A} & \frac{\partial u_R}{\partial \psi} \\ \frac{\partial u_I}{\partial A} & \frac{\partial u_I}{\partial \psi} \end{vmatrix} = A. \tag{11.12}$$

We therefore have a probability density in terms of amplitude and phase as

$$p(A, \psi) = \frac{A}{2\pi\sigma^2} \exp\left[-\frac{A^2}{2\sigma^2}\right]. \tag{11.13}$$

It is to be noted that this function does not explicitly depend on ψ at all. It is trivially possible to separate the probability densities for amplitudes and phases in the form

$$p(A) = \frac{A}{\sigma^2} \exp\left[-\frac{A^2}{2\sigma^2}\right], \tag{11.14}$$

$$p(\psi) = \frac{1}{2\pi}. \tag{11.15}$$

When integrated over their respective ranges, namely, $0 \leq A < \infty$ and $-\pi \leq \psi < \pi$, we find that each density integrates to unity. The phase is uniformly distributed over all possible values; this is unsurprising, as we had basically built this into our initial phasor description. A speckle field with a uniform phase probability distribution is known as *fully developed speckle*; we will for simplicity restrict our attention to such cases.

We may also determine the probability density function for the intensity of the total field, where $I = A^2$. The relationship between $p(A)$ and $p(I)$ comes from the simple expression

$$|p(A)dA| = |p(I)dI|, \tag{11.16}$$

from which we may write

$$p(I) = p(A = \sqrt{I}) \left| \frac{dA}{dI} \right|, \tag{11.17}$$

or

$$p(I) = \frac{1}{2\sigma^2} \exp\left[-I/2\sigma^2\right]. \tag{11.18}$$

If we calculate the mean intensity of the field, we find that $\bar{I} = 2\sigma^2$, so we may finally write

$$p(I) = \frac{1}{\bar{I}} \exp\left[-I/\bar{I}\right]. \tag{11.19}$$

The probability density function for the intensity of the field is a negative exponential, as illustrated in Figure 11.2. One can also readily demonstrate that $\overline{I^2} = 2\bar{I}^2$, so that the variance σ_I^2 of the intensity is given by

$$\sigma_I^2 = \overline{I^2} - \bar{I}^2 = \bar{I}^2. \tag{11.20}$$

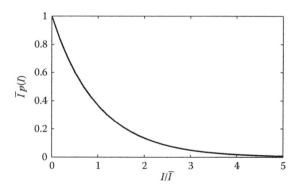

FIGURE 11.2 The negative exponential probability density function.

It is perhaps worth noting that Equation 11.19 immediately suggests that the likelihood of getting an exact zero of intensity is vanishingly small, as the probability of getting any *precise* value of intensity is zero. This does not necessarily conflict with our intuition of optical vortices, however, as we have seen that such singularities are isolated points in the plane. The likelihood of choosing a random point in the cross section of a wavefield and finding zero intensity is vanishingly small.

The characteristic function of the intensity distribution may be defined as

$$M_I(\omega) \equiv \int_0^\infty p(I) \exp[iI\omega] dI. \tag{11.21}$$

This is simply found to be

$$M_I(\omega) = \frac{1}{1 - i\omega I}. \tag{11.22}$$

We can then find the average moments of the intensity from the formula

$$\overline{I^n} = \frac{1}{i^n} \frac{\partial^n}{\partial \omega^n} M_I(\omega)|_{\omega=0}. \tag{11.23}$$

11.2 SPECKLE STATISTICS AT MULTIPLE POINTS

We now turn to correlations of the speckle field at two or more points in space. As the field must have Gaussian statistics at any individual point, we expect that the probability density function for the field at multiple points will also be of Gaussian form. Assuming that we are interested in the field at N different points, with associated complex random variables $u_1, u_2, \ldots, u_N$, we introduce a characteristic function of the form

$$M(\omega) = \exp\left[-\frac{1}{4}\omega^\dagger \cdot \mathbf{W} \cdot \omega\right], \tag{11.24}$$

where ω is a complex vector of N components itself, and $\mathbf{W}$ is known as the *covariance matrix*. The properties of this covariance matrix, and the detailed form of the probability density function, are elaborated upon in Section 11.7; for now, we take it as given and consider the implications.

The characteristic function is defined as the Fourier transform of the probability density function $p(\mathbf{u})$, that is,

$$M(\omega) = \int p(\mathbf{u}) \exp\left[i\mathrm{Re}\{\omega^* \cdot \mathbf{u}\}\right] d^{2N}u, \tag{11.25}$$

which may also be written as

$$M(\omega) = \int p(\mathbf{u}) \exp\left[i(\omega^* \cdot \mathbf{u} + \omega \cdot \mathbf{u}^*)/2\right] d^{2N}u. \qquad (11.26)$$

The real part in the exponent leaves only products for each complex random variable u_j in the form $u_{jR}\omega_{jR} + u_{jI}\omega_{jI}$, which means that we take the Fourier transform with respect to the real and imaginary part of each random variable. We may treat ω_{jR} and ω_{jI} as independent variables, or alternatively use ω_j and ω_j^*. In the latter case, we readily find that we may write

$$\overline{u_j} = \frac{2}{i} \frac{\partial}{\partial \omega_j^*} M(\omega)|_{\omega=0} = 0, \qquad (11.27)$$

where the latter result follows directly, and that

$$\overline{u_i^* u_j} = \frac{2^2}{i^2} \frac{\partial}{\partial \omega_i} \frac{\partial}{\partial \omega_j^*} M(\omega)|_{\omega=0} = \int u_i^* u_j p(\mathbf{u}) d^{2N}u = W_{ij}. \qquad (11.28)$$

Referring back to Equation 9.29 of Chapter 9, we see that W_{ij} is simply the cross-spectral density of the field with respect to the two points i and j. Let us assume, for simplicity, that the spectral density at every point is the same, namely,

$$\overline{u_j^* u_j} = 2\sigma^2. \qquad (11.29)$$

This amounts to the real and imaginary parts of each random process having variance σ^2. We may then write the coefficients of the matrix as

$$W_{ij} = 2\sigma^2 \mu_{ij}, \qquad (11.30)$$

where μ_{ij} is the spectral degree of coherence between the points i and j.

We have therefore, through our definition of $M(\omega)$ in Equation 11.24, introduced correlations between pairs of points in our random speckle field. But what is the value of μ_{ij}? Our general statistical description does not give us any indication of how to progress; so, we must turn to the details of a specific optical system to learn more.

We consider the scattering of coherent light off of a random rough surface; the scattered light is then measured on an observation plane a

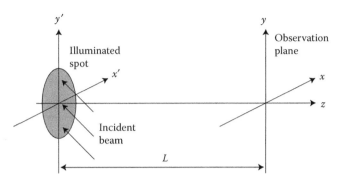

FIGURE 11.3   The geometry used in calculating correlations in a speckle field.

distance L away. The system is illustrated in Figure 11.3. This system is representative, for instance, of the simple case of shining a laser pointer off of a wall.

The field $U(\mathbf{r}, L)$ at a point in the observation plane, with $\mathbf{r} = (x, y)$, may be determined from the scattered field $U_0(\mathbf{r}')$ in the surface plane, with $\mathbf{r}' = (x', y')$, by use of the Fresnel diffraction formula [Goo96, Section 4.2]

$$U(\mathbf{r}, L) = \frac{e^{ikL}}{i\lambda L} s(\mathbf{r}, L) \int U_0(\mathbf{r}') s(\mathbf{r}', L) e^{-ik\mathbf{r}\cdot\mathbf{r}'/z} d^2 r', \qquad (11.31)$$

where for brevity we have introduced

$$s(\mathbf{r}, L) \equiv \exp\left[i\frac{k}{2L} |\mathbf{r}|^2 \right], \qquad (11.32)$$

which is a complex spherical wave in the Fresnel approximation. If we take the product of the field at two points $\mathbf{r}_1$ and $\mathbf{r}_2$, we may calculate the average of this product using the space–frequency ensemble, as discussed in Section 9.1, to get

$$W(\mathbf{r}_1, \mathbf{r}_2, L) = \langle U^*(\mathbf{r}_1, L) U(\mathbf{r}_2, L) \rangle_\omega. \qquad (11.33)$$

The cross-spectral density of the field on the observation plane is therefore

$$W(\mathbf{r}_1, \mathbf{r}_2, L) = \frac{s^*(\mathbf{r}_1, L) s(\mathbf{r}_2, L)}{\lambda^2 z^2} \int \int s^*(\mathbf{r}'_1, L) s(\mathbf{r}'_2, L) W_0(\mathbf{r}'_1, \mathbf{r}'_2)$$
$$\times e^{ik\mathbf{r}_1\cdot\mathbf{r}'_1/L} e^{-ik\mathbf{r}_2\cdot\mathbf{r}'_2/L} d^2 r'_1 d^2 r'_2, \qquad (11.34)$$

where we have also used

$$W_0(\mathbf{r}_1', \mathbf{r}_2') = \langle U_0^*(\mathbf{r}_1') U_0(\mathbf{r}_2') \rangle_\omega. \tag{11.35}$$

To proceed further, we need some sort of model for the effect of the surface roughness on the illuminating field. We will make the simple assumption

$$W_0(\mathbf{r}_1', \mathbf{r}_2') = \alpha S_0(\mathbf{r}_1') \delta^{(2)}(\mathbf{r}_2' - \mathbf{r}_1'), \tag{11.36}$$

where $S_0(\mathbf{r})$ is the spectral density of the illumination, α is a positive constant, and $\delta^{(2)}$ is the two-dimensional Dirac delta function. The form of Equation 11.36 suggests that the surface has height variations larger than a wavelength, thus effectively washing out the phase of the incident field, and also that it has very small spatial correlations along the surface. The constant α has dimensions of squared length, and is there solely to make Equation 11.36 dimensionally consistent.

On substitution, we arrive at the expression

$$W(\mathbf{r}_1, \mathbf{r}_2, L) = \alpha \frac{s^*(\mathbf{r}_1, L)s(\mathbf{r}_2, L)}{\lambda^2 z^2} \int S_0(\mathbf{r}') e^{-i\mathbf{k}\mathbf{r}\cdot\mathbf{r}'/L} d^2 r', \tag{11.37}$$

where we have introduced $\mathbf{r} \equiv \mathbf{r}_2 - \mathbf{r}_1$. To a good approximation, then, the cross-spectral density of the field in the observation plane depends only on the spatial separation of the observation points. We may further introduce the spectral degree of coherence in the observation plane, which follows from Equation 9.32 as

$$\mu(\mathbf{r}_1, \mathbf{r}_2, L) = s^*(\mathbf{r}_1, L)s(\mathbf{r}_2, L) \frac{\int S_0(\mathbf{r}') e^{-i\mathbf{k}\mathbf{r}\cdot\mathbf{r}'/L} d^2 r'}{\int S_0(\mathbf{r}') d^2 r'}. \tag{11.38}$$

This result is equivalent to a classic result of coherence theory, the *van Cittert–Zernike theorem* [Wol07, Section 3.2], which indicates that the correlation properties of light far from an incoherent source are determined by the Fourier transform of the intensity distribution in the source plane.

We are now in a position to make some quantitative estimates. Let us take the illumination spot to be circular, with radius a. Using properties of Bessel functions, we can quickly determine that

$$|\mu(\mathbf{r}_1, \mathbf{r}_2, L)| = \left| \frac{2J_1(kra/L)}{kra/L} \right|, \tag{11.39}$$

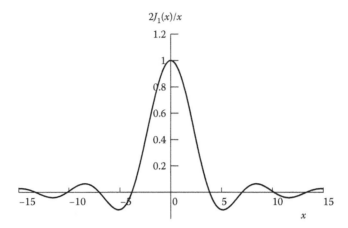

FIGURE 11.4   The Bessel-related function that appears in speckle correlations.

where $J_1(x)$ is the Bessel function of order 1. This Bessel-dependent function appears quite often in optics, especially when dealing with circular apertures or sources. It is plotted in Figure 11.4. The first zero of this function appears at $x \approx 3.83$.

This correlation function can be directly applied to the characteristic function of Equation 11.24 to determine the covariance matrix for any pair of points. For our purposes, however, it will be more useful to use it to determine the effective "correlation area" of the speckle pattern. More precisely, we may phrase this as the following question: within what area is the intensity of light in the speckle correlated?

As will be proven in Section 11.7, a multivariate Gaussian random process satisfies the moment theorem

$$\overline{U_1^* U_2^* U_3 U_4} = W_{42} W_{31} + W_{41} W_{32}. \tag{11.40}$$

Let us consider the case when $U_3 = U_1$ and $U_4 = U_2$; we then have

$$\overline{I_1 I_2} = W_{22} W_{11} + W_{12} W_{21} = \overline{I}^2 \left[1 + |\mu_{12}|^2 \right], \tag{11.41}$$

where in the last step, we have assumed that the intensities are the same at the two points. Introducing a normalized intensity covariance function of the form

$$C_I(\mathbf{r}) \equiv \frac{\overline{I^2(\mathbf{r})} - \overline{I}^2}{\overline{I}^2} = |\mu(\mathbf{r}, L)|^2, \tag{11.42}$$

we may then define a correlation area of the speckle as

$$A_c \equiv \int |\mu(\mathbf{r}, L)|^2 d^2 r. \tag{11.43}$$

The integral in question is nontrivial, but can be evaluated exactly using Fourier properties of Bessel functions. The result is

$$A_c = \frac{(\lambda L)^2}{\pi a^2}. \tag{11.44}$$

The denominator is simply the area A of the original illuminated patch. Also, noting that L^2/A is approximately the solid angle Ω subtended by the patch as observed from the observation plane, we may also write

$$A_c = \frac{\lambda^2}{\Omega}. \tag{11.45}$$

This result is, in essence, the qualitative result of the van Cittert–Zernike theorem: as the distance L from the source increases, the solid angle Ω decreases, and the correlation area A_c of the speckle increases.

We have done a lot of work here to derive a very simple result in the form of Equation 11.45! What we have found, however, is the most important characteristic length associated with a speckle pattern. We will see in the next section that this length is directly related to vortex statistics.

11.3 STATISTICS OF VORTICES IN RANDOM WAVEFIELDS

At long last, we can return to a discussion of vortices in random wavefields. Our first task is to calculate the density of vortices in any cross section of a speckle field. Ideally, this would be accomplished by directly counting the number of zeros in a section of the field of unit area. As we have seen, a zero of the field is a point where $u_R = u_I = 0$. Naively, then, it would seem that we could count the number of zeros in an area by integrating over the delta-based distribution

$$D(u_R, u_I) = \delta(u_R)\delta(u_I), \tag{11.46}$$

as each zero of the field will contribute a unit value to the integral.

This is not quite right, however—the function $D(u_R, u_I)$ in fact has dimensions of inverse field squared, not inverse area. We need to do a coordinate transform from (u_R, u_I) to x, y, with a corresponding Jacobian $|\mathbf{J}| = |u_{Rx}u_{Iy} - u_{Ry}u_{Ix}|$, where the x, y subscripts indicate partial derivatives. In terms of the complex field u, we may also write this as $|\mathbf{J}| = |\text{Im}\{u_x^* u_y\}|$. To count the number of vortices in an area A, we therefore need to perform the following integral:

$$N_A = \int \delta(u_R)\delta(u_I)|\text{Im}\{u_x^* u_y\}|dx\,dy, \tag{11.47}$$

or, if we want to determine the number of vortices per unit area n, just look at the quantity

$$n = \delta(u_R)\delta(u_I)|\text{Im}\{u_x^* u_y\}|. \tag{11.48}$$

Of course, we are interested in the average properties, so, we must properly average this over the appropriate probability density function

$$\bar{n} = \overline{\delta(u_R)\delta(u_I)|\text{Im}\{u_x^* u_y\}|}. \tag{11.49}$$

But now we have a problem. We already know the function $p(u)$ for the complex field at a single point, given by Equation 11.2, but this function depends on the derivatives u_x and u_y of this field, as well. Before evaluating the average, then, we must determine the correlation properties between u, u_x, and u_y.

We will discuss these correlation properties in a very qualitative manner; a detailed discussion can be found in Goodman [Goo07, Section 4.7]. First, we note that we expect u_x and u_y to also be Gaussian random processes, as they are linear functions of u, which is itself a Gaussian process. It is also reasonable to expect that u_x and u_y will not be correlated with u, as the value of a wavefield does not in general have any relationship to the local slope of the wave. Finally, provided that the illuminated patch has rotational symmetry, we expect that there will be no correlations between u_x and u_y.

With these observations, we expect that we may write a characteristic function $M(\omega)$ as

$$M(\omega) = \exp\left[-\frac{1}{4}\omega^\dagger \cdot \mathbf{M} \cdot \omega\right], \tag{11.50}$$

where $\omega = (\omega_0, \omega_x, \omega_y)$ and

$$\mathbf{M} = \begin{bmatrix} 2\sigma^2 & 0 & 0 \\ 0 & 2\alpha_x^2 & 0 \\ 0 & 0 & 2\alpha_y^2 \end{bmatrix}, \tag{11.51}$$

where σ^2 is again the variance of u, and α_x^2, α_y^2 are the variances of u_x, u_y. Following the results of Section 11.7, the probability density function is then of the form

$$p(\mathbf{u}) = \frac{1}{\pi^3} \frac{1}{8\sigma^2 \alpha_x^2 \alpha_y^2} \exp\left[-\mathbf{u}^\dagger \cdot \mathbf{M}^{-1} \cdot \mathbf{u}\right], \tag{11.52}$$

where $\mathbf{u} = (u, u_x, u_y)$ and

$$\mathbf{M}^{-1} = \begin{bmatrix} 1/2\sigma^2 & 0 & 0 \\ 0 & 1/2\alpha_x^2 & 0 \\ 0 & 0 & 1/2\alpha_y^2 \end{bmatrix}. \tag{11.53}$$

Writing the exponent in explicit form and setting up the integrations, our average becomes

$$\bar{n} = \frac{1}{\sigma^2 \alpha_x^2 \alpha_y^2} \frac{1}{(2\pi)^3} \int\int \left|\frac{u_x^* u_y - u_y^* u_x}{2i}\right| e^{-|u_x|^2/2\alpha_x^2} e^{-|u_y|^2/2\alpha_y^2} d^2 u_x \, d^2 u_y. \tag{11.54}$$

We have already performed the delta function integrals over u_R and u_I. If we write $u_x = |u_x| e^{i\phi_x}$, with a similar expression for u_y, and express our integrals in polar form, we have

$$\bar{n} = \frac{1}{\sigma^2 \alpha_x^2 \alpha_y^2} \frac{1}{(2\pi)^3} \int\int |u_x|^2 |u_y|^2 |\sin(\phi_y - \phi_x)| e^{-|u_x|^2/2\alpha_x^2} e^{-|u_y|^2/2\alpha_y^2}$$

$$\times du_x du_y d\phi_x d\phi_y. \tag{11.55}$$

We may evaluate the angular integrals by replacing the variables ϕ_y, ϕ_x with $\phi \equiv \phi_y - \phi_x$ and ϕ_x. We finally have

$$\bar{n} = \frac{1}{\sigma^2 \alpha_x^2 \alpha_y^2} \frac{1}{(2\pi)^2} \left[\int_{-\pi}^{\pi} |\sin\phi| d\phi\right] \left[\int_0^{\infty} u_x^2 e^{-u_x^2/2\alpha_x^2} du_x\right]$$

$$\times \left[\int_0^{\infty} u_y^2 e^{-u_y^2/2\alpha_y^2} du_y\right]. \tag{11.56}$$

These three independent integrals can be readily evaluated, and the result is

$$\bar{n} = \frac{\alpha_x \alpha_y}{2\pi\sigma^2}. \tag{11.57}$$

But we are still not done, as we need physical expressions for α_x and α_y! From our definition of the matrix **M**, we note that we may write

$$2\alpha_x^2 = \overline{\partial_x u^*(\mathbf{r}_1)\partial_x u(\mathbf{r}_2)}|_{\mathbf{r}_1=\mathbf{r}_2}, \tag{11.58}$$

$$2\alpha_y^2 = \overline{\partial_y u^*(\mathbf{r}_1)\partial_y u(\mathbf{r}_2)}|_{\mathbf{r}_1=\mathbf{r}_2}. \tag{11.59}$$

The two expressions are formally the same, only differing in the derivative. Using Equation 11.37, we may write

$$2\alpha_x^2 = \frac{\alpha}{\lambda^2 z^2}\frac{\partial}{\partial x_1}\frac{\partial}{\partial x_2}\int S_0(\mathbf{r})e^{-i k \mathbf{r}\cdot\mathbf{r}'/z}d^2 r'\bigg|_{\mathbf{r}_1=\mathbf{r}_2}. \tag{11.60}$$

Here, we have ignored the spherical wave terms $s(\mathbf{r}, L)$; provided we are sufficiently far away from the source, these terms should be slowly varying and their derivatives should be negligible compared to those of the integral. The derivatives of the integral may be easily evaluated, and the result is

$$2\alpha_x^2 = \frac{\alpha}{\lambda^2 z^2}\frac{k^2}{z^2}\int x'^2 S_0(\mathbf{r}')d^2 r'. \tag{11.61}$$

To evaluate this for our circular illuminated patch, we note that we may write

$$2\alpha_x^2 + 2\alpha_y^2 = \frac{\alpha(2\pi)^2}{\lambda^4 z^4}\int S_0(\rho)\rho^3 d\rho d\phi = \frac{(2\pi)^3}{4\lambda^4 z^4}a^4. \tag{11.62}$$

For a circular patch, we expect $\alpha_x = \alpha_y$, so that we finally have

$$\alpha_x^2 = \alpha_y^2 = \frac{S_0\alpha\pi^3 a^4}{2\lambda^4 z^4}, \tag{11.63}$$

where S_0 is the average intensity of the source. We may also use Equation 11.37 to evaluate $2\sigma^2$, and readily find that

$$2\sigma^2 = \overline{u_j^* u_j} = \frac{\alpha S_0}{\lambda^2 z^2}\pi a^2. \tag{11.64}$$

Finally, we may substitute from Equations 11.63 and 11.64 into Equation 11.57, to get the result

$$\overline{n} = \frac{1}{2A_c}. \tag{11.65}$$

We have done a large amount of mathematics to derive, in the end, a very simple formula, but it is a very significant formula. It tells us that one vortex appears, on average, in every two correlation areas of the speckle pattern. For the first time, we have a statistical result related to the *structure* of light, as opposed to the one related to correlations of the complex fields.

The earliest calculation of this form was apparently done by Baranova et al. [BZM⁺81], who verified their results with experimental measurements. An extremely detailed discussion of vortex statistics in speckle was performed by Berry [Ber78], and later elaborated upon by Freund, Shvartsman, and Freilikher [FSF93] and Freund [Fre94].

We have already noted that the correlation area A_c increases as the field propagates further from the illumination patch; this implies that the density of vortices decreases on propagation. This may at first glance seem paradoxical, considering the conservation of topological charge. In the propagation regime we are considering, however, the field is already primarily a spherical wave, which means that the field is locally propagating in a radial direction and we expect that the vortex lines will follow a similar path. The lines will not generally be created or destroyed, but will become less dense as the field propagates. The illumination field was assumed to be coherent and vortex free; the speckle vortices are therefore created by interference of the scattered field in the near zone of the surface or, if the surface is relatively smooth, somewhat further away in the intermediate zone.

There is one simple modification we can make to our calculation that is of some interest. If we consider the interference of our speckle field with a coherent plane wave, we can examine the density of vortices as the plane wave amplitude increases. Taking our plane wave to be normally incident to the observation plane and real valued in that plane, the probability density of Equation 11.52 takes on the modified form

$$p(\mathbf{u}) = \frac{1}{\pi^3} \frac{1}{8\sigma^2 \alpha_x^2 \alpha_y^2} \exp\left[-((u_R - A_0)^2 + u_I^2)/2\sigma^2\right]$$
$$\times \exp\left[-|u_x|^2/2\alpha_x^2\right] \exp\left[-|u_y|^2/2\alpha_y^2\right], \qquad (11.66)$$

where A_0 is the amplitude of the plane wave. The entire calculation is the same as before, except that the integral over u_R brings out an explicit Gaussian term. We have, in the end,

$$\bar{n} = \frac{e^{-I_0/2\sigma^2}}{2A_c}. \qquad (11.67)$$

As the intensity $I_0 = A_0^2$ of the plane wave increases, the density of the vortices decreases, becoming negligible once I_0 becomes larger than the average intensity of the speckle, $2\sigma^2$. From our knowledge of vortex reactions, we expect that the vortices annihilate as the plane wave intensity increases.

11.4 CORRELATIONS OF VORTICES IN RANDOM WAVEFIELDS

Perhaps unnoticed in the barrage of mathematics of the preceding section is a rather surprising and quite novel paradigm shift in the way we study random fields. Throughout Chapters 9 and 10, and the first two sections of this chapter, we concerned ourselves with correlations between fields or between intensities, studying quantities such as the cross-spectral density

$$W(\mathbf{r}_1, \mathbf{r}_2, \omega) = \langle U^*(\mathbf{r}_1, \omega) U(\mathbf{r}_2, \omega) \rangle_\omega. \tag{11.68}$$

When looking at such field correlations alone, a speckle field would appear, in a sense, to be the least interesting possible random field or, to put it in another way, the most random field we can create. A speckle field is locally homogeneous and isotropic, and by our calculations, we have found that it possesses field correlations only over an area A_c.

But our calculation of the vortex density $\bar{n}$ is the first of another set of possibilities: studies of the *structural* correlations of a random field. This vortex density could not be calculated from the field averages alone, and suggests that our emphasis on such averages in general will obscure other features of interest. This was described very elegantly by Isaac Freund,[*] who said of field averages,

> All this leads directly to the conventional view of a Gaussian speckle pattern: inside a coherence area the fields remain correlated because they can scarcely change, while outside this area when they do change they change randomly, rapidly driving all correlators to zero. From this point of view the apparent randomness of speckle patterns reflects the true nature of the fields, and so what one sees is what one really gets—pure randomness.
>
> Since conventional correlators are equivalent to spatial averages, however, we need to be concerned that such correlators may

---

[*] See his paper on "1001 correlations in random wave fields" [Fre98], which I found to be quite a revelation in my understanding of random waves.

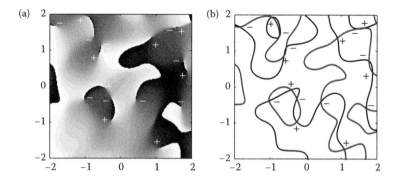

FIGURE 11.5 Illustrating the (a) phase and (b) real and imaginary zeros of a complex random field.

> average local positive and negative correlations to zero, thereby hiding significant local order.

In other words, the averaging process used to derive the traditional correlation functions almost certainly washes out important *structural* correlations in the wavefield. The result of the previous section is the first illustration of this; we now consider other structural properties of seemingly random wavefields.

Figure 11.5a shows the phase in a small cross section of a random field generated from 60 plane waves with $|k_x|, |k_y| \leq \pi$ and each plane wave being assigned a uniformly distributed phase $|\psi| \leq 32\pi$. A variety of phase singularities are clearly visible in the figure: for convenience, the sign of the topological charge is labeled for each.

One curious property is immediately obvious from the figure: looking at any given vortex, the topological charge of the nearest neighbor is almost always of the opposite sign. Through numerical simulations, Shvartsman and Freund [SF94a,SF94b] have demonstrated that 90% of vortices in a Gaussian random wavefield have nearest neighbors of opposite sign, and this result was confirmed through laboratory experiments. They further showed that 58% of next nearest neighbors are also of opposite sign.

This statistical effect is, in fact, grounded in a topological property of wavefields, as can be seen in Figure 11.5b, which shows the zeros of the real and imaginary parts of the wavefield. Starting at any crossing of real and imaginary lines and following one or the other to the next crossing, one finds that the neighboring singularity along any real or imaginary line is *always* of opposite sign. This is known as the *sign principle* [FS94]; it is a deterministic rule that is based on the topology of wavefields. It results in a

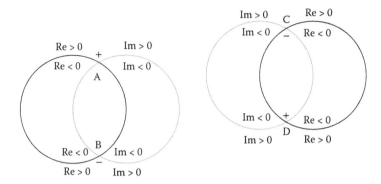

FIGURE 11.6 Illustration of the sign principle, where black lines indicate $\text{Re}\{U\} = 0$ and gray lines indicate $\text{Im}\{U\} = 0$.

strong correlation between the signs of nearest-neighbor vortices, but does not result in a deterministic nearest-neighbor rule because the zero lines can become quite convoluted.

To see how the sign principle arises, we consider a hypothetical system of zero crossings for the real and imaginary parts of the wavefield in Figure 11.6. Before we begin, we note the (obvious) statement that the sign of these components changes across any zero line. Let us assume that the sign of the vortex at point A is positive. Just by making that statement, we determine, to within an absolute overall sign, the signs of $\text{Re}\{U\}$ and $\text{Im}\{U\}$ across their respective zero lines. For instance, we know that when the phase of the wave $\theta = 0$, then $\text{Re}\{U\} > 0$ and $\text{Im}\{U\} = 0$; similarly, when $\theta = \pi/2$, then $\text{Re}\{U\} = 0$ and $\text{Im}\{U\} > 0$. We choose them as shown in the figure, and then consider the sign of the vortex at point B. Because the signs of the real and imaginary parts are fixed, we automatically find that the vortex at point B must be negative.

Remarkably, we find that all the vortex charges are now determined, even if they don't lie upon the same real and imaginary zero lines. For instance, point C has the signs on either side of its zero lines determined from the values neighboring A and it is found to be negative. Finally, point D can be determined to be positive by using A, B, or C as a reference.

This astonishing result suggests that, once a single vortex charge in a wavefield is determined, all others are determined from the topology of the wavefield itself, even for all future times. Any new vortices will be created by new intersections of real and imaginary zero lines, and their signs will be determined by the vortices already in existence.[*]

---

[*] Freund has dubbed this "the genesis principle" [Fre98].

The knowledge of the strong anticorrelation of nearest-neighbor vortices is of potential use in phase retrieval problems. In Section 6.3, we noted that phase retrieval is generally hampered by the presence of vortices, as vortices of opposite sign are indistinguishable from their intensity patterns. Simply knowing that vortices are generally of first order, however, reduces the amount of computational guessing one must try to deduce the correct phase, and an awareness of the anticorrelation properties of vortices suggests that further reduction is possible by using educated guesses.

11.5 QUASICRYSTALLINE ORDER

If we extend our search to include structural features other than vortices, we can find even more surprising correlations in wavefields. Decomposing the complex field into its real and imaginary parts again, that is, $u = u_R + iu_I$, we consider the stationary points of each of these components. The stationary points are, of course, maxima, minima, and saddles, and can be found at points for which the individual components are stationary. For instance, stationary points of the real part of the field are found when

$$\frac{\partial u_R}{\partial x} = \frac{\partial u_R}{\partial y} = 0. \tag{11.69}$$

Using the same technique as the last section, we generate a random wavefield and look at the zero crossings of these derivatives. An example is shown in Figure 11.7. Following either $u_{Rx} = 0$ or $u_{Ry} = 0$, it is readily apparent that there is a strong tendency for either maxima to alternate with saddle points or minima to alternate with saddle points. Furthermore, it is to be

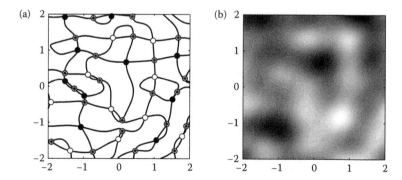

FIGURE 11.7 Illustration of (a) the zero crossings of the derivatives of u_R and (b) the real part of the field u_R. A white circle indicates a maximum, a black circle a minimum, and a star a saddle point.

noted that the zero lines tend to be elongated along the direction opposite to the derivative taken; that is, zero lines of u_{Ry} tend to be horizontal while zero lines of u_{Rx} tend to be vertical. Their intersection locally forms a crystal-like lattice, as can be clearly seen in the upper half of Figure 11.7a.

This phenomenon is now referred to as *quasicrystalline order*, and was first noted by Freund and Shvartsman [FS95]. It should be clear that a similar order appears for the imaginary part of the field as well. It is also to be noted that the spacing of the zero lines in both the horizontal and vertical directions is surprisingly regular, making the system appear quite periodic. The spacing has been shown to be on the order of $\sqrt{A_c}$, which indicates that the quasicrystalline structure generally extends over many coherence areas of the field.

A simple illustration of an extended region of such a crystalline pattern is shown in Figure 11.8, just to provide a feeling for the overall structure. On average, one expects roughly equal amounts of maxima and minima, and twice as many saddle points as there are maxima.

Even now, we have hardly touched upon all the possible correlations in random wavefields! The reader is pointed to Freund [Fre98] to learn more.

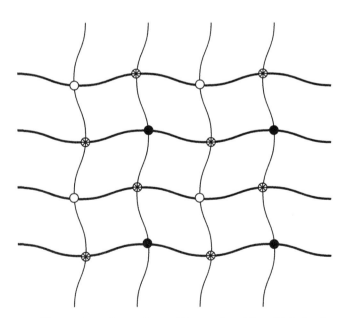

FIGURE 11.8 Illustration of quasicrystalline order. The thick horizontal lines represent zeros of u_{Ry}, while the thin vertical lines represent zeros of u_{Rx}.

11.6 POLARIZATION SINGULARITIES IN RANDOM WAVEFIELDS

At this point, it will not be surprising to learn that the techniques used in the previous sections can also be adapted to the study of polarization singularities in random electromagnetic fields. The topic was first broached by Berry and Dennis [BD01c], and later studied in extensive detail by Dennis [Den02]. The calculations are even more elaborate than those of Section 11.3; for this reason, we only provide a sketch here, followed by the main results; the reader can refer to the original papers for more details.

The first natural question to consider is the density of polarization singularities in a random electromagnetic speckle field. Here, we hardly need to any work at all: the transverse polarization may be decomposed in any plane into left- and right-circularly polarized components, and as in Section 11.3, we expect each of these components to have vortices with a density $\bar{n} = 1/2A_c$, or one vortex for every two correlation areas. Provided the polarization components are uncorrelated, we therefore expect that the density $\bar{n}_C$ of C-points in the total field is simply twice this, or

$$\bar{n}_C = \frac{1}{A_c}. \tag{11.70}$$

We would also expect, without doing any calculation, that the densities of left-handed and right-handed C-points are equal.

Less trivial is the determination of the density of specific types of C-points, namely, lemons, stars, and monstars. We first recall that a C-point is defined as a point where $S_1 = S_2 = 0$; this suggests that we may find the total density $\bar{n}_C$ of C-points by calculating the ensemble average

$$\bar{n}_C = \langle \delta(S_1)\delta(S_2)|S_{1x}S_{2y} - S_{1y}S_{2x}|\rangle, \tag{11.71}$$

where S_{1x} is the x partial derivative of S_1, and so forth. However, from Section 7.5, we know that the index of a C-point is given by the sign of

$$D_I = S_{1x}S_{2y} - S_{1y}S_{2x}, \tag{11.72}$$

and the number of separatrices is given by the sign of

$$D_L = \left[(2S_{1y} + S_{2x})^2 - 3S_{2y}(2S_{1x} - S_{2y})\right]\left[(2S_{1x} - S_{2y})^2 \right. $$
$$\left. + 3S_{2x}(2S_{1y} + S_{2x})\right] - (2S_{1x}S_{1y} + S_{1x}S_{2x} - S_{1y}S_{2y} + 4S_{2x}S_{2y})^2, \tag{11.73}$$

with $D_L > 0$ indicating three lines and $D_L < 0$ indicating one line. To the average quantity in Equation 11.71, then, we may add step functions that isolate a particular topological index and particular number of separatrices, thus leaving us with the density of a particular type of polarization singularity.

As discussed in Dennis [Den02], it is in fact more convenient to work with a modified set of parameters instead of the Stokes parameters. In any case, it is found that

$$\bar{n}_{star} = 0.5, \tag{11.74}$$

$$\bar{n}_{monstar} = 0.05279, \tag{11.75}$$

$$\bar{n}_{lemon} = 0.44721. \tag{11.76}$$

It is reassuring to note that the density of index $n = -1/2$ singularities, namely, stars, is equal to the density of $n = +1/2$ singularities, lemons, and monstars. As one might expect, the density of monstars is quite low, as it represents a transition singularity that only appears in the process of creation/annihilation.

These results were confirmed experimentally by Flossmann et al. [FODP08]. The vector field was generated by splitting a single laser beam into a pair of orthogonally polarized fields using a Wollaston prism; these orthogonal beams then illuminate separate regions of a ground glass screen, and are recombined to form the complete random electromagnetic field.

11.7 MULTIVARIATE COMPLEX CIRCULAR GAUSSIAN RANDOM PROCESSES

In this section, we derive results related to the so-called *multivariate complex circular Gaussian random process*. More details, and more general forms of such processes, are described in Goodman and Picinbono [Goo63,Pic96]. "Multivariate" refers to our consideration of multiple random variables simultaneously. "Circular" indicates that the variance of each random variable is the same for the real and imaginary parts. In such a case, we can write the probability density function in terms of the complex random variables directly, instead of writing it in terms of the real and imaginary parts of these variables separately. We assume that we are interested in N complex random variables of zero mean; the characteristic function in this case is of the form

$$M(\omega) = \exp\left[-\frac{1}{4}\omega^\dagger \cdot \mathbf{W} \cdot \omega\right], \tag{11.77}$$

where ω is a complex vector of N components, and $\mathbf{W}$ is the so-called covariance matrix. This characteristic function is the $2N$-variable Fourier transform of the probability density function, that is,

$$M(\omega) = \int p(\mathbf{u}) \exp\left[i(\omega^* \cdot \mathbf{u} + \omega \cdot \mathbf{u}^*)\right] d^{2N}u, \tag{11.78}$$

or

$$M(\omega) = \int p(\mathbf{u}) \exp\left[i(u_{iR}\omega_{iR} + u_{iI}\omega_{iI})/2\right] d^{2N}u. \tag{11.79}$$

It is to be noted that $\omega^\dagger$ is a vector with components ω_i^*, with $i = 1,\ldots,N$. If we treat ω_i and ω_i^* as independent variables, we may then derive the moments of this multivariate process through formulas such as

$$\overline{u_i^* u_j} = \frac{2^2}{i^2}\frac{\partial}{\partial\omega_i}\frac{\partial}{\partial\omega_j^*} M(\omega)|_{\omega=0} = \int u_i^* u_j p(\mathbf{u}) d^{2N}u = W_{ij}. \tag{11.80}$$

The latter result comes from directly taking the derivatives of Equation 11.77. Clearly, however, this implies that W_{ij} is simply the cross-spectral density of the field evaluated at a pair of points i and j, as

$$W_{ij} = \overline{u_i^* u_j} = \sqrt{S_i}\sqrt{S_j}\mu_{ij}, \tag{11.81}$$

where we have introduced the spectral density S_i and the spectral degree of coherence μ_{ij}. We therefore note that the matrix $\mathbf{W}$ must be nonnegative definite and Hermitian. If we assume that each random process has the same variance $S_i = 2\sigma^2$ (the variance of the real and imaginary parts are each taken to be σ^2), we may then write

$$W_{ij} = 2\sigma^2 \mu_{ij}. \tag{11.82}$$

Our primary interest in this section is to calculate the probability density function $p(\mathbf{u})$ from the characteristic function. To do so, we first note that there exists a unitary matrix Θ that diagonalizes $\mathbf{W}$; the diagonal form $\mathbf{D}$ is defined by a similarity transformation

$$\Theta \mathbf{W} \Theta^\dagger = \mathbf{D}. \tag{11.83}$$

We may also introduce a pair of new complex vectors

$$\mathbf{w} = \Theta \cdot \omega, \tag{11.84}$$

$$\mathbf{v} = \Theta \cdot \mathbf{u}. \tag{11.85}$$

In terms of $\mathbf{w}$, then, the characteristic function is

$$M(\mathbf{w}) = \exp\left[-\frac{1}{4}\mathbf{w}^\dagger \cdot \mathbf{D} \cdot \mathbf{w}\right], \tag{11.86}$$

and in terms of $\mathbf{v}$ the probability density is

$$p(\mathbf{v}) = \frac{1}{(2\pi)^{2N}} \int M(\mathbf{w}) \exp\left[i\mathrm{Re}\{\mathbf{w}^* \cdot \mathbf{v}\}\right] d^{2N}v. \tag{11.87}$$

It should be noted that, because the determinant of a unitary matrix has unit magnitude, that the coordinate transformation does not change the form of the Fourier transform.

In diagonal form, the characteristic function simplifies to the form

$$M(\mathbf{w}) = \exp\left[-\frac{1}{4}(w_{1R}^2 + w_{1I}^2)\lambda_1\right] \cdots \exp\left[-\frac{1}{4}(w_{NR}^2 + w_{NI}^2)\lambda_N\right], \tag{11.88}$$

where λ_i is the ith eigenvalue of the matrix. We may determine the Fourier transform for each complex variable w_i independently, then, and the result for the probability density function is

$$p(\mathbf{v}) = \frac{1}{\pi^N} \frac{1}{\lambda_i \cdots \lambda_N} \exp\left[-\sum_{i=1}^{N}(v_{iR}^2 + v_{iI}^2)/\lambda_i\right]. \tag{11.89}$$

The product of all the eigenvalues is simply the determinant of $\mathbf{D}$, which in turn is simply the determinant of $\mathbf{W}$. Furthermore, the exponent may be written as a diagonal quadratic form of $\mathbf{v}$ whose diagonal elements are the inverse eigenvalues of $\mathbf{D}$; this implies that the matrix is simply $\mathbf{D}^{-1}$. We may write

$$p(\mathbf{v}) = \frac{1}{\pi^N} \frac{1}{\text{Det}[\mathbf{W}]} \exp\left[\mathbf{v}^\dagger \cdot \mathbf{D}^{-1} \cdot \mathbf{v}\right]. \tag{11.90}$$

Substituting our definition of $\mathbf{v}$ back into this expression, we have

$$p(\mathbf{u}) = \frac{1}{\pi^N} \frac{1}{\text{Det}[\mathbf{W}]} \exp\left[\mathbf{u}^\dagger \cdot \Theta^\dagger \mathbf{D}^{-1}\Theta \cdot \mathbf{u}\right]. \tag{11.91}$$

But we may use the identity

$$\mathbf{W}^{-1} = \Theta^\dagger \mathbf{D}^{-1}\Theta, \tag{11.92}$$

to arrive at our final expression

$$p(\mathbf{u}) = \frac{1}{\pi^N} \frac{1}{\text{Det}[\mathbf{W}]} \exp\left[-\mathbf{u}^\dagger \cdot \mathbf{W}^{-1} \cdot \mathbf{u}\right]. \tag{11.93}$$

One important result that arises for such a random process is the moment theorem, which can be readily demonstrated from the use of the characteristic function. We consider the moment

$$\overline{u_1^* u_2^* u_3 u_4} = 16 \frac{\partial}{\partial w_1} \frac{\partial}{\partial w_2} \frac{\partial}{\partial w_3^*} \frac{\partial}{\partial w_4} M(w)|_{w=0}. \tag{11.94}$$

Taking the first two derivatives, with respect to w_4^* and w_3^*, we get

$$\overline{u_1^* u_2^* u_3 u_4} = \frac{\partial}{\partial w_1} \frac{\partial}{\partial w_2} \left[W_{4j}w_j W_{3j}w_j M(w)\right]_{w=0}. \tag{11.95}$$

If either of the remaining two derivatives act on $M(w)$ directly, they will bring down additional powers of w_i^* and all these terms will vanish when we set $w = 0$. The final derivatives may only act on the terms outside of the exponent; we then have

$$\overline{u_1^* u_2^* u_3 u_4} = W_{42} W_{31} + W_{41} W_{32}. \tag{11.96}$$

We find that a fourth-order moment can be expressed in terms of products of second-order moments. For a Gaussian random process, it follows that all average properties are determined by the second-order moment alone.

11.8 EXERCISES

1. Extend the Gaussian moment theorem to evaluate the average $\overline{u_1^* u_2^* u_3^* u_4 u_5 u_6}$, by taking the appropriate derivatives of the characteristic function.

2. Prove that Equation 11.8 holds for the particular example of a Gaussian random process by comparing the results of the expression to the explicit moment calculation using Equation 11.2.

3. Prove Equation 11.20 by explicitly calculating the appropriate moments of the negative exponential distribution.

4. Derive Equation 11.67 for the density of vortices in the interference of a speckle field with a plane wave by repeating the steps of Section 11.3. Describe how the calculation changes if the plane wave has an arbitrary constant phase; does the result change?

CHAPTER 12

Unusual Singularities and Topological Tricks

I N THIS FINAL CHAPTER, we take a look at a number of special and unusual topics in singular optics, many delving further into the topology of light than we have so far considered.

12.1 BESSEL–GAUSS VORTEX BEAMS

Throughout this book, we have primarily focused on Laguerre–Gauss vortex beams, as discussed in detail in Chapter 2. However, other classes of vortex beams exist, and one of these classes, the Bessel–Gauss beams, have a number of additional interesting properties worth considering. In this section, we look at the derivation, generation, and behavior of Bessel–Gauss beams, and some of the mischief they have caused in the scientific literature.

Owing to the wave nature of light, we generally expect beams of light to spread on propagation, no matter how directional and localized they are to begin with. We may now ask, however, if this is always the case! We return to the Helmholtz equation 2.22 for a scalar monochromatic wavefield $U(\mathbf{r})$, which is given again below:

$$\nabla^2 U(\mathbf{r}) + k^2 U(\mathbf{r}) = 0. \tag{12.1}$$

We seek solutions of this wave equation whose cross sections are invariant along the direction of propagation z. We explore a trial solution of the form

$$U(\mathbf{r}) = u(x, y)e^{i\beta z}, \tag{12.2}$$

where β is a real, positive propagation constant and $u(x, y)$ is independent of z. On substitution into the Helmholtz equation, we find that $u(x, y)$ satisfies

$$\nabla_\perp^2 u(x, y) + \alpha^2 u(x, y) = 0, \tag{12.3}$$

with $\nabla_\perp^2$ the Laplacian in x and y and

$$\alpha^2 = k^2 - \beta^2. \tag{12.4}$$

We take $\beta^2 < k^2$ so that α is also a real-valued quantity. Equation 12.3 is, in essence, a two-dimensional Helmholtz equation, which we know will have nontrivial solutions. With an eye toward designing vortex beams, we now attempt a further separation of the field in polar coordinates in the form

$$u(x, y) = u(\rho, \phi) = R(\rho)e^{im\phi}, \tag{12.5}$$

with m an integer. On substitution of this expression into Equation 12.3, we find that $R(\rho)$ satisfies

$$\rho^2 \frac{\partial^2 R}{\partial \rho^2} + \rho \frac{\partial R}{\partial \rho} + (\alpha^2 \rho^2 - m^2)R = 0. \tag{12.6}$$

If we introduce $x \equiv \alpha \rho$, our differential equation reduces to the final form

$$x^2 \frac{\partial^2 R}{\partial x^2} + x \frac{\partial R}{\partial x} + (x^2 - m^2)R = 0. \tag{12.7}$$

This is Bessel's equation,[*] and the solutions are the Bessel functions $J_m(x)$, or

$$R(\rho) = J_m(\alpha \rho). \tag{12.8}$$

---

[*] See [Gbu11], Chapter 16, from which we will take all Bessel-related results.

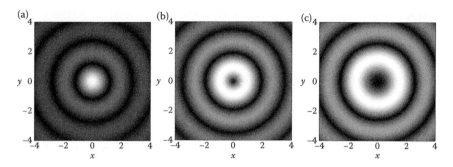

FIGURE 12.1 Amplitude cross section $|U(\mathbf{r})|$ of Bessel beams with (a) $m = 0$, (b) $m = 1$, and (c) $m = 2$. For all, $\alpha = 2\,\text{mm}^{-1}$.

(We immediately discard the Neumann function solutions $N_m(\alpha\rho)$, which diverge at $\rho = 0$ and are therefore unphysical.) We have therefore found a new set of solutions to the Helmholtz equation, given by[*]

$$U(\mathbf{r}) = U_0 J_{|m|}(\alpha\rho)e^{im\phi}e^{i\beta z}. \tag{12.9}$$

The most striking thing about these solutions, referred to as *Bessel beams*, is that their shape is unchanged on propagation. That is, the intensity of the field, $|U(\mathbf{r})|^2$, is independent of z. For this reason, such beams are also referred to as *nondiffracting beams*, though this latter term encompasses more than just Bessel types, as we will see.

The amplitude cross sections $|U(\mathbf{r})|$ of several Bessel beams are illustrated in Figure 12.1. For all beams, except the $m = 0$ beam, there is an intensity null in the beam's center; this is a screw dislocation, as evident from the $e^{im\phi}$ phase factor in Equation 12.9. As it is known that, for $u \ll 1$, $J_m(u) \sim u^m$, it can be seen that near the beam axis the field has the familiar local form $(x \pm iy)^m$.

The possibility of nondiffracting beams was first explored by Durnin [Dur87], and soon after followed up experimentally by Durnin, Miceli, and Eberly [DJE87]. Since that time, they have been investigated out of both practical and scientific interests; Bouchal [Bou03] has written a review of the subject.

We can readily determine the physics behind the nondiffracting nature of such beams. It is well known that a monochromatic wavefield may always be expressed as a superposition of plane waves and evanescent waves in

---

[*] We use the absolute value of *m* in this formula for convenience in describing both positive and negative vortex states.

what is known as an *angular spectrum representation* [Gbu11, Section 11.9] of the form

$$U(x, y, z) = \int A(\kappa_x, \kappa_y) e^{i(\kappa_x x + \kappa_y y + \kappa_z z)} d\kappa_x d\kappa_y, \qquad (12.10)$$

where $\kappa_z^2 = k^2 - \kappa_x^2 - \kappa_y^2$ and $A(\kappa_x, \kappa_y)$ is the *spectral amplitude* of the field in the plane $z = 0$. When $\kappa_x^2 + \kappa_y^2 > k^2$, κ_z is imaginary and the wave is exponentially decaying in the z-direction: it is evanescent. One physical result that arises from the identification of evanescent waves in Equation 12.10 is the observation that the evanescent waves decay away in free propagation; beyond a few wavelengths from a source, all waves are effectively bandlimited to the range $\kappa_x^2 + \kappa_y^2 \leq k^2$.

In polar coordinates such that $(\kappa_x, \kappa_y) = (\kappa, \phi_\kappa)$, Equation 12.10 may be written in the form

$$U(\rho, \phi, z) = \int A(\kappa, \phi_\kappa) e^{i\kappa\rho \cos(\phi - \phi_\kappa)} \kappa d\kappa d\phi_\kappa. \qquad (12.11)$$

We now consider the special case for which

$$A(\kappa, \phi_\kappa) = e^{im\phi_\kappa} \delta(\kappa - \alpha), \qquad (12.12)$$

where $\delta(\kappa)$ represents a one-dimensional Dirac delta function. This choice of spectral amplitude results in a total field that is a superposition of plane waves whose orientations lie on a cone of angle θ, where

$$\tan\theta = \frac{\alpha}{\beta}. \qquad (12.13)$$

Such a cone is illustrated in Figure 12.2.

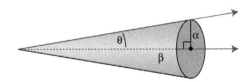

FIGURE 12.2 Cone of plane wave directions for a Bessel beam.

On substitution from Equations 12.12 into 12.11, we may integrate over κ to find

$$U(\rho, \phi, z) = \alpha e^{i\beta z} \int_0^{2\pi} e^{im\phi_\kappa} e^{i\alpha\rho \cos(\phi-\phi_\kappa)} d\phi_\kappa. \tag{12.14}$$

We now change the variable of integration to $\Phi \equiv \phi - \phi_\kappa$, which results in the field being of the form

$$U(\rho, \phi, z) = -\alpha e^{i\beta z} e^{im\phi} \int_0^{2\pi} e^{-im\Phi} e^{i\alpha\rho \cos(\Phi)} d\Phi. \tag{12.15}$$

The integral is a standard representation of the Bessel function of order m, namely,

$$J_m(x) = \frac{1}{2\pi i^m} \int_0^{2\pi} e^{-im\Phi} e^{ix \cos(\Phi)} d\Phi. \tag{12.16}$$

This tells us that the field is of the form

$$U(\rho, \phi, z) = -2\pi i^m \alpha e^{i\beta z} e^{im\phi} J_m(\alpha\rho), \tag{12.17}$$

which is, within a trivial constant, the set of nondiffracting fields of Equation 12.9.

Because each plane wave has the same propagation constant along the z-direction, they remain in phase on propagation and therefore the shape of the beam does not change. This observation conversely allows us to think generally of wave diffraction as resulting from a change in phase relations between the constituent plane waves of a beam.

This plane wave representation also immediately suggests a simple method to create Bessel beams. We imagine the system shown in Figure 12.3, in which a coherent plane wave illuminates an (ideally) infinitely thin ring in an aperture. This ring is placed in the front focal plane of a lens; the field $U_1(x, y)$ in the rear focal plane is then proportional to the Fourier transform of the field $U_0(x', y')$ in the front focal plane:

$$U_1(x, y) \sim \int U_0(x', y') \exp[-2\pi i(xx' + yy')/\lambda f] dx' dy'. \tag{12.18}$$

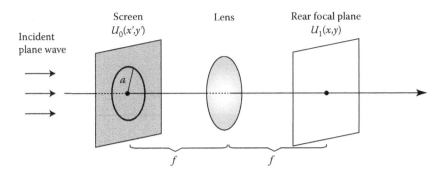

FIGURE 12.3 A simple experimental method to generate Bessel beams.

We may approximate the field in a thin ring aperture by a delta function, namely,

$$U_0(x', y') \approx \delta(\rho' - a). \tag{12.19}$$

On substitution, we readily find that the field in the rear focal plane is given by

$$U_1(x, y) \sim a \int_0^{2\pi} e^{-2\pi i a \rho \cos(\phi - \phi')/\lambda f} d\phi', \tag{12.20}$$

and from Equation 12.16, we find that the output field is proportional to $J_0(a\rho/\lambda f)$, and is a Bessel field with $\alpha = a/\lambda f$. If an azimuthal phase mask of order m is placed immediately before or after the ring aperture, the output field will instead be an mth-order Bessel beam.

The use of a thin ring aperture is highly inefficient, however, due to the loss of light. More efficient generation methods include the use of computer-generated holograms [VTF89] or axicons [AD00].

It is important to note that it is not possible to generate an ideal Bessel beam. This is already clear from Figure 12.3, as any real aperture will have a finite width and Equation 12.19 will only approximately apply. That it is not possible, even in principle, to create a perfect Bessel beam can be seen by considering the integrated intensity E in any cross section of the field, given by

$$E = \int |U(x, y)|^2 dx dy \sim 2\pi \int |J_m(\alpha\rho)|^2 \rho d\rho. \tag{12.21}$$

Asymptotically,

$$J_m(x) \sim \sqrt{\frac{2}{\pi x}} \cos[x - m\pi/2 - \pi/4] \text{ as } x \to \infty. \quad (12.22)$$

The integrand of Equation 12.21 therefore does not approach zero as $x \to \infty$ and the integral diverges: the energy of an ideal Bessel beam is infinite. In fact, the energy in each radial ring of a Bessel beam is comparable in size to the energy of the innermost peak or ring.

Bessel fields can therefore only be approximately generated, and will only possess their nondiffracting ability over a finite propagation distance. The most commonly considered realization is the Bessel–Gauss beam, which for lowest order has the form

$$U(\rho, z = 0) = U_0 J_0(\alpha\rho) \exp[-\rho^2/w_0^2]. \quad (12.23)$$

This represents an ideal Bessel field truncated by a Gaussian envelope. The Fresnel diffraction formula for such a beam can be evaluated analytically [GGP87], and the result is

$$U(\rho, z) = \frac{U_0 w_0}{w(z)} \exp[i(k - \alpha^2/2k)z - \Phi(z)] J_0[\alpha\rho/(1 + iz/z_0)]$$
$$\times \exp\{[-1/w^2(z) + ik/2R(z)](\rho^2 + \alpha^2 z^2/k^2)\}, \quad (12.24)$$

where $w(z)$, $\Phi(z)$, $R(z)$, and z_0 are defined as for a Gaussian beam in Section 2.2.

Under what conditions does such a beam fairly reproduce a nondiffracting Bessel beam? We first note that, like an ideal Bessel beam, a Bessel–Gauss beam may be considered a combination of constituent waves arranged on a cone of angle θ; in this case, however, the constituent waves are Gaussians of width w_0.

There are therefore two quantities that characterize the propagation of a Bessel-Gauss beam: the angular divergence θ of the Bessel function, and the diffractive spreading of a Gaussian beam, which can be shown to have an angular divergence $\theta_G = \lambda/\pi w_0$ far from the waist plane. The ratio of these two factors is approximately given by

$$\theta/\theta_G \approx \frac{\alpha w_0}{2}. \quad (12.25)$$

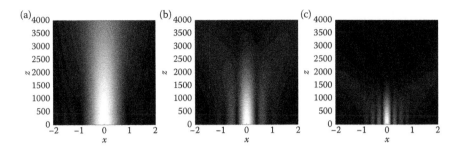

FIGURE 12.4 Amplitude of Bessel–Gauss beams on propagation for (a) $\theta = 2 \times 10^{-4}$, (b) $\theta = 5 \times 10^{-4}$, and (c) $\theta = 1 \times 10^{-3}$. Here, we have $w_0 = 1\,\text{mm}$, $\lambda = 500\,\text{nm}$, and $\theta_G = 1.6 \times 10^{-4}$. All distances on the plot are in mm.

When this factor is small, that is, $\theta/\theta_G < 1$, the spreading of the constituent Gaussians is much stronger than the Bessel divergence, and our field will look not much different than a Gaussian. In the case where $\theta/\theta_G > 1$, we expect Bessel effects to survive only while the constituent Gaussian beams continue to overlap. At a distance z, we expect the center of an individual Gaussian to have moved a distance l from the central axis, where $l = \theta/z$. A conservative estimate of the position at which "Besselness" begins to break down is the distance z at which the constituent Gaussian beams are a distance w_0 from the axis, or

$$z = w_0/\theta. \tag{12.26}$$

An illustration of these effects is shown in Figure 12.4. When θ is small, the beam appears primarily Gaussian; for larger θ, the beam appears more Bessel-like, although its effective range of diffraction-free propagation is inversely related to θ. This observation is in good agreement with Equation 12.26.

Looking at the figure, the reader may wonder at this point whether there is any advantage to a Bessel beam at all, as the leftmost Gaussian beam appears to maintain its shape and intensity over a longer distance than the Bessel fields with the same Gaussian width. It should be noted, however, that the Bessel–Gauss beams are effectively much narrower than the comparable Gaussian; the use of Bessel beams allows one to break the usual assumption that a narrower beam must necessarily spread much faster.

Bessel fields have caused quite a lot of mischief in recent years, most notably in claims of superluminal ("faster-than-light") propagation, even in vacuum [MRR00]. At a particular frequency, the field of a zeroth-order

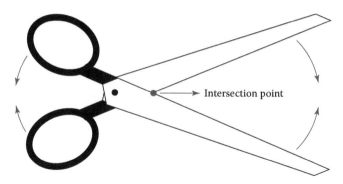

FIGURE 12.5 The intersection point of the blades of scissors.

Bessel beam has the form

$$U(\rho, z, t) = J_0(\rho k \sin \theta) \exp[i(kz \cos \theta - \omega t)], \qquad (12.27)$$

and this implies a wavenumber along the z-direction of the form $\kappa_z = k \cos \theta$, and a corresponding group velocity v_g of the form

$$v_g = \frac{1}{d\kappa_z/d\omega} = c/\cos\theta, \qquad (12.28)$$

which can apparently be made arbitrarily high!

The resolution of this seeming violation of Einstein's special relativity comes from special relativity itself. It has long been appreciated that the intersection point of a pair of ordinary scissors (illustrated in Figure 12.5) moves increasingly fast as the blades close (resulting in the characteristic "snip" sound they make) and that, with a large enough pair of scissors, and blades nearly parallel, it is possible to make the intersection point move arbitrarily fast. It would seem, at first glance, that one could use these scissors to send a message superluminally, moving the handles and cutting a distant string at the tip of the blades.

However, this assumes that the blades move instantaneously when the handles are squeezed. In reality, the force applied to the handle will be propagated, at the speed of light, along the blades until they are completely in motion. The superluminal motion of the intersection point will therefore always lag behind the speed-of-light motion of the blades themselves.

An analogous effect occurs in pulsed Bessel beams. Let us consider an extremely simple model of such a beam, consisting of only two angled sources of finite transverse extent, illustrated in Figure 12.6a. The Bessel

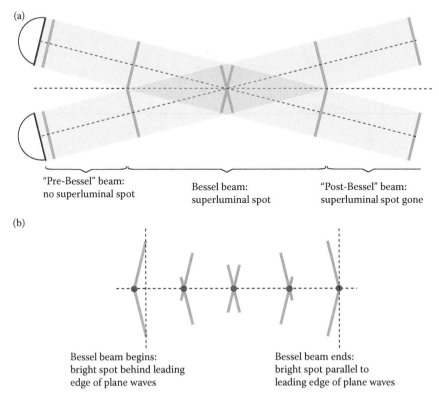

FIGURE 12.6 (a) Simple model of a pulsed Bessel beam. (b) Illustration showing how the superluminal bright spot lags behind the front of the wave.

pulse only forms in the region where the individual beams intersect, and the intersection of these creates a superluminal bright spot along the axis. In Figure 12.6b, we trace the motion of this bright spot. When the Bessel pulse begins, the bright spot is at the rear of the two beams. It then moves superluminally, and ends up at the front of the two beams only when the overall Bessel beam ends. Just as in the scissors case, the superluminal Bessel pulse is always lagging behind the leading edge of the signal, which moves at less than the speed of light. In fact, it is now acknowledged that pulsed Bessel beams actually travel *slower* than the vacuum speed of light in free space [GRP+15].

This noncausal propagation of a light spot in a Bessel beam also provides a simple explanation for one of the most intriguing properties of such beams: the ability to "heal" even when partially obstructed by an obstacle [BWC98,Bou02]. As illustrated in Figure 12.7, the angled plane waves

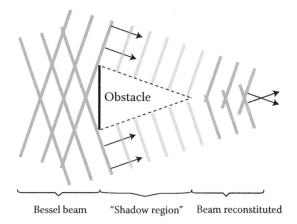

Bessel beam   "Shadow region"   Beam reconstituted

FIGURE 12.7   Self-healing of a Bessel beam on encountering an obstruction. The constituent plane waves of the Bessel beam (shown at large angles for clarity) are clipped by the obstruction, but the unobstructed parts continue on to reconstitute the beam.

of a Bessel beam never directly hit an obstruction along the line of propagation; rather, they are truncated. Ignoring diffraction edge effects for simplicity, we note that the constituent plane waves (or Gaussian beams) of the Bessel beam will then begin to interfere some distance downstream from the obstruction, and will appear just as the original field. Curiously, even ordinary Laguerre–Gauss vortex beams show some degree of self-reconstitution upon obstruction [VMS00], though the origin of this effect appears to be related to the conservation of angular momentum.

It is to be noted that nondiffracting partially coherent beams also exist [TVF91,BP02]. An example of such a beam is one that is Bessel-correlated, that is, $\mu(\mathbf{r}_1, \mathbf{r}_2, \omega) = J_0(\alpha|\mathbf{r}_2 - \mathbf{r}_1|)$. It can be synthesized in a simple manner by illuminating the experimental arrangement of Figure 12.3 with incoherent light rather than coherent light. Bessel-correlated fields have been shown to have interesting properties, such as a minimum of intensity at the geometric focus [GV03a].

Nondiffracting beams do not have to be necessarily of a simple Bessel form. Any field whose spectral amplitude has a delta ring of the form $A(\kappa, \phi_\kappa) = A(\phi_\kappa)\delta(\kappa - \alpha)$ will produce a nondiffracting solution. The function $A(\phi_\kappa)$ can be decomposed in Fourier series to express such a field as a superposition of fundamental Bessel beams.

One additional surprising class of beams are worth mentioning that are closely related to Bessel beams. In 1979, Berry and Balazs [BB79]

showed that it is possible to construct pulsed solutions of the free particle Schrödinger equation with a temporal profile in the form of an Airy
function that not only maintain their shape on propagation but also freely
accelerate. In 2007, taking advantage of the mathematical analogy between
the Schrödinger equation and the paraxial wave equation, it was demonstrated [SBDC07] that it is possible to make beams with an Airy function
cross section that transversely accelerate on propagation. The Airy function
$Ai(x)$ is defined by the integral

$$Ai(x) = \frac{1}{\pi} \int_0^\infty \cos(t^3/3 + xt)dt. \tag{12.29}$$

Assuming a paraxial wave whose form in the $z = 0$ plane is $U(x, 0) = Ai(x/x_0)$, with x_0 a length parameter, it has been shown that the solution
on propagation may be written as

$$U(\zeta, s) = Ai[s - (\zeta/2)^2] \exp[i(s\zeta/2) - i(\zeta^3/12)], \tag{12.30}$$

where $s \equiv x/x_0$ and $\zeta = z/kx_0^2$. An illustration of the propagation of an Airy
beam is shown in Figure 12.8.

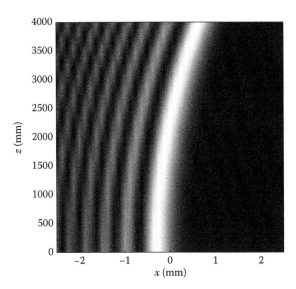

FIGURE 12.8 Propagation and acceleration of an Airy beam with $x_0 = 0.3$ mm
and $\lambda = 500$ nm.

As suggested earlier, Airy functions, and consequently Airy beams, may be expressed in terms of Bessel and modified Bessel functions, depending on the sign of the argument:

$$Ai(x) = \frac{1}{\pi}\sqrt{\frac{x}{3}}K_{1/3}\left(\frac{2}{3}x^{3/2}\right), \tag{12.31}$$

$$Ai(-x) = \sqrt{\frac{x}{9}}\left[J_{1/3}\left(\frac{2}{3}x^{3/2}\right) + J_{-1/3}\left(\frac{2}{3}x^{3/2}\right)\right], \tag{12.32}$$

with $x > 0$. Like their Bessel counterparts, perfect Airy beams are not realizable in practice because they possess infinite energy. Finite-energy Airy beams, truncated by an exponential function, have been shown to accelerate over finite distances [SC07].

At this point, our discussion has wandered well away from the singular optics of Bessel beams, so we end it here. Bessel beams should always be kept in mind, however, as an additional class of vortex-carrying beams.

12.2 RIEMANN–SILBERSTEIN VORTICES

One striking aspect of the vortices considered throughout this book is that they are not relativistically invariant—that is, a vortex that exists in one inertial frame of reference will not, in general, exist in another such frame. This can be readily understood simply by looking at the relativistic transformation equations that relate the electric field $\mathbf{E}(\mathbf{R})$ and magnetic field $\mathbf{B}(\mathbf{R})$ in one inertial frame with those in a frame moving with velocity $\mathbf{v}$, denoted by primes:

$$\mathbf{E}'_{\|} = \mathbf{E}_{\|}, \tag{12.33}$$

$$\mathbf{E}'_{\perp} = \gamma(\mathbf{E}_{\perp} + \mathbf{v} \times \mathbf{B}), \tag{12.34}$$

$$\mathbf{B}'_{\|} = \mathbf{B}_{\|}, \tag{12.35}$$

$$\mathbf{B}'_{\perp} = \gamma(\mathbf{B}_{\perp} - \frac{1}{c^2}\mathbf{v} \times \mathbf{E}), \tag{12.36}$$

where "$\|$" represents the component of the field parallel to the velocity $\mathbf{v}$ and "$\perp$" represents the perpendicular component, and $\gamma \equiv 1/\sqrt{1 - v^2/c^2}$. We have also used $\mathbf{R} \equiv (t, \mathbf{r})$ to represent the position four-vector of spacetime. These transformations suggest that a field that has a zero transverse electric field $\mathbf{E}_{\perp}$ in one reference frame will not necessarily have such a zero in another, due to the mixing of the electric and magnetic fields in the relativistic transformation formulas. (We noted in Section 7.3 that singularities of the electric and magnetic fields are not, generally, coincident.)

The nonrelativistic nature of vortices is not particularly problematic, as optical experiments rarely (if ever) involve multiple observers moving at relativistic speeds. It is interesting, however, to ask if there are any relativistically invariant line singularities of the electromagnetic field. Such singularities were introduced in 2003 by Bialynicki-Birula and Bialynicka-Birula [BBBB03], and have been dubbed *Riemann–Silberstein vortices* (RS vortices) after early work by Bernhard Riemann and Ludwik Silberstein on relativistically appropriate quantities.

We introduce the *Riemann–Silberstein vector* (RS vector), of the form

$$\mathbf{F}(\mathbf{r}, t) \equiv \mathbf{E}(\mathbf{r}, t) + ic\mathbf{B}(\mathbf{r}, t). \tag{12.37}$$

It is to be noted that this is a complex field quantity constructed from real-valued, time-dependent fields $\mathbf{E}(\mathbf{r}, t)$ and $\mathbf{B}(\mathbf{r}, t)$. We do not assume monochromaticity. This vector is significant because we can use it to replace the four real-valued, free-space Maxwell's equations given by Equations 2.7 to 2.10 with two complex-valued free-space equations for $\mathbf{F}$:

$$\nabla \cdot \mathbf{F} = 0, \tag{12.38}$$

$$\frac{i}{c}\frac{\partial \mathbf{F}}{\partial t} = \nabla \times \mathbf{F}. \tag{12.39}$$

No field information is lost in this change, as taking the real and imaginary parts of this pair of equations immediately reproduce Maxwell's equations.

The vector $\mathbf{F}$ was reintroduced in recent years as a quantity that might be interpreted as the quantum wave function of a photon [BB94]. Then, in analogy with vortices that appear in the Schrödinger equation [BBBBS00], it was quite natural to look for vortex properties of this presumed photon wave function.

The immediate difficulty is one we have encountered before: the complex vector $\mathbf{F}$ in principle has three phases, one associated with each vector component. We instead consider the complex square of this vector:

$$\mathbf{F}^2 = \mathbf{F} \cdot \mathbf{F} = (\mathbf{E}^2 - c^2\mathbf{B}^2) + 2ic\mathbf{E} \cdot \mathbf{B}, \tag{12.40}$$

which is itself a complex scalar. The quantity $\mathbf{F}^2$ will be zero when its real and imaginary parts are simultaneously zero; we have two constraints in three-dimensional space, and expect that the zero manifolds will be lines. However, our fields **E** and **B** are time dependent, so we also expect that these lines will generally be in motion; we elaborate on this momentarily.

A particularly striking aspect of the quantity $\mathbf{F}^2$ is that the real and imaginary parts, $S \equiv \mathbf{E}^2 - c^2\mathbf{B}^2$ and $P \equiv \mathbf{E} \cdot \mathbf{B}$, are relativistically invariant quantities. In other words, we may readily show that

$$[\mathbf{E}']^2 - c^2[\mathbf{B}']^2 = \mathbf{E}^2 - c^2\mathbf{B}^2, \quad \mathbf{E}' \cdot \mathbf{B}' = \mathbf{E} \cdot \mathbf{B}. \tag{12.41}$$

This may be proven by the use of Equations 12.33 to 12.36, which we leave as an exercise. The line singularities defined by $\mathbf{F}^2 = 0$ therefore manifest in every inertial frame of reference, though their location in space and time will vary in accordance with the Lorentz transformations.

Much like we did for ordinary optical vortices in Section 3.2, it is possible to find vectors $\mathbf{F}(\mathbf{r}, t)$ that are exact solutions of Maxwell's equations possessing RS vortices. One example from [BBBB03] is given by

$$\mathbf{F}(\mathbf{r}, t) = (y + t, a - i(z + a + t), x + it), \tag{12.42}$$

which can readily be shown to satisfy Equations 12.38 and 12.39, with a a real scale parameter. The quantities S and P may be found via Equation 12.40, with the result

$$S = x^2 + y^2 + 2yt + (z + a + t)^2 + a^2, \tag{12.43}$$
$$P = -2a(z + a + t) + 2xt. \tag{12.44}$$

We may solve the expressions $S = 0$, $P = 0$ as $x(z, t)$, $y(z, t)$, in the form

$$x(z, t) = a(a + z)/t - a, \tag{12.45}$$
$$y(z, t) = -t \pm \sqrt{t^2 - a^2}\sqrt{a^2 - 2at + 2t^2 + 2az - 2zt + z^2}/t. \tag{12.46}$$

We may now examine the distribution and evolution of the singularities by treating these equations as parametric functions of z. A plot of the result is illustrated in Figure 12.9. For $t < a$, no singularities exist. At $t = a$, a pair of vortex lines manifest as a single straight line, which then breaks into two vortices that separate away from one another. Because of the symmetry of Equations 12.46, two lines also exist for $t < -a$, and they come together and annihilate for $t = -a$.

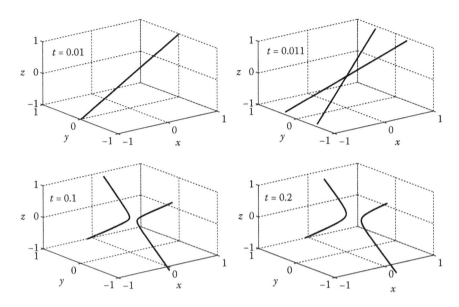

FIGURE 12.9 Creation and separation of RS vortices, for $a = 0.01$, and the RS vector as given in Equation 12.42.

We may readily verify that these singularities of the squared RS vector are vortices by plotting the argument of $\mathbf{F}^2$ in a plane of constant z. This is done in Figure 12.10, and it can be clearly seen that a pair of first-order vortices of opposite handedness have been created.

The previous example may be said to be only a local solution as, for a fixed time, the RS vector increases without bound in all directions. It is not difficult, however, to create a well-behaved vector containing singularities from a small number of plane waves. As one might expect from earlier discussions in this book (particularly Section 3.1, which discusses Young's three-pinhole experiment), one requires a minimum of three plane waves to see generic behavior. We again use an example from [BBBB03], in which a set of orthogonal unit vectors $\hat{\mathbf{u}}, \hat{\mathbf{v}}, \hat{\mathbf{w}}$ are directions of propagation of a trio of monochromatic plane waves. Each wave is taken to have right-hand circular polarization, so that

$$\mathbf{E}_1(\mathbf{r}, t) = \hat{\mathbf{u}} \cos(k\hat{\mathbf{w}} \cdot \mathbf{r} - \omega t) - \hat{\mathbf{v}} \sin(k\hat{\mathbf{w}} \cdot \mathbf{r} - \omega t), \tag{12.47}$$

$$\mathbf{E}_2(\mathbf{r}, t) = \hat{\mathbf{v}} \cos(k\hat{\mathbf{u}} \cdot \mathbf{r} - \omega t) - \hat{\mathbf{w}} \sin(k\hat{\mathbf{u}} \cdot \mathbf{r} - \omega t), \tag{12.48}$$

$$\mathbf{E}_3(\mathbf{r}, t) = \hat{\mathbf{w}} \cos(k\hat{\mathbf{v}} \cdot \mathbf{r} - \omega t) - \hat{\mathbf{u}} \sin(k\hat{\mathbf{v}} \cdot \mathbf{r} - \omega t). \tag{12.49}$$

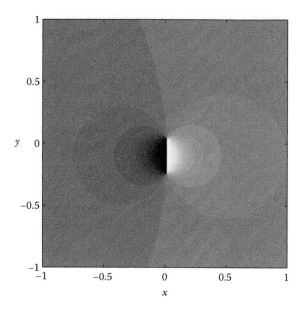

FIGURE 12.10   Phase of $\mathbf{F}^2$ for the RS vector given by Equation 12.42, with $z = 0$ and $t = 0.1$.

It is straightforward to construct the RS vectors of each of these fields, first by finding the magnetic field from Faraday's law and then combining the electric and magnetic fields as in Equation 12.37. We have

$$F_1(\mathbf{r}, t) = (\hat{\mathbf{u}} + i\hat{\mathbf{v}})e^{i(kw-\omega t)}, \tag{12.50}$$

$$F_2(\mathbf{r}, t) = (\hat{\mathbf{v}} + i\hat{\mathbf{w}})e^{i(ku-\omega t)}, \tag{12.51}$$

$$F_3(\mathbf{r}, t) = (\hat{\mathbf{w}} + i\hat{\mathbf{u}})e^{i(kv-\omega t)}. \tag{12.52}$$

On calculating the vector $\mathbf{F}^2$, we readily find that the time dependence $\exp[-2i\omega t]$ completely factors out; for this case, the singularities of the RS vector are stationary. The phase of $\mathbf{F}^2$ is illustrated in Figure 12.11. It can be seen that we have two shifted hexagonal arrays of vortices, one with positive charge, one with negative charge.

Though the singularities in the plane wave example are stationary, we generally expect RS singularities to evolve in time. For a monochromatic field of frequency ω, the quantity $\mathbf{F}^2$ will usually have a nontrivial oscillation at frequency 2ω; at optical frequencies, this motion is too fast to observe. However, as noted by Berry [Ber04c], zeros can still be found in

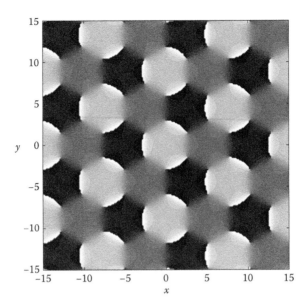

FIGURE 12.11 Phase of $\mathbf{F}^2$ in the interference of three orthogonal plane waves with electric fields given by Equations 12.49, for $t = 0$, $z = 0$.

the *cycle-averaged* RS vector $\langle \mathbf{F} \cdot \mathbf{F} \rangle$; defining the scalar field $\psi(\mathbf{r})$ as

$$\psi(\mathbf{r}) = 2\langle \mathbf{F} \cdot \mathbf{F} \rangle, \tag{12.53}$$

we can quickly find that

$$\psi(\mathbf{r}) = (\mathbf{E}^* + ic\mathbf{B}^*) \cdot (\mathbf{E} + ic\mathbf{B}), \tag{12.54}$$

where $\mathbf{E}$ and $\mathbf{B}$ are now the complex representations of the monochromatic fields. This complex quantity (note that the two vectors are not complex conjugates) will now possess stationary vortices.

It is interesting to note that there is a mathematical analogy between the definition of singularities of $\psi(\mathbf{r})$ and those of correlation functions in Chapter 9. In both cases, it is found that the "instantaneous" vortices in time are reflected, in some sense, in the average properties of the field.

Considering only paraxial fields, Berry demonstrated that the scalar field $\psi(\mathbf{r})$ satisfies the relation

$$\psi(\mathbf{r}) \approx \frac{1}{2k^2} \left[\left(\frac{\partial^2}{\partial x^2} - \frac{\partial^2}{\partial y^2} + 2i\frac{\partial^2}{\partial x \partial y} \right) \left(|E_x|^2 - |E_y|^2 - 2i\mathrm{Re}\{E_x^* E_y\} \right) \right]. \tag{12.55}$$

If we introduce the left- and right-circular polarization states, $E_L = (E_x + iE_y)/\sqrt{2}$, $E_R = (E_x - iE_y)/\sqrt{2}$ and the complex coordinates $\zeta = x + iy$, $\bar{\zeta} = x - iy$, it can be shown that the paraxial relation for $\psi(\mathbf{r})$ reduces to the form

$$\psi(\mathbf{r}) \approx \frac{4}{k^2} \frac{\partial^2}{\partial \bar{\zeta}^2} (E_L^* E_R). \tag{12.56}$$

This connection between the quantity $\psi(\mathbf{r})$ and the circular polarization states seems to suggest a close relationship between polarization singularities and the RS singularities. In fact, as noted by both Berry [Ber04c] and Kaiser [Kai04], the conditions for RS vortices and polarization singularities are structurally similar. The former type of singularities appear when $\mathbf{E}^2 - \mathbf{B}^2 = 0$ and $\mathbf{E} \cdot \mathbf{B} = 0$, while the latter appear when $\mathbf{p}^2 - \mathbf{q}^2 = 0$ and $\mathbf{p} \cdot \mathbf{q} = 0$, with $\mathbf{p}$ and $\mathbf{q}$ defined in Equation 7.6. There is not, however, a direct link between the classes, as the presence of the second derivative in Equation 12.56 indicates that $\psi(\mathbf{r})$ can be nonzero even when one (or both) of E_L and E_R are zero at a point.

Interesting connections nevertheless exist, as noted by Kaiser [Kai04]. To see this, we first decompose the RS vector in frequency using a Fourier expansion, that is,

$$\mathbf{F}(\mathbf{r}, t) = \int_{-\infty}^{\infty} e^{-i\omega t} \mathbf{F}_\omega(\mathbf{r}) d\omega, \tag{12.57}$$

where $\mathbf{F}_\omega(\mathbf{r})$ is the RS vector at frequency ω. We may, of course, break this into positive and negative frequency terms:

$$\mathbf{F}(\mathbf{r}, t) = \mathbf{F}_+(\mathbf{r}, t) + \mathbf{F}_-(\mathbf{r}, t), \tag{12.58}$$

where

$$\mathbf{F}_\pm(\mathbf{r}, t) = \int_0^\infty e^{\mp i\omega t} \mathbf{F}_{\pm\omega}(\mathbf{r}) d\omega \tag{12.59}$$

represent those parts of the vector due to the positive and negative frequencies alone.

These components are not related, however, because the vector $\mathbf{F}(\mathbf{r}, t)$ is complex. The quantities $\mathbf{F}_+(\mathbf{r}, t)$ and $\mathbf{F}_-(\mathbf{r}, t)$ are in fact independent solutions of Maxwell's equations:

$$\nabla \cdot \mathbf{F}_\pm = 0, \quad \nabla \times \mathbf{F}_\pm = \frac{i}{c}\frac{\partial \mathbf{F}_\pm}{\partial t}, \tag{12.60}$$

and may be considered separately.

If we consider only monochromatic fields, that is,

$$\mathbf{F}(\mathbf{r}, t) = e^{-i\omega t}\mathbf{F}_\omega(\mathbf{r}) + e^{i\omega t}\mathbf{F}_{-\omega}(\mathbf{r}), \tag{12.61}$$

we further find that $\mathbf{F}_{\pm\omega}$ are independent solutions to Maxwell's equations, as well, and Maxwell's equations may be reduced to a single expression for positive or negative ω:

$$\nabla \times \mathbf{F}_\omega = \frac{\omega}{c}\mathbf{F}_\omega. \tag{12.62}$$

Let us explicitly consider a plane wave solution, where

$$\mathbf{E}_\omega = \mathbf{E}_\omega^0 e^{i\mathbf{k}\cdot\mathbf{r}}, \quad \mathbf{B}_\omega = \mathbf{B}_\omega^0 e^{i\mathbf{k}\cdot\mathbf{r}}, \tag{12.63}$$

and ω is allowed to be positive or negative. The monochromatic form of Maxwell's equations then becomes

$$\mathbf{k} \times \mathbf{E}_\omega^0 = \omega\mathbf{B}_\omega^0, \tag{12.64}$$

$$-c\mathbf{k} \times \mathbf{B}_\omega^0 = \frac{\omega}{c}\mathbf{E}_\omega^0. \tag{12.65}$$

For $\omega > 0$, we find that the triplet of vectors $(\mathbf{E}_0, \mathbf{B}_0, \mathbf{k})$ form a right-handed system; for $\omega < 0$, they form a left-handed system. Explicitly, if we set $\mathbf{k} = k\hat{\mathbf{z}}$ and $\mathbf{E}_\omega^0 = \hat{\mathbf{x}}$ for simplicity, we have $\mathbf{B}_\omega^0 = \mathrm{sign}(\omega)\hat{\mathbf{y}}$; our RS vector has the form

$$\mathbf{F}_\omega = \left[\hat{\mathbf{x}} + i\,\mathrm{sign}(\omega)\hat{\mathbf{y}}\right]e^{ikz-\omega t}. \tag{12.66}$$

This is mathematically analogous to a circularly polarized plane wave, where the handedness of the vector $\mathbf{F}_\omega$ is determined by the sign of the frequency ω. Kaiser has referred to this circulation of a monochromatic plane wave as the *helicity* of the wave, and has suggested that the general temporal instability of RS vortices is due to a "tug of war" between the positive and negative circulations of the positive and negative frequency components. This is at least consistent with our three plane wave example of Equations 12.52, which were constructed in such a way that the RS vectors all have a single positive frequency component. The singularities in that case are, as we have seen, stationary.

12.3 FRACTIONAL CHARGE VORTEX BEAMS

In Section 4.1, it was noted that one of the most straightforward methods used to produce vortex beams is a spiral phase plate, which is a ramped optical element designed to impart a phase exp[$im\phi$] on a transmitted beam. This gives the beam a net topological charge m, which should be an integer due to continuity requirements of the wavefield.

However, a phase plate can be designed with any size ramp, and therefore can impart any helical phase, with a transmission function

$$t(\phi) = \exp[i\alpha\phi], \tag{12.67}$$

where α may be integer or fractional. As such beams cannot carry a fractional topological charge, it is of interest to see what happens when a "fractional beam" is generated, and how such beams evolve during a continuous change in α.

Before considering a theoretical model, it is worthwhile to consider what we might expect for a fractional vortex beam. For a value of α between two integer values, it is reasonable to expect that the total topological charge must take on one value or another, and that there is some critical value at which the total charge "jumps" discontinuously. This jump must somehow arise from the creation/annihilation of vortices. Furthermore, it should be noted that the $\phi = 0$ line in the source plane is now a line where the field is discontinuous. We expect strong diffraction effects to appear along this line as a fractional vortex beam propagates.

The first investigation of a fractional vortex was done by Vasnetsov, Basistiy, and Soskin [VBS98], who considered an effective charge $\alpha = 1/2$ beam both theoretically and experimentally. They observed the creation of vortex pairs along the $\phi = 0$ line, but could not investigate changes in α.

For a theoretical model, we follow the work of Berry [Ber04b], who examined in detail the behavior of a fractional vortex field for varying α. For simplicity, he considered a plane wave, propagating in the positive z-direction, incident upon an infinite fractional spiral phase plate in the plane $z = 0$ with transmission function given by Equation 12.67. To determine the field for $z > 0$, let us use again the angular spectrum representation of Section 12.1, which indicates that the field for positive z is given by

$$U(x, y, z) = \int A(\kappa_x, \kappa_y) e^{i(\kappa_x x + \kappa_y y + \kappa_z z)} d\kappa_x d\kappa_y, \tag{12.68}$$

where

$$\kappa_z = \sqrt{k^2 - \kappa_x^2 - \kappa_y^2}, \tag{12.69}$$

and k is again the free-space wavenumber. It is to be noted that κ_z is imaginary when $\kappa_x^2 + \kappa_y^2 > k^2$, which implies a field exponentially decaying in the z-direction: an evanescent wave. The angular spectrum of the field can be readily determined by setting $z = 0$ in Equation 12.68, which results in a Fourier relationship between the field in the source plane $U_0(x, y)$ and the angular spectrum, that is,

$$U_0(x, y) = \int A(\kappa_x, \kappa_y) e^{i(\kappa_x x + \kappa_y y)} d\kappa_x d\kappa_y, \tag{12.70}$$

or, inverted,

$$A(\kappa_x, \kappa_y) = \frac{1}{(2\pi)^2} \int U_0(x, y) e^{-i(\kappa_x x + \kappa_y y)} dx dy. \tag{12.71}$$

This expression illustrates that the properties of a wave for all values of $z > 0$ can be found from its properties in the source plane.

Let us first consider the behavior of a phase plate of arbitrary integer order n. In this case, the angular spectrum $A_n(\kappa_x, \kappa_y)$ may be written in polar coordinates in the form

$$A_n(\kappa, \phi_\kappa) = \frac{1}{(2\pi)^2} \int_0^\infty \int_0^{2\pi} \rho e^{-i\kappa\rho\cos(\phi - \phi_\kappa)} e^{in\phi} d\rho d\phi, \tag{12.72}$$

where $\kappa = \sqrt{\kappa_x^2 + \kappa_y^2}$ and ϕ_κ is the azimuthal angle of the vector κ. Introducing a new variable $\phi' = \phi - \phi_\kappa$, we may write

$$A_n(\kappa, \phi_\kappa) = \frac{1}{2\pi} e^{in\phi_\kappa} \int_0^\infty \rho \frac{1}{2\pi} \int_0^{2\pi} e^{-i\kappa\rho\cos\phi'} e^{in\phi'} d\phi'. \tag{12.73}$$

From Equation 12.16, we may immediately write

$$A_n(\kappa, \phi_\kappa) = \frac{i^{|n|}}{2\pi} e^{in\phi_\kappa} \int_0^\infty J_{|n|}(\kappa\rho) \rho d\rho. \tag{12.74}$$

The integral over ρ can be found from standard integral tables [GR07, Section 6.561, Equation 14], and the final result is

$$A_n(\kappa_x, \kappa_y) = \frac{|n|i^{|n|}}{2\pi\kappa^2}e^{in\phi_\kappa}. \tag{12.75}$$

To propagate the field, we substitute from Equation 12.75 into 12.68. However, the resulting integrals are not in general evaluable. To simplify things, we neglect the evanescent waves and consider only paraxial components (propagating near z-axis) of the field. In this approximation, we may write

$$\sqrt{k^2 - \kappa^2} \approx k - \frac{\kappa^2}{2k}, \tag{12.76}$$

so that the field $U_n(x, y, z)$ of an nth-order vortex mask is given by

$$U_n(\rho, \phi, z) = e^{ikz}\int A_n(\kappa_x, \kappa_y)e^{i\kappa\rho\cos(\phi_\kappa-\phi)}e^{-i\kappa^2 z/2k}\kappa d\kappa d\phi_\kappa, \tag{12.77}$$

where we have already written the expression in polar coordinates. We now substitute our angular spectrum, Equation 12.75, into this expression; the integral over ϕ_κ becomes another Bessel function and we have

$$U_n(\rho, \phi, z) = e^{ikz}e^{in\phi}|n|\int_0^\infty \frac{J_{|n|}(\kappa\rho)}{\kappa}e^{-i\kappa^2 z/2k}d\kappa. \tag{12.78}$$

Here, we must be a little clever. Let us first define $b \equiv z/2k$ for brevity. We then rewrite our integral in the form

$$U_n(\rho, \phi, z) = \frac{|n|\rho e^{ikz}e^{in\phi}}{2}\int_0^\infty \frac{2J_{|n|}(\rho\kappa)}{\rho\kappa}e^{-ib\kappa^2}d\kappa. \tag{12.79}$$

We now use the standard Bessel identity

$$\frac{2nJ_n(x)}{x} = J_{n-1}(x) + J_{n+1}(x), \tag{12.80}$$

and may write

$$U_n(\rho, \phi, z) = \frac{\rho e^{ikz}e^{in\phi}}{2}\int_0^\infty \left[J_{|n|-1}(\rho\kappa) + J_{|n|+1}(\rho\kappa)\right]e^{-ib\kappa^2}d\kappa. \tag{12.81}$$

At this point, we must turn to the integral tables once more. From [GR07, Section 6.618, Equation 1], we have

$$\int_0^\infty e^{-\alpha^2 x^2} J_\nu(\beta x)\,dx = \frac{1}{2}\sqrt{\frac{\pi}{\alpha}}e^{-\beta^2/8\alpha}I_{\nu/2}(\beta^2/8\alpha),\tag{12.82}$$

with $\mathrm{Re}(\alpha) > 0$, $\beta > 0$ and $\mathrm{Re}(\nu) > -1$, and $I_\mu(x)$ is a modified Bessel function. Each of the integrals of Equation 12.81 may be evaluated in this form with imaginary α. Further applying the relation,

$$I_m(-ix) = (-i)^m J_m(x),\tag{12.83}$$

we may after some manipulation arrive at the result

$$U_n(\rho,\phi,z) = e^{ikz}e^{in\phi}e^{ik\rho^2/4z}\sqrt{\frac{\pi}{2}}(-i)^{|n|/2}\sqrt{\frac{k\rho^2}{4z}}$$

$$\times \left[J_{\frac{1}{2}(|n|-1)}\left(\frac{k\rho^2}{4z}\right) - iJ_{\frac{1}{2}(|n|+1)}\left(\frac{k\rho^2}{4z}\right)\right].\tag{12.84}$$

This expression represents the field created by a vortex mask with integer charge. To find the field for a fractionally charged mask, we next use the Fourier series representation of the phase:

$$e^{i\alpha\phi} = \frac{e^{i\pi\alpha}\sin(\pi\alpha)}{\pi}\sum_{n=-\infty}^{\infty}\frac{e^{in\phi}}{\alpha - n}.\tag{12.85}$$

On substitution from this into the angular spectrum formula and subsequent formula for the propagated field $U(\mathbf{r})$, we readily find that the field due to a fractional vortex mask is given by

$$U(\mathbf{r}) = \frac{e^{i\pi\alpha}\sin(\pi\alpha)}{\pi}\sum_{n=-\infty}^{\infty}\frac{U_n(\mathbf{r})}{\alpha - n}.\tag{12.86}$$

We now have a model that we may use to study fractional vortex beams. Noting that the field depends entirely on the scaled coordinates,

$$\xi = \sqrt{\frac{k}{2z}}x, \quad \eta = \sqrt{\frac{k}{2z}}y,\tag{12.87}$$

we will plot the phase of the field in a transverse plane in these coordinates.

The evolution of the phase is illustrated in Figure 12.12 for a transition between vortex charge $\alpha = 3$ and $\alpha = 4$. For $\alpha = 3$, we expect to have a pure charge-3 vortex field. As the charge increases, the third-order vortex breaks into three charge-1 vortices, as in Figure 12.12a. As α approaches

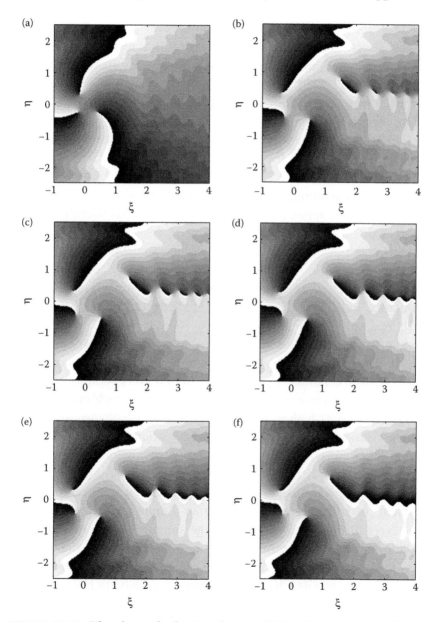

FIGURE 12.12 The phase of a fractional vortex field, with (a) $\alpha = 3.1$, (b) $\alpha = 3.46$, (c) $\alpha = 3.5$, (d) $\alpha = 3.54$, (e) $\alpha = 3.55$, and (f) $\alpha = 3.6$.

3.5, we find that pairs of oppositely charged vortices begin to form from the origin outward, roughly coinciding with the $\phi = 0$ line discontinuity, as seen in Figure 12.12b. It can be shown that, for $\alpha = 3.5$, there is an infinite line of singularities along this line; the plot near the origin is shown in Figure 12.12c. Finally, as α increases further, pairs of vortices annihilate from the outside in, eventually leaving an unpaired positive charge that coalesces into the charge-4 beam when $\alpha = 4$. To put it in a slightly different way: vortices are formed in "$(+, -)$" pairs as $\alpha \to 3.5$, and then annihilate in "$(-, +)$" pairs as $\alpha \to 4$.

How do we interpret this creation of an unbalanced charge in a system that is, effectively, undergoing a small perturbation? We have previously noted in Section 3.3 that unbalanced singularities may be created at the boundary of a system; in the case of a crystal, the boundary is the edge of the crystal, while in the case of a light beam, the boundary is at the transverse point at infinity. In Section 9.2, we observed an unbalanced correlation singularity appearing from infinity as the spatial coherence of light is decreased.

In this case, however, it appears that we have something very different, even quite remarkable, occurring: the unbalanced charge appears through the peculiar arithmetic of transfinite, that is, infinite, numbers (as discussed, for instance, in [Bre06]). The evolution of Figure 12.12 is, in fact, a near-perfect realization of the famous example [Gam88, page 17] of "Hilbert's Hotel," often used to illustrate the weirdness of infinity. We imagine a hotel consisting of a countably infinite number of rooms, each labeled with one of the natural numbers: $0, 1, 2, 3, \ldots$. Even if every room is occupied, we can always accommodate a new guest by asking each current guest to move to the next-highest room; Hilbert's Hotel simultaneously has no vacancies and an arbitrarily large number of vacancies! In our fractional vortex field, we may associate positive charges with "rooms" and negative charges with "guests." As α approaches a half-integer value, pairs of positive and negative charges are created: each guest steps out of his or her room. At half-integer α, there are an infinite number of pairs. As α increases further, each negative charge annihilates with its rightmost neighbor: the guests enter the next-highest room, leaving in the end one empty room, that is, positive charge. This analogy and its implications were explored further by Gbur [Gbu16]; a generalized Hilbert's Hotel example serves as the cover image for this book. A quite different, but still vortex-based, optical realization of Hilbert's Hotel was undertaken by Potoček et al. [PMM+15].

In the same paper from which the preceding results are derived, Berry showed numerically [Ber04b] that the total topological charge of the field jumps discontinuously when the phase mask has half-integer values. We may concisely write the topological charge t_α of a beam produced by a fractional mask as

$$t_\alpha = \text{int}(\alpha + 1/2), \tag{12.88}$$

where int(x) represents the closest integer to x, rounded down. The jump therefore occurs when the number of vortices in the field is infinite and the topological charge is ambiguous.

Experimental work by Leach, Yao, and Padgett [LYP04] has confirmed the above predictions; the researchers used a spatial light modulator to generate the fractional vortex masks. At about the same time, Tao, Lee, and Yuan [TLY04] generated a fractional Bessel vortex beam holographically, seeing similar effects. A half-integer spiral phase plate for optical frequencies has also been fabricated by Oemrawsingh et al. [OEW+04].

Fractional beams still carry orbital angular momentum, and they can be used to trap and rotate particles, as demonstrated by Tao et al. [TYL+05]. The dark line of vortices along the $\phi = 0$ line can hinder the rotation, but can also be used to guide and transport particles.

12.4 KNOTS, BRAIDS, AND LINKED VORTICES

We have seen that optical phase singularities typically manifest as lines in three-dimensional space, around which the phase has a circulating or helical behavior. But lines, and collections of lines, can themselves take on nontrivial structures, in the form of knots, links, and braids. In all our examples so far, vortex lines have been quite simple, usually either closed loops or lines that stretch to infinity; it is natural to ask whether vortex lines of light can form more complicated behaviors. In this section, we show that they can.

The first strategy to do so was introduced by Berry and Dennis [BD01a]. We imagine constructing a field that possesses an axial vortex line of charge m, which threads a vortex loop of charge n; this is illustrated in Figure 12.13. Such an arrangement of nongeneric vortices is unstable, and when perturbed by another field we expect the higher-order vortex loop to decompose into a collection of first-order vortices of some sort. Because of the twist in the central dislocation, we expect that it will be disrupted in a very asymmetric manner, possibly leading to a knotted singularity.

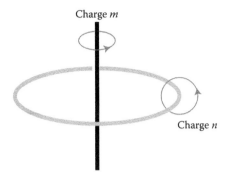

FIGURE 12.13 The unstable vortex arrangement used to create knots and links. The smaller circles indicate the sense in which the topological charge is defined.

To construct such a vortex arrangement, we use a sum of higher-order Bessel beams, of the type described in Section 12.1. To create an axial vortex of charge m, we require that all components be Bessel beams of order m, but with different effective widths b_j. Letting $k = 1$ for simplicity, we may write

$$U_{m,j}(\rho, \phi, z) = J_m(b_j\rho)e^{im\phi}e^{iz\sqrt{1-b_j^2}}. \tag{12.89}$$

Locally, the field at a loop dislocation of order n must vanish, and all the derivatives perpendicular to the loop up to order $n - 1$ must vanish as well, and the derivatives of order n must *not* vanish. We therefore have the set of $n(n + 1)/2$ conditions:

$$\partial_\rho^q \partial_z^{p-q} U(\rho_0, \phi, 0) = 0, \quad 0 \le q \le p, \quad 0 \le p \le n - 1, \tag{12.90}$$

$$\partial_\rho^q \partial_z^{n-q} U(\rho_0, \phi, 0) \ne 0, \quad 0 \le q \le n. \tag{12.91}$$

We will need $n(n + 1)/2$ Bessel fields in superposition in order to match these conditions, so we try a field of the form

$$U(\rho, \phi, z) = \sum_{j=1}^{n(n+1)/2} a_j U_{m,j}(\rho, \phi, z). \tag{12.92}$$

We may choose the set of b_j with some freedom, as well as the value of ρ_0; the values of a_j must then be found numerically. After Berry and Dennis [BD01a], we use $b_1 = 1$, $b_2 = 1/3$, $b_3 = 2/3$. Then with $m = 3, n = 2$, we have

$$a_1 = 1, \quad a_2 = 10.0302, \quad a_3 = -3.18960, \tag{12.93}$$

and with $m = 2, n = 2$, we have

$$a_1 = 1, \quad a_2 = 4.73341, \quad a_3 = -2.70176. \tag{12.94}$$

These unstable arrangements are then perturbed by adding a zeroth-order Bessel beam, of the form

$$U_0(\rho, \phi, z) = \epsilon J_0(\rho/4)e^{i\sqrt{15}z/4}, \tag{12.95}$$

where ϵ is the perturbation parameter.

What sort of evolution do we expect under this perturbation? The phase in a path around the loop, as in Figure 12.13, increases by $2\pi n$, which is another way of saying that there are n full phase cycles along the path. It is helpful to think of a cross-section of the vortex loop as a ratcheted gear with n teeth. Because of the axial singularity of order m, though, the phase along any closed path parallel to the loop must increase by $2\pi m$. Each 2π factor change moving parallel to the loop is equivalent to the phase pattern around the loop rotating by $2\pi/n$, or the gear ratcheting by the same amount. In passing around the axial singularity, the gear ratchets by a total $2\pi m/n$.

Now we imagine perturbing the field. If m and n are coprime—that is, they have no common divisor except unity—then it is not possible for the loop to break into n first-order loops, because each loop would have to have the same $2\pi m/n$ ratcheting after traversing the axial singularity, but with a gear with a single tooth, leaving it in a different position after its full circuit. We therefore expect one continuous line, looping n times around the singularity. It can be shown that this single line must twist m times around the original loop, which leads to a torus knot of order (m, n).

A simulation of the case $m = 3, n = 2$ is shown in Figure 12.14. The decomposition of the second-order vortex loop into a simple knot is clearly seen; the axial third-order vortex line breaks into three first-order helical lines.

For the case where m and n are not coprime, but share a common multiple such that $(m, n) = (Nm_0, Nn_0)$, with m_0 and n_0 coprime, it can be shown that the loop evolves into N linked loops, each of which consists of an (m_0, n_0) knot. The simplest case for $m = 2, n = 2$ is shown in Figure 12.15.

It was initially believed that such vortex knots could not be formed with paraxial wavefields, as it can be shown that a field satisfying the paraxial wave equation cannot form a dislocation loop higher than order 1. However, it was soon demonstrated that paraxial waves can form knots, by a method similar to that described above, that manages to sidestep

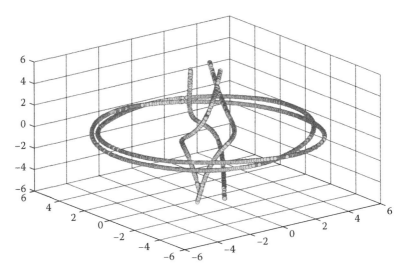

FIGURE 12.14 A $(3, 2)$ torus knot, threaded by three helical vortex lines, formed by a perturbation $\epsilon = 0.02$. Zero points were found by a zero-crossing algorithm, with a few missed points filled in manually.

the prohibition by using additional creation events [BD01b]. This allows the use of standard Laguerre–Gauss beams for knot creation instead of Bessel beams, the latter of which we have seen can only be approximately generated in the laboratory.

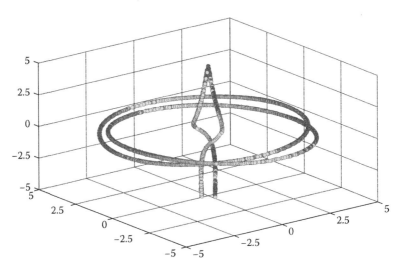

FIGURE 12.15 A $(2, 2)$ configuration resulting in a pair of linked loops, threaded by two helical vortex lines, formed by a perturbation $\epsilon = 0.02$. Zero points were found by a zero-crossing algorithm, with a few missed points filled in manually.

Following Berry and Dennis [BD01b], we introduce a set of Laguerre–Gauss beams with a special normalization and with $k = 1$, of the form

$$\psi_{nm}(x, y, z) = \left[\frac{w(-z)}{w(z)}\right]^n \frac{(x + iy)^m}{w(z)^{m+1}} \exp\left[-\rho^2/2w(z)\right] L_n^m(\rho^2/|w(z)|^2),$$

$$(12.96)$$

where $w(z) = 1 + iz$ and $\rho = \sqrt{x^2 + y^2}$. For simplicity, we have assumed that $m \geq 0$.

The following combination results in a knot-creation event:

$$\psi(x, y, z) = \psi_{03}(x, y, z) - \frac{8}{13}\psi_{13}(x, y, z) + \frac{2}{13}\psi_{23}(x, y, z) + \epsilon,$$

where ϵ represents a perturbative plane wave propagating along the z-direction. An overall $\exp[ikz]$ dependence of all paraxial waves is suppressed.

The evolution of the system as ϵ is increased is shown in Figure 12.16. Instead of the strict set of boundary conditions chosen for the Bessel fields, here the fields were constructed by requiring the field and its first derivative to vanish at a particular radius in the waist plane. This results in three zero rings, two of charge 1 above and below the waist, and one nongeneric, charge-0 ring in the plane, as can be seen in Figure 12.16a. As the perturbation is increased, the charge-0 ring breaks into a trio of crescent-shaped rings, as seen in Figure 12.16b. Finally, when a critical perturbation level is reached, $\epsilon = 0.000972409$, the different rings reconnect into the trefoil knot, as in Figure 12.16c.

Such an evolution was demonstrated experimentally by Leach et al. [LDCP05]. A spatial light modulator was used to synthesize the appropriate combination of beam phases and intensities. The image at different cross sections in the knot region was measured with a CCD camera, and the phase was measured by interfering the knot field with a plane wave.

Knots and links can also be generated without a central threading axial vortex. Using algebraic topology, Dennis et al. [DKJ+10] devised methods for constructing such isolated vortex knots, and confirmed the results experimentally.

Other nontrivially twisted vortex structures can be imagined. Again using superpositions of Bessel beams, Dennis [Den03] showed that it is possible to braid vortex lines. In this case, a pair of transversely shifted, counterpropagating Bessel beams were interfered with an orthogonal plane

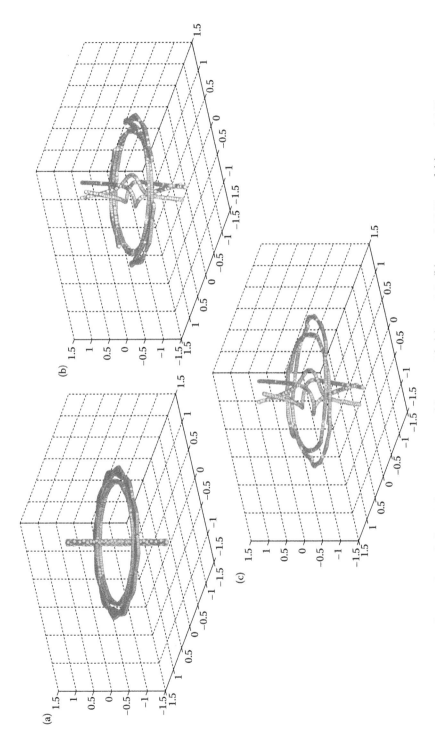

FIGURE 12.16 Simulation of a trefoil (3, 2) knot in a paraxial beam, with (a) $\epsilon = 0.0$, (b) $\epsilon = 0.0008$, and (c) $\epsilon = 0.002$.

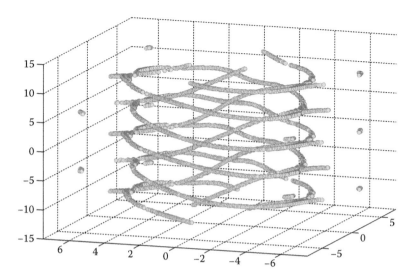

FIGURE 12.17 Simulation of a Bessel braid, using the parameters of Dennis [Den03].

wave. The result is what is known as the Borromean braid; the optical result is simulated in Figure 12.17.

For those interested in learning more about the mathematics of knots, many books are available; see, for instance, Crowell and Fox [CF63].

12.5 CASCADES OF SINGULARITIES

Though we have seen many different types of singularities throughout this book, we have also seen time and again that these singularities are closely related to one another. Decreasing the spatial coherence of an optical vortex beam, for instance, results in the embedded vortex evolving into a correlation singularity, as seen in Section 9.2. Also, a polarization singularity arises when an electromagnetic wave has a phase singularity in one component of a circular polarization basis, as noted in Section 7.3. With this in mind, it is of some interest to consider a single optical system in which different types of singularities evolve into each other in a continuous "cascade."

This was first done by Visser and Schoonover [VS08], who considered a cascade of singularities in Young's two-pinhole experiment. Though the singularities in this case are not generic, as was noted in Section 3.1, this example nevertheless illustrates how multiple classes of singularities can be observed and transformed in a single system.

As Young's interference pattern is essentially two-dimensional over the paraxial regime, we consider an explicit two-dimensional (double-slit)

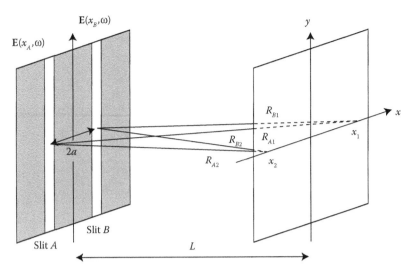

FIGURE 12.18   Illustration of the geometry for demonstrating a cascade of singularities in Young's experiment.

configuration as illustrated in Figure 12.18; the system is invariant along the y-axis and we only need consider displacements along x. The fields incident upon slits A and B are expressed as

$$\mathbf{E}(x_A, \omega) = U(x_A, \omega)\hat{\mathbf{x}}, \tag{12.97}$$

$$\mathbf{E}(x_B, \omega) = U(x_B, \omega)\left[\hat{\mathbf{x}}\cos\alpha + \hat{\mathbf{y}}\sin\alpha\right], \tag{12.98}$$

where $U(x, \omega)$ represents a partially coherent scalar wavefield. We take the cross-spectral density between the two slits to be of the form

$$\langle U^*(x_A, \omega)U(x_B, \omega)\rangle_\omega = S_0\mu_0, \tag{12.99}$$

and the spectral density at each slit is given by

$$S(x_A, \omega) = S(x_B, \omega) = S_0. \tag{12.100}$$

We assume that the degree of coherence μ_0 is real-valued and tunable, as is the angle of linear polarization α of the second pinhole.

Let us begin with a system for which the field is partially coherent, that is, $\mu_0 < 1$, and the polarization is the same at the two pinholes, that is, $\alpha = 0$. As discussed in Section 9.1, we do not expect to see any phase singularities, but we do expect correlation singularities. The cross-spectral density between pairs of points on the observation screen may be determined from the space–frequency representation of the field, using

$$W(x_1, x_2, \omega) = \left\langle \left[E(x_A, \omega) \frac{e^{ikR_{A1}}}{R_{A1}} + E(x_B, \omega) \frac{e^{ikR_{B1}}}{R_{B1}} \right]^* \right.$$

$$\times \left[E(x_A, \omega) \frac{e^{ikR_{A2}}}{R_{A2}} + E(x_B, \omega) \frac{e^{ikR_{B2}}}{R_{B2}} \right] \right\rangle, \quad (12.101)$$

where $R_{A1} = \sqrt{|x_A - x_1|^2}$ is the distance from the Ath pinhole to the observation point 1, and so forth. Evaluating the ensemble averages, we get the following expression:

$$W(x_1, x_2, \omega) = S_0 \frac{e^{ik(R_{A2} - R_{A1})}}{R_{A2} R_{A1}} + S_0 \frac{e^{ik(R_{B2} - R_{B1})}}{R_{B2} R_{B1}} + \mu_0 S_0 \frac{e^{ik(R_{B2} - R_{A1})}}{R_{B2} R_{A1}}$$

$$+ \mu_0 S_0 \frac{e^{ik(R_{A2} - R_{B1})}}{R_{A2} R_{B1}}. \quad (12.102)$$

Within the paraxial approximation, we may approximate all $R_{\alpha\beta}$'s in the denominators of the preceding expression by L, and in the exponents may make the following set of approximations:

$$R_{\alpha\beta} \approx L - \frac{x_\alpha x_\beta}{L}, \quad (12.103)$$

where $\alpha = A, B$ and $\beta = 1, 2$.

The pinholes are taken to be symmetric about the y-axis, that is, $x_A = -a$, $x_B = +a$, which leads us to a relatively simple expression for the cross-spectral density on the observation screen:

$$W(x_1, x_2, \omega) = \frac{S_0}{L^2} \left[e^{-\frac{ik}{L}(x_A x_2 - x_A x_1)} + e^{-\frac{ik}{L}(x_B x_2 - x_B x_1)} \right.$$

$$\left. + \mu_0 e^{-\frac{ik}{L}(x_B x_2 - x_A x_1)} + \mu_0 e^{-\frac{ik}{L}(x_A x_2 - x_B x_1)} \right]. \quad (12.104)$$

Even this expression is still quite difficult to understand, as there are still two degrees of freedom, namely, the two observation positions x_1 and x_2. Let us only consider points on the observation screen symmetric with respect to the x-axis, that is, $x_1 = +x$, $x_2 = -x$. Our result then immediately simplifies to

$$W(x, -x, \omega) = \frac{2S_0}{L^2} \left[\cos\left(\frac{2kax}{L} \right) + \mu_0 \right], \quad (12.105)$$

and the spectral density is given by

$$S(x, \omega) = \frac{2S_0}{L^2}\left[1 + \mu_0 \cos\left(\frac{2kax}{L}\right)\right]. \tag{12.106}$$

It can be seen from Equation 12.106 that, as expected, the spectral density possesses no zeros for any $\mu_0 < 1$, as this condition would require

$$\cos\left(\frac{2kax}{L}\right) = -\frac{1}{\mu_0}. \tag{12.107}$$

The cross-spectral density, however, does possess zeros; they appear at symmetric points such that

$$\cos\left(\frac{2kax}{L}\right) = -\mu_0. \tag{12.108}$$

In the limit that $\mu_0 \to 1$, the preceding two equations are identical: the zeros of the correlation functions evolve into zeros of the spectral density.

This evolution, and the locations of the singularities, are illustrated in Figure 12.19. A pair of correlation singularities appear on either side of the position of the eventual phase singularity; this circumstance can be compared with the behavior of correlation vortices as shown in Section 9.2.

From coherent phase singularities, we may then rotate the polarization of pinhole B in order to see the evolution of polarization singularities. As we know from Section 7.3 that a C-point is characterized in terms of the normalized Stokes parameters by $s_3 = \pm 1$ and an L-point is characterized by $s_3 = 0$, we may find such points in a field simply by plotting the parameter s_3. Now the complex field components on the observation screen are

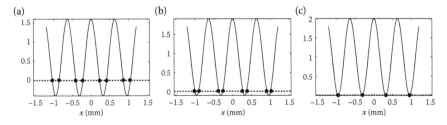

FIGURE 12.19 Solutions of Equation 12.108 for (a) $\mu_0 = 0.6$, (b) $\mu_0 = 0.9$, and (c) $\mu_0 = 1.0$. The dots represent positions for which singularities exist.

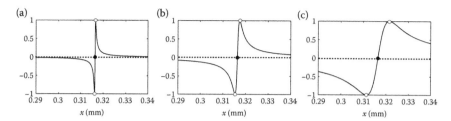

FIGURE 12.20 Evolution of the Stokes parameter s_3 when (a) $\alpha = 0.002$, (b) $\alpha = 0.01$, and (c) $\alpha = 0.05$. The white circles indicate C-points, while the black circles indicate L-points.

given by

$$E_x(x) = E_0 \left[\frac{e^{ikR_A}}{R_A} + \cos\alpha \frac{e^{ikR_B}}{R_B} \right], \tag{12.109}$$

$$E_y(x) = E_0 \sin\alpha \frac{e^{ikR_B}}{R_B}, \tag{12.110}$$

where R_A is the distance from the slit A to the observation point, and so forth, and $E_0 = \sqrt{S_0}$ is the field amplitude.

From these equations, the Stokes parameter s_3 can be calculated. The result of this calculation is shown in Figure 12.20, where we have focused our attention on the first phase singularity on the right of Figure 12.19. As soon as the angle α is increased, the phase singularity unfolds into a trio of polarization singularities, an L-point with C-points of opposite handedness on either side. As the angle is increased, the C-points move further apart from one another. This figure is particularly illuminating because it illustrates how an L-point must come between two C-points of opposite handedness as long as s_3 is continuous.

It is also possible to study a similar cascade with generic singularities, for instance, by using a 3-pinhole interferometer. This was done by Pang, Gbur, and Visser [PGV15], and they showed relationships between coherent and partially coherent scalar and electromagnetic singularities. The calculations are much more involved, however, and we will not consider them here.

12.6 LISSAJOUS SINGULARITIES

In the discussion of polarization singularities of Chapter 7, we almost entirely considered the properties of monochromatic wavefields. Even the discussion of partial polarization in Section 7.10 was restricted to quasi-monochromatic fields, for which the behavior of the electric field vector could be decomposed into a fully polarized and a fully unpolarized part.

Though sufficient for many optical systems, a monochromatic approximation is by no means the only time dependence possible. In 2003, Kessler and Freund [KF03] considered the polarization properties of a field possessing two frequency components, ω and 2ω, such as might be produced in the process of second-harmonic generation. Not long after, Freund [Fre03] undertook a study of such systems in more detail. With two frequencies present, the electric field vector no longer traces out a simple polarization ellipse, but instead traces a figure known as a Lissajous figure.[*] Just like the case of elliptical polarization, however, it is possible to define singularities of handedness and orientation. We consider such *Lissajous singularities* in this section; they serve as an illustrative example of how existing singular optics can be adapted to more complicated systems.

We describe the transverse electric field of a bichromatic field of frequencies ω and 2ω by the expression

$$\mathbf{E} = \mathrm{Re}\left\{\mathbf{e}e^{-i\omega t} + \mathbf{f}e^{-i2\omega t}\right\}, \tag{12.111}$$

where $\mathbf{e}$ and $\mathbf{f}$ are complex vectors with x and y components. Because these vectors are complex, a complete characterization of the path of the electric field vector requires eight real numbers, as opposed to the four required for monochromatic fields, and the polarization figures produced are correspondingly more complex. Following Freund, we write these parameters in two pairs, as follows:

$$\{\mathbf{e}, \mathbf{f}\} = [e_{x,R}, e_{x,I}, f_{x,R}, f_{x,I}], [e_{y,R}, e_{y,I}, f_{y,R}, f_{y,I}]. \tag{12.112}$$

In Figure 12.21, we illustrate some of the variety of figures possible by showing symmetric examples, which were determined by Freund through "exhaustive analysis." It can be seen that the first and third figures are linear and elliptical polarization, respectively, which are special cases for which the amplitude of the second frequency component is zero. It is to be noted that, in general, a Lissajous figure may cross itself 0, 1, 2, or 3 times.

With single-frequency elliptical polarization, we noted that linear polarization represented singularities of handedness, while circular polarization

---

[*] These figures are named after the mathematician Jules Antoine Lissajous (1822–1880), who undertook the first detailed investigation of them.

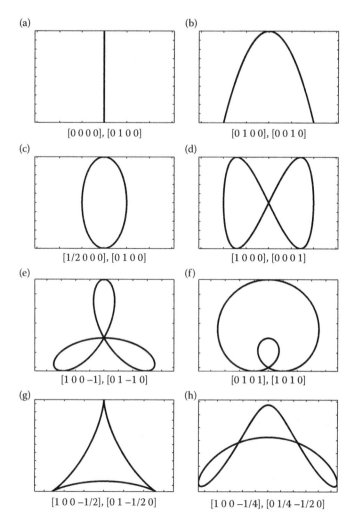

FIGURE 12.21   Set of symmetric Lissajous figures, as found by Freund [Fre03].

represented singularities of orientation. Naively, we would expect that (a) and (b) of Figure 12.21 might represent more generalized handedness singularities, while (e) and (g) might represent orientation singularities. We would like to make this idea quantitative, however, and also be able to account for asymmetric figures such as those shown in Figure 12.22.

We will find it useful to introduce the Stokes parameters for the bichromatic field; as the Stokes parameters are simply cycle-averaged products of fields, we can readily find that the total Stokes parameters S_j are given by

$$S_j = S_j(\omega) + S_j(2\omega), \quad j = 0, 1, 2, 3, \tag{12.113}$$

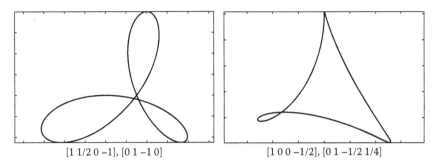

[1 1/2 0 −1], [0 1 −1 0]   [1 0 0 −1/2], [0 1 −1/2 1/4]

FIGURE 12.22   A pair of asymmetric Lissajous figures.

where $S_j(\omega)$ is the jth Stokes parameter of the component of the field at frequency ω, that is,

$$S_0(\omega) = e_{x,R}^2 + e_{x,I}^2 + e_{y,R}^2 + e_{y,I}^2, \tag{12.114}$$

$$S_1(\omega) = e_{x,R}^2 + e_{x,I}^2 - e_{y,R}^2 - e_{y,I}^2, \tag{12.115}$$

$$S_2(\omega) = 2(e_{x,R}e_{y,R} + e_{x,I}e_{y,I}), \tag{12.116}$$

$$S_3(\omega) = 2(e_{x,R}e_{y,I} - e_{y,R}e_{x,I}). \tag{12.117}$$

Now let us consider the handedness of a bichromatic field. In Section 7.3, we noted that the orientation of nontransverse polarization ellipses could be defined as the vector area of the ellipse, in accordance with the right-hand rule. For transverse polarization ellipses, the handedness is then the signed area of the ellipse; we try an analogous definition for a bichromatic field. For an infinitesimal path element ds, the infinitesimal area element dA may be given by $dA = |\mathbf{E}|ds$; in polar coordinates, the path element may be written as $ds = |\mathbf{E}|d\theta$. We then find that the area element may be written as

$$dA = |\mathbf{E}|^2 d\theta. \tag{12.118}$$

The angle θ in $E_x - E_y$-coordinates, however, is simply given by

$$\theta = \arctan(E_y/E_x). \tag{12.119}$$

With some simple manipulation, one can readily find that

$$dA = E_x dE_y - E_y dE_x, \tag{12.120}$$

or, if we write it as the rate of change of the area in time,

$$\frac{dA}{dt} = E_x\frac{dE_y}{dt} - E_y\frac{dE_x}{dt}. \tag{12.121}$$

We may then consider the average area swept out during a complete cycle of the field:

$$\langle A \rangle = \frac{\omega}{2\pi} \int\limits_0^{2\pi/\omega} \left[E_x\frac{dE_y}{dt} - E_y\frac{dE_x}{dt} \right] dt. \tag{12.122}$$

We may evaluate this expression by substituting from Equation 12.111; the result is of the form

$$\langle A \rangle = S_3(\omega) + 2S_3(2\omega). \tag{12.123}$$

A positive value of $\langle A \rangle$ roughly indicates a net counterclockwise motion of the electric field, while a negative value indicates net clockwise motion. We therefore define the handedness h of a Lissajous figure as

$$h \equiv \text{sign}[S_3(\omega) + 2S_3(2\omega)]. \tag{12.124}$$

One can easily verify that (a) and (b) of Figure 12.21 are singularities of h, in directly analogy with the L-lines of monochromatic fields. However, it is also straightforward to demonstrate that part (h) of the same figure is also a singularity of h, even though the electric field traces out a definite region of space and has a definite direction at every point on the curve.

We may also introduce a generalized orientation by analogy with monochromatic fields. We have seen that the orientation of a polarization ellipse can be defined by the matrix $\mathbf{Q}$, which in terms of Stokes parameters has the form

$$\mathbf{Q} = \frac{2}{S_3^2}\left[\begin{array}{cc} S_0 - S_1 & -S_2 \\ -S_2 & S_0 + S_1 \end{array} \right]. \tag{12.125}$$

We directly apply this same matrix to Lissajous figures, substituting the total Stokes parameters of the bichromatic field into the expression. The orientation angle ψ of the figure is then given by

$$\tan(2\psi) = \frac{S_2}{S_1}, \tag{12.126}$$

and the orientation will be undefined when

$$S_1 = S_2 = 0. \tag{12.127}$$

This is the same condition that makes $\mathbf{Q}$ proportional to the identity matrix. Referring again to Figure 12.21, we find that (b), (d), (e), (f), and (g) are all of undefined orientation. Some of these are obvious, such as (e) and (g); others are less so.

In analogy with monochromatic fields, then, we find that there is a single condition, $h = 0$, that a field must satisfy to be a singularity of handedness, or h-singularity. There are two conditions, $S_1 = S_2 = 0$, that a field must satisfy to be a singularity of orientation, or α-singularity. We therefore expect h-singularities to be lines in a transverse plane, separating regions of opposite handedness, and α-singularities to be points.

To illustrate this, we consider the example of second harmonic generation from Kessler and Freund [KF03]. The field is taken to have a C-point at the origin, by setting, in dimensionless coordinates,

$$e_x = 1 + (x + iy), \tag{12.128}$$

$$e_y = i[1 - (x + iy)], \tag{12.129}$$

and the components of the frequency doubled field are

$$f_x = -e_x e_y, \tag{12.130}$$

$$f_y = (e_y^2 - e_x^2)/2. \tag{12.131}$$

Figure 12.23a shows the behavior of the ω-component of the field alone in space as well as the combined field. The ω-component has a C-point at the origin, and a circle of linear polarization surrounding it at a radius $r = 1$. There is a corresponding Lissajous singularity at the origin, a symmetric trefoil, as shown in Figure 12.23b.

The Lissajous case has additional features, however. A look at the orientation of the Lissajous figures, in Figure 12.23c, in the region $r < 1$ suggests that the singularity is a lemon with a topological index $n = +1/2$; however, looking at the orientation of the figures for $r > 1$ suggests an index $n = -1$. By use of Equation 12.127, we can find that there are three additional α-singularities on the circle $r = 0$, namely, at $(x, y) = (1, 0)$, $(-1/2, \sqrt{3}/2)$, and $(-1/2, -\sqrt{3}/2)$. Each of these is a star with a topological index $n =$

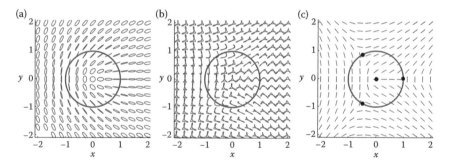

FIGURE 12.23 (a) Polarization ellipses for a monochromatic wavefield, and (b) Lissajous figures for the corresponding bichromatic field. (c) Orientation of the Lissajous figures, showing clearly the α-singularities, marked as black dots. The circles indicate a line that is a singularity of handedness.

$-1/2$, bringing us to a total index for the region of $n = -1$. It is notable that we are allowed to have simultaneous α- and h-singularities at the same point in space for Lissajous figures.

Even more characteristics of bichromatic Lissajous fields can be found in the paper by Freund [Fre03]. The evolution of Lissajous singularities on propagation has also been studied by Yan and Lü [YL10], showing pair creation and annihilation as well as reversal of the handedness of singularities.

12.7 OPTICAL MÖBIUS STRIPS

In Chapter 7, our discussion of electromagnetic polarization singularities was primarily restricted to the transverse polarization, that is, the components of the field perpendicular to the propagation axis. We now lift that restriction and consider, in a limited manner, how polarization singularities manifest in fully three-dimensional fields.

To help understand what we will encounter, we reproduce a typical transverse polarization singularity in Figure 12.24, again constructed by the superposition of a vortex beam and a plane wave (to be discussed again below). This singularity is a lemon, as discussed in Section 7.4. Though they have no physical significance in this context, the arrows are included to point out that the polarization ellipses make only a π rotation in a full circuit around the singularity.

If we perturb our system, the major axes of the polarization ellipses in a closed path around the singularity will form a surface in three-dimensional space, one that must maintain that same π-twist. This sort of object—a

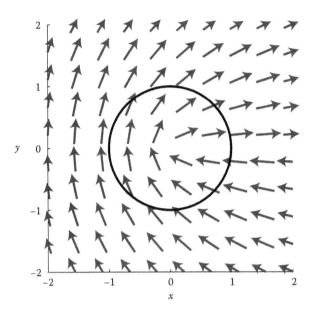

FIGURE 12.24   A lemon polarization singularity, with arrows drawn to emphasize the half-turn of the major axis in a path around the singularity.

surface with a half-twist—is well known in topology as a Möbius strip, a one-sided surface. We therefore expect that the polarization ellipses will take the form of a Möbius strip around a C-line.

We may test this by using a simple model. We take the combination of a right-hand circularly polarized vortex of topological charge $t = 1$, of the form

$$\mathbf{E}_1(x, y, z) = (\hat{\mathbf{x}} - i\hat{\mathbf{y}})(x + iy)e^{ikz}, \qquad (12.132)$$

and a left-hand circularly polarized plane wave

$$\mathbf{E}_0(x, y, z) = (\hat{\mathbf{x}} + i\hat{\mathbf{y}})e^{ikz}. \qquad (12.133)$$

However, we rotate the direction of propagation of the plane wave about the y-axis by an angle θ, which leaves it in the form

$$\mathbf{E}_0(x, y, z) = [\cos\theta\hat{\mathbf{x}} + i\hat{\mathbf{y}} + \sin\theta\hat{\mathbf{z}}]e^{ik(-x\sin\theta + z\cos\theta)}. \qquad (12.134)$$

When $\theta = 0$, the polarization of these two fields is transverse, and we reproduce the lemon pattern of Figure 12.24 in the $z = 0$ plane.

What happens when $\theta \neq 0$? Here, we must be careful, as the major and minor axes, as well as the vector area of the ellipse, must be specified by vectors for nontransverse fields. Following Berry [Ber04a], we introduce explicit vector formulas for the major axis α, the minor axis β, and the vector area $\mathbf{N}$:

$$\alpha = \frac{1}{|\sqrt{\mathbf{E} \times \mathbf{E}}|} \mathrm{Re} \left\{ \mathbf{E}^* \sqrt{\mathbf{E} \times \mathbf{E}} \right\}, \qquad (12.135)$$

$$\beta = \frac{1}{|\sqrt{\mathbf{E} \times \mathbf{E}}|} \mathrm{Im} \left\{ \mathbf{E}^* \sqrt{\mathbf{E} \times \mathbf{E}} \right\}, \qquad (12.136)$$

$$\mathbf{N} = \mathrm{Im} \left\{ \mathbf{E}^* \times \mathbf{E} \right\}. \qquad (12.137)$$

It is to be noted that the quantity in the square root is complex, and not proportional to the intensity of the electromagnetic wave. We will not derive these formulas, other than to note that the electric field vector $\mathbf{E}$ defines the vectors $\mathbf{p}$ and $\mathbf{q}$ of Section 7.1, which define the plane of polarization. The major and minor axes must lie in this plane, and may be related to $\mathbf{p}$ and $\mathbf{q}$ by

$$\mathbf{E} = \mathbf{p} + i\mathbf{q} = \exp[i\gamma](\alpha + i\beta), \qquad (12.138)$$

where γ is a "rectification angle" that brings the ellipse axes into alignment with $\mathbf{p}$ and $\mathbf{q}$. This angle is given by the square root terms in the above equations.

We consider the orientation of the major axis of the ellipse around a circular path of radius r in the $z = 0$ plane for the sum $\mathbf{E}_0 + \mathbf{E}_1$; the evolution is shown in Figure 12.25 for several values of r. For small values of r, the evolution of the major axis possesses a single half-twist; as the radius increases, additional pairs of half-twists are born, resulting in multitwist Möbius strips of increasing complexity. These additional twists are created from the transverse phase of the field, in this case from the tilted plane wave. Because it is always orthogonal to the major axis, the minor axis of the polarization ellipse also undergoes similar Möbius evolution.

In a closed circuit around an ordinary point, the major axis will make an even number of half-twists. This is shown in Figure 12.26, where a plane wave rotated by an angle θ_1 about the y-axis is superimposed with a plane wave rotated by an angle θ_2 about the x-axis. It can be seen that the major axis makes an even number of half-twists about the origin, with the number of twists increasing with radius.

To properly describe the complex topology of the polarization field in three dimensions, a number of new topological indices must be introduced,

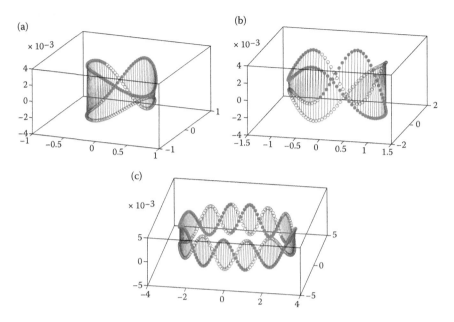

FIGURE 12.25 The major axes of optical Möbius strips, with (a) $r = 0.5$, (b) $r = 1.0$, and (c) $r = 3.0$. In all examples, $\lambda = 1$ and $\theta = 0.05 \times (2\pi)$. The spheres are used to clearly illustrate the evolution of orientation.

and even then an astounding complexity of behavior exists. Much has been written on the subject, starting with the pioneering work of Freund [Fre05, Fre10a,Fre10b,Fre14]. The appearance of such Möbius strips in intersecting Laguerre–Gauss beams has been predicted [Fre11], and such strips were recently observed experimentally [BBK+15].

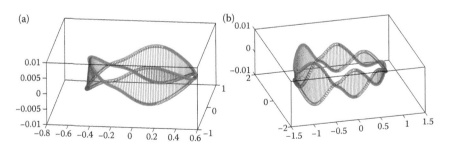

FIGURE 12.26 The major axes of optical ribbons, with (a) $r = 0.5$ and (b) $r = 1.0$. In all examples, $\lambda = 1$, $\theta_1 = 0.05 \times (2\pi)$, and $\theta_2 = 0.1 \times (2\pi)$. The spheres are used to clearly illustrate the evolution of orientation.

12.8 SUPEROSCILLATORY FIELDS

Fourier analysis tells us that any signal can be decomposed into a sum of monochromatic components of different frequencies. If we let $f(x)$ be our signal, we may then write

$$f(x) = \int_{-\infty}^{\infty} \tilde{f}(k)e^{ikx}dk, \tag{12.139}$$

where $\tilde{f}(k)$ is the Fourier transform (or spectrum) of $f(x)$, given by

$$\tilde{f}(k) = \frac{1}{2\pi} \int_{-\infty}^{\infty} f(x)e^{-ikx}dx. \tag{12.140}$$

Similar transforms may be defined for multivariable functions. The Fourier transform forms the basis of signal processing and is of fundamental importance in optics and any field of physics that deals with waves.

There are, of course, many insights to be gained from Fourier analysis. One seemingly straightforward one comes from the study of bandlimited functions, that is, functions for which

$$\tilde{f}(k) = 0, \quad |k| > \Delta, \tag{12.141}$$

where Δ represents the bandlimit. The Fourier representation of the function is then given simply by

$$f(x) = \int_{-\Delta}^{\Delta} \tilde{f}(k)e^{ikx}dk. \tag{12.142}$$

This formula suggests that the fastest at which the function $f(x)$ can oscillate in any region is with a frequency Δ, which is the highest-frequency component with nonzero amplitude. That is, a single cycle of the function $f(x)$ should occur over a region no shorter than $\delta x = 2\pi/\Delta$. This is, qualitatively, why it is possible to reconstruct a bandlimited signal from samples spaced, at most, a distance π/Δ, in what is known as the Nyquist–Shannon sampling theorem [Gbu11, Section 13.1].

It is now, however, known that this restriction on the rate of oscillations is not generally true. It is in fact possible to create a bandlimited signal that possesses oscillations of arbitrarily high frequency that extend over an arbitrarily long duration. Such peculiar oscillations are known as *super-oscillations*, and we will see that they are intimately connected to singular optics.

Superoscillations appear to have been first noted in physics by Aharonov et al. [AAPV90] in the context of quantum mechanics; not long after, Berry [Ber95] brought them to the attention of the broader physics community in a paper with the compelling title "Faster than Fourier." Since then, there has been a flurry of work done on studying the possibilities and limitations of such oscillations as well as their possible application to breaking resolution limits in optical systems.

Several different techniques have been introduced to construct super-oscillatory waves, though these methods tend to require somewhat compli-cated mathematics. Berry, for instance, considered the asymptotic analysis of a complex-valued integral, while Ferreira, Kempf, and Reis [FKR07] used a reconstruction technique based on the sampling theorem. However, there is a much simpler way to create superoscillatory functions of arbitrary length and frequency, as was shown by Chremmos and Fikioris [CF15].

An example is given in Figure 12.27. Part (a) shows the spectrum $\tilde{g}(k)$, which is bandlimited such that $\Delta = 1$. In part (b), we see a large range of the function $g(x)$, and it can be seen that it conforms to our expectation that a single cycle occurs over a range roughly 2π. Near the origin, however, the function appears almost flat; we zoom into this region in part (c). It can now be seen that there are rapid, low-amplitude oscillations in this area with a period of 2, much faster than the bandlimit $\delta x = 2\pi$ seems to allow. These are the superoscillations.

The construction of such a superoscillatory function is surprisingly straightforward. We start with a function $f(x)$, which is assumed to be

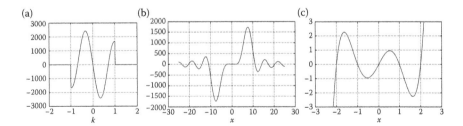

FIGURE 12.27 (a) Spectrum $\tilde{g}(k)$ of a superoscillatory field, for $a = 1$. (b) Expanded view of the function $g(x)$. (c) Zoomed-in view of $g(x)$, showing super-oscillations.

bandlimited to the range $-\Delta \le k \le \Delta$. We then construct a new function

$$g(x) = h(x)f(x), \tag{12.143}$$

where $h(x)$ is a polynomial of the form

$$h(x) = \sum_{l=0}^{n} a_l x^l. \tag{12.144}$$

We may then determine the spectrum of $g(x)$. The answer follows directly from an elementary property of Fourier transforms, namely,

$$\frac{1}{2\pi} \int_{-\infty}^{\infty} x^l f(x) e^{-ikx} dx = i^l \frac{\partial^l \tilde{f}(k)}{\partial k^l}. \tag{12.145}$$

With this, we may write the spectrum of $g(x)$ as

$$\tilde{g}(k) = \left[\sum_{l=0}^{n} a_l \frac{\partial^l}{\partial k^l} \right] \tilde{f}(k). \tag{12.146}$$

Because $\tilde{f}(k)$ is bandlimited, the function $g(k)$ must be as well, as the derivatives of $\tilde{f}(k)$ will not extend its range. We further require that the function $\tilde{f}(k)$ and its derivatives up to $(n-1)$th order be continuous in order that the spectrum of $\tilde{g}(k)$ not possess any delta function-like singularities.

Because the form of the polynomial $h(x)$ has no effect at all on the bandwidth of the function $g(x)$, we can choose it to have as many zeros as we like and we may place them as arbitrarily close together as we like. For the example of Figure 12.27, we used

$$h(x) = x(x^2 - a^2)(x^2 - 4a^2), \tag{12.147}$$

with $a = 1$, and

$$\tilde{f}(k) = \begin{cases} [\cos(\pi k/2\Delta)]^5, & |k| \le \Delta, \\ 0, & |k| > \Delta. \end{cases} \tag{12.148}$$

We therefore have five zeros, spaced by $a = 1$; we may however take a as small as desired. In Figure 12.28, a second example is shown with $a = 0.1$.

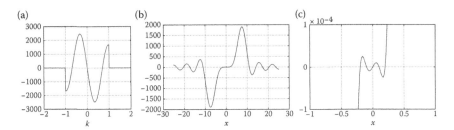

FIGURE 12.28 (a) Spectrum $\tilde{g}(k)$ of a superoscillatory field, for $a = 0.1$. (b) Expanded view of the function $g(x)$. (c) Zoomed-in view of $g(x)$, showing superoscillations.

This image shows that superoscillation comes with a price: a larger frequency comes with a dramatic reduction in amplitude, in this case four orders of magnitude less for a single order of magnitude decrease in period. The spectra of Figures 12.27 and 12.28 look nearly identical, and it can be shown that the a-dependent terms are much smaller than what we might call the carrier wave of the overall signal.

The creation of superoscillations by closely packing zeros practically cries out for identification with singular optics. We noted in Section 12.1, by use of the angular spectrum representation, that waves propagating sufficiently far from any source are bandlimited; that is, their plane wave components all satisfy the restriction $k_x^2 + k_y^2 \leq k^2$. Nevertheless, even a single vortex in such a beam oscillates arbitrarily fast around its core. As we well know by now, the phase increases by a minimum of 2π in a closed path around a vortex center; if we make the radius of that path arbitrarily small, the rate of change of the phase correspondingly grows arbitrarily large. To quantify the superoscillatory nature of wavefields with vortices, Dennis, Hamilton, and Courtial [DHC08] introduced the local wavenumber of the field. If we write the wave as $U(\mathbf{r}) = U_R(\mathbf{r}) + iU_I(\mathbf{r})$, the local wavenumber is defined as

$$k(\mathbf{r}) \equiv \mathrm{Im}\left\{\nabla \log[U(\mathbf{r})]\right\} = \frac{|U_R(\mathbf{r})\nabla U_I(\mathbf{r}) - U_I(\mathbf{r})\nabla U_R(\mathbf{r})|}{|U(\mathbf{r})|^2}. \quad (12.149)$$

They found that a remarkably high fraction of a speckle pattern is superoscillatory, 1/5 of the total area of the speckle pattern for a uniform distribution of plane waves. Not long after, Berry and Dennis [BD09] demonstrated that similar results hold for superoscillations in a wavefield of any number of dimensions.

Quite extreme vortex-based superoscillations may be constructed using a method similar to that given above for the signal $f(x)$. For a two-dimensional complex field $f(x, y)$, let us assume that its spectrum $\tilde{f}(k_x, k_y)$ is bandlimited such that $|\mathbf{k}| \leq \Delta$, that is, $\tilde{f}(k_x, k_y) = 0$ for wavenumbers outside this region. We now multiply $f(x, y)$ by a polynomial $h(z)$, that is,

$$h(z) = \sum_{l=0}^{n} a_l z^l. \qquad (12.150)$$

Introducing a new function

$$g(x, y) = h(z)f(x, y), \qquad (12.151)$$

we expect that $g(x, y)$ will be bandlimited as well, and will possess generic left-handed vortices at points where $h(z) = 0$. To demonstrate that the function is bandlimited, we use the Fourier result

$$\frac{1}{(2\pi)^2} \int (x + iy)^l f(x, y) e^{-i(k_x x + k_y y)} dx dy = i^l \left[\partial_{k_x} + i \partial_{k_y} \right]^l \tilde{f}(k_x, k_y). \qquad (12.152)$$

As in the earlier case, provided our function $\tilde{f}(k_x, k_y)$ is also taken to have continuous nth partial derivatives, the spectrum of $g(x, y)$ will be well behaved and bandlimited to the same region Δ. This is discussed by Smith and Gbur [SGa].

Superoscillations have captured much attention as a possible way to beat traditional Fourier limits, though as we have noted they come with an exponentially increasing cost. Berry [Ber95] noted that it is possible to encode Beethoven's 9th Symphony on a 1 Hz bandwidth signal using super-oscillations; however, this would require a surrounding signal, such as the peaks in the above figure, that are 10^{19} times stronger than a conventional oscillation!

Nevertheless, some researchers have successfully applied superoscilla-tions to beat conventional diffraction limits, albeit by a relatively small amount. Huang et al. [HCdAZ07] demonstrated that a quasi-periodic array of nanoholes in a metal screen can produce subwavelength spots in the far-field, and attribute these spots to the presence of superoscillations. Wong and Eleftheriades [WE11] have applied superoscillations to pro-duce a single subwavelength focal spot. Superoscillations have even been encoded in diffractionless beams [GSH+13], possibly providing a way to carry subwavelength information over long distances.

12.9 EXERCISES

1. For the following Riemann–Silberstein vector, (a) confirm that it satisfies Maxwell's equations, (b) calculate the quantities S and P, (c) find parametric equations $x(z, t)$ and $y(z, t)$ for the zeros, and (d) plot the results for some illustrative times:

$$\mathbf{F} = (y + it, z - a + i(a + t), x + it).$$

2. For the following Riemann–Silberstein vector, (a) confirm that it satisfies Maxwell's equations, (b) calculate the quantities S and P, (c) find parametric equations $x(z, t)$ and $y(z, t)$ for the zeros, and (d) plot the results for some illustrative times:

$$\mathbf{F} = (z^2 + t^2 - iat, a^2 - i(2zt + a^2 + ax), a(y - t)).$$

3. Starting with the plane waves of Equations 12.49, change the polarization of the third plane wave from left-handed to right-handed circular polarization, and calculate the square of the Riemann–Silberstein vector. Describe the locations of the Riemann–Silberstein vortices in this case, and how they evolve in time.

4. Derive the paraxial equation for $\psi(\mathbf{r})$, Equation 12.56, from Equation 12.55.

5. Derive the general formulas for the major axis α and minor axis β for a general polarization ellipse in three dimensions, as given by Equations 12.135 and 12.136.

References

[AAPV90]   Y. Aharonov, J. Anandan, S. Popescu, and L. Vaidman. Super-positions of time evolutions of a quantum system and a quantum time-translation machine. *Phys. Rev. Lett.*, 64:2965–2968, 1990.

[ABB15]   M. Andersson, E. Berglind, and G. Björk. Orbital angular momentum modes do not increase the channel capacity in communication links. *New J. Phys.*, 17:043040, 2015.

[ABdAC12]   M.E. Anderson, H. Bigman, L.E.E. de Araujo, and J.L. Chaloupka. Measuring the topological charge of ultrabroadband optical-vortex beams with a triangular aperture. *J. Opt. Soc. Am. B*, 29:1968–1976, 2012.

[ABGS94]   D. Ambrosini, V. Bagini, F. Gori, and M. Santarsiero. Twisted Gaussian Schell-model beams: A superposition model. *J. Mod. Opt.*, 41:1391–1399, 1994.

[ABP94]   L. Allen, M. Babiker, and W.L. Power. Azimuthal Doppler shift in light beams with orbital angular momentum. *Opt. Commun.*, 112:141–144, 1994.

[ABP03]   L. Allen, S.M. Barnett, and M.J. Padgett, editors. *Optical Angular Momentum*. IOP Publishing. Bristol, 2003.

[Abr09]   M. Abraham. Zur Elektrodynamik bewegter Körper. *Rend. Circ. Matem. Palermo*, 28:1–28, 1909.

[ABSW92]   L. Allen, M.W. Beijersbergen, R.J.C. Spreeuw, and J.P. Woerdman. Orbital angular momentum of light and the transformation of Laguerre-Gaussian laser modes. *Phys. Rev. A*, 45:8185–8189, 1992.

[AD73]   A. Ashkin and J.M. Dziedzic. Radiation pressure on a free liquid surface. *Phys. Rev. Lett.*, 30:139–142, 1973.

[AD00]   J. Arlt and K. Dholakia. Generation of high-order Bessel beams by use of an axicon. *Opt. Commun.*, 177:297–301, 2000.

[ADAP99]   J. Arlt, K. Dholakia, L. Allen, and M.J. Padgett. Parametric down-conversion for light beams possessing orbital angular momentum. *Phys. Rev. A*, 59:3950–3952, 1999.

[ADBC86]   A. Ashkin, J.M. Dziedzic, J.E. Bjorkholm, and S. Chu. Observation of a single-beam gradient force optical trap for dielectric particles. *Opt. Lett.*, 11:288–290, 1986.

[AFN+01]   L.J. Allen, H.M.L. Faulkner, K.A. Nugent, M.P. Oxley, and D. Paganin. Phase retrieval from images in the presence of first-order vortices. *Phys. Rev. E*, 63:037602, 2001.

[AG06]   G.S. Agarwal and G. Gbur. Rotational frequency shifts for electromagnetic fields of arbitrary states of coherence and polarization. *Opt. Lett.*, 31:3080–3082, 2006.

[AGK+90]   V.I. Adzhalov, M.A. Golub, S.V. Karpeev, I.N. Sisakyan, and V.A. Soĭfer. Multichannel computer-optics components matched to mode groups. *Sov. J. Quantum Electron.*, 20:136–140, 1990.

[AGP05]   D. Ambrosini, F. Gori, and D. Paoletti. Destructive interference from three partially coherent point sources. *Opt. Commun.*, 254:30–39, 2005.

[AGR81]   A. Aspect, P. Grangier, and G. Roger. Experimental tests of realistic local theories via Bell's theorem. *Phys. Rev. Lett.*, 47:460–463, 1981.

[AMLS09]   S. Albaladejo, M.I. Marqués, M. Laroche, and J.J. Sáenz. Scattering forces from the curl of the spin angular momentum of a light field. *Phys. Rev. Lett.*, 102:113602, 2009.

[AMP+02]   J. Arlt, M. MacDonald, L. Paterson, W. Sibbett, K. Dholakia, and K. Volke-Sepulveda. Moving interference patterns created using the angular Doppler effect. *Opt. Exp.*, 10:844–852, 2002.

[ANV08]   J.A. Anguita, M.A. Neifeld, and B.V. Vasic. Turbulence-induced channel crosstalk in an orbital angular momentum-multiplexed free-space optical link. *Appl. Opt.*, 47:2414–2429, 2008.

[AOEW05]   A. Aiello, S.S.R. Oemrawsingh, E.R. Eliel, and J.P. Woerdman. Non-locality of high-dimensional two-photon orbital angular momentum states. *Phys. Rev. A*, 72:052114, 2005.

[APS12]   M. Alonso, G. Piquero, and J. Serna. Proposals for the generation of angular momentum from non-uniformly polarized beams. *Opt. Commun.*, 285:1631–1635, 2012.

[Ara11]   D.F.J. Arago. Mémoire sur une modification remarquable qu'éprouvent les rayons lumineux dans leur passage à traverse certains corps diaphanes, et sur quelques autres nouveaux phénomènes d'optique. *Mém. de l'Inst.*, 12:93–134, 1811.

[Ara92]   P.K. Aravind. A simple proof of Pancharatnam's theorem. *Opt. Commun.*, 94:191–196, 1992.

[Ash70]   A. Ashkin. Acceleration and trapping of particles by radiation pressure. *Phys. Rev. Lett.*, 24:156–159, 1970.

[Ash78]   A. Ashkin. Trapping of atoms by resonance radiation pressure. *Phys. Rev. Lett.*, 40:729–732, 1978.

[Ash92]   A. Ashkin. Forces of a single-beam gradient laser trap on a dielectric sphere in the ray optics regime. *Biophys. J.*, 61:569–582, 1992.

[AV91]   E. Abramochkin and V. Volostnikov. Beam transformations and nontransformed beams. *Opt. Commun.*, 83:123–135, 1991.

[AW01]   G.B. Arfken and H.J. Weber. *Mathematical Methods for Physicists*. Harcourt Press, Amsterdam and Boston, 5th edition, 2001.

[Bar02] S.M. Barnett. Optical angular-momentum flux. *J. Opt. B*, 4:S7–S16, 2002.

[Bar08] S.M. Barnett. On the quantum core of an optical vortex. *J. Mod. Opt.*, 55:2279–2292, 2008.

[Bar10] S.M. Barnett. Resolution of the Abraham–Minkowski dilemma. *Phys. Rev. Lett.*, 104:070401, 2010.

[Bat03] R.W. Batterman. Falling cats, parallel parking, and polarized light. *Stud. Hist. Phil. Mod. Phys.*, 34:527–557, 2003.

[BAvdVW93] M.W. Beijersbergen, L. Allen, H.E.L.O. van der Veen, and J.P. Woerdman. Astigmatic laser mode converters and transfer of orbital angular momentum. *Opt. Commun.*, 96:123–132, 1993.

[BB79] M.V. Berry and N.L. Balazs. Nonspreading wave packets. *Am. J. Phys.*, 47:264–267, 1979.

[BB94] I. Bialynicki-Birula. On the wave function of the photon. *Acta Phys. Pol. A*, 86:97–116, 1994.

[BBA10] A.M. Beckley, T.G. Brown, and M.A. Alonso. Full Poincaré beams. *Opt. Exp.*, 18:10777–10785, 2010.

[BBA12] A.M. Beckley, T.G. Brown, and M.A. Alonso. Full Poincaré beams II: Partial polarization. *Opt. Exp.*, 20:9357–9362, 2012.

[BBBB97] I. Bialynicki-Birula and Z. Bialynicka-Birula. Rotational frequency shift. *Phys. Rev. Lett.*, 78:2539–2542, 1997.

[BBBB03] I. Bialynicki-Birula and Z. Bialynicka-Birula. Vortex lines of the electromagnetic field. *Phys. Rev. A*, 67:062114, 2003.

[BBBBŚ00] I. Bialynicki-Birula, Z. Bialynicka-Birula, and C. Śliwa. Motion of vortex lines in quantum mechanics. *Phys. Rev. A*, 61:032110, 2000.

[BBK$^+$15] T. Bauer, P. Banzer, E. Karimi, S. Orlov, A. Rubano, L. Marrucci, E. Santamato, R.W. Boyd, and G. Leuchs. Observation of optical polarization Möbius strips. *Science*, 347:964–966, 2015.

[BBSV93] I.V. Basistiy, V. Yu. Bazhenov, M.S. Soskin, and M.V. Vasnetsov. Optics of light beams with screw dislocations. *Opt. Commun.*, 103:422–428, 1993.

[BCKW94] M.W. Beijersbergen, R.P.C. Coerwinkel, M. Kristensen, and J.P. Woerdman. Helical-wavefront laser beams produced with a spiral phaseplate. *Opt. Commun.*, 112:321–327, 1994.

[BD01a] M.V. Berry and M.R. Dennis. Knotted and linked phase singularities in monochromatic waves. *Proc. Roy. Soc. Lond. A*, 457:2251–2263, 2001.

[BD01b] M.V. Berry and M.R. Dennis. Knotting and unknotting of phase singularities: Helmholtz waves, paraxial waves and waves in 2+1 spacetime. *J. Phys. A*, 34:8877–8888, 2001.

[BD01c] M.V. Berry and M.R. Dennis. Polarization singularities in isotropic random vector waves. *Proc. Roy. Soc. Lond. A*, 457:141–155, 2001.

[BD04] M.V. Berry and M.R. Dennis. Quantum cores of optical phase singularities. *J. Opt. A*, 6:S178–S180, 2004.

[BD09]   M.V. Berry and M.R. Dennis. Natural superoscillations in
            monochromatic waves in D dimensions. *J. Phys. A*, 42:022003,
            2009.

[BDL04]   M.V. Berry, M.R. Dennis, and R.L. Lee Jr. Polarization singularities
            in the clear sky. *New J. Phys.*, 6:162, 2004.

[BDW67]   A. Boivin, J. Dow, and E. Wolf. Energy flow in the neighborhood of
            the focus of a coherent beam. *J. Opt. Soc. Am.*, 57:1171–1175, 1967.

[Bel64]   J.S. Bell. On the Einstein Podolsky Rosen paradox. *Physics*, 1:195–
            200, 1964.

[Ben01]   I. Bendixson. Sur les courbes définies par des équations différen-
            tielles. *Acta Math.*, 24:1–88, 1901.

[Ber78]   M.V. Berry. Disruption of wavefronts: Statistics of dislocations in
            incoherent Gaussian random waves. *J. Phys. A*, 11:27–37, 1978.

[Ber84]   M.V. Berry. Quantal phase factors accompanying adiabatic
            changes. *Proc. Roy. Soc. Lond. A*, 392:45–57, 1984.

[Ber87]   M.V. Berry. The adiabatic phase and Pancharatnam's phase for
            polarized light. *J. Mod. Opt.*, 34:1401–1407, 1987.

[Ber95]   M.V. Berry. Faster than Fourier. In J.S. Anandan and J.L. Safko,
            editors, *Proceedings of the International Conference on Fundamen-
            tal Aspects of Quantum Theory*, pages 55–65, World Scientific,
            Singapore, 1995.

[Ber98]   M. Berry. Paraxial beams of spinning light. *Proc. SPIE*, 3487:6–11,
            1998.

[Ber02a]   M.V. Berry. Coloured phase singularities. *New. J. Phys.*, 4:66.1–
            66.14, 2002.

[Ber02b]   M.V. Berry. Exploring the colours of dark light. *New. J. Phys.*,
            4:74.1–74.14, 2002.

[Ber04a]   M.V. Berry. Index formulae for singular lines of polarization. *J. Opt.
            A*, 6:675–678, 2004.

[Ber04b]   M.V. Berry. Optical vortices evolving from helicoidal integer and
            fractional phase steps. *J. Opt. A*, 6:259–268, 2004.

[Ber04c]   M.V. Berry. Riemann–Silberstein vortices for paraxial waves. *J.
            Opt. A*, 6:S175–S177, 2004.

[Ber09]   M.V. Berry. Optical currents. *J. Opt. A*, 11:094001, 2009.

[Bet36]   R.A. Beth. Mechanical detection and measurement of the angular
            momentum of light. *Phys. Rev.*, 50:115–125, 1936.

[BEZ00]   D. Bouwmeester, A. Ekert, and A. Zeilinger. *The Physics of Quan-
            tum Information*. Springer, Berlin, 2000.

[BFP+03]   G.V. Bogatyryova, C.V. Fel'de, P.V. Polyanskii, S.A. Ponomarenko,
            M.S. Soskin, and E. Wolf. Partially coherent vortex beams with a
            separable phase. *Opt. Lett.*, 28:878–880, 2003.

[BGV04]   M.A. Bandres and J.C. Gutiérrez-Vega. Ince–Gaussian beams. *Opt.
            Lett.*, 29:144–146, 2004.

[BH77]   M.V. Berry and J.H. Hannay. Umbilic points on Gaussian random
            surfaces. *J. Phys. A*, 10:1809–1821, 1977.

[BH83]   C.F. Bohren and D.R. Huffman. *Absorption and Scattering of Light by Small Particles*. Wiley, New York, 1983.

[BK96]   M.V. Berry and S. Klein, Geometric phases from stacks of crystal plates. *J. Mod. Opt.* 43:165–180, 1996.

[BL52]   W. Braunbek and G. Laukien. Einzelheiten zur Halbebenen-Beugung. *Optik*, 9:174, 1952.

[BL66]   B.R. Brown and A.W. Lohmann. Complex spatial filtering with binary masks. *Appl. Opt.*, 5:967–969, 1966.

[BL10]   C. Baxter and R. Loudon. Radiation pressure and the photon momentum in dielectrics. *J. Mod. Opt.*, 57:830–842, 2010.

[BLC$^+$10] G.C.G. Berkhout, M.P.J. Lavery, J. Courtial, M.W. Beijersbergen, and M.J. Padgett. Efficient sorting of orbital angular momentum states of light. *Phys. Rev. Lett.*, 105:153601, 2010.

[BLPB11]   G.C.G. Berkhout, M.P.J. Lavery, M.J. Padgett, and M.W. Beijersbergen. Measuring orbital angular momentum superpositions of light by mode transformation. *Opt. Lett.*, 36:1863–1865, 2011.

[BMMJ09]   E. Brasselet, N. Murazawa, H. Misawa, and S. Juodkazis. Optical vortices from liquid crystal droplets. *Phys. Rev. Lett.*, 103:103903, 2009.

[BMP$^+$97] A. Beržanksis, A. Matijošius, A. Piskarskas, V. Smilgevičius, and A. Stabinis. Conversion of topological charge of optical vortices in a parametric frequency converter. *Opt. Commun.*, 140:273–276, 1997.

[BMR92]   S.L. Braunstein, A. Mann, and M. Revzen. Maximal violation of Bell inequalities for mixed states. *Phys. Rev. Lett.*, 68:3259–3261, 1992.

[BN47]   L. Bragg and J.F. Nye. A dynamical model of crystal structure. *Proc. Roy. Soc. Lond. A*, 190:474–481, 1947.

[BO05]   L. Basano and P. Ottonello. Complete destructive interference of partially coherent sources of acoustic waves. *Phys. Rev. Lett.*, 94:173901, 2005.

[Bop63]   F. Bopp. Ist der Poynting-Vektor beobachtbar? *Ann. Physik*, 466:35–51, 1963.

[Bou54]   C.J. Bouwkamp. Diffraction theory. *Rep. Prog. Phys.*, 17:35–100, 1954.

[Bou02]   Z. Bouchal. Resistance of nondiffracting vortex beam against amplitude and phase perturbations. *Opt. Commun.*, 210:155–164, 2002.

[Bou03]   Z. Bouchal. Nondiffracting optical beams: Physical properties, experiments, and applications. *Czech. J. Phys.*, 53:537–578, 2003.

[Boy08]   R.W. Boyd. *Nonlinear Optics*. Academic Press, Amsterdam, 3rd edition, 2008.

[BP89]   S.M. Barnett and D.T. Pegg. On the Hermitian optical phase operator. *J. Mod. Opt.*, 36:7–19, 1989.

[BP90]   S.M. Barnett and D.T. Pegg. Quantum theory of rotation angles. *Phys. Rev. A*, 41:3427–3435, 1990.

[BP02]   Z. Bouchal and J. Peřina. Non-diffracting beams with controlled spatial coherence. *J. Mod. Opt.*, 49:1673–1689, 2002.

[BRBSG08]   M.M. Bernitsas, K. Raghavan, Y. Ben-Simon, and E.M.H. Garcia. VIVACE (Vortex Induced Vibration Aquatic Clean Energy): A new concept in generation of clean and renewable energy from fluid flow. *J. Offshore Mech. Arct. Eng.*, 130:041101, 2008.

[Bre06]   J. Breuer. *Introduction to the Theory of Sets*. Dover Publications, New York, 2006.

[Bro08]   T.G. Brown. Unconventional polarization states: Beam propagation, focusing, and imaging. In E. Wolf, editor, *Progress in Optics*, volume 56, pages 81–129, Elsevier, Amsterdam, 2008.

[Bru02]   D. Bruß. Characterizing entanglement. *J. Math. Phys.*, 43:4237–4251, 2002.

[Bry74]   O. Bryngdahl. Geometrical transformations in optics. *J. Opt. Soc. Am.*, 64:1092–1099, 1974.

[BS07]   A. Ya. Bekshaev and M.S. Soskin. Transverse energy flows in vectorial fields of paraxial beams with singularities. *Opt. Commun.*, 271:332–348, 2007.

[Bur40]   J.M. Burgers. Geometrical considerations concerning the structural irregularities to be assumed in a crystal. *Proc. Phys. Soc.*, 52:23–33, 1940.

[BW65]   A. Boivin and E. Wolf. Electromagnetic field in the neighborhood of the focus of a coherent beam. *Phys. Rev.*, 138:B1561–B1565, 1965.

[BW70]   D.C. Burnham and D.L. Weinberg. Observation of simultaneity in parametric production of optical photon pairs. *Phys. Rev. Lett.*, 25:84–87, 1970.

[BW99]   M. Born and E. Wolf. *Principles of Optics*. Cambridge University Press, Cambridge, 7th edition, 1999.

[BWC98]   Z. Bouchal, J. Wagner, and M. Chlup. Self-reconstruction of a distorted nondiffracting beam. *Opt. Commun.*, 151:207–211, 1998.

[BYR⁺13]   N. Bozinovic, Y. Yue, Y. Ren, M. Tur, P. Kristensen, H. Huang, A.E. Willner, and S. Ramachandran. Terabit-scale orbital angular momentum mode division multiplexing in fibers. *Science*, 340:1545–1548, 2013.

[BZM⁺81]   N.B. Baranova, B. Ya. Zel'dovich, A.V. Mamaev, N.F. Pilipetskii, and V.V. Shkukov. Dislocation of the wavefront of a speckle-inhomogeneous field (theory and experiment). *JETP Lett.*, 33:195–199, 1981.

[CDM04]   K. Crabtree, J.A. Davis, and I. Moreno. Optical processing with vortex-producing lenses. *Appl. Opt.*, 43:1360–1367, 2004.

[CDR+98]   J. Courtial, K. Dholakia, D.A. Robertson, L. Allen, and M.J. Pad-
             gett. Measurement of the rotational frequency shift imparted to a
             rotating light beam possessing orbital angular momentum. *Phys.
             Rev. Lett.*, 80:3217–3219, 1998.

[CF63]   R.H. Crowell and R.H. Fox. *Introduction to Knot Theory*. Ginn and
             Company, Boston, 1963.

[CF15]   I. Chremmos and G. Fikioris. Superoscillations with arbitrary
             polynomial shape. *J. Phys. A*, 48:265204, 2015.

[CG03]   J.E. Curtis and D.G. Grier. Structure of optical vortices. *Phys. Rev.
             Lett.*, 90:133901, 2003.

[CGL+02]   D. Collins, N. Gisin, N. Linden, S. Massar, and S. Popescu. Bell
             inequalities for arbitrarily high-dimensional systems. *Phys. Rev.
             Lett.*, 88:040404, 2002.

[CHSH69]   J.F. Clauser, M.A. Horne, A. Shimony, and R.A. Holt. Proposed
             experiment to test local hidden-variable theories. *Phys. Rev. Lett.*,
             23:880–884, 1969.

[Cir80]   B.S. Cirel'son. Quantum generalizations of Bell's inequality. *Lett.
             Math. Phys.*, 4:93–100, 1980.

[CK83]   D. Colton and R. Kress. *Integral Equation Methods in Scattering
             Theory*. John Wiley and Sons, New York, 1983.

[CL95]   P.M. Chaikin and T.C. Lubensky. *Principles of Condensed Matter
             Physics*. Cambridge University Press, Cambridge, 1995.

[CNV00]   P.C. Chaumet and M. Nieto-Vesperinas. Time-averaged total force
             on a dipolar sphere in an electromagnetic field. *Opt. Lett.*, 25:1065–
             1067, 2000.

[CP76]   D. Casasent and D. Psaltis. Position, rotation, and scale invariant
             optical correlation. *Appl. Opt.*, 15:1795–1799, 1976.

[CP78]   D. Casasent and D. Psaltis. Multiple-invariant space-variant optical
             processors. *Appl. Opt.*, 17:655–659, 1978.

[CRD+98]   J. Courtial, D.A. Robertson, K. Dholakia, L. Allen, and M.J. Pad-
             gett. Rotational frequency shift of a light beam. *Phys. Rev. Lett.*,
             81:4828–4830, 1998.

[CRO07]   M. Chen, F.S. Roux, and J.C. Olivier. Detection of phase singulari-
             ties with a Shack–Hartmann wavefront sensor. *J. Opt. Soc. Am. A*,
             24:1994–2002, 2007.

[CTDL77]   C. Cohen-Tannoudji, B. Diu, and F. Laloë. *Quantum Mechanics*.
             Wiley, New York and London, 1977.

[CWY88]   B. Campbell, G.A.H. Walker, and S. Yang. A search for substellar
             companions to solar-type stars. *Astrophys. J.*, 331:902–921, 1988.

[dAA11]   L.E.E. de Araujo and M.E. Anderson. Measuring vortex charge
             with a triangular aperture. *Opt. Lett.*, 36:787–789, 2011.

[DCLZ98]   R. Dum, J.I. Cirac, M. Lewenstein, and P. Zoller. Creation of dark
             solitons and vortices in Bose–Einstein condensates. *Phys. Rev.
             Lett.*, 80:2972–2975, 1998.

[Del94]   T. Delmarcelle. *The Visualization of Second-Order Tensor Fields*. PhD thesis, Stanford University, 1994.

[Den02]   M.R. Dennis. Polarization singularities in paraxial vector fields: Morphology and statistics. *Opt. Commun.*, 213:201–221, 2002.

[Den03]   M.R. Dennis. Braided nodal lines in wave superpositions. *New J. Phys.*, 5:134, 2003.

[Den08]   M.R. Dennis. Polarization singularity anisotropy: Determining monstardom. *Opt. Lett.*, 33:2572–2574, 2008.

[DH94]   T. Delmarcelle and L. Hesselink. The topology of symmetric, second-order tensor fields. In *VIS '94: Proceedings of the Conference on Visualization '94*, pages 140–147, IEEE Computer Society Press, 1994.

[DHC08]   M.R. Dennis, A.C. Hamilton, and J. Courtial. Superoscillation in speckle patterns. *Opt. Lett.*, 33:2976–2978, 2008.

[Dir31]   P.A.M. Dirac. Quantised singularities in the electromagnetic field. *Proc. Roy. Soc. A*, 133:60–72, 1931.

[DJE87]   J. Durnin, J.J. Miceli Jr., and J.H. Eberly. Diffraction-free beams. *Phys. Rev. Lett.*, 58:1499–1501, 1987.

[DKJ$^+$10]   M.R. Dennis, R.P. King, B. Jack, K. O'Holleran, and M.J. Padgett. Isolated optical vortex knots. *Nat. Phys.*, 6:118–121, 2010.

[DKZ01]   T. Durt, D. Kaszlikowski, and M. Zukowski. Violations of local realism with quantum systems described by n-dimensional Hilbert spaces up to $n = 16$. *Phys. Rev. A*, 64:024101, 2001.

[DMCC00]   J.A. Davis, D.E. McNamara, D.M. Cottrell, and J. Campos. Image processing with the radial Hilbert transform: Theory and experiments. *Opt. Lett.*, 25:99–101, 2000.

[DQL03]   R. Dorn, S. Quabis, and G. Leuchs. Sharper focus for a radially polarized light beam. *Phys. Rev. Lett.*, 91:233901, 2003.

[DSPA96]   K. Dholakia, N.B. Simpson, M.J. Padgett, and L. Allen. Second-harmonic generation and the orbital angular momentum of light. *Phys. Rev. A*, 54:R3742–R3745, 1996.

[Dur87]   J. Durnin. Exact solutions for nondiffracting beams. I. The scalar theory. *J. Opt. Soc. Am. A*, 4:651–654, 1987.

[DV04]   D.W. Diehl and T.D. Visser. Phase singularities of the longitudinal field components in high-aperture systems. *J. Opt. Soc. Am. A*, 21:2103–2108, 2004.

[EG79]   M. Eisenberg and R. Guy. A proof of the hairy ball theorem. *Am. Math. Monthly*, 86:572–574, 1979.

[EJ12]   O. Edfors and A.J. Johansson. Is orbital angular momentum (OAM) based radio communication an unexploited area? *IEEE Trans. Ant. Prop.*, 60:1126–1131, 2012.

[EKH$^+$92]   T. Erdogan, O. King, G.W. Hicks, D.G. Hall, E.H. Anderson, and M.J. Rooks. Circularly symmetric operation of a concentric-circle-grating, surface-emitting, AlGaAs/GaAs quantum-well semiconductor laser. *Appl. Phys. Lett.*, 60:1921–1923, 1992.

[EL08]   A. Einstein and J. Laub. Über die im elektromagnetischen Felde
              auf ruhende Körper ausgeübten ponderomotorischen Kräfte. *Ann.
              Phys. (Leipzig)*, 26:541–550, 1908.

[ELG⁺98]   T.W. Ebbesen, H.J. Lezec, H.F. Ghaemi, T. Thio, and P.A. Wolff.
              Extraordinary optical transmission through sub-wavelength hole
              arrays. *Nature*, 391:667–669, 1998.

[EPR35]   A. Einstein, B. Podolsky, and N. Rosen. Can quantum-mechanical
              description of physical reality be considered complete? *Phys. Rev.*,
              47:777–780, 1935.

[Erd55]   A. Erdélyi. *Asymptotic Expansions*. Dover Publications, New York,
              1955.

[Eri90]   F. Eriksson. On the measure of solid angles. *Math. Mag.*, 63:184–
              187, 1990.

[FA01a]   G.W. Forbes and M.A. Alonso. Consistent analogs of the Fourier
              uncertainty relation. *Am. J. Phys.*, 69:1091–1095, 2001.

[FA01b]   G.W. Forbes and M.A. Alonso. Measures of spread for periodic dis-
              tributions and the associated uncertainty relations. *Am. J. Phys.*,
              69:340–347, 2001.

[FABPA02]   S. Franke-Arnold, S.M. Barnett, M.J. Padgett, and L. Allen. Two-
              photon entanglement of orbital angular momentum states. *Phys.
              Rev. A*, 65:033823, 2002.

[FABY⁺04]   S. Franke-Arnold, S.M. Barnett, E. Yao, J. Leach, J. Courtial, and
              M. Padgett. Uncertainty principle for angular position and angular
              momentum. *New J. Phys.*, 6:103, 2004.

[FC72]   S.J. Freeman and J.F. Clauser. Experimental test of local hidden-
              variable theories. *Phys. Rev. Lett.*, 28:938–941, 1972.

[FERDH96]   M.E.J. Friese, J. Enger, H. Rubinsztein-Dunlop, and N.R. Hekcen-
              berg. Optical angular-momentum transfer to trapped absorbing
              particles. *Phys. Rev. A*, 54:1593–1596, 1996.

[FHPW61]   P.A. Franken, A.E. Hill, C.W. Peters, and G. Weinreich. Generation
              of optical harmonics. *Phys. Rev. Lett.*, 7:118–119, 1961.

[FJBRM05]   S. Fürhapter, A. Jesacher, S. Bernet, and M. Ritsch-Marte. Spi-
              ral phase contrast imaging in microscopy. *Opt. Exp.*, 13:689–694,
              2005.

[FKR07]   P.J.S.G. Ferreira, A. Kempf, and M.J.C.S. Reis. Construction of
              Aharonov–Berry's superoscillations. *J. Phys. A*, 40:5141–5147,
              2007.

[FLP⁺12]   R. Fickler, R. Lapkiewicz, W.N. Plick, M. Krenn, C. Schaeff,
              S. Ramelow, and A. Zeilinger. Quantum entanglement of high
              angular momenta. *Science*, 338:640–643, 2012.

[FLS64]   R. Feynman, R.B. Leighton, and M.L. Sands. *The Feynman Lectures
              on Physics*, volume 2. Addison-Wesley, Reading, MA, 1964.

[FNHRD98]   M.E.J. Friese, T.A. Nieminen, N.R. Heckenberg, and
              H. Rubinsztein-Dunlop. Optical alignment and spinning of
              laser-trapped microscopic particles. *Nature*, 394:348–350, 1998.

[FODP08]   F. Flossmann, K. O'Holleran, M.R. Dennis, and M.J. Padgett. Polarization singularities in 2D and 3D speckle fields. *Phys. Rev. Lett.*, 100:203902, 2008.

[FPGS05]   G. Foo, D.M. Palacios, and G.A. Swartzlander Jr. Optical vortex coronagraph. *Opt. Lett.*, 30:3308–3310, 2005.

[Fra58]   F.C. Frank. On the theory of liquid crystals. *Discuss. Faraday Soc.*, 25:19–28, 1958.

[FRDG$^+$01] M.E.J. Friese, H. Rubinsztein-Dunlop, J. Gold, P. Hagberg, and D. Hanstorp. Optically driven micromachine elements. *Appl. Phys. Lett.*, 78:547–549, 2001.

[Fre68]   A.P. French. *Special Relativity*. W.W. Norton, New York, 1968.

[Fre94]   I. Freund. Optical vortices in Gaussian random wave fields: Statistical probability densities. *J. Opt. Soc. Am. A*, 11:1644–1652, 1994.

[Fre98]   I. Freund. "1001" correlations in random wave fields. *Waves Random Media*, 8:119–158, 1998.

[Fre01]   I. Freund. Poincaré vortices. *Opt. Lett.*, 26:1996–1998, 2001.

[Fre03]   I. Freund. Bichromatic optical Lissajous fields. *Opt. Commun.*, 226:351–376, 2003.

[Fre05]   I. Freund. Cones, spirals, and Möbius strips, in elliptically polarized light. *Opt. Commun.*, 249:7–22, 2005.

[Fre10a]   I. Freund. Optical Möbius strips in three-dimensional ellipse fields: I. Lines of circular polarization. *Opt. Commun.*, 283:1–15, 2010.

[Fre10b]   I. Freund. Optical Möbius strips in three-dimensional ellipse fields: II. Lines of linear polarization. *Opt. Commun.*, 283:16–28, 2010.

[Fre11]   I. Freund. Möbius strips and twisted ribbons in intersecting Gauss–Laguerre beams. *Opt. Commun.*, 284:3816–3845, 2011.

[Fre14]   I. Freund. Optical Möbius strips and twisted ribbon cloaks. *Opt. Lett.*, 39:727–730, 2014.

[FS82]   A.T. Friberg and R.J. Sudol. Propagation parameters of Gaussian Schell-model beams. *Opt. Commun.*, 41:383–387, 1982.

[FS94]   I. Freund and N. Shvartsman. Wave-field phase singularities: The sign principle. *Phys. Rev. A*, 50:5164–5172, 1994.

[FS95]   I. Freund and N. Shvartsman. Structural correlations in Gaussian random wave fields. *Phys. Rev. E*, 51:3770–3773, 1995.

[FSF93]   I. Freund, N. Shvartsman, and V. Freilikher. Optical dislocation networks in highly random media. *Opt. Commun.*, 101:247–264, 1993.

[FV04]   D.G. Fischer and T.D. Visser. Spatial correlation properties of focused partially coherent light. *J. Opt. Soc. Am. A*, 21:2097–2102, 2004.

[GA79]   B.A. Garetz and S. Arnold. Variable frequency shifting of circularly polarized laser radiation via a rotating half-wave retardation plate. *Opt. Commun.*, 31:1–3, 1979.

[Gab48]   D. Gabor. A new microscopic principle. *Nature*, 161:777–778, 1948.

[GAF10]   G. Grynberg, A. Aspect, and C. Fabre. *Introduction to Quantum Optics*. Cambridge University Press, Cambridge, 2010.

[Gam88]   G. Gamow. *One Two Three... Infinity*. Dover Publications, New York, 1988.

[Gar81]   B.A. Garetz. Angular Doppler effect. *J. Opt. Soc. Am.*, 71:609–611, 1981.

[Gbu]   G. Gbur. Falling cats and geometric phases, in preparation.

[Gbu11]   G.J. Gbur. *Mathematical Methods for Optical Physics and Engineering*. Cambridge University Press, Cambridge, 2011.

[Gbu13]   G. Gbur. Invisibility physics: Past, present and future. In E. Wolf, editor, *Progress in Optics*, volume 58, pages 65–114, Elsevier, Amsterdam, 2013.

[Gbu16]   G. Gbur. Fractional vortex Hilbert's Hotel. *Optica*, 3:222–225, 2016.

[GCMP$^+$03] V. Garcés-Chávez, D. McGloin, M.J. Padgett, W. Dultz, H. Schmitzer, and K. Dholakia. Observation of the transfer of the local angular momentum density of a multiringed light beam to an optically trapped particle. *Phys. Rev. Lett.*, 91:093602, 2003.

[GCP$^+$04]   G. Gibson, J. Courtial, M.J. Padgett, M. Vasnetsov, V. Pas'ko, S.M. Barnett, and S. Franke-Arnold. Free-space information transfer using light beams carrying orbital angular momentum. *Opt. Exp.*, 12:5448–5456, 2004.

[GDGE13]   Y. Gorodetski, A. Drezet, C. Genet, and T.W. Ebbesen. Generating far-field orbital angular momentum from near-field optical chirality. *Phys. Rev. Lett.*, 110:203906, 2013.

[GG07]   C.H. Gan and G. Gbur. Phase and coherence singularities generated by the interference of partially coherent fields. *Opt. Commun.*, 280:249–255, 2007.

[GG12]   Y. Gu and G. Gbur. Measurement of atmospheric turbulence strength by vortex beam. *Opt. Commun.*, 283:1209–1212, 2012.

[GG13]   Y. Gu and G. Gbur. Scintillation of nonuniformly correlated beams in atmospheric turbulence. *Opt. Lett.*, 38:1395–1397, 2013.

[GGG$^+$02]   D. Ganic, X. Gan, M. Gu, M. Hain, S. Somalingam, S. Stankovic, and T. Tschudi. Generation of doughnut laser beams by use of a liquid-crystal cell with a conversion efficiency near 100%. *Opt. Lett.*, 27:1351–1353, 2002.

[GGP87]   F. Gori, G. Guattari, and C. Padovani. Bessel–Gauss beams. *Opt. Commun.*, 64:491–495, 1987.

[GGS08]   G. Gbur and G.A. Swartzlander Jr. Complete transverse representation of a correlation singularity of a partially coherent field. *J. Opt. Soc. Am. B*, 25:1422–1429, 2008.

[GH47]   F. Goos and H. Hänchen. Ein neuer und fundamentaler Versuch zur Totalreflexion. *Ann. Physik*, 436:333–346, 1947.

[GH96]   P.L. Greene and D.G. Hall. Diffraction characteristics of the azimuthal Bessel–Gauss beam. *J. Opt. Soc. Am. A*, 13:962–966, 1996.

[GHXD06]   C.-S. Guo, Y.-J. Han, J.-B. Xu, and J. Ding. Radial Hilbert transform with Laguerre–Gaussian spatial filters. *Opt. Lett.*, 31:1394–1396, 2006.

[GHZ89]   D.M. Greenberger, M.A. Horne, and A. Zeilinger. Going beyond Bell's theorem. In M. Kafatos, editor, *Bell's Theorem and the Conception of the Universe*, pages 69–72, Kluwer Academic, Dordrecht, 1989.

[GK04]   C.C. Gerry and P.L. Knight. *Introductory Quantum Optics*. Cambridge University Press, Cambridge, 2004.

[GKG09]   Y. Gu, O. Korotkova, and G. Gbur. Scintillation of nonuniformly polarized beams in atmospheric turbulence. *Opt. Lett.*, 34:2261–2263, 2009.

[GKK⁺83]   M.A. Golub, S.V. Karpeev, S.G. Krivoshlykov, A.M. Prokhorov, I.N. Sisakyan, and V.S. Soĭfer. Experimental investigation of spatial filters separating transverse modes of optical fields. *Sov. J. Quantum Electron.*, 13:1123–1124, 1983.

[GKK⁺84]   M.A. Golub, S.V. Karpeev, S.G. Krivoshlykov, A.M. Prokhorov, I.N. Sisakyan, and V.S. Soĭfer. Spatial filter investigation of the distribution of power between transverse modes in a fiber waveguide. *Sov. J. Quantum Electron.*, 14:1255–1256, 1984.

[GKK⁺88]   M.A. Golub, S.V. Karpeev, N.L. Kazanskii, A.V. Mirzov, I.N. Sisakyan, and V.S. Soĭfer. Spatial phase filters matched to transverse modes. *Sov. J. Quantum Electron.*, 18:392–393, 1988.

[GKK⁺96]   M.A. Golub, E.L. Kaganov, A.A. Kondorov, V.A. Soĭfer, and G.V. Usplen'ev. Experimental investigation of a multibeam holographic optical element matched to Gauss–Laguerre modes. *Quantum Electron.*, 26:184–186, 1996.

[GKLU07]   A. Greenleaf, Y. Kurylev, M. Lassas, and G. Uhlmann. Electromagnetic wormholes and virtual magnetic monopoles from metamaterials. *Phys. Rev. Lett.*, 99:183901, 2007.

[Gla63]   R.J. Glauber. The quantum theory of optical coherence. *Phys. Rev.*, 130:2529–2539, 1963.

[GO01]   P. Galajdá and P. Ormos. Complex micromachines produced and driven by light. *Appl. Phys. Lett.*, 78:249–251, 2001.

[Gol65]   A.S. Goldhaber. Role of spin in the monopole problem. *Phys. Rev.*, 140:B1407–B1414, 1965.

[Goo63]   N.R. Goodman. Statistical analysis based on a certain multivariate complex Gaussian distribution (an introduction). *Ann. Math. Stat.*, 34:152–177, 1963.

[Goo96]   J.W. Goodman. *Introduction to Fourier Optics*. McGraw-Hill, New York, 2nd edition, 1996.

[Goo07]   J.W. Goodman. *Speckle Phenomena in Optics*. Roberts and Company, Englewood, CO, 2007.

[Gor73]   J.P. Gordon. Radiation forces and momenta in dielectric media. *Phys. Rev. A*, 8:14–21, 1973.

[GP92]   N. Gisin and A. Peres. Maximal violation of Bell's inequality for arbitrarily large spin. *Phys. Lett. A*, 162:15–17, 1992.

[GPK+06]   O. Guyon, E.A. Pluzhnik, M.J. Kuchner, B. Collins, and S.T. Ridgway. Theoretical limits on extrasolar terrestrial planet detection with coronagraphs. *Astrophys. J. Suppl. Ser.*, 167:81–99, 2006.

[GPSS82]   M.A. Golub, A.M. Prokhorov, I.N. Sisakyan, and V.A. Soĭfer. Synthesis of spatial filters for investigation of the transverse mode composition of coherent radiation. *Sov. J. Quantum Electron.*, 12:1208–1209, 1982.

[GR81]   A. Garuccio and V.A. Rapisarda. Bell's inequalities and the four-coincidence experiment. *Nuovo Cimento*, 65:269–297, 1981.

[GR07]   I.S. Gradshteyn and I.M. Ryzhik. *Table of Integrals, Series, and Products*. Academic Press, Burlington, MA, 7th edition, 2007.

[Gri95]   D.J. Griffiths. *Introduction to Quantum Mechanics*. Prentice-Hall, Upper Saddle River, NJ, 1995.

[Gri13]   D.J. Griffiths. *Introduction to Electrodynamics*. Pearson, Boston, 4th edition, 2013.

[GRN95a]   T.E. Gureyev, A. Roberts, and K.A. Nugent. Partially coherent fields, the transport-of-intensity equations, and phase uniqueness. *J. Opt. Soc. Am. A*, 12:1942–1946, 1995.

[GRN95b]   T.E. Gureyev, A. Roberts, and K.A. Nugent. Phase retrieval with the transport-of-intensity equation: Matrix solution with use of Zernike polynomials. *J. Opt. Soc. Am. A*, 12:1932–1941, 1995.

[GRP+15]   D. Giovannini, J. Romero, V. Potoček, G. Ferenczi, F. Speirits, S.M. Barnett, D. Faccio, and M.J. Padgett. Spatially structured photons that travel in free space slower than the speed of light. *Science*, 347:857–860, 2015.

[GS06]   G.A. Swartzlander Jr. Achromatic optical vortex lens. *Opt. Lett.*, 31:2042–2044, 2006.

[GS15]   F. Gori and M. Santarsiero. Twisted Gaussian Schell-model beams as series of partially coherent modified Bessel–Gauss beams. *Opt. Lett.*, 40:1587–1590, 2015.

[GS96]   K.T. Gahagan and G.A. Swartzlander Jr. Optical vortex trapping of particles. *Opt. Lett.*, 21:827–829, 1996.

[GS98]   K.T. Gahagan and G.A. Swartzlander Jr. Trapping of low-index microparticles in an optical vortex. *J. Opt. Soc. Am. B*, 15:524–534, 1998.

[GSH$^+$13]   E. Greenfield, R. Schley, I. Hurwitz, J. Nemirovsky, K.G. Markis, and M. Segev. Experimental generation of arbitrarily shaped diffractionless superoscillatory optical beams. *Opt. Exp.*, 21:13425–13435, 2013.

[GSoU89]   M.A. Golub, I.N. Sisakyan, V.A. Soĭfer, and G.V. Uvarov. Optical components for the analysis and formation of the transverse mode composition. *Sov. J. Quantum Electron.*, 19:543–549, 1989.

[GST$^+$09]   R.M. Gomes, A. Salles, F. Toscano, P.H. Souto Ribeiro, and S.P. Walborn. Observation of a nonlocal optical vortex. *Phys. Rev. Lett.*, 103:033602, 2009.

[GT08]   G. Gbur and R.K. Tyson. Vortex beam propagation through atmospheric turbulence and topological charge conservation. *J. Opt. Soc. Am. A*, 25:225–230, 2008.

[GV03a]   G. Gbur and T.D. Visser. Can spatial coherence effects produce a local minimum of intensity at focus? *Opt. Lett.*, 28:1627–1629, 2003.

[GV03b]   G. Gbur and T.D. Visser. Coherence vortices in partially coherent beams. *Opt. Commun.*, 222:117–125, 2003.

[GVW02a]   G. Gbur, T.D. Visser, and E. Wolf. Anomalous behavior of spectra near phase singularities of focused waves. *Phys. Rev. Lett.*, 88:013901, 2002.

[GVW02b]   G. Gbur, T.D. Visser, and E. Wolf. Singular behavior of the spectrum in the neighborhood of focus. *J. Opt. Soc. Am. A*, 19:1694–1700, 2002.

[GVW04a]   G. Gbur, T.D. Visser, and E. Wolf. Complete destructive interference of partially coherent fields. *Opt. Commun.*, 239:15–23, 2004.

[GVW04b]   G. Gbur, T.D. Visser, and E. Wolf. "Hidden" singularities in partially coherent fields. *J. Opt. A*, 6:S239–S242, 2004.

[Haj87a]   J.V. Hajnal. Singularities in the transverse fields of electromagnetic waves. I. Theory. *Proc. Roy. Soc. Lond. A*, 414:433–446, 1987.

[Haj87b]   J.V. Hajnal. Singularities in the transverse fields of electromagnetic waves. II. Observations on the electric field. *Proc. Roy. Soc. Lond. A*, 414:447–468, 1987.

[Hal96]   D.G. Hall. Vector-beam solutions of Maxwell's wave equation. *Opt. Lett.*, 21:9–11, 1996.

[Han98]   J.H. Hannay. The Majorana representation of polarization, and the Berry phase of light. *J. Mod. Opt.*, 45:1001–1008, 1998.

[Har96]   P. Hariharan. *Optical Holography*. Cambridge University Press, Cambridge, 2nd edition, 1996.

[HBD$^+$15]   B. Hensen, H. Bernien, A.E. Dréau, A. Reiserer, N. Kalb, M.S. Blok, J. Reitenberg, R.F.L. Vermeulen, R.N. Schouten, C. Abellán, W. Amaya, V. Pruneri, M.W. Mitchell, M. Markham, D.J. Twitchen,

D. Elkouss, S. Wehner, T.H. Taminiau, and R. Hanson. Loophole-free Bell inequality violation using electron spins separated by 1.3 kilometres. *Nature*, 526:682–686, 2015.

[HBS⁺02] G. Horváth, B. Bernáth, B. Suhai, A. Barta, and R. Wehner. First observation of the fourth neutral polarization point in the atmosphere. *J. Opt. Soc. Am. A*, 19:2085–2099, 2002.

[HCdAZ07] F.M. Huang, Y. Chen, F.J. Garcia de Abajo, and N.I. Zheludev. Optical super-resolution through super-oscillations. *J. Opt. A*, 9:S285–S288, 2007.

[HCP74] J.O. Hirschfelder, A.C. Christoph, and W.E. Palke. Quantum mechanical streamlines. I. Square potential barrier. *J. Chem. Phys.*, 61:5435–5455, 1974.

[HDD87] W.J. Hossack, A.M. Darling, and A. Dahdouh. Coordinate transformations with multiple computer-generated optical elements. *J. Mod. Opt.*, 34:1235–1250, 1987.

[HdDL16] B.C. Hiesmayr, M.J.A. de Dood, and W. Löffler. Observation of four-photon orbital angular momentum entanglement. *Phys. Rev. Lett.*, 116:073601, 2016.

[HFHRD95] H. He, M.E.J. Friese, N.R. Heckenberg, and H. Rubinsztein-Dunlop. Direct observation of transfer of angular momentum to absorptive particles from a laser beam with a phase singularity. *Phys. Rev. Lett.*, 75:826–829, 1995.

[HFSCC10] J.M. Hickmann, E.J.S. Fonseca, W.C. Soares, and S. Chávez-Cerda. Unveiling a truncated optical lattice associated with a triangular aperture using light's orbital angular momentum. *Phys. Rev. Lett.*, 105:053904, 2010.

[HGB74] J.O. Hirschfelder, C.J. Goebel, and L.W. Bruch. Quantized vortices around wavefunction nodes. II. *J. Chem. Phys.*, 61:5456–5459, 1974.

[HHRD95] H. He, N.R. Heckenberg, and H. Rubinsztein-Dunlop. Optical particle trapping with higher-order doughnut beams produced using high efficiency computer generated holograms. *J. Mod. Opt.*, 42:217–223, 1995.

[HHTV94] M. Harris, C.A. Hill, P.R. Tapster, and J.M. Vaughan. Laser modes with helical wave fronts. *Phys. Rev. Lett.*, 49:3119–3122, 1994.

[HLSK06] M. Hautakorpi, J. Lindberg, T. Setälä, and M. Kaivola. Rotational frequency shifts in partially coherent optical fields. *J. Opt. Soc. Am. A*, 23:1159–1163, 2006.

[HM85] C.K. Hong and L. Mandel. Theory of parametric frequency down conversion of light. *Phys. Rev. A*, 31:2409–2418, 1985.

[HMR99] P. Hariharan, S. Mujumdar, and H. Ramachandran. A simple demonstration of the Pancharatnam phase as a geometric phase. *J. Mod. Opt.*, 46:1443–1446, 1999.

[HMS⁺92] N.R. Heckenberg, R. McDuff, C.P. Smith, H. Rubinsztein Dunlop, and M.J. Wegener. Laser beams with phase singularities. *Opt. Quantum Electron.*, 24:S951–S962, 1992.

[HMSW92]   N.R. Heckenberg, R. McDuff, C.P. Smith, and A.G. White. Generation of optical phase singularities by computer-generated holograms. *Opt. Lett.*, 17:221–223, 1992.

[HOB67]   S.E. Harris, M.K. Oshman, and R.L. Byer. Observation of tunable optical parametric fluorescence. *Phys. Rev. Lett.*, 18:732–734, 1967.

[HOM87]   C.K. Hong, Z.Y. Ou, and L. Mandel. Measurement of subpicosecond time intervals between two photons by interference. *Phys. Rev. Lett.*, 59:2044–2046, 1987.

[Jac75]   J.D. Jackson. *Classical Electrodynamics*. Wiley, New York, 2nd edition, 1975.

[Jer54]   H.G. Jerrard. Transmission of light through birefringent and optically active media: The Poincaré sphere. *J. Opt. Soc. Am.*, 44:634–640, 1954.

[JFAM^{+}08]   G.A. Swartzlander Jr., E.L. Ford, R.S. Abdul-Malik, L.M. Close, M.A. Peters, D.M. Palacios, and D.W. Wilson. Astronomical demonstration of an optical vortex coronagraph. *Opt. Exp.*, 16:10200–10207, 2008.

[JFBRM05]   A. Jesacher, S. Fürhapter, S. Bernet, and M. Ritsch-Marte. Shadow effects in spiral phase contrast microscopy. *Phys. Rev. Lett.*, 94:233902, 2005.

[JH94]   R.H. Jordan and D.G. Hall. Free-space azimuthal paraxial wave equation: The azimuthal Bessel–Gauss beam solution. *Opt. Lett.*, 19:427–429, 1994.

[JHA07]   G.A. Swartzlander Jr. and R.I. Hernandez-Aranda. Optical Rankine vortex and anomalous circulation of light. *Phys. Rev. Lett.*, 99:163901, 2007.

[JL63]   D. Judge and J.T. Lewis. On the commutator $[l_z, \phi]$. *Phys. Lett.*, 5:190, 1963.

[Jon41]   R.C. Jones. New calculus for the treatment of optical systems. *J. Opt. Soc. Am.*, 31:488–493, 1941.

[JT04]   T. Jarvis and J. Tanton. The hairy ball theorem via Sperner's lemma. *Am. Math. Monthly*, 111:599–603, 2004.

[Jud63]   D. Judge. On the uncertainty relation for l_z and ϕ. *Phys. Lett.*, 5:189, 1963.

[Kai04]   G. Kaiser. Helicity, polarization and Riemann–Silberstein vortices. *J. Opt. A*, 6:S243–S245, 2004.

[KBvDW97]   G.P. Karman, M.W. Beijersbergen, A. van Duijl, and J.P. Woerdman. Creation and annihilation of phase singularities in a focal field. *Opt. Lett.*, 22:1503–1505, 1997.

[KBW94]   M. Kristensen, M.W. Beijersbergen, and J.P. Woerdman. Angular momentum and spin-orbit coupling for microwave photons. *Opt. Commun.*, 104:229–233, 1994.

[Kel62]   J.B. Keller. Geometrical theory of diffraction. *J. Opt. Soc. Am.*, 52:116–130, 1962.

[Kem15]   Brandon A. Kemp. Macroscopic theory of optical momentum. In E. Wolf, editor, *Progress in Optics*, volume 60, pages 437–488, Elsevier, Amsterdam, 2015.

[KESC94]   P.G. Kwiat, P.H. Eberhard, A.M. Steinberg, and R.Y. Chiao. Proposal for a loophole-free Bell inequality experiment. *Phys. Rev. A*, 49:3209–3220, 1994.

[KF03]   D.A. Kessler and I. Freund. Lissajous singularities. *Opt. Lett.*, 28:111–113, 2003.

[KFF⁺14]   M. Krenn, R. Fickler, M. Fink, J. Handsteiner, M. Malik, T. Scheidl, R. Ursin, and A. Zeilinger. Communication with spatially modulated light through turbulence air across Vienna. *New J. Phys.*, 16:113028, 2014.

[KG12]   S.M. Kim and G. Gbur. Angular momentum conservation in partially coherent wave fields. *Phys. Rev. A*, 86:043814, 2012.

[KGZ⁺00]   D. Kaszlikowski, P. Gnaciński, M. Zukowski, W. Miklaszewski, and A. Zeilinger. Violations of local realism by two entangled n-dimensional systems are stronger than for two qubits. *Phys. Rev. Lett.*, 85:4418–4421, 2000.

[KMW⁺95]   P.G. Kwiat, K. Mattle, H. Weinfurter, A. Zeilinger, A.V. Sergienko, and Y. Shih. New high-intensity source of polarization-entangled photon pairs. *Phys. Rev. Lett.*, 75:4337–4341, 1995.

[KS69]   T.R. Kane and M.P. Scher. A dynamical explanation of the falling cat phenomenon. *Int. J. Solids Structures*, 5:663–670, 1969.

[KSG⁺13]   O. Korech, U. Steinitz, R.J. Gordon, I.Sh. Averbukh, and Y. Prior. Observing molecular spinning via the rotational Doppler effect. *Nat. Photon.*, 7:711–714, 2013.

[LBCP11]   M.P.J. Lavery, G.C.G. Berkhout, J. Courtial, and M.J. Padgett. Measurement of the light orbital angular momentum spectrum using an optical geometric transformation. *J. Opt.*, 13:064006, 2011.

[LDCP05]   J. Leach, M.R. Dennis, J. Courtial, and M.J. Padgett. Vortex knots in light. *New J. Phys.*, 7:55, 2005.

[Leo14]   U. Leonhardt. Abraham and Minkowski momenta in the optically induced motion of fluids. *Phys. Rev. A*, 90:033801, 2014.

[LFJS06]   J.H. Lee, G. Foo, E.G. Johnson, and G.A.S. Swartzlander Jr. Experimental verification of an optical vortex coronagraph. *Phys. Rev. Lett.*, 97:053901, 2006.

[LFP⁺07]   S.C. Lowry, A. Fitzsimmons, P. Pravec, D. Vokrouhlický, H. Boehnhardt, P.A. Taylor, J.-L. Margot, A. Galád, M. Irwin, J. Irwin, and P. Kusnirák. Direct detection of the asteroidal YORP effect. *Science*, 316:272–274, 2007.

[LG04]   K. Ladavac and D.G. Grier. Microoptomechanical pumps assembled and driven by holographic optical vortex arrays. *Opt. Exp.*, 12:1144, 2004.

[LGBV13]   A. Lubk, G. Guzzinati, F. Börrnert, and J. Verbeeck. Transport of intensity phase retrieval of arbitrary wave fields including vortices. *Phys. Rev. Lett.*, 111:173902, 2013.

[LJR+09]   J. Leach, B. Jack, J. Romero, M. Ritsch-Marte, R.W. Boyd, A.K. Jha, S.M. Barnett, S. Franke-Arnold, and M.J. Padgett. Violation of a Bell inequality in two-dimensional orbital angular momentum state-spaces. *Opt. Exp.*, 17:8287–8293, 2009.

[LMN72]   F. Landstorfer, H. Meinke, and G. Niedermair. Ringförmiger Energiewirbel im Nahfeld einer Richtantenne. *Nachr. Tech. Z.*, 25:537–576, 1972.

[LON88]   L.A. Lugiato, C. Oldano, and L.M. Narducci. Cooperative frequency locking and stationary spatial structures in lasers. *J. Opt. Soc. Am. B*, 5:879–887, 1988.

[Lou83]   R. Loudon. *The Quantum Theory of Light*. Oxford University Press, Oxford, 2nd edition, 1983.

[LRB+12]   M.P.J. Lavery, D.J. Robertson, G.C.G. Berkhout, G.D. Love, M.J. Padgett, and J. Courtial. Refractive elements for the measurement of the orbital angular momentum of a single photon. *Opt. Exp.*, 20:2110–2115, 2012.

[LSBP13]   M.P.J. Lavery, F.C. Speirits, S.M. Barnett, and M.J. Padgett. Detection of a spinning object using light's orbital angular momentum. *Science*, 341:537–540, 2013.

[LSPL13]   Y. Liu, S. Sun, J. Pu, and B. Lü. Propagation of an optical vortex beam through a diamond-shaped aperture. *Opt. Laser Tech.*, 45:473–479, 2013.

[LTPL11]   Y. Liu, H. Tao, J. Pu, and B. Lü. Detecting the topological charge of vortex beams using an annular triangle aperture. *Opt. Laser Tech.*, 43:1233–1236, 2011.

[LU62]   E.N. Leith and J. Upatnieks. Reconstructed wavefronts and communication theory. *J. Opt. Soc. Am.*, 52:1123–1130, 1962.

[LU63]   E.N. Leith and J. Upatnieks. Wavefront reconstruction with continuous-tone objects. *J. Opt. Soc. Am.*, 53:1377–1381, 1963.

[LWG+08]   J. Leach, A.J. Wright, J.B. Götte, J.M. Girkin, L. Allen, S. Franke-Arnold, S.M. Barnett, and M.J. Padgett. "Aether drag" and moving images. *Phys. Rev. Lett.*, 100:153902, 2008.

[Lyo39]   M.B. Lyot. A study of the solar corona and prominences without eclipses. *Mon. Not. R. Astron. Soc.*, 99:580–594, 1939.

[LYP04]   J. Leach, E. Yao, and M.J. Padgett. Observation of the vortex structure of a non-integer vortex beam. *New J. Phys.*, 6:71, 2004.

[MA07]   A.V. Martin and L.J. Allen. Phase imaging from a diffraction pattern in the presence of vortices. *Opt. Commun.*, 277:288–294, 2007.

[MAH+99]   M.R. Matthews, B.P. Anderson, P.C. Haljan, D.S. Hall, C.E. Wieman, and E.A. Cornell. Vortices in Bose–Einstein condensate. *Phys. Rev. Lett.*, 83:2498–2501, 1999.

[Maj32]   E. Majorana. Atomi orientati in campo magnetico variabile. *Nuovo Cim.*, 9:43–50, 1932.

[Mar94]   É.J. Marey. Des mouvements que certains animaux exécutent pour retomber sur leurs pieds, lorsquils sont précipités dun lieu élevé. *Comptes Rendus*, 119:714–717, 1894.

[Max92]   J.C. Maxwell. *A Treatise on Electricity and Magnetism.* Clarendon Press, Oxford, 3rd edition, 1892.

[Mil78]   J. Milnor. Analytic proofs of the "hairy ball theorem" and the Brouwer fixed point theorem. *Am. Math. Monthly*, 85:521–524, 1978.

[Mil06]   K.A. Milton. Theoretical and experimental status of magnetic monopoles. *Rep. Prog. Phys.*, 69:1637–1711, 2006.

[Mil08]   D.A.B. Miller. *Quantum Mechanics for Scientists and Engineers.* Cambridge University Press, Cambridge, 2008.

[Min08]   H. Minkowski. Die Grundgleichungen für die elektromagnetischen Vorgänge in bewegten Körpern. *Nachr. Ges. Wiss. Göttingen*, 53–111, 1908.

[MM67]   D. Magde and H. Mahr. Study in ammonium dihydrogen phosphate of spontaneous parametric interaction tunable from 4400 to 16000 Å. *Phys. Rev. Lett.*, 18:905–907, 1967.

[MMSB13]   M. Mirhosseini, M. Malik, Z. Shi, and R.W. Boyd. Efficient separation of the orbital angular momentum eigenstates of light. *Nat. Commun.*, 4:2781, 2013.

[MPMS04]   I.D. Maleev, D.M. Palacios, A.S. Marathay, and G.A. Swartzlander Jr. Spatial correlation vortices in partially coherent light: Theory. *J. Opt. Soc. Am. B*, 21:1895–1900, 2004.

[MPP10]   M.L. Marasinghe, M. Premaratne, and D.M. Paganin. Coherence vortices in Mie scattering of statistically stationary partially coherent fields. *Opt. Exp.*, 18:6628–6641, 2010.

[MPPA12]   M.L. Marasinghe, M. Premaratne, D.M. Paganin, and M.A. Alonso. Coherence vortices in Mie scattered nonparaxial partially coherent beams. *Opt. Exp.*, 20:2858–2875, 2012.

[MRAS05]   D. Mawet, P. Riaud, O. Absil, and J. Surdej. Annular groove phase mask coronagraph. *Astrophys. J.*, 633:1191–1200, 2005.

[MRP98]   C.H. Monken, P.H. Souto Ribeiro, and S. Pádua. Transfer of angular spectrum and image formation in spontaneous parametric down-conversion. *Phys. Rev. A*, 57:3123–3126, 1998.

[MRR00]   D. Mugnai, A. Ranfagni, and R. Ruggeri. Observation of superluminal behaviors in wave propagation. *Phys. Rev. Lett.*, 84:4830–4833, 2000.

[MvdS99]   H.J. Metcalf and P. van der Straten. *Laser Cooling and Trapping.* Springer, New York, 1999.

[MVWZ01]   A. Mair, A. Vaziri, G. Weihs, and A. Zeilinger. Entanglement of the orbital angular momentum states of photons. *Nature*, 412:313–316, 2001.

[MW95]   L. Mandel and E. Wolf. *Optical Coherence and Quantum Optics.* Cambridge University Press, Cambridge, 1995.

[MZW97]   K.-P. Marzlin, W. Zhang, and E.M. Wright. Vortex couple for atomic Bose–Einstein condensates. *Phys. Rev. Lett.,* 79:4728–4731, 1997.

[NB74]   J.F. Nye and M.V. Berry. Dislocations in wave trains. *Proc. Roy. Soc. A,* 336:165–190, 1974.

[NB09]   A.V. Novitsky and L.M. Barkovsky. Poynting singularities in optical dynamic systems. *Phys. Rev. A,* 79:033821, 2009.

[NBYB01]   L. Novotny, M.R. Beversluis, K.S. Youngworth, and T.G. Brown. Longitudinal field modes probed by single molecules. *Phys. Rev. Lett.,* 86:5251–5254, 2001.

[NH87]   J.F. Nye and J.V. Hajnal. The wave structure of monochromatic electromagnetic radiation. *Proc. Roy. Soc. Lond. A,* 409:21–36, 1987.

[NHH88]   J.F. Nye, J.V. Hajnal, and J.H. Hannay. Phase saddles and dislocations in two-dimensional waves such as the tides. *Proc. Roy. Soc. Lond. A,* 417:7–20, 1988.

[Nie96]   G. Nienhuis. Doppler effect induced by rotating lenses. *Opt. Commun.,* 132:8–14, 1996.

[NP00]   K.A. Nugent and D. Paganin. Matter-wave phase measurement: A noninterferometric approach. *Phys. Rev. A,* 61:063614, 2000.

[Nye83a]   J.F. Nye. Lines of circular polarization in electromagnetic wave fields. *Proc. Roy. Soc. Lond. A,* 389:279–290, 1983.

[Nye83b]   J.F. Nye. Polarization effects in the diffraction of electromagnetic waves: The role of disclinations. *Proc. Roy. Soc. Lond. A,* 387:105–132, 1983.

[Nye99]   J.F. Nye. *Natural Focusing and the Fine Structure of Light.* Institute of Physics Publishing, Bristol and Philadelphia, 1999.

[OEW+04]   S.S.R. Oemrawsingh, E.R. Eliel, J.P. Woerdman, E.J.K. Verstegen, J.G. Kloosterboer, and G.W. 't Hooft. Half-integral spiral phase plates for optical wavelengths. *J. Opt. A,* 6:S288–S290, 2004.

[OMAP02]   A.T. O'Neil, I. MacVicar, L. Allen, and M.J. Padgett. Intrinsic and extrinsic nature of the orbital angular momentum of a light beam. *Phys. Rev. Lett.,* 88:053601, 2002.

[OMMB12]   M.N. O'Sullivan, M. Mirhosseini, M. Malik, and R.W. Boyd. Near-perfect sorting of orbital angular momentum and angular position states of light. *Opt. Exp.,* 20:24444–24449, 2012.

[OMV+05]   S.S.R. Oemrawsingh, X. Ma, D. Voigt, A. Aiello, E.R. Eliel, G.W. 't Hooft, and J.P. Woerdman. Experimental demonstration of fractional orbital angular momentum entanglement of two photons. *Phys. Rev. Lett.,* 95:240501, 2005.

[O'N56]   E.L. O'Neill. Spatial filtering in optics. *IRE Trans. Inf. Theory,* 2:56–65, 1956.

[Opr95]   J. Oprea. Geometry and the Foucault pendulum. *Am. Math. Monthly*, 102:515–522, 1995.

[Oro34]   E. Orowan. Zur Kristallplastizitat. III. Uber den Mechanismus des Gleitvorganges. *Z. Phys.*, 89:634–659, 1934.

[Pan56]   S. Pancharatnam. Generalized theory of interference, and its applications. *Proc. Indian Acad. Sci. Sect. A*, 44:247–262, 1956.

[Pap88]   C.H. Papas. *Theory of Electromagnetic Wave Propagation*. Dover Publications, New York, 1988.

[PASA96]   M. Padgett, J. Arlt, N. Simpson, and L. Allen. An experiment to observe the intensity and phase structure of Laguerre–Gaussian laser modes. *Am. J. Phys.*, 64:77–82, 1996.

[PBZ$^+$05] D.T. Pegg, S.M. Barnett, R. Zambrini, S. Franke-Arnold, and M. Padgett. Minimum uncertainty states of angular momentum and angular position. *New J. Phys.*, 7:62, 2005.

[PC77]   D. Psaltis and D. Casasent. Deformation invariant optical processors using coordinate transformations. *Appl. Opt.*, 16:2288–2292, 1977.

[PCNS15]   J. Prat-Camps, C. Navau, and A. Sanchez. A magnetic wormhole. *Sci. Rep.*, 5:12488, 2015.

[PD24]   J. Proudman and A.T. Doodson. The principal constituent of the tides of the North Sea. *Phil. Trans. Roy. Soc. Lond. A*, 224:185–219, 1924.

[Per00]   M.A.C. Perryman. Extra-solar planets. *Rep. Prog. Phys.*, 63:1209–1272, 2000.

[PGV15]   X. Pang, G. Gbur, and T.D. Visser. Cycle of phase, coherence and polarization singularities in Young's three-pinhole experiment. *Opt. Exp.*, 23:34093–34108, 2015.

[Pic96]   B. Picinbono. Second-order complex random vectors and normal distribution. *IEEE Trans. Signal Process.*, 44:2637–2640, 1996.

[PMM$^+$04] D.M. Palacios, I.D. Maleev, A.S. Marathay, and G.A. Swartzlander Jr. Spatial correlation singularity of a vortex field. *Phys. Rev. Lett.*, 92:143905, 2004.

[PMM$^+$15] V. Potoček, F.M. Miatto, M. Mirhosseini, O.S. Magaña-Loaiza, A.C. Liapis, D.K.L. Oi, R.W. Boyd, and J. Jeffers. Quantum Hilbert Hotel. *Phys. Rev. Lett.*, 115:160505, 2015.

[PNHRD07] R.N.C. Pfeifer, T.A. Nieminen, N.R. Heckenberg, and H. Rubinsztein-Dunlop. Colloquium: Momentum of an electromagnetic wave in dielectric media. *Rev. Mod. Phys.*, 79:1197–1216, 2007.

[Pol34]   M. Polanyi. Über eine art gitterstörung, die einen Kristall plastich machen könnte. *Z. Phys.*, 89:660–664, 1934.

[Poy09]   J.H. Poynting. The wave motion of a revolving shaft, and a suggestion as to the angular momentum in a beam of circularly polarized light. *Proc. Roy. Soc. A*, 82:560–567, 1909.

[PP02]   A. Papoulis and S.U. Pillai. *Probability, Random Variables and Stochastic Processes*. McGraw-Hill, New York, 4th edition, 2002.

[PS01]   B.C. Platt and R. Shack. History and principles of Shack–Hartmann wavefront sensing. *J. Refract. Surg.*, 17:S573–S577, 2001.

[QLHE15]   X.-F. Qian, B. Little, J.C. Howell, and J.H. Eberly. Shifting the quantum-classical boundary: Theory and experiment for statistically classical optical fields. *Optica*, 2:611–615, 2015.

[Rae85]   H. Raether. *Surface Plasmons*. Springer-Verlag, Berlin, 1985.

[Ray71]   Lord Rayleigh. On the light from the sky, its polarization and colour. *Phil. Mag.*, 61:107–120 and 274–279, 1871.

[RG04]   J. Rudnick and G. Gaspari. *Elements of the Random Walk*. Cambridge University Press, Cambridge, 2004.

[RGF+12]   G. Ropars, G. Gorre, A. Le Floch, J. Enoch, and V. Lakshminarayanan. A depolarizer as a possible precise sunstone for Viking navigation by polarized skylight. *Proc. Roy. Soc. A*, 468:671–684, 2012.

[RGG12]   C. Rosenbury, Y. Gu, and G. Gbur. Phase singularities, correlation singularities, and conditions for complete destructive interference. *J. Opt. Soc. Am. A*, 29:410–416, 2012.

[RHWT99]   M. Reicherter, T. Haist, E.U. Wagemann, and H.J. Tiziani. Optical particle trapping with computer generated holograms written on a liquid-crystal display. *Opt. Lett.*, 24:608–610, 1999.

[Rig63]   W.W. Rigrod. Isolation of axi-symmetric optical-resonator modes. *Appl. Phys. Lett.*, 2:51–53, 1963.

[Rin91]   W. Rindler. *Introduction to Special Relativity*. Oxford University Press, Oxford, 2nd edition, 1991.

[RLM+12]   B. Rodenburg, M.P.J. Lavery, M. Malik, M.N. O'Sullivan, M. Mirhosseini, D.J. Robertson, M. Padgett, and R.W. Boyd. Influence of atmospheric turbulence on states of light carrying orbital angular momentum. *Opt. Lett.*, 37:3735–3737, 2012.

[Rob29]   H.P. Robertson. The uncertainty principle. *Phys. Rev.*, 34:163–164, 1929.

[RSS+15]   G.J. Ruane, G.A. Swartzlander Jr., S. Slussarenko, L. Marrucci, and M.R. Dennis. Nodal areas in coherent beams. *Optica*, 2:147–150, 2015.

[RSV12]   S.B. Raghunathan, H.F. Schouten, and T.D. Visser. Correlation singularities in partially coherent electromagnetic beams. *Opt. Lett.*, 37:4179–4181, 2012.

[RSV13]   S.B. Raghunathan, H.F. Schouten, and T.D. Visser. Topological reactions of correlation functions in partially coherent electromagnetic beams. *J. Opt. Soc. Am. A*, 30:582–588, 2013.

[RW59]   B. Richards and E. Wolf. Electromagnetic diffraction in optical systems II. Structure of the image field in an aplanatic system. *Proc. Roy. Soc. A*, 253:358–379, 1959.

[SA03]   A.V. Smith and D.J. Armstrong. Generation of vortex beams by an image-rotating optical parametric oscillator. *Opt. Exp.*, 11:868–873, 2003.

[SAST00]   B.E.A. Saleh, A.F. Abouraddy, A.V. Sergienko, and M.C. Teich. Duality between partial coherence and partial entanglement. *Phys. Rev. A*, 62:043816, 2000.

[SBDC07]   G.A. Siviloglou, J. Broky, A. Dogariu, and D.N. Christodoulides. Observation of accelerating Airy beams. *Phys. Rev. Lett.*, 99:213901, 2007.

[SC04]   C.J.R. Sheppard and A. Choudhury. Annular pupils, radial polarization, and superresolution. *Appl. Opt.*, 43:4322–4327, 2004.

[SC07]   G.A. Siviloglou and D.N. Christodoulides. Accelerating finite energy Airy beams. *Opt. Lett.*, 32:979–981, 2007.

[SC10]   D. Sarid and W. Challener. *Modern Introduction to Surface Plasmons*. Cambridge University Press, Cambridge, 2010.

[Sch30]   E. Schrödinger. Zum heisenbergschen unschärfeprinzip. *Sitzungsberichte der Preussischen Akademi der Wissenschaften, Physikalisch-mathematische Klasse*, 14:296–303, 1930.

[Sch35]   E. Schrödinger. Die gegenwärtige Situation in der Quantenmechanik. *Naturwissenschaften*, 23:807–812, 823–828, 844–849, 1935.

[Sch61]   A.C. Schell. *The Multiple Plate Antenna*. PhD thesis, Massachusetts Institute of Technology, 1961.

[Sch80]   E. Schrödinger. The present situation in quantum mechanics. *Proc. Am. Phil. Soc.*, 124:323–338, 1980.

[SDAP97]   N.B. Simpson, K. Dholakia, L. Allen, and M.J. Padgett. Mechanical equivalence of spin and orbital angular momentum of light: An optical spanner. *Opt. Lett.*, 22:52–54, 1997.

[SF94a]   N. Shvartsman and I. Freund. Vortices in random wavefields: Nearest neighbor anticorrelations. *Phys. Rev. Lett.*, 72:1008–1011, 1994.

[SF94b]   N. Shvartsman and I. Freund. Wave-field phase singularities: Near-neighbor correlations and anticorrelations. *J. Opt. Soc. Am. A*, 11:2710–2718, 1994.

[SGa]   M.K. Smith and G. Gbur. Construction of arbitrary vortex and superoscillatory fields, submitted.

[SGb]   C. Stahl and G. Gbur. Analytic calculation of vortex diffraction by a triangular aperture. *J. Opt. Soc. Am. A*, 33:1175–1180, 2016.

[SG14]   C.S.D. Stahl and G. Gbur. Complete representation of a correlation singularity in a partially coherent beam. *Opt. Lett.*, 39:5985–5988, 2014.

[SGV+97]   M.S. Soskin, V.N. Gorshkov, M.V. Vasnetsov, J.T. Malos, and N.R. Heckenberg. Topological charge and angular momentum of light beams carrying optical vortices. *Phys. Rev. A*, 56:4064–4075, 1997.

[SGV⁺03]   H.F. Schouten, G. Gbur, T.D. Visser, D. Lenstra, and H. Blok. Creation and annihilation of phase singularities near a sub-wavelength slit. *Opt. Exp.*, 11:371–380, 2003.

[SGVW03]   H.F. Schouten, G. Gbur, T.D. Visser, and E. Wolf. Phase singularities of the coherence functions in Young's interference pattern. *Opt. Lett.*, 28:968–970, 2003.

[Sha80]   R. Shankar. *Principles of Quantum Mechanics*. Plenum Press, New York and London, 1980.

[SII91]   S. Sato, M. Ishigure, and H. Inaba. Optical trapping and rotational manipulation of microscopic particles and biological cells using higher-order mode Nd:YAG laser beams. *Electron. Lett.*, 27:1831–1832, 1991.

[SKO83]   Y. Saito, S. Komatsu, and H. Ohzu. Scale and rotation invariant real time optical correlator using computer generated hologram. *Opt. Commun.*, 47:8–11, 1983.

[SM62]   A. Savage and R.C. Miller. Measurements of second harmonic generation of the ruby laser line in piezoelectric crystals. *Appl. Opt.*, 1:661–664, 1962.

[SM93]   R. Simon and N. Mukunda. Twisted Gaussian Schell-model beams. *J. Opt. Soc. Am. A*, 10:95–109, 1993.

[SMB10]   E. Serabyn, D. Mawet, and R. Burruss. An image of an exoplanet separated by two diffraction beamwidths from a star. *Nature*, 464:1018–1020, 2010.

[Som96]   A. Sommerfeld. Mathematisch Theorie der Diffraction. *Math. Ann.*, 47:317–374, 1896.

[SPAGR10]   G.A. Swartzlander Jr., T.J. Peterson, A.B. Artusio-Glimpse, and A.D. Raisanen. Stable optical lift. *Nat. Photon.*, 5:48–51, 2010.

[SPO09]   G. Situ, G. Pedrini, and W. Osten. Spiral phase filtering and orientation-selective edge detection/enhancement. *J. Opt. Soc. Am. A*, 26:1788–1797, 2009.

[Spr98]   R.J.C. Spreeuw. A classical analogy of entanglement. *Found. Phys.*, 28:361–374, 1998.

[SS04]   G.A. Swartzlander Jr. and J. Schmit. Temporal correlation vortices and topological dispersion. *Phys. Rev. Lett.*, 93:093901, 2004.

[SSM93a]   R. Simon, K. Sundar, and N. Mukunda. Twisted Gaussian Schell-model beams. I. Symmetry structure and normal-mode spectrum. *J. Opt. Soc. Am. A*, 10:2008–2016, 1993.

[SSM93b]   K. Sundar, R. Simon, and N. Mukunda. Twisted Gaussian Schell-model beams. II. Spectrum analysis and propagation characteristics. *J. Opt. Soc. Am. A*, 10:2017–2023, 1993.

[Sta86]   J.J. Stamnes. *Waves in Focal Regions*. Adam Hilger, Bristol and Boston, 1986.

[Sto52]   G.G. Stokes. On the composition and resolution of streams of polarized light from different sources. *Trans. Camb. Phil. Soc.*, 9:399–416, 1852.

[STS+14]   A. Shanker, L. Tian, M. Sczyrba, B. Connolly, A. Neureuther, and L. Waller. Transport of intensity phase imaging in the presence of curl effects induced by strongly absorbing photomasks. *Appl. Opt.*, 53:J1–J6, 2014.

[SV98]   M.S. Soskin and M.V. Vasnetsov. Nonlinear singular optics. *Pure Appl. Opt.*, 7:301–311, 1998.

[SVG+04a]   H.F. Schouten, T.D. Visser, G. Gbur, D. Lenstra, and H. Blok. Diffraction of light by narrow slits in plates of different materials. *J. Opt. A*, 6:S277–S280, 2004.

[SVG+04b]   H.F. Schouten, T.D. Visser, G. Gbur, D. Lenstra, and H. Blok. On the connection between phase singularities and the radiation pattern of a slit in a metal plate. *Phys. Rev. Lett.*, 93:173901, 2004.

[SVL04]   H.F. Schouten, T.D. Visser, and D. Lenstra. Optical vortices near sub-wavelength structures. *J. Opt. B: Quantum Semiclass. Opt.*, 6:S404–S409, 2004.

[SVLB03]   H.F. Schouten, T.D. Visser, D. Lenstra, and H. Blok. Light transmission through a subwavelength slit: Waveguiding and optical vortices. *Phys. Rev. E*, 67:036608, 2003.

[Swa09]   G.A. Swartzlander Jr. The optical vortex coronagraph. *J. Opt. A*, 11:094022, 2009.

[SWPO10]   G. Situ, M. Warber, G. Pedrini, and W. Osten. Phase contrast enhancement in microscopy using spiral phase filtering. *Opt. Commun.*, 283:1273–1277, 2010.

[SYF08]   W. She, J. Yu, and R. Feng. Observation of a push force on the end face of a nanometer silica filament exerted by outgoing light. *Phys. Rev. Lett.*, 101:243601, 2008.

[Tam31]   Ig. Tamm. Die verallgemeinerten Kugelfunktionen und die Wellenfunktionen eines Elektrons im Felde eines Magnetpoles. *Zeit. Phys.*, 71:141–150, 1931.

[Tam88]   Chr. Tamm. Frequency locking of two transverse optical modes of a laser. *Phys. Rev. A*, 38:5960–5963, 1988.

[Tay34]   G.I. Taylor. The mechanism of plastic deformation of crystals. Part I. Theoretical. *Proc. Roy. Soc. Lond. A*, 145:362–387, 1934.

[TCN78]   A.S. Thorndike, C.R. Cooley, and J.F. Nye. The structure and evolution of flow fields and other vector fields. *J. Phys. A*, 11:1455–1490, 1978.

[TCPC12]   M. Tamagnone, C. Craeye, and J. Perruisseau-Carrier. Comment on "Encoding many channels on the same frequency through radio vorticity: First experimental test". *New J. Phys.*, 14:118001, 2012.

[Tea82]   M.R. Teague. Irradiance moments: Their propagation and use for unique retrieval of phase. *J. Opt. Soc. Am.*, 72:1199–1209, 1982.

[Tea83]   M.R. Teague. Deterministic phase retrieval: A Green's function solution. *J. Opt. Soc. Am.*, 73:1434, 1983.

[Tho09]   J.J. Thomson. *Elements of the Mathematical Theory of Electricity and Magnetism*. Cambridge University Press, London, 4th edition, 1909.

[TLCP12] H. Tao, Y. Liu, Z. Chen, and J. Pu. Measuring the topological charge of vortex beams by using an annular ellipse aperture. *Appl. Phys. B*, 106:927–932, 2012.

[TLY04] S.H. Tao, W.M. Lee, and X. Yuan. Experimental study of holographic generation of fractional Bessel beams. *Appl. Opt.*, 43:122–126, 2004.

[TMs⁺12a] F. Tamburini, E. Mari, A. Sponselli, B. Thidé, A. Bianchini, and F. Romanato. Encoding many channels on the same frequency through radio vorticity: First experimental test. *New J. Phys.*, 14:033001, 2012.

[TMs⁺12b] F. Tamburini, E. Mari, A. Sponselli, B. Thidé, A. Bianchini, and F. Romanato. Reply to Comment on "Encoding many channels on the same frequency through radio vorticity: First experimental test". *New J. Phys.*, 14:118002, 2012.

[TRS⁺96] G.A. Turnbull, D.A. Robertson, G.M. Smith, L. Allen, and M.J. Padgett. The generation of free-space Laguerre-Gaussian modes at millimetre-wave frequencies by use of a spiral phaseplate. *Opt. Commun.*, 127:183–188, 1996.

[TVF91] J. Turunen, A. Vasara, and A.T. Friberg. Propagation invariance and self-imaging in variable-coherence optics. *J. Opt. Soc. Am. A*, 8:282–289, 1991.

[TW90] Chr. Tamm and C.O. Weiss. Bistability and optical switching of spatial patterns in a laser. *J. Opt. Soc. Am. B*, 7:1034–1038, 1990.

[TWD03] B.M. Terhal, M.M. Wolf, and A.C. Doherty. Quantum entanglement: A modern perspective. *Phys. Today*, April, 46–52, 2003.

[TYL⁺05] S.H. Tao, X.-C. Yuan, J. Lin, X. Peng, and H.B. Niu. Fractional optical vortex beam induced rotation of particles. *Opt. Exp.*, 13:7726–7731, 2005.

[VBL99] T.D. Visser, H. Blok, and D. Lenstra. Theory of polarization-dependent amplification in a slab waveguide with anisotropic gain and losses. *IEEE J. Quantum Electron.*, 35:240–249, 1999.

[VBS98] M.V. Vasnetsov, L.V. Basistiy, and M.S. Soskin. Free-space evolution of monochromatic mixed screw-edge wavefront dislocations. *Proc. SPIE*, 3487:29–33, 1998.

[VMS00] M.V. Vasnetsov, I.G. Marienko, and M.S. Soskin. Self-reconstruction of an optica vortex. *JETP Lett.*, 71:192–196, 2000.

[VP90] J.A. Vaccaro and D.T. Pegg. Physical number-phase intelligent and minimum-uncertainty states of light. *J. Mod. Opt.*, 37:17–39, 1990.

[VPJ⁺03] A. Vaziri, J.-W. Pan, T. Jennewein, G. Weihs, and A. Zeilinger. Concentration of higher dimensional entanglement: Qutrits of photon orbital angular momentum. *Phys. Rev. Lett.*, 91:27902, 2003.

[VS08] T.D. Visser and R.W. Schoonover. A cascade of singular field patterns in Young's interference experiment. *Opt. Commun.*, 281:1–6, 2008.

[VTF89]   A. Vasara, J. Turunen, and A.T. Friberg. Realization of general non-diffracting beams with computer-generated holograms. *J. Opt. Soc. Am. A*, 6:1748–1754, 1989.

[VWZ02]   A. Vaziri, G. Weihs, and A. Zeilinger. Experimental two-photon, three-dimensional entanglement for quantum communication. *Phys. Rev. Lett.*, 89:240401, 2002.

[Wat44]   G.N. Watson. *A Treatise on the Theory of Bessel Functions*. Cambridge University Press, Cambridge, 2nd edition, 1944.

[WDH+06]   W. Wang, Z. Duan, S.G. Hanson, Y. Miyamoto, and M. Takeda. Experimental study of coherence vortices: Local properties of phase singularities in a spatial coherence function. *Phys. Rev. Lett.*, 96:073902, 2006.

[WdOTM04]   S.P. Walborn, A.N. de Oliveira, R.S. Thebaldi, and C.H. Monken. Entanglement and conservation of orbital angular momentum in spontaneous parametric down-conversion. *Phys. Rev. A*, 69:023811, 2004.

[WE11]   A.M.H. Wong and G.V. Eleftheriades. Sub-wavelength focusing at the multi-wavelength range using superoscillations: An experimental demonstration. *IEEE Trans. Ant. Prop.*, 59:4766–4776, 2011.

[Whe33]   W. Whewell. Essay towards a first approximation to a map of cotidal lines, *Proc. Roy. Soc. Lond.* 122:147–236, 1833.

[Wie90]   O. Wiener. Stehende Lichtwellen und die Schwingungsrichtung polarisirten Lichtes. *Ann. Phys. Chem.*, 38:203–243, 1890.

[Wil48]   H.A. Wilson. Note on Dirac's theory of magnetic poles. *Phys. Rev.*, 75:309, 1948.

[Wol50]   H. Wolter. Zur Frage des Lichtweges bei Totalreflexion. *Z. Naturforsch. A*, 5:276–283, 1950.

[Wol82]   E. Wolf. New theory of partial coherence in the space-frequency domain. Part 1: Spectra and cross-spectra of steady-state sources. *J. Opt. Soc. Am.*, 72:343–351, 1982.

[Wol83]   E. Wolf. Recollections of Max Born. *Opt. News*, 9:10–16, 1983.

[Wol03]   E. Wolf. Unified theory of coherence and polarization of random electromagnetic beams. *Phys. Lett. A*, 312:263–267, 2003.

[Wol07]   E. Wolf. *Introduction to the Theory of Coherence and Polarization of Light*. Cambridge University Press, Cambridge, 2007.

[Wol08]   E. Wolf. Can a light beam be considered to be the sum of a completely polarized and a completely unpolarized beam? *Opt. Lett.*, 33:642–644, 2008.

[Wol09]   H. Wolter. Concerning the path of light upon total reflection. *J. Opt. A*, 11:090401, 2009.

[WYF+12]   J. Wang, J.-Y. Yang, I.M. Fazal, N. Ahmed, Y. Yan, H. Huang, Y. Ren, Y. Yue, S. Dolinar, M. Tur, and A.E. Willner. Terabit free-space data transmission employing orbital angular momentum multiplexing. *Nat. Photon.*, 6:488–496, 2012.

[YB00a]   K.S. Youngworth and T.G. Brown. Focusing of high numerical aperture cylindrical vector beams. *Opt. Exp.*, 7:77–87, 2000.

[YB00b]   K.S. Youngworth and T.G. Brown. Inhomogeneous polarization in scanning optical microscopy. *Proc. SPIE*, 3919:75–85, 2000.

[YL10]   H. Yan and B. Lü. Dynamical evolution of Lissajous singularities in free-space propagation. *Phys. Lett. A*, 374:3695–3700, 2010.

[You07]   T. Young. *A Course of Lectures on Natural Philosophy and the Mechanical Arts*, volume 1st. Joseph Johnson, London, 1807.

[Zha04]   Q. Zhan. Trapping metallic Rayleigh particles with radial polarization. *Opt. Exp.*, 12:3377–3382, 2004.

[ZSPL15]   L. Zhang, W. She, N. Peng, and U. Leonhardt. Experimental evidence for Abraham pressure of light. *New J. Phys.*, 17:053035, 2015.

Index

A

Abraham–Minkowski controversy,
        136–137
    Ashkin and Dziedzic
        experiment, 145
    difference between Abraham and
        Minkowski formulations, 137
    Lorentz force of fields, 141
    macroscopic Maxwell's equations,
        138–139
    matter force density, 139–140
    Maxwell's equations, 136–137
    Minkowski field momentum density,
        142–144
    resolution of Abraham–Minkowski
        dilemma, 145–146
    SYF08, 144–145
ADP crystal, see Ammonium dihydrogen
        phosphate crystal
Allen, Babiker, and Power theory, 181
Ammonium dihydrogen phosphate crystal
        (ADP crystal), 364
Ampére–Maxwell law, 10
Amphidromic points, 4
Angular Doppler effect, 180
Angular momentum, 355, 399; see also
        Entanglement of angular
        momentum states
    difference-frequency
        experiment, 363
    flux density, 118
    OAM, 361
    parametric processes, 357
    quartz prism spectrometer, 359
    second-harmonic signal, 360

second-order polarization, 356, 362
    SHG types, 358
Angular momentum in wavefields, 112
    analogous calculation, 116–117
    Cartesian coordinate, 115–116
    divergence theorem, 117–118
    field quantities, 113–114
    momentum conservation law, 113
    momentum density of electromagnetic
        field, 114–115
Angular momentum of light, 118
    angular momentum of
        Laguerre–Gaussian beams,
        122–127
    intrinsic and extrinsic angular
        momentum, 127–131
    momentum in matter and
        Abraham–Minkowski
        controversy, 136–146
    orbital and spin angular momentum,
        118–122
    trapping forces, 131–136
Angular spectrum representation, 456
Annihilation events, 54
    computationally model effect, 60
    focusing configuration, 58
    idealized visualization of vortex, 55
    intensity of field, 59
    measurement plane, 56
    phase singularities, 57
    real and imaginary parts, 55
Anti-Sun, 233
Associated Laguerre equation, 29
Axis-finder, 80
Azimuthal beams, 222–223

B

Babinet point, 234
Beamsplitter, 352, 353
Beam wander
    model, 307
    vortex beams, 330
Bell's inequalities for angular momentum
      states, 380
    CHSH case, 387
    classical entanglement, 389
    correlation function, 383
    hypothetical experiment, 381
    nonlocal quantum theory, 388
    probability density, 382
    quantum mechanicsm, 385
    similarity transformation of matrix, 384
    single-photon state, 390
    Vaziri, Weihs, and Zeilinger
      experiment, 386
Bessel beams, 455
Bessel–Gauss vortex beams, 453; *see also*
      Fractional charge vortex beams
    airy functions, 465
    amplitude cross section, 455
    amplitude of Bessel–Gauss beams on
      propagation, 460
    constituent waves, 459
    experimental method to generate
      Bessel beams, 458
    integral, 457
    intersection point of blades
      of scissors, 461
    propagation and acceleration, 464
    real-valued quantity, 454
    self-healing, 463
    simple model of pulsed Bessel
      beam, 462
    spectral amplitude, 456
Bialynicki-Birula and Bialynicka-Birula
      work, 181
Bose–Einstein condensates, 335
Braids vortices, 479–485
Brewster point, 234–235
Burgers circuit, 48, 49
Burgers vector, 48, 49

C

Cartesian coordinate system, 187
Cascades of singularities, 485–489; *see also*
      Lissajous singularities
Catastrophe theory, 252
Caustics, 33–34
CDM04, *see* Crabtree, Davis, and Moreno
      theory
Center in dynamic systems, 282
Circular polarization, 79, 193–194,
      201–202
Classical correlations, 349
Clear sky singularities, 227, 233
    center singularities in clear sky, 234
    light scattering, 233
    polarization lines and singularities of
      sky, 235
    Rayleigh scattering in atmosphere, 234
    total topological index on sky sphere,
      233–234
Coherence singularities, 293; *see also*
      Polarization singularities
    electromagnetic correlation
      singularities, 321–324
    experiments and applications, 324–326
    OAM and rankine vortices, 331–334
    optical coherence, 294–304
    phase singularities in partially coherent
      fields, 317–321
    singularities of correlation functions,
      305–312
    structure of correlation singularity,
      312–317
    tGSM beams, 326–331
Coherent/coherence, 299
    complex degree, 298
    mode representation, 304
    states, 342
    time, 299
Complex analytic field, 390
Complex analytic signal, 295
Complex degree of coherence, 298
Computer-generated holograms, 72
    Bessel functions, 99–100
    binary spiral zone plates, 74
    complex transmission function,
      96–97

fork diagram, 76
four-channel modan, 98–99
Fourier series representation, 76–77
Fraunhofer integral, 95–96
Gabor in-line hologram, 72
helical wavefront, 73
Leith–Upatnieks hologram, 94–95
and modans, 94
multichannel modans, 100–101
portion of plane, 74–75
transmission function of multichannel
modan, 97–98
Continuous-wave lasers (CW lasers), 294
Contour property, 251
Cooperative frequency locking, 79
Coronograph, 166
Bessel functions, 167–169
Fourier optics, 169–170
Lyot-style coronagraph, 166–167
"ring of fire" simulation, 169
vortex coronagraphs, 170
Correlation functions, singularities of, 305
behavior of spectrum, 312
coherent superposition, 310–311
cross-correlation function, 309
phase contours, 308
position of correlation vortices as
observation point, 309
properties of correlation vortices, 306
spectral density, 305–306
universal color pattern, 312
Young's two-pinhole experiment, 305
Correlation singularities, 309, 312, 326
basic hyperbolic feature of, 317
cross-spectral density, 315, 316
methods of producing optical
vortices, 312
nature, 317
notation for linear optical system, 313
projections, 316–317
Correlation vortices, 305, 306, 320, 325
properties, 306
in random wavefields, 442–445
Covariance matrix, 432
C-points, 202
characterizing of polarized light, 207
identification as singularities, 207
lines, 203

as phase singularities, 203
of projection of polarization ellipse, 204
transverse, 205
of transverse electric field, 204
Crabtree, Davis, and Moreno theory
(CDM04), 173
Creation operators, 339, 341
Cross-spectral density, 300, 326, 442
Crystal dislocations, 46
burgers circuit, 49
crystal dislocations, 50
edge and screw dislocations
in crystals, 48
formation and motion of dislocation, 48
incorrectmodel for plastic
deformation, 47
types of liquid crystal disclinations, 51
Cubic potential, 360
CW lasers, *see* Continuous-wave lasers
Cycle-averaged Poynting vector, 263
Cylindrical vector beams, 218

D

Degree of polarization, 231
Density operator, 348
Destruction, 339, 341
Dichroic azimuthal linear polarizer, 80
Difference-frequency generation, 357
Diffraction-based method, 87–89
Dirac's magnetic monopole, 416
continuous distribution, 418
cylindrical coordinates, 419
displacement of singularity path, 420
electromagnetic angular momentum
and charge quantization, 423
geometric phase, 422
magnetostatic fields, 425
obscure vector, 424
vector potential, 417, 421
Directional wave, 40
Direct laser generation, 77–79
Divergence theorem, 118
Domain integral equation method, 287
Doppler effect, 176
Dove prisms (DP), 325
Dove prism technique, 87
DP, *see* Dove prisms

Duality, 377
Dynamic phase, 236, 243

E

Edge detection, 170
  calibrated image of HR 8799
    system, 171
  filtering operation, 172–173
  Fourier plane, 171–172
  interior cut-out, 173–175
  optical systems, 175
Edge dislocation, 42–44
Eigenvectors, 208
Electric energy density, 263
Electric fields, 268, 465
Electromagnetic correlation singularities,
    321–324; see also Polarization
    singularities
Electromagnetic cross-spectral density
    matrix, 321
Electromagnetic degree of
    coherence, 322
Electromagnetic energy density, 262
Electromagnetic fields, 10, 111, 227, 229,
    246, 395
  average energy density, 346
  law of angular momentum
    conservation, 116–117
  momentum density, 114–115
  monochromatic, 208
  quantization of, 336–343
  time-averaged spin density, 135
  transverse, 13, 185
Electron–positron pair, 350
Ellipticity, 192
Elliptic streamlines, 209
Energy
  density, 264
  flow, 266
Ensemble, 296
  average, 296
  monochromatic, 306, 331
  space–frequency, 313, 434
Entanglement, 347
  of angular momentum states, 363–375
  applications, 355
  beamsplitter, 352, 353

coherence theory, 349
  density operator, 348
  electron–positron pair, 350
  Hong–Ou–Mandel experiment, 354
  macroscopic system, 351
Entanglement of angular momentum
    states, 363; see also Angular
    momentum
  Bell's inequalities, 380–390
  cone of photons emitted in type I
    SPDC, 368
  cones of photons emitted in type II
    SPDC, 370
  corresponding density matrix, 373
  detect OAM entanglement
    of photons, 371
  noncolinear phasematching
    geometry, 364
  OAM modes, 374–375
  optics of anisotropic media, 365
  polarization-entangled pair
    of photons, 369
  principal velocities, 366
  projection operator, 372
  uniaxial crystal, 367
Ergodic field, 296
Eta singularities, 323
Extrinsic angular momentum, 127
  Earth–Sun system, 127–128
  intrinsic quantity, 128–129
  spin angularmomentum, 131
  vortex beam, 130

F

Faraday's law, 10, 119
  application, 261
First quantization, 336
Fluid vortices, 4, 7
Foo, Palacios, and Swartzlander
    strategy, 167
Fractional charge vortex beams, 473; see
    also Bessel–Gauss vortex beams
  angular spectrum, 474
  Bessel function, 475
  Fourier series representation
    of phase, 476
  fractional mask, 479

peculiar arithmetic of transfinite, 478
phase of fractional vortex field, 477
Fractional spiral filter, 175
Fraunhofer patterns, 90
Full Poincaré beams, 223

G

Gabor in-line hologram, 72
Gauge freedom, 336
Gaussian Schell-model beams (GSM
    beams), 326; *see also* Twisted
    Gaussian Schell-model beams
    (tGSM beams)
Gauss's law, 10, 121–122
Geometrical mode separation, 101
    log-polar transformation, 101
    method of stationary phase, 103–104
    mode separator, 106
    PM, 102–103
    SLMs, 105
Geometric phases, 236
Goos–Hänchen shift, 269
Gouy phase shift, 15–16
Gradient force, 134
Green's function techniques, 162–163
GSM beams, *see* Gaussian
    Schell-model beams

H

Hairy ball theorem, 8, 233–234, 252–256
Hamiltonian operator, 339
Harmonic generation, 359
Helicity, 192
Helmholtz equation, 44
Helmholtz's theorem, 164
Hermite equation, 26–27
Hermite–Gauss beams, 16, 27
Hermite–Gauss mode, 68–69, 378
Hermitian property, 228
High-pass filter, 172
Higher-order Gaussian beams, 23
    associated Laguerre equation, 30–31
    Hermite–Gauss beams, 31–32
    Hermite–Gauss modes, 27–29
    independent differential equations,
        26–27
    Laguerre polynomials, 29–30

radius of curvature, 24–26
RMS, 23
Higher-order polarization singularities,
    216–218
Hologram, 72
Hong–Ou–Mandel experiment, 354
Hydrodynamic model, 393
Hyperbolic streamlines, 209

I

Ince–Gaussian beams, 32
Incoherent field, 299
Intensity distribution, 432
Interference-based methods, 84
    interference patterns, 84–85
    interference patterns, 87
    Mach–Zender interferometer, 85–86
    pattern recognition, 87
Interferometry, 84
Intrinsic angular momentum, 127
    Earth–Sun system, 127–128
    intrinsic quantity, 128–129
    spin angularmomentum, 131
    vortex beam, 130
Inverse fourier transforms, 377

J

Jacobian of transformation, 430

K

KBW94, *see* Kristensen, Beijersbergen, and
    Woerdman work
KDP crystal, *see* Potassium dihydrogen
    phosphate crystal
Kernel of optical system, 313
Knots vortices, 479–485
Kristensen, Beijersbergen, and Woerdman
    work (KBW94), 65
Kronecker delta, 115
KTP, *see* Potassium titanyl phosphate

L

Laguerre–Gauss beams, 416
Laguerre–Gaussian beams, angular
    momentum of, 122

Laguerre–Gaussian beams, angular
    momentum of (*Continued*)
  paraxial beams of light, 122–123
  polar coordinates, 125–126
  Poynting vector, 123–124
  quantitative connection, 126–127
  topological charge of beam, 127
Laguerre–Gauss mode, 161, 178, 374
  vortex mode, 332
Laguerre–Gauss paraxial beams, 14
  Gouy phase shift, 15–16
  integrated cross-sectional intensity,
    16–17
  intensity and phase of, 18–19
  intensity and zero crossings of, 21–22
  paraxial wave equation, 14–15
  phase of complex wavefield, 19–20
  phase problem, 22
  topological charge of vortex beam,
    17–18
Laguerre polynomials, 29–30
Laplace's equation, 41
LBO crystal, *see* Lithium triborate crystal
LCDs, *see* Liquid crystal displays
Leith–Upatnieks hologram, 74
Lemon, 211, 217, 220
Levi–Civita tensor, 117, 119
Linear optics, 355
Linear polarization, 192–193, 203, 240
Linear refractive index, 358
Linked vortices, 479–485
Liquid crystal displays (LCDs), 83
Lissajous singularities, 489–495; *see also*
    Cascades of singularities
Lithium triborate crystal
    (LBO crystal), 361
L-lines, 202, 203
  of electric and magnetic fields, 207
  identification as singularities, 207
  point on equator of Poincaré
    sphere, 207
  of projection of polarization ellipse, 204
  relationship between an L-surface, 205
  surfaces, 203
  transverse, 205
Log-polar transform, 101
Lorentz force, 141
  law, 112, 260

Low-pass filter, 172
Lugiato, Oldano, and Narducci model, 79

M

Mach–Zender interferometer, 81, 85–86
Macroscopic system, 351
Magnetic energy density, 263
Magnetic fields, 10, 133, 206, 268, 284, 397,
    465, 469
Magnetic force, 132–133, 260
Maxwell–Ampére law, 365
Maxwell's equations, 9, 113, 336
  curl equations, 263
  paraxial wave equation, 13–14
  scalar Helmholtz equation, 13
  single frequency wavefields, 11–12
  in SI units, 9–10
  source–free regions, 10–11
Maxwell stress tensor, 115
Method of stationary phase, 103, 107–109
Minkowski field momentum density,
    142–143
Mixed edge-screw dislocations, 44–46
Momentum density, 116
Momentum flux density, 116
Momentum in matter, 136
  Ashkin and Dziedzic experiment, 145
  difference between Abraham and
    Minkowski formulations, 137
  Lorentz force of fields, 141
  macroscopic Maxwell's equations,
    138–139
  matter force density, 139–140
  Maxwell's equations, 136–137
  Minkowski field momentum density,
    142–144
  resolution of Abraham–Minkowski
    dilemma, 145–146
  SYF08, 144–145
Momentum in wavefields, 112
  analogous calculation, 116–117
  angular momentum of light, 118
  Cartesian coordinate, 115–116
  divergence theorem, 117–118
  field quantities, 113–114
  momentum conservation law, 113

momentum density of electromagnetic field, 114–115

Monochromatic electromagnetic fields, 208

Monstar, 211

Multiple points, speckle statistics at, 432–437

Multivariate complex circular Gaussian random process, 448–451

Mutual coherence function, 295, 297

N

Node in dynamical systems, 282

Noncolinear phasematching geometry, 364

Nondiffracting beams, 455

Nonlinear optics, 355
  difference-frequency experiment, 363
  OAM, 361
  parametric processes, 357
  quartz prism spectrometer, 359
  second-harmonic signal, 360
  second-order polarization, 356, 362
  types of SHG, 358

Nonlocal optical vortex, 375
  experiment to measure, 380
  gaussian factors, 379
  Hermite–Gauss modes, 378
  inverse fourier transforms, 377
  total output wavefunction, 376

Nonuniformly polarized beams, 218–222

Nonuniform polarization, 79–82
  dichroic azimuthal linear polarizer, 80–81
  generating vortex beams, 82
  uniform polarization, 81–82

Normalization, 412

N-pinhole interferometer, 320–321

Number operator, 340

O

OAM, *see* Orbital angular momentum

OPO, *see* Optical parametric oscillator

Optical angular momentum, 112

Optical coherence, 294
  advantages, 301
  coherent mode representation, 304
  complex analytic signal, 295

degree of coherence, 299

Fourier transform of Wolf equations, 300–301

Michelson interferometer, 298

mutual coherence function, 295–296

oscillations of optical fields, 294–295

spectral density, 303

wave properties of mutual coherence function, 297

Young's experiment with filtered light, 302

Optical coherence theory, 293

Optical communications, 156
  average topological charge, 157–158
  free-space data transmission, 160–161
  OAM beams for communication, 159–160
  optical vortex beams, 156–157
  positive topological charge, 159
  value of vortex beams, 161

Optical möbius strips, 495–498

Optical parametric oscillator (OPO), 82

Optical phase singularities, 8

Optical rectification, 356

Optical spanner, 154

Optical tweezing, 112

Optical vortices, 1, 7, 268
  asymmetric beam, 70–71
  beams, 156–157
  computer-generated holograms, 72–77, 94–101
  coronograph, 166–170
  detection of, 84
  diffraction-based method, 87–89
  direct laser generation, 77–79
  electric field, 272
  generation, 63
  geometrical mode separation, 101–106
  geometry using by Wolter, 270
  Hermite–Gauss mode, 68–69
  incident wave, 275
  interference-based methods, 84–87
  location of Poynting singularities, 274
  method of stationary phase, 107–109
  micromanipulation, spanning, and trapping, 149
  mode conversion, 68
  nonuniform polarization, 79–82

Optical vortices (*Continued*)
  optical communications, 156–161
  optical gear arrangement, 154
  optical OAM, 152–153
  optical rotation, 156
  other methods, 82–84
  output beam, 71–72
  phase of E_y with Poynting vector, 273
  phase retrieval, 161–166
  polarizability of particle, 151–152
  Poynting vector, 272
  quantum cores, 343–347
  reflected zero line propagation, 271
  rotational doppler shifts, 176–182
  Shack–Hartmann method, 89–94
  signal processing and edge detection,
    170–175
  spiral phase plate, 64–68
  time-lapse composite, 155
  top view of cogwheels, 155
  traditional boundary value, 150–151
  two plane waves, 269
Optical wavefields, polarization in, 186
  circular polarization, 193–194
  left-handed and right-handed elliptical
    polarization, 192
  linear polarization, 192–193
  polarization ellipse, 188–189
Orbital angular momentum (OAM), 112,
    118, 331–334, 361
  Gauss's law, 121–122
  paraxial wavefield, 120–121
  separation of angular momentum
    density, 122
  spin part, 118–119
  state, 156

P

Pancharatnam phase, 235
  challenges, 239–240
  example, 236
  experiment, 244
  experimental measurement, 243
  Pancharatnam's connection, 236
  pendulums at different latitudes, 241
  Poincaré sphere in, 237
Parabolic streamlines, 209

Parallel transport of vector, 240
Parametric fluorescence, 364
Paraxial beams, 9
  paraxial wave equation, 13–14, 40
  scalar Helmholtz equation, 13
  single frequency wavefields, 11–12
  in SI units, 9–10
  source-free regions, 10–11
Partially coherent fields, 293
Partially polarized light, 227
  characterizing polarization properties,
    227–228
  electromagnetic field to, 229
  fully polarized light, 227
  Poincaré sphere, 232
  polarization matrix, 229, 231
  polarized field, 228–229
Phase mask (PM), 102, 325
Phase problem, 22
Phase retrieval, 161
  Green's function techniques, 162–163
  modified TIE, 164–165
  TIE for, 165–166
  vortex circulation, 163–164
Phase singularities, 6, 34, 275
  amplitude of pinholes, 39
  binomial expansions, 36
  caustics, 33–34
  creation and annihilation events, 54–60
  crystal dislocations and wavefronts,
    46–51
  cubic equations, 319
  designing, 317
  intensity and phase, 38
  N-pinhole interferometer, 320–321
  paradoxical result, 321
  in partially coherent fields, 317
  of partially coherent fields, 325–326
  phase singularities, 34–39
  properties of, 22–23
  spectral density, 318–319
  systems, 321
  typical forms of wave dislocations,
    40–46
  Young's interferometer, 34–39
  Young's two-pinhole experiment, 35, 37
  Young's three-pinhole experiment, 318
Phasors, 428

PM, *see* Phase mask
Poincaré beams, 222–225
Poincaré sphere, 194, 197, 235
   highlighting nature of linear and
      circular polarization, 200
   position on, 200
   relationship with Stokes
      parameters, 200
   stereographic projection, 198
   unit vectors, 201
Poincaré vortices, 225–227
Point-spread function, 313
Polarization
   converter, 81
   of partially coherent field, 323
Polarization singularities; *see also*
      Coherence singularities
   C-points and L-lines, 202–207
   features, 207–212
   hairy ball theorem, 252–256
   higher-order, 216–218
   nonuniformly polarized beams,
      218–222
   Pancharatnam phase, 235–244
   partially polarized light and
      singularities of clear sky,
      227–235
   Poincaré beams, 222–225
   Poincaré sphere, 194–202
   Poincaré vortices, 225–227
   polarization in optical wavefields,
      186–194
   in random wavefields, 447–448
   second-order tensor fields and their
      singularities, 244–252
   stokes parameters, 194–202
   topological reactions, 212–216
   types, 185–186
Polarization state, 201
Potassium dihydrogen phosphate crystal
      (KDP crystal), 359
Potassium titanyl phosphate (KTP), 82
Power flow in electromagnetic waves, 260
   calculating for simple electromagnetic
      waves, 265
   change in mechanical energy
      equation, 261
   energy flow, 266

   magnetic force, 260
   Poynting vector as flux of energy per
      unit area, 262
   time dependence, 263
Power flow of beam, 8
Poynting singularities, observations of,
      275–279
Poynting vector, 123–124, 261
   generic singularities of transverse, 283
   streamlines of, 282
Poynting vector singularities,
      259, 279, 286
   absorption, 282
   angles of rotation for planes waves, 285
   annihilating in pairs of vortices, 285
   components, 279
   conceptual difficulties with, 266–268
   example, 285
   first optical vortex, 268–275
   observations, 275–279
   power flow in electromagnetic waves,
      260–266
   properties, 279–286
   singularities of vector field, 280
   singularities, transmission, and
      radiation, 286–292
   topological index, 283
   transverse Poynting vector direction
      and amplitude, 286
Principal dielectric constants, 365
Probability density, 382
Projection operator, 372
Pump photon, 357

Q

Quantization of electromagnetic field, 336
   destruction and creation operators,
      339, 341
   normally ordered component, 343
   number operator, 340
   real-valued functions, 338
   RMS deviation, 342
   time-dependent components, 337
Quantum cores of optical vortices, 343
   electromagnetic field, 346
   short-wavelength electromagnetic
      waves, 347

Quantum cores of optical vortices
(*Continued*)
    thin spherical shell, 345
    total electric field, 344
Quantum correlations, 349
Quantum optics, singularities
            and vortices in
    Bell's inequalities for angular
            momentum states, 380–390
    Dirac's magnetic monopole, 416–425
    entanglement, 347–355, 363–375
    nonlinear optics and angular
            momentum, 355–363
    nonlocal optical vortex, 375–380
    quantization of electromagnetic field,
            336–343
    quantum cores of optical vortices,
            343–347
    in Schrödinger's equation, 390–399
Quartz prism spectrometer, 359
Quasi-homogeneous source, 377
Quasicrystalline order, 445–446
Qutrit states, 375

R

Radial beams, 222–223
Radial Hilbert transform, 173
Radiation, 286–292
    pressure force, 134
Random wavefields, vortices in
    correlations of vortices, 442–445
    multivariate complex circular Gaussian
            random process, 448–451
    polarization singularities, 447–448
    quasicrystalline order, 445–446
    speckle statistics at multiple points,
            432–437
    speckle statistics at single point,
            428–432
    statistics of vortices, 437–442
Rankine vortices, 331–334
Rayleigh particle, 150
Rayleigh range, 14
Riemann–Silberstein vector (RS vector),
            285, 466
Riemann–Silberstein vortices, 465
    Lorentz transformations, 467

monochromatic form, 472
orthogonal plane waves, 470, 471
phase of F^2, 469
RS vector, 466, 468
"Ring of fire" simulation, 169, 183
RMS, *see* Root mean square
Root mean square (RMS), 23, 342
Rotational Doppler shifts, 176
    angular Doppler effect, 180–181
    Cartesian coordinates, 178–179
    OAM modes, 182
    ordinary Doppler effect, 179–180
    transformation of vortex beam,
            177–178
    translational Doppler effect, 176–177
Rotational frequency shift, 181
RS vector, *see* Riemann–
            Silberstein vector

S

Scalar Helmholtz equation, 13
Scattering force, 134
Schell-model, 313, 327
Schrödinger's equation, 335
    electromagnetic field, 395
    electron wavefunction of hydrogen
            atom, 392
    evolution in time of ring vortex, 397
    hydrodynamic model, 393
    monochromatic fields, 391
    probability density, 394
    uniform magnetic field, 398–399
    vortices in, 390
    wavefunctions, 396
Schrödinger uncertainty relation, 401
Screw dislocation, 41–42
Second-harmonic generation (SHG), 356,
            358
Second-order polarization effects, 356
Second-order tensor fields and
            singularities, 244–252
    closed path encircling region with
            singularities, 247
    roots of cubic function, 251
    structure of polarization
            singularities, 252
    symmetric and real-valued, 245

theorems to topological index, 246
   types of tensor singularities, 247–278
Second quantization, 33
Separatrix, 209, 211
   monster, 215
Shack–Hartmann method, 89
   averaging process, 94
   phase function, 93–94
   phase singularity, 89–90
   wavefront sensor, 90–91
Shear forces, 136
SHG, *see* Second-harmonic generation
Signal photon, 357
Signal processing, 170
   calibrated image of HR 8799
      system, 171
   filtering operation, 172–173
   Fourier plane, 171–172
   interior cut-out, 173–175
   optical systems, 175
Single point, speckle statistics at, 428–432
Singularities, 286–292
Singular optics, 1
Sink in singular optics, *see* Node in
   dynamical systems
SLM, *see* Spatial light modulator
Spatial coherence, 303
Spatial correlation singularities, 310
Spatial light modulator (SLM), 83
SPDC, *see* Spontaneous parametric down
   conversion
Speckle statistics
   at multiple points, 432–437
   at single point, 428–432
Spectral degree of coherence, 302
Sperner's lemma, 252–256
Spin angular momentum, 118
   Gauss's law, 121–122
   paraxial wavefield, 120–121
   separation of angular momentum
      density, 122
   spin part, 118–119
Spin force, 135
Spinor representation, 201
Spiral phase filtering, 173
Spiral phase plate, 64, 473
   Gaussian beam, 64–65
   phase plates, 65–66

   Taylor expansion, 66–67
Spiral point, 282
Spontaneous parametric down conversion
   (SPDC), 355
Star, 211, 217
Statistically stationary, 294
Statistics of vortices in random wavefields,
   437
   circular patch, 440
   exponent in explicit form, 439
   plane wave increases, 442
   probability density function, 438
   structure of light, 441
Stereographic projection, 198
Stimulated process, 363
Stokes parameters, 194
   experimental arrangement for
      measuring, 195
   from measuring intensities, 196
   relationship with Poincaré sphere, 200
   set of four parameters, 194
   set of normalized, 197
Stokesparameters, 213
Stokes' theorem, 92, 421
Sum-frequency generation, 357
Superoscillatory fields, 499–503
Surface plasmons, 83
Surfaces of constant phase, 4

T

Temporal coherence, 299
Temporal correlation vortices, 310
Tensor field, second-order, 208
tGSM beams, *see* Twisted Gaussian
   Schell-model beams
TIE, *see* Transport-of-intensity equation
Topological charge
   and index, 51–53
   of vortex beam, 17–18
Topological index, 214, 215
   of radial and azimuthal beams, 219–220
Topological reactions of polarization
   singularities, 212–216
Total electric field, 344
Translational Doppler effect, 176–177
Transmission, 286–292
Transport-of-intensity equation (TIE), 162

Trapping forces, 131
  gradient beam trap, 131
  gradient force, 134–135
  magnetic component of Lorentz force
    law, 132–133
  shear forces, 136
  time-averaged spin density, 135–136
  vector calculus, 133–134
Twisted Gaussian Schell-model beams
    (tGSM beams), 326
  construction by incoherent
    superposition of vortex
    modes, 331
  difference with beam wander vortex
    beams, 330
  expression for, 328
  Gaussian modes to constructing, 329
  physics of, 328

U

Uncertainty principle for angular
    momentum, 399, *see* Quantum
    optics, singularities and
    vortices in
  angle operator, 409
  arbitrary state, 410
  commutator, 407
  diffraction theory, 404
  expectation value of
    anti-commutator, 401
  function distributed over
    unit circle, 415
  Gaussian beam, 414
  Laguerre–Gauss beams, 416
  mean angular position, 406
  minimum uncertainty state, 402
  momentum space, 403
  normalization, 412
  phase operators, 413
  physical states, 411
  Schwartz inequality, 400
  uniform distribution, 405
  zero-angle state, 408
Uniform polarization, 185
Unusual singularities
  Bessel–Gauss vortex beams, 453–465
  cascades of singularities, 485–489

fractional charge vortex beams,
    473–479
knots, braids, and linked vortices,
    479–485
Lissajous singularities, 489–495
optical möbius strips, 495–498
Riemann–Silberstein vortices, 465–472
superoscillatory fields, 499–503

V

van Cittert–Zernike theorem, 435
Vector beams, 218, 220
Vector Helmholtz equation, 12
Vector phase mask, 170
Vector potential, 337
VIV, *see* Vortex-induced vibration
von Karman vortex streets, 2, 3
Vortex-induced vibration (VIV), 3
Vortex beams, 17
  higher-order Gaussian beams, 23–32
  Laguerre–Gauss paraxial beams, 14–22
  Maxwell's equations and paraxial
    beams, 9–14
  properties of phase singularities, 22–23
Vortices, 1
  complex scalar field in complex plane, 6
  example of vortex creation, 3
  fluids, 3, 4, 7–8
  optical, 1
  optical waves, 8
  phase, 84
  of power flow, 259
  production, 2
  in singular optics, 282
  technical drawings of collection of, 1–2
  wave, 4
  wingtip vortex of passing aircraft visible
    in smoke, 4

W

Wave dislocations typical forms, 40
  edge dislocation, 42–44
  mixed edge-screw dislocations, 44–46
  screw dislocation, 41–42
Wavefield, generic features of, 34
Wavefronts, 46
  burgers circuit, 49

crystal dislocations, 50
edge and screw dislocations
    in crystals, 48
formation and motion of dislocation, 48
incorrectmodel for plastic
    deformation, 47
types of liquid crystal disclinations, 51
Wavefunctions, 350, 377, 396
    bound-state, 392
    position, 377
    total output, 376
    two-photon, 375
Wave–particle duality of light, 335
Wave vortices, 4
Wiener–Khintchine theorem, 298

Wiener's 1890 experiment, 206
Wolf equations, 296

Y

Young's interferometer, 34
    amplitude of pinholes, 39
    binomial expansions, 36
    intensity and phase, 38
    Young's two-pinhole experiment, 35,
        37, 485
    Young's three-pinhole experiment, 318

Z

Zero of electric field, 212, 273–274

Printed and bound by CPI Group (UK) Ltd, Croydon, CR0 4YY

01/11/2024

01782622-0016